LIPOFUSCIN AND
CEROID PIGMENTS

ADVANCES IN EXPERIMENTAL MEDICINE AND BIOLOGY

Recent Volumes in this Series

LIPOFUSCIN AND CEROID PIGMENTS

Edited by

Eduardo A. Porta

University of Hawaii
Honolulu, Hawaii

SPRINGER SCIENCE+BUSINESS MEDIA, LLC

Library of Congress Cataloging-in-Publication Data

International Symposium on Lipofuscin and Ceroid Pigments: State of
  the Art (3rd : 1989 : Wailea, Maui, Hawaii)
   Lipofuscin and ceroid pigments / edited by Eduardo A. Porta.
      p.   cm. -- (Advances in experimental medicine and biology ; v.
  266)
    "Proceedings of the Third International Symposium on Lipofuscin
  and Ceroid Pigments--1989: State of the Art, held August 20-23,
  1989, in Maui, Hawaii"--T.p. verso.
    Includes bibliographical references.
    Includes index.
    ISBN 978-1-4899-5341-4   ISBN 978-1-4899-5339-1 (eBook)
    DOI 10.1007/978-1-4899-5339-1

    1. Lipofuscins--Congresses. 2. Neuronal ceroid-lipofuscinosis-
  -Congresses.   I. Porta, Eduardo A.   II. Title.   III. Series.
    [DNLM: 1. Pigments--metabolism--congresses.   W1 AD559 v. 266 / QU
  110 1623L 1989]
  QP671.L52I57 1989
  612'.01528--dc20
  DLC
  for Library of Congress   90-7337
                                                                CIP

Proceedings of the Third International Symposium on
Lipofuscin and Ceroid Pigments — 1989: State of the Art,
held August 20-23, 1989, in Maui, Hawaii

© 1989 Springer Science+Business Media New York
Originally published by Plenum Press, New York in 1989
Softcover reprint of the hardcover 1st edition 1989

To the memory of my son, Eduardo Dante

PREFACE

These biennial symposia were first established in 1985, in Italy, in response to the clear need to concentrate, interchange and disseminate the rapid advances on the biogenesis, properties and nosologic implications of lipofuscin and ceroid pigments. The first international meeting in Vico Equense (Naples) was organized by E. Aloj-Totaro, F. Pisanti and P. Glees, and the second took place in Debrecen, Hungary, and was organized by I. Zs.-Nagy in 1987.

This third international symposium was held in the Wailea area of the island of Maui, Hawaii, and I was fortunate to count on the help of the members of the International Organizing Committee: L.S. Wolfe (Canada), I. Zs.-Nagy (Hungary), E. Aloj-Totaro (Italy), R.D. Jolly (New Zealand), and D. Harman (USA).

The field of lipofuscin and ceroid pigments is rapidly expanding and is noted for its diversity in thought and technology. Our goal has been to provide a balanced forum for the most recent advances in this multidisciplinary area of autofluorescent lipopigments. The criteria for selection of invited speakers was based on scientific excellence, nondiscriminatory factors, and representation from diverse geographical areas of the world.

The main sessions of the symposium were: I. Lipofuscin (Age Pigment), II. Ceroid Pigments in Nutritional and Environmental Pathology, III. Ceroid in Genetic Disorders, and IV. In Vitro Systems of Lipopigment Formation. There was also a poster session. Each presentation was followed by discussions from the audience. This volume contains the complete manuscripts submitted by the invited speakers, the summaries of poster presentations, and the discussions following each presentation, including those following a brief oral presentation of the posters' contents. Finally, we have also included a general discussion on the needs for future research lines and approaches, as perceived by the attendants.

We are particularly indebted to the support given to this symposium by the Vice-President of the University of Hawaii, Dr. David Yount (Research and Graduate Education), and by the Dean of the Medical School, University of Hawaii, Dr. Christian L. Gulbrandsen, and his secretary Mrs. Gayle Gilbert. Our gratitude is also extended to Miss Valerie Song for the transcription and typing of the discussions, to Mrs. Gwen Helliker, from On-Line Hawaii, for the coordination of trips and lodgings, and to Ms. Melanie Yelity, Editor, Plenum Publishing Corporation, for her patience and help in editing this book.

Eduardo A. Porta, M.D.
The Editor

ACKNOWLEDGMENTS

The sponsorship and financial support of the following organizations are gratefully acknowledged

Division of Research and Graduate Education, University of Hawaii

John A. Burns, School of Medicine, University of Hawaii

Center on Aging, University of Hawaii

Pacific Islands Geriatric Education Center, Hawaii

American Aging Association

Children's Brain Diseases Foundation, U.S.A.

Hokama-Yagawa Memorial Fund

Verzar International Laboratory for Experimental Gerontology, Italy-Hungary

Alzheimer's Association, Honolulu Chapter

Center of Experimental Pathology, Univ. Buenos Aires, Argentina

Pfizer Pharmaceuticals

CONTENTS

LIPOFUSCIN (AGE PIGMENT)

CEROID PIGMENTS IN NUTRITIONAL AND ENVIRONMENTAL PATHOLOGY

CEROID IN GENETIC DISORDERS

IN VITRO SYSTEMS OF LIPOPIGMENT FORMATION

POSTER PRESENTATIONS

LIPOFUSCIN (AGE PIGMENT)

LIPOFUSCIN AND CEROID FORMATION: THE CELLULAR RECYCLING SYSTEM

Denham Harman

University of Nebraska
College of Medicine
Omaha, Nebraska 68105

SUMMARY

Lipofuscin, age pigment, is a dark pigment with a strong auto-fluorescence seen with increasing frequency with advancing age in the cytoplasm of postmitotic cells. By bright-field light microscopy lipofuscin appears as irregular yellow to brown granules ranging in size from 1-2mm in diameter. The fluorescent spectra of lipofuscin in situ generally show excitation maxima at about 360nm and a yellowish emission maxima at 540-650nm. Ultrastructurally the granules, localized in residual body-type lysosomes, are extremely heterogeneous and vary from one cell type to another, and frequently within a single cell. The pigment granules usually contain numerous liquid droplets embedded in an electron-dense matrix. The granules stain positively for neutral lipids but are not soluble in polar or non-polar lipid solvents. Lipofuscin contains about 50 percent by weight of proteinaceous substances, a lesser fraction of lipid-like material, and probably less than one percent by weight fluorophore(s); it is enriched in metals such as Al, Cu, and Fe, and in dolichols.

Free radical reactions and the proteolytic system are implicated in lipopigment formation. Thus the rate of lipopigment formation is increased by vitamin E deficiency and by increased intake of polyunsaturated fatty acids as well as by protease inhibitors such as leupeptin. Free radical reactions and proteolysis are involved in the continual turnover of cellular components. Cellular damage from free radical reactions, and others such as hydrolysis, has been present since the beginning of life. The evolution of more complex cells necessitated development of defenses - DNA repair processes, antioxidants, etc. - against damaging reactions as well as the removal and replacement of altered parts, and of those no longer needed by the cells. Proteins "marked" for disposal by oxidation damage, or other means such as conjugation with ubiquitin, are apparently rendered more hydrophobic so that they are "recognized" for degradation by the lysosomes and the proteinases and peptidases of the cytosol and mitochondria. Oxidatively altered lipids are removed by enzymes such as phospholipase A_2. The products of the degradation processes are reused by the cells.

Normally the recycling of damaged components works extremely well. There may be some slow slippage with advancing age as the rate of free radical damage increases while protease activity decreases. As a result a gradually increasing fraction of lysosomal"food" may be converted to non-digestible forms, lipofuscin, before it can be broken down to reusable components. Ceroid is apparently formed when the disposal system is "overloaded" or impaired. Considering the complexity of this system there could be many genetic defects that result in ceroid formation in addition to those involved in Batten s disease.

KEY WORDS: Lipofuscin, ceroid, proteolysis, free radicals, evolution.

INTRODUCTION

Lipopigments can be categorized into those that occur normally,lipofuscin and neuromelanin, and the ceroids that are experimentally produced or associated with some pathological disorders. Lipofuscin was first observed in 1842, in dissected human neurons (Hannover, 1842), and neuromelanin in 1865 (Luys, 1865) in the human substantia nigra. Apparently ceroid was described first in 1936, in the uterus and kidneys of rats on a vitamin E free diet (Martin and Moore, 1936). Interest in lipofuscin was increased in 1866 when it was recognized that the pigment accumulated with age (Koneff, 1886). The voluminous literature that has accumulated on lipofuscin (Sohal, 1981; Totaro et al., 1987; Zs.-Nagy, 1988) is in part due to the hope that further knowledge of this lipopigment might contribute to an understanding of the aging process. Neuromelanin also accumulates with age; it is not as widely distributed as lipofuscin and is now considered to be melanized lipofuscin (Barden and Brizzee, 1987). Ceroid studies are prompted by the possibility that they will help in determining the mechanism of lipofuscin formation, for the two lipopigments have some properties in common, as well as by their presence in specific disorders. Included under the term, ceroid-lipofuscinosis are a group of autosomal recessive lysosomal storage diseases of humans and animals (Koppang, 1973/1974; Palmer, et al., 1986) associated with brain atrophy, blindness, dementia, and premature death. Human disorders include the Hermansky-Pudlak syndrome (Hermansky and Pudlak, 1959; Witkup et al., 1988) and the four subgroups of Batten s disease (Armstrong et al., 1982) infantile, late infantile, juvenile and adult (Kufs disease) (Berkovic, et al., 1988).

PROPERTIES OF LIPOPIGMENTS

Distribution

Lipofuscin has been found in the postmitotic cells of all species studied, from neurospora (Munkres, 1987) nematodes (Zuckerman, 1987; Epstein and Gershon, 1972) and flies (Miquel et al., 1977; Sohal, 1981) to humans (Brizzee and Ordy, 1981; Porta and Hartroft, 1969). The distribution in the human brain is uneven, being related to the level of oxidative enzymes (Friede, 1962, Friede et al., 1962), while it is uniform throughout the heart (Strehler et al., 1959).

Neuromelanin is not a "universal" pigment like lipofuscin. It is found only in mammals in the primates, carnivores, horse, and giraffe. Neuromelanin accumulates in several areas of the human brain stem, particularly the substantia nigra and the locus ceruleus as well as in the dorsal root ganglia of the spinal cord (Barden, 1981).

Ceroid can accumulate in both mitotic and post mitotic cells.

In situ Studies

__Light Microscopy__. Lipofuscin, neuromelanin, and the ceroids all appear about the same, as irregular yellow to brown cytoplasmic granules about 1-20 μm in diameter (Porta and Hartroft, 1969; Wolman, 1980; Barden, 1981). These pigments all display the same physical and biochemical characteristics at some time in their evolution (Porta and Hartroft, 1969; Miyagishi et al., 1967).

__Electron Microscopy__. The most prominent feature of the lipofuscin granules is their polymorphic internal configuration and the almost constant presence of a single membrane envelope (Porta and Hartroft, 1969). The granules contain varying numbers of: 1) translucent or almost translucent vacuoles, 2) compact, rounded or coarse bodies of variable electron opacity and osmiophilia, usually very dense, and 3) laminated bodies of various shapes; they have been categorized into four groups on the basis of these characteristics (Miyagishi et al., 1967). The electron dense component of the granules increase with age (Siakotos et al., 1977).

The granules of ceroid are somewhat similar to those of lipofuscin. They accumulate in lysomes as rounded subgranules of 0.2μm to several microns in diameter and often contain mixtures of small granular dense bodies, droplets of fat of moderate osmiophilia, empty vacuoles or vacuoles with a peripheral rim of fat and sometimes laminated areas (Porta and Hartroft, 1969).

__Histochemical__. The lipopigments are relatively insoluble in polar or non-polar solvents. They stain positively for the presence of neutral lipids with oil red O, Sudan black or osmium. They are positive for polysaccharides with the periodic acid - Schiff reaction, are acid fast and exhibit moderate to strong basophilia. They are usually positive for acid phosphatase, a marker enzyme for lysomes (Oliver, 1981; Porta and Hartroft, 1969; 1981).

__Autofluoresence__. Autofluorescence is one of the most consistent features of the lipopigments; this property was first observed in 1911 (Stubel, 1911). The lipopigments exhibit a organ-yellow fluorescent when excited by near visible ultraviolet light, about 360nm (Wolman, 1980).

Isolated Lipopigments

Studies on the isolation and characterization of the lipopigments are few in comparison to the in situ ones, largely because the low concentrations of the lipopigments in the tissues make them difficult to isolate and characterize. Because of this lipopigments are generally quantified by the spectrofluorimetic method (Hammer and Braun, 1988). Isolation of lipofuscin for the first time was in 1901, by peptic digestion of human intestinal mucosa (Rosenfeld, 1901). Later procedures employed acid hydrolysis and enzymatic degradation (Moore and Wang, 1947), and sonification (Hendly et al., 1963). Less drastic separation procedures were subsequently developed which involved tissue homogenization and density gradient centrifugation that apparently produce little or no change in the lipofuscin or ceroid pigments (Siakotos et al., 1970; 1981; Palmer et al., 1986; Ng Ying Kin et al., 1983).

The purity of isolated lipopigments is usually assessed by phase-contrast and fluorescent microscopy; "pure" being the absence of debris and non-fluorescent particles; electron microscopy is also frequently employed to evaluate purity. These procedures yield about 100-200 mg of lipofuscin per normal 40-50 year old human brain (1.2-1.4 kg) while the ceroid from the brain of a patient with Batten's disease may constitute around 5 percent of the total brain weight (Siakotos et al., 1970). Ceroid separated from the liver, kidney, pancreas and brain of 12 to 24 month old sheep with ceroid lipofuscinosis was about 1.4 mg dry mass of ceroid per gram of wet weight of tissue (Palmer et al., 1986). The fraction of lipopigment present in situ that is extracted by these procedures is apparently unknown.

The fluorophore(s) in lipofuscin is probably less than 1 percent by weight of the lipopigment mass (Wolfe, 1988).

Electron microscopy of isolated heart and liver lipofuscin shows a homogeneous population of characteristic lipopigments, unique for each organ system, while brain lipofuscin is heterogeneous. Ceroid fractions from pathologic human and canine brain show significant morphological differences according to the type of neuronal ceroid-lipofuscin (Siakotos and Munkres, 1981; Palmer, et al, 1986).

The properties of isolated lipopigments have been extensively discussed (Elleder, 1981). They have generally been reported to contain 40-70 percent by weight of proteinaceous material (Bjorkerud et al., 1964; Palmer et al., 1986) the remainder being lipid with small amounts of several metals - the levels of copper and iron are higher in ceroid than in lipofuscin (Siakotos and Koppang, 1973). Lipids that have been isolated include the lysomal marker bis(monoacylglycero)phosphate, free fatty acids, cholesterol, ubiquinone, dolichol, and dolichol esters. Reflecting the absence of standardized isolation procedures, and the need for them to be reevaluated, was the finding in one study that over 80 percent of the lipofuscin was polymalonaldehyde (Siakotos and Munkres, 1982).

POSSIBLE MECHANISM FOR LIPOFUSCIN FORMATION

Free radical reactions are involved in lipopigment formation (Wolman, 1980, 1981). This was indicated first in 1928 when a granular, acid-fast material resembling age pigment was observed (Pinkertan, 1928) in the pulmonary parenchyma of rabbits after the intratracheal injection of cod liver oil; this oil contains high concentrations of readily oxidized polyunsaturated lipids. A similar pigment was noted in the myometrium of vitamin E deficient rats in 1936; the investigators named the pigment "ceroid" (Martin and Moore, 1936) - lipofuscin was named in 1912 (Hueck, 1912). Further studies on the effect of dietary unsaturated lipids (Danse and Verschuren, 1978; Reddy et al., 1973; Miyagishi et al., 1967), of vitamin E deficiency (Reddy et al., 1973; Miyagishi et al., 1967), and of the intraperineal injection of mice with 1-methyl-4-phenyl-1,2,3,6-tetrahydropyridine on the formation of autofluorescent lipopigments, as well as the formation of fluorophores during the in vitro oxidation of lipids (Gutteridge, 1984; Kikugawa, 1988; Porta et al., 1988) contributed to the current acceptance of the role of peroxidation in the genesis of lipopigments. Also contributing to this acceptance are cell culture studies (Thaw et al., 1984; Ball et al., 1988) and experiments which demonstrate that vitamin E decreases the rate of lipopigment formation (Epstein et al., 1972; Tappel et al., 1973; Blackett and Hall, 1981).

In spite of numerous studies the structure(s) of the fluorophore(s) in the lipopigments is unknown (Eldred and Katz, 1988; Kikugawa, 1988; Elleder, 1981); the suggestion (Tappel et al., 1973; Feeney-Burns et al., 1980) that the fluorophore(s) are conjugated Schiff bases, $R_1N=CH-CH=CH-NH-R_2$, resulting from the condensation of malonaldehyde produced during lipid peroxidation with amino groups of proteins and amino acids is unlikely (Eldred and Katz., 1988; Palmer et al., 1986; Kikugawa, 1988) in part because their fluorescence emission is in the range of 420-470nm whereas that of lipopigments is about 450-650nm (Porta et al., 1988). More recently it has been suggested the autofluorescence of the pigment is a property of the protein/lipid complex, rather than a chemical species significant to the pathogenesis of the pigment (Dalefield et al., 1988), while evidence has been reported to support the possibility that retinoyl complexes may be the source of the autofluorescence in Batten's disease (Wolfe et al., 1977) and lipofuscin (Eldred and Katz, 1988).

Proteolysis may also be implicated in lipopigment formation as administration of proteinase inhibitors such as leupeptin enhance ceroid formation in experimental animals (Ivy and Gurd, 1988; Ivy et al., 1984). Leupeptin is an oligopeptide endopeptidase inhibitor produced by the actinomycetes; it has a low toxicity and inhibits the most important thiol proteinases, namely cathepsins B, H, and L. The terminal aminoaldehyde group of leupeptin inhibits thiol and serine proteinases by binding semireversibly with their thiol groups (Marzella and Glaumann, 1987; Baugh and Schnebli, 1980).

Intracellular proteins are continuously synthesized and degraded (Schoenheimer, 1946; Glaumann and Ballard, 1987); turnover rates differ and are modulated by the nutritional state (Goldberg and Dice., 1974; Pine, 1972). The rate of intracellular degradation may be high. In liver, for example, the rate may vary from 0.3 to 4.5 per cent per hour (Marzella and Glaumann, 1987; the lower-limit range, 0.3-1.5 per cent, is operationally defined as basal degradation. Possible reasons for this phenomena include (Marzella and Glaumann, 1987): 1) elimination of abnormal and denatured proteins and proteins resulting from synthetic errors, 2) modulation of important metabolic pathways by removal of enzymes not needed by the cell or synthesized in excess of actual demand, for example, phenobarbital-induced cytochrome P-450 dependent enzymes are degraded in lysosomes following cessation of phenobarbital treatment, 3) post-translational processing to activate enzymes or hormones, 4) modulation of cell growth in cooperation with synthesis, 5) provision of amino acids for synthesis of proteins and for gluconeogenesis when the exogenous supply is limited, 6) involution of organs like the uterus and the mammary glands, and 7) phenotypic alterations of cells, for example, pancreatic acinus cells can lose their secretory machinery by lysosomal degradation and may undergo transformation into hepatocytes.

There are at least two major proteolytic systems in eukaryotic cells, the lysosomal system and the soluble ATP-dependent pathway (Goldberg, 1987). The latter is responsible for the rapid elimination of cytosolic proteins with abnormal structures, such as those that might result from mutation, biosynthetic errors or postsynthetic damage (Hershko and Ciechanover, 1982) while the lysosomes are apparently responsible for the most part for the degradation of long-lived proteins, fragments of membranes, and intact organelles such as mitochondria. There are at least two different types of ATP-dependent systems, the ubiquitin system in the cytosol and the vanadate-sensitive system in the mitochondria (Goldberg, 1987). The ATP-dependent proteases are most active at 7.8 pH; they have molecular weight of around 500,000. In contrast the proteases of the lysosomes are most active at acid pH's; they are glycoproteins with molecular weights of 20,000 to 40,000. It has been estimated that 30-70 percent of endogenous cell protein breakdown occurs outside the lysosomes (Ballard, 1977).

Components to be degraded are "marked" in at least one of eight ways (Stadtman, 1986) for selective removal by lysosomal and/or non-lysosomal pathways. "Marking" increases the hydrophobicity or otherwise alters (Bohley et al., 1985; Cervera and Levine, 1986; Roseman and Levine, 1987) a component so that it is "recognized" by a protease and degraded. For example, exposure of glutamine synthetase to an oxidizing system causes inactivation, a loss of one histidine residue, and the introduction of carbonyl groups. This form of the enzyme is more hydrophilic than the native form and it is not a substrate for proteolysis. On continued exposure to the oxidizing system, a second histidine residue is lost and there is a further increase in carbonyl content. This form is significantly more hydrophobic than the native form and is very susceptible to proteolytic attack (Cervera and Levine, 1986; Roseman and Levine, 1987). The selective capacity of lysosomes for "marked" proteins, membranes and organelles may be increased by dolichols. These long chain hydrophobic polyprenols containing an unsaturated isoprene unit have a high affinity for membrane structures. They are synthesized in the microsomes and then transported for the most part to the lysosomes (Wong et al., 1982a,b). The only known function of dolichol, in the form of dolichol phospate, is its participation as an active "carrier" of glycosyl groups in the synthesis of N-linked glycoproteins (Struck and Lennarz, 1980). The phosphorylated forms of dolichol amount to only a small fraction of total cellular dolichol. The remaining dolichol and dolichol esters may possibly serve to increase the hydrophobicity of the lysosomal surface so as to enhance recognition and uptake of cellular components marked for degration.

In addition to "normal marking", such as by ubiquitin conjugation (Hershko, 1985), by the oxidation of amino acid residues by mixed-function oxidation systems, or by the phosphorylation of serine or threonine residues (Stadtman, 1986; 1988), cellular components may be "marked" in a more-or-less random manner by reacting with free radicals arising from non-enzymatic sources or "leaking" from enzymatic free radical reactions. Such free radical reactions may make a significant contribution to the rate of "marking" of cellular components as they go on continuously throughout the cells and tissues. The level of these reactions increases with the metabolic rate and with age (Harman, 1986; Noy et al., 1985). These free radical reactions may be partly responsible for the observed increases with age of altered proteins in the cells (Gershon and Gershon, 1970; Rothstein, 1982) - inactivated, but not readily proteolyzed, as well as for decreases in neutral protease activity with age (Stadtman, 1988); the latter in turn could contribute to the increases in altered proteins with age.

Oxidatively altered proteins are largely proteolyzed by cytosolic proteases (Stadtman, 1988). Single mitochondrial proteins may be degraded in the mitochondria (Desautels and Goldberg, 1982) while the bulk turnover of the entire mitochondria takes place is lysosomes (Glaumann et al., 1981; Pfeifer, 1978). Similarly, oxidized membrane lipids are "recognized" and removed, at least in part, by the action of phospholipase A_2 (Sevanian and Muakkassah-Kelly, 1983).

Cellular damage from free radical reactions has been present from the beginning of life (Harman, 1986). The evolution of more complex and longer living cells and organisms apparently occurred through the gradual selection and development of 1) defenses against deleterious chemical reactions, particularly against the constantly present free radical reactions - during the early part of evolution these reactions were largely initiated by ionizing radiation from the sun, later, after formation of the ozone shield and the shift to aerobic metabolism, they arose for the most part endogenously, and 2) means to repair or replace cellular components that were rendered defective by free radical reactions, or others such as hydrolysis. These processes ensured preservation of beneficial cellular activities while still permitting changes conducive to further evolution.

Defenses that have evolved to minimize the rate of production of free radical damage include antioxidants, such as the tocopherols and carotenes, glutathione peroxidase, and superoxide dismutases, as well as cellular components more resistant to free radical attack (Joenje et al., 1985).

Measures which evolved to remove and degrade cellular components damaged by free radical reactions, or others, include the DNA repair systems, the lysosomal system, and the proteinases and peptidases of the cytosol and mitochondria. As cells become more complex the proteolytic system apparently expanded to serve other purposes as indicated above while enzymes such as phospholipase A_2 appeared to aid in the removal of oxidatively damaged lipids. Compounds resulting from these measures were then available for reuse by the cell.

Normally this recycling system works extremely well. However, there could be some slow slippage with advancing age. The rate of free radical reaction damage increases, possibly exponentially judging from the exponential increase in the chance of death with advancing age, while protease activity decreases - due, at least in part to increasing free radical damage. As a result a gradually increasing fraction of lysosomal "food" may be converted to non-digestible forms (Davies, 1988), lipofuscin, before it can be broken down to reusable components. The pigment serves as a reminder of the extreme efficiency of the cellular recycling system.

Ceroid apparently is formed when the disposal system is "overloaded" or impaired. Overloaded, for example, by increased intake of polyunsaturated fatty acids (these probably increase the rate of free radical damage to cellular components), or by a genetic defect that results in an increased rate of "marking" of one or more components of the cells. The system may be impaired, e.g., by protease inhibitors such as leupeptin or genetic defects involving lysome formation, protease activity, etc. The reason for the accumulation in ceroid-lipofuscinosis is not clear. It may be due to an increased susceptibility to lipid peroxidation. Abnormalities in the fatty acid composition of phospholipids have been reported in these diseases. Thus, in a case of the infantile syndrome (Svennerholm, 1976) the content of $22:4\omega6$ and of $22:6\omega3$ in the phospholipids of the gray matter of the brain was much lower than in the controls while that of $18:1\omega9$ and $20:4\omega6$ were proportionally much higher. Another early onset case had a 30 percent decrease in the $22:6\omega3$ content of phosphatidylserine, this was replaced by $18:1\omega9$ (Jervis and Pullarkat, 1976). A study of infantile, late infantile and adult ceroid-lipofuscinosis (Pullarkat et al., 1982) brain gray matter found that the phosphatidylserine fraction had a 35-90 percent reduction in docosahexaeonic acid ($22:6\omega3$). Studies such as these, the recent report of increased levels of 4-hydroxynonenal (Siakotos et al., 1988), an oxidation product of $22:4\omega6$, in a canine model of ceroid-lipofuscinosis and the increased iron content reported in this disorder (Johansson et al., 1984) all suggest that there is some genetic defect(s) in this disorder(s) which results in increased lipid peroxidation damage and/or abnormality in phosphatidylserine which predisposes it to oxidative damage. Considering the complexity of the cellular disposal system there could be many genetic defects that result in ceroid formation in addition to those involved in the above disorders.

REFERENCES

Armstrong, D., Koppang, N., and Rider, J. A., editors: ceroid-
    lipofuscinosis (Batten's disease), New York, Elsevier Biomedical
    Press, 1982.
Ball, R. Y., Carpenter, K. L. H., and Mitchinson, M. J.: Ceroid
    accumulation by murine periotoneal macrophages exposed to artificial
    lipoproteins: ultrastructural observations. Brit. J. Exper. Path.,
    69:43-56, 1988.
Ballard, F. J.: Intracellar protein degradation, in Essays in
    Biochemistry, Vol. 13, edited by Campbell, P. N., and Aldridge, W.
    N., New York, Academic Press, 1977, pp. 1-37.
Barden, H. and Brizzee, K. R.: The histochemistry of lipofuscin and
    neuromelanin, in Advances in the Biosciences, Vol. 64, Advances in
    Age Pigments Research, Totaro, E. A., Glees, P., and Pisanti, F. A.,
    editors, New York, Pergamon Press, 1987, pp. 339-392.
Barden, H.: The biology and chemistry of neuromelanin, in Age Pigments,
    edited by Sohal, R. S., New York, Elsevier/North Holland Biomedical
    Press, 1981, pp. 155-175.
Baugh, R. J., and Schnebli, H. P.: Role and potential therapeutic value
    of proteinase inhibitors in tissue distruction, in Monograph Series
    of the European Organization for Research on Treatment of Cancer,
    Vol. 6, edited by Strauli, P., Barett, A. J., and Baici, A., New
    York, Raven Press, 1980, pp. 59-67.
Berkovic, S. F., Carpenter, S., Andermann, F., Andermann, E., and Wolfe,
    L. S.: Kufs disease: a critical reappraisal. Brain, 111: 27-62,
    1988.
Bjorkerud, S.: Isolated lipofuscin granules - a survey of a new field, in
    Advances in Gerontological Research, Vol. 1, edited by Strehler, B.
    L., New York, Academic Press, 1964, pp. 257-288.
Blackett, A. D., and Hall, D. A.: Tissue vitamin E levels and lipofuscin
    accumulation with age in the mouse. J. Gerontol., 36:529-533, 1981.
Bohley, P., Hieke, C., Kirschke, H., and Schaper, S.: Protein degradation
    in rat liver cells, in Progress in Clinical and Biological Research,
    Vol. 180, Intracellular Protein Catabolism, edited by Khairallak, E.
    A., Bond, J. S., and Bird, J. W. C., New York, Alan R. Liss, Inc.,
    1985, pp. 447-455.
Brizzee, K.R., and Ordy, J.M.: Cellular features, regional accumulation,
    and prospects of modification of age pigments in mammals, in Age
    Pigments, edited by Sohal, R.S., New York, Elsevier/North Holland
    Biomedical Press, 1981, pp. 101-154.
Cervera, J., and Levine, R. L.: Mixed function oxidation of glutamine
    synthetase modulates the hydrophobicity of the protein and controls
    its susceptibility to proteolyses. Fed. Proc., 45:1597, 1986.
Dalefield, R. R., Jolly, R. D., Craig, A. S., Martinus, R. D., and
    Palmer, D. N.: Age pigment in the thyroid of aged horses, in
    Lipofuscin-1987: State of the Art, edited by Zs.-Nagy, I., New York,
    Elsevier Science Publishers, 1988, pp. 213-226.
Danse, L. H. W. C., Verschuren, P. M.: Fish oil-induced yellow fat
    disease in rats. I. Histological changes. Vet. Pathol., 15:114-
    124, 1978.
Davies, K. J. A.: Protein oxidation, protein cross-linking, and
    proteolysis in the formation of lipofuscin: rationale and methods
    for the measurement of protein degradation, in Lipofuscin-1987:
    State of the Art, edited by Zs.-Nagy, I., New York, Elsevier Science
    Publishers, 1988, pp. 109-132.
Desautels, M., and Goldberg, A. L.: Demonstration of an ATP-dependent,
    vanadate-sensitive endoprotease in the matrix of rat liver
    mitochondria. J. Biol. Chem., 257:11673-11679, 1982.
Eldred, G. E. and Katz, M. L.: Fluorophores of the human retinal pigment
    epithelium: Separation and spectral characterization. Exper. Eye
    Res., 47:71-86, 1988.

10

Elleder, M.: Chemical characterization of age pigments, in Age Pigments, edited by Sohal, R. S., New York, Elsevier/North Holland Biomedical Press, 1981. pp. 203-241.

Epstein, J., and Gershon, D.: Studies on ageing in nematodes. IV. The effect of antioxidants on cellular damage and life span. Mech. Age. Dev., 1:257-264, 1972.

Feeney-Burns, L., Berman, E. R., and Rothman, H.: Lipofuscin of human retinal pigment epithelium. Amer. J. Ophthalmol., 90:783-791, 1980.

Friede, R. L.: The relation of the formation of lipofuscin to the distribution of oxidative enzymes in the human brain. Acta Neuropathologica, 2:113-125, 1962.

Friede, R. L., and Fleming, L. M.: A mapping of oxidative enzymes in the human brain. J. Neurochem., 9:179-198, 1962.

Gershon, H., and Gershon, D.: Detection of inactive enzyme molecules in ageing organisms. Nature, 227:1214-1217, 1970.

Glaumann, H., Ericsson, J. L. E., and Marzella, L.: Mechanisms of intraliposomal degradation with special reference to autophagocytosis and heterophagocytosis of cell organelles. Intl. Rev. Cytol., 73:149-182, 1981.

Glaumann, H., and Ballard, F. J., editors. Lysomes: Their Role in Protein Breakdown. New York, Academic Press, 1987.

Goldberg, A. L.: The ATP-dependent pathway for protein degradation in mitochondria, in Lysosomes: Their Role in Protein Breakdown, edited by Glaumann, H., and Ballard, F. J., New York, Academic Press, 1987, pp. 715-722.

Goldberg, A. L. and Dice, T. F.: Intracellular protein turnover in mammalian and bacterial cells. Ann. Rev. Biochem., 48:835-869, 1974.

Gutteridge, J. M. C.: Age pigments: role of iron and copper salts in the formation of fluorescent lipid complexes. Mech. Ageing Dev., 25:205-214, 1984.

Hadjiconstantinou, M., Tijoe, S., Alho, H., Miller, C., and Neff, N. H.: 1-Methyl-4-phenyl-1,2,3,6-tetrahydropyridine (MPTP) accelerates the accumulation of lipofuscin in mouse adrenal gland. Neuroscience Letters, 83:1-6, 1987.

Hammer, C., and Braun, E.: Mini-review: Quantification of age pigments (lipofuscin). Comp. Bichem. Physiol., 90B:7-17, 1988.

Hamperl, H.: Die fluoreszenzmikroskopie menschlicher gewebe. Virchows Arch. Pathol. Anat. Physiol., 292:1-51, 1934.

Hannover, A.: Mikroskopische undersogelser of nervensystemf. Kgl. Danske Videns Kabernes Selskabs. C. Naturv. og Math. Afh, Copenhagen, 10:1-112, 1842.

Harman, D.: Free radical theory of aging: role of free radicals in the origination and evolution of life, aging, and disease processes, in Free Radicals, Aging, and Degenerative Diseases, edited by Johnson, Jr., J. E., Walford, R., Harman, D., and Miquel, J., New York, Alan R. Liss, 1986, pp. 3-49.

Hendley, D.D., Mildvan, A. S., Reporter, M. C., and Strehler, B. L.: The properties of isolated human cardiac age pigment. I. Preparation and physical properties. J. Gerontol., 18:144-150, 1963.

Hermansky, F. and Pudlak, P.: Albinism associated with hemorrhagic diathesis and unusual pigmented reticular cells in the bone marrow: Report of two cases with histochemical studies. Blood, 14:162-169, 1959.

Hershko, A, and Ciechanover, A.: Mechanisms of intracellular protein breakdown. Ann. Rev. Biochem., 51:335-364, 1982.

Hershko, A.: The ATP-ubiquitin proteolytic pathway, in Progress in Clinical and Biological Research, Vol. 180, Intracellular Protein Catabolism, edited by Khairallah, E. A., Bond, J. S., and Bird, J. W. C., New York, Alan R. Liss, Inc., 1985, pp. 17-31.

Hueck, W.: Pigmenstudien. Beitrag. Path. Anat., 54:68-232, 1912.

Ivy, G. O., Schottler, F., Wenzel, J., Baudry, M., and Lynch, G.:

Inhibitors of lysosomal enzymes: accumulation of lipofuscin-like dense bodies in the brain Science, 226:985-987, 1984.

Ivy, G. O., and Gurd, J. W.: A proteinase inhibitor model of lipofuscin formation, in Lipofuscin-1987: State of the Art, edited by Zs.-Nagy, I., New York, Elsevier Science Publishers, 1988, pp. 83-106.

Jervis, G. A., and Pullarkat, R. K.: Pigment variant of lipofuscinosis. Neurology, 28:500-503, 1978.

Joenje, H., Gille, J. J. P., Oostra, A. B., van der Valk, P.: Some characteristics of hyperoxia-adapted Hela cells. Lab. Invest., 52:420-428, 1985.

Johansson, E., Lindh, U., Alanen, T., Westermarck, T., Heiskala, H., Santavuori, P., and Elovaara, I.: Elemental profiles of blood cells in certain neurological disorders, Medical Biol., 62:139-142, 1984.

Kikugawa, K.: Involvement of lipid oxidation products in the formatin of lipofuscin, in Lipofuscin-1987: State of the Art, edited by Zs.-Nagy, I., New York, Excerpta Medica, 1988, pp. 51-68.

Koneff, H.: Beitrage zur kenntniss der nervonzellenin den peripheren ganglien. Mitth. d. Naturf. Gesellsch. Bern, 44:13-14, 1886.

Koppang, N.: Canine ceroid-lipofuscinosis - a model for human neuronal ceroid-lipofuscinosis and aging. Mech. Ageing Dev., 2:421-445, 1973/1974.

Levine, R. L., Oliver, C. N., Fulks, R. M., and Stadtman, E. R.: Turnover of bacterial glutamine synthetase: Oxidative inactivation precedes proteolysis. Proc. Natl. Acad. Sci. USA, 78:2120-2124, 1981.

Levine, R. L.: Oxidative modification of glutamine synthetase. I. Inactivation is due to loss of one histidine residue. J. Biol. Chem., 258:11823-11827, 1983.

Luys, J.: Recerces sur Le Systeme Nerveux Cerebro-Spinal, Sa Structure, Ses Fonetions Et Ses Maladies. J.-B. Bailliere Et Fils, Paris, 1865.

Martin, A. J. P. and Moore, T.: Changes in the uterus and kidneys in rats kept on a vitamin E free diet. Chem. Ind., 55:236, 1936.

Marzella, L., and Glaumann, H.: Autophagy, microautophagy and crinophagy as mechanisms for protein degradation, in Lysomes: Their Role in Protein Breakdown, edited by Glaumann, H., and Ballard, F. J., New York, Academic Press, 1987, pp. 319-366.

Miquel, J., Oro, J., Bensch, K. G., and Johnson, Jr., J. E.: Lipofuscin: fine structural and biochemical studies, in Free Radicals in Biology, Vol. 3, edited by Pryor, W. A., New York, Academic Press, 1977, pp. 133-182.

Miyagishi, T., Takahata, N., and Iizuka, R.: Electron microscopic studies on the lipo-pigments in the cerebral cortex nerve cells of senile and vitamin E deficient rats. Acta Neuropathologica, 9:7-17, 1967.

Moore, T., and Wang, Y. L.: Fluorescent pigment and vitamin E. Brit. J. Nutrit., 1:53-64, 1947.

Munkres, K. D.: Neurospora age pigments resemble malonaldehyde polymers, in Advances in the Biosciences, Vol. 64. Advances in Age Pigments Research, edited by Totaro, E. A., Glees, P., and Pisanti, F. A., New York, Pergamon Press, 1987, pp. 165-184.

Nagy, I. Zs.-, editor: Lipofuscin-1987: State of the Art, New York, Excerpta Medica, 1988.

Ng Ying Kin, N. M. K., Palo, J., Haltia, M., and Wolfe, L. S.: High levels of brain dolichols in neuronal ceroid-lipofuscinosis and senescence J. Neurochem., 40:1465-1473, 1983.

Nichols, W. W., and Murphy, D. G., editors. DNA Repair Process. Miami, Miami Symposium Specialties, 1977.

Noy, N., Schwartz, H., and Gafni, A.: Age-related changes in the redox status of rat muscle cells and their role in enzyme-aging. Mech. Ageing Devel., 29:63-69, 1985.

Oliver, C.: Lipofuscin and ceroid accumulation in experimental animals, in Age Pigments, edited by Sohal, R. S., New York, Elsevier/North

Holland Biomedical Press, 1981, pp. 335-350.

Oliver, C. N., Levine, R. L., and Stadtman, E. R.: A role of mixed-function oxidation reactions in the accumulation of altered enzyme forms during aging. J. Amer. Geriatrics Soc., 35:947-956, 1987.

Palmer, D. N., Barns, G., Husbands, D. R., and Jolly, R. D.: Ceroid lipofuscinosis in sheep. II. The major component of the lipopigment in liver, kidney, pancreas, and brain is low molecular weight proteins. J. Biol. Chem., 261:1773-1777, 1986.

Pfeifer, W.: Inhibition by insulin of the formation of autophagic vacuoles in rat liver. J. Cell Biol., 78:152-167, 1978.

Pine, M. F.: Turnover of intracellular proteins. Ann. Rev. Microbiol., 26:103-126, 1972.

Pinkerton, H.: The reaction to oil and fats in the lung. A.M.A. Arch. Pathol., 5:380-401, 1928.

Porta, E.A.: Tissue lipoperoxidation and lipofuscin accumulation as influenced by age, type of dietary fat and levels of vitamin E in rats, in Advances in the Biosciences, Vol. 64, Advances in Age Pigments Research, edited by Totaro, E. H., Glees, P., and Pisanti, F. A., New York, Pergamon Press, 1987, pp. 37-74.

Porta, E. A., Mower, H. F., Moroye, M., Lee Ch., and Palumbo, N. E. Differential features between lipofuscin (age pigment) and various experimentally produced "ceroid" pigments, in Lipofuscin-1987: State of the Art, edited by Zs.-Nagy, I., New York, Exerpta Medica, 1988, pp. 341-372.

Porta, E. A., and Hartroft, W. S.: Lipid pigments in relation to aging and dietary factors (lipofuscins), in Pigments in Pathology, edited by Wolman, M., New York, Academic Press, 1969, pp. 191-235.

Pullarkat, R., Reha, H., Patel, V. K., and Goebel, H. H.: Docosahexaenoic acid levels in brains of various forms of ceroid-lipofuscinosis, in Ceroid-lipofucinosis (Batten's disease), edited by Armstrong, D., Koppang, N., and Rider, J. A., New York, Elsevier Biomedical Press, 1982, pp. 335-342.

Reddy, K., Fletcher, B., Tappel, A., Tappel, A. L.: Measurement and spectral characteristics of fluorescent pigments in the tissues of rats as a function of polyunsaturated fats and vitamin E. J. Nutr., 103:908-915, 1973.

Roseman, J. E., and Levine, R. L.: Purification of a protease from Escherichia coli with specificity for oxidized glutamine synthetase. J. Biol. Chem., 262:2101-2110, 1987.

Rosenfeld, M.: Weber das pigment der hamochromatose des darmes. Arch. Exper. Pathol. Pharmak., 45:46-50, 1901.

Rothstein, M.: Enzymes and altered protein. Biochemical Approaches to Aging, New York, Academic Press, 1982, pp. 213-255.

Schoenheimer, R.: The Dynamic State of Body Constituents. Boston, Harvard University Press, 1946.

Sevanian, A., Muakkassah-Kelly, S. F., and Montestrusque, S.: The influence of phospholipase A_2 and glutathione on the elimination of membrane lipid peroxides. Arch. Biochem. Biophys., 223:441-452, 1983.

Siakotos, A. N., Bray, B., Dratz, E., van Kuyk, F., Sevanian, A., and Koppang, N.: 4-Hydroxynonenal: a specific indictor for canine neuronal-retinal ceroidosis. Amer. J. Med. Genetics Supplement, 5:171-181, 1988.

Siakotos, A. N., and Munkres, K. D.: Recent developments in the isolation and properties of autofluorescent lipopigments, in Ceroid-lipofuscinosis (Batten s disease), edited by Armstrong, D., Koppang, N., and Rider, J. A., New York, Elsevier Biomedical Press, 1982, pp. 167-178.

Siakotos, A. N., and Munkres, K. D.: Purification and properties of age pigments, in Age Pigments, edited by Sohal, R. S., New York,

Elsevier/North-Holland Biomedical Press, 1981, pp. 181-202.

Siakotos, A. N., Armstrong, D., Koppang, N., and Mulles, J.: Biochemical significance of age pigment in neurons, in The Aging Brain and Senile Dementia, edited by Nandy, K., and Sherwin, I., New York, Plenum Press, 1977, pp. 99-118.

Siakotos, A. N., and Koppang, N.: Procedures for the isolation of lipopigments from brain, heart, and liver, and their properties: a review. Mech. Ageing Dev., 2:177-200, 1973.

Siakotos, A. N., Watanabe, I., Saito, A., and Fleischer, S.: Procedures for the isolation of two distinct lipopigments from human brain: lipofuscin and ceroid. Biochemical Med., 4:361-375, 1970.

Sohal, R. S., editor: Age Pigments, New York, Elsevier/North Holland Biomedical Press, 1981.

Sohal, R. S.: Metabolic rate, aging, and lipofuscin accumulation, in Age Pigments, edited by Sohal, R. S., New York, Elsevier/North-Holland Biomedical Press, 1981, pp. 303-316.

Stadtman, E. R.: Oxidation of proteins by mixed-function oxidation systems: implication in protein turnover, ageing and neutrophil function. Trends Biochem. Sci., 11:11-12, 1986.

Stadtman, E. R.: Biochemical markers of aging. Exper. Gerontol., 23:327-347, 1988.

Strehler, B. L., Mark, D. D., Mildvan, A. S., and Gee, M. V.: Rate and magnitude of age pigment accumulation in the human myocardium. J. Gerontol., 14:430-439, 1959.

Struck, D. K., and Lennarz, W. G.: in The Biochemistry of Glycoproteins and Proteoglycans, edited by Lennarz, W. J., New York, Plenum Press, 1980, pp. 35-83.

Stubel, H.: Die fluoreszenz tierischer gewebe in ultraviolettem licht. Pfluegers Arch. Gesamte Physiol. Menschen Tiere, 142:1-14, 1911.

Svennerholm, L.: Polyunsaturated fatty acid lipidosis: a new nosological entity, in Current Trends in Sphingolipidoses and Allied Disorders, edited by Volk, B. W. and Schneck, L., New York, Plenum Press, 1976, pp. 389-402.

Tappel, A. L., Fletcher, B., and Deamer, D.: Effect of antioxidants and nutrients on lipid peroxidation fluorescent products and aging parameters in the mouse. J. Gerontol., 28:415-424, 1973.

Thaw, H. H., Collins, V. P., and Brunk, U. T.: Influence of oxygen tension, pro-oxidants and antioxidants on the formation of lipid peroxidative products (lipofuscin) in individual cultivated human glial cells. Mech. Ageing Dev., 24:211-223, 1984.

Totaro, E. A., Glees, P., and Pisanti, F. A.: Advances in the Biosciences, Vol. 64. Advances in Age Pigments Research, New York, Pergamon Press, 1987.

Witkop, C. J.,White J. G., Townsend D., Sedano, H. O., Cal, S. X., Babcock, M., Krumwiede, M., Keenan, K., Love, J. E., and Wolfe, L. S.: Ceroid storage disease in Hermansky-Pudlak syndrome: Induction in animal models, in Lipofuscin-1987: State of the Art, edited by Zs.-Nagy, I., New York, Excerpta Medica, 1988, pp. 413-435.

Wolfe, L. S.: Comment, in Lipofuscin-1987: State of the Art, edited by Zs.-Nagy, I., New York, Excerpta Medica, 1988, p. 181.

Wolfe, L. S., Ng Ying Kin, N. M. K., Baker, R. R., Carpenter, S., and Andermann, F. A.: Identification of retinoyl complexes as the autofluorescent component of the neuronal storage material in Batten s disease. Science, 195:1360-1362, 1977.

Wolman, M.: Factors affecting lipid pigment formation, in Age Pigments, edited by R. S. Sohal, New York, Elsevier/North-Holland Biomedical Press., 1981, pp. 265-281.

Wolman, M.: Lipid pigments (chromolipids): their origin, nature, and significance, in Pathobiology Annual, Vol. 10, edited by Ioachim, H.

L., New York, Raven Press, 1980, pp. 253–267.

Wong, T. K., Decker, G. L., and Lennard, W. J.: Localization of dolichols in the lysosomal fraction of rat liver. J. Biol. Chem. 257:6614–6618, 1982a.

Wong, T. K., and Lennard, W. J.: The site of synthesis and intracellular deposition of dolichol in rat liver. J. Biol. Chem., 257:6619–6624, 1982b.

Zuckerman, B. M.: The nematode Caenorhabiditis elegans as a model for rapid evaluation of cellular aging, in Advances in the Biosciences, Vol. 64, Advances in Age Pigment Research, edited by Totaro, E. A., Glees, P., and Pisanti, F. A., New York, Pergamon Press, 1987, pp.155–163.

DISCUSSION

KATZ: One of the points you emphasized was the probable central role of oxidation in the formation of age-pigment and ceroid pigments. I concur that oxidation probably does play some role because in our in vivo experiments with animals deficient in antioxidant nutrients we have observed acceleration of autofluorescent pigment build-up. However, we have done some in vitro experiments oxidizing the same tissues where we have seen lipopigment accumulation in vivo, and we have seen different fluorophores than those you see in age-pigment. I think, therefore, that the oxidation story might be more complex than simple chemical oxidation that you can reproduce in the test tube.

HARMAN: I think that this depends in part on what is in the cell at the time the process is taking place. In fact, when I was reviewing the literature I began to wonder how much of what is attributed to peroxidation is actually real or an artifact. For example, it is commonly believed that peroxides are present in the atherosclerotic plaques, but these findings, as well as yours, have been done after death. However, if you extrapolate the contents of peroxides from the time of analysis back to the moment of death, you would find almost a linear regression to zero. In other words, the steady-state concentration of peroxides in atherosclerotic plaques may be essentially zero. I don't know for sure how much or what percentage of what people are seen and measuring in these autofluorescent pigments is actually taking place in the cells in vivo, and what is taking place as a function of the extraction procedures. What I would agree is that you may get a lot of different fluorophores depending upon what has been oxidized or undergoing changes at a given time in a cell. You have to keep in mind that in analyzing a tissue you usually work under relatively high concentration of oxygen as compared with the real conditions in vivo, so this will make a big difference in the relative contributions of the components to the final formation of fluorophores.

KATZ: I agree with what you are saying, but I think that in order to prove a mechanism we eventually need to reproduce it in the test tube, and up to now we have not been able to do it. If we are going to demonstrate that these yellow fluorescent pigments are oxidation products, we eventually have to be able to synthesize them, and prove that oxidative reactions are responsible for their formation.

HARMAN: I agree with your last comments.

LIPOFUSCIN AS AN INDICATOR OF OXIDATIVE STRESS AND AGING

R.S. Sohal[1] and U.T. Brunk[2]

[1]Department of Biological Sciences, Southern Methodist University, Dallas, Texas 75275
and [2]Department of Pathology, Linköping University, S-581 85 Linköping, Sweden

SUMMARY

There is a considerable body of evidence indicating that oxidative stress is a causal factor both in lipofuscinogenesis as well as in aging. Studies on the effects of pro-oxidants and antioxidants on lipofuscin accumulation in cultured rat cardiac myocytes and human glial cells indicated that pro-oxidants accelerate while antioxidants retard the rate of lipofuscin accumulation. Lipofuscin was measured by microspectrofluorometry; the reliability of this method was independently validated by comparison with electron microscopical morphometry. *In vivo* studies on hibernating mammals and on insects indicate that rate of lipofuscin is enhanced while life span is shortened by elevation in metabolie rate. The increase in metabolic rate has been shown to enhance the rate of lipid peroxidation, measured as n-pentane exhalation *in vivo*. Overall, it seems that there is enough evidence at hand to reasonably infer that lipofuscin can be used as a marker of oxidative stress and aging.

INTRODUCTION

It is now well established that a greenish-yellowish fluorescent material accumulates progressively during aging in the cytoplasm of several post-mitotic cell types in a broad variety of animal species[1-6]. The chemical composition of this substance, often referred to as lipofuscin, is presently poorly defined and there appears to be considerable variation in its constituents detectable by cytochemical staining or by biochemical analysis[1]. However, lipofuscin invariably contains polymerized residues of peroxidized lipids and proteins incorporated into the secondary lysosomes by autophagocytosis of cellular constituents. Ever since its discovery more than a century ago, lipofuscin has retained the interest of gerontologists apparently in the hope that this marker of cellular aging may provide clues to the underlying mechanism of aging. Accordingly, a variety of experimental and comparative studies have been conducted to identify the factors that influence lipofuscinogenesis and link it to the aging process. Although we are still in the midst of this quest, a considerable amount of knowledge has been gained to lead to the postulation of

KEY WORDS: Lipofuscin, Oxygen, Free radicals, Aging, Age pigment, Cardiac myocytes, Glial cells, Cellular senescence

Lipofuscin and Ceroid Pigments
Edited by E. A. Porta
Plenum Press, New York, 1990

two hypotheses regarding the causes of lipofuscin accumulation and its role in the aging process. One hypothesis proposes oxidative stress, i.e., the ratio of pro-oxidants to anti-oxidants as the main factor[1,3,7-10], while the other hypothesis regards a decline in the degradative ability of cells to be primarily responsible for lipofuscin formation[11-13]. The two hypotheses are not necessarily mutually incompatible as it is possible that oxidative reactions involving lipid and protein peroxidation, among others, may induce both a decline in the activity of lytic enzymes as well as an increase in the production of modified molecules that are precursors of lipofuscin. Nevertheless, the key difference between the two hypotheses is that which of the two processes plays the most dominant role in lipofuscinogenesis. It is fair to say that in general the first hypothesis is more popular among gerontologists while the second is favored by pathologists and some experimentalists dealing with the accumulation of "ceroid", a related pigmented substance, which seems to be an early, somewhat immature, variety of lipofuscin.

We have conducted a series of investigations examining the role of oxygen free radicals in cellular aging. The objective of this paper is to provide a synthetic overview of the evidence supporting the concept that oxidative stress is a causal factor in lipofuscinogenesis as well as in aging. Our overall hypothesis is that lipofuscin can be used as an indicator of both oxidative stress and of rate of aging.

TERMINOLOGY AND QUANTIFICATION OF LIPOFUSCIN

Unfortunately, understanding the mechanism of lipofuscinogenesis has been hampered by a confusion in the terminology, the purported chemical nature, as well as the methods employed for the quantification of lipofuscin (see references 10 and 14 for detailed discussion). Most of the confusion can be traced to the lack of knowledge of the chemical nature of lipofuscin as well as the absence of any specific compound of known chemical composition that can be diagnostically associated with lipofuscin. Classically, the term lipofuscin has been used to refer to the material emitting greenish to yellowish autofluorescence when excited with blue light. This material is present in membrane-bound organelles that can be identified as secondary lysosomes and has the cytochemical characteristics of a lipoproteinous material associated with carbohydrate groups[3,4]. The fluorescent material is resistant to extraction in aqueous or organic solvents. The chemical nature of the material that is specifically responsible for the characteristic fluorescence is presently completely unknown. Because of the lack of such information and to avoid confusion in the future, it is imperative to reach a certain consensus about terminology. Presently, lipofuscin is being identified and measured by four different methods:

(1) The most reliable method measures the autofluorescence of lipofuscin granules and employs microspectrofluorometry of cells *in situ*.

(2) Another method uses cytochemical staining and morphometry of sectioned material and can only provide semi-quantitative measurements. The drawback of this method is that it cannot determine if all the granules under study exhibit fluorescence, and if they do of what intensity?

(3) Another morphological method employs electron microscopy and morphometry of sectioned material. It suffers from the handicap that it cannot specifically identify the relative fluorescent content of the secondary lysosomes.

(4) One method that has been frequently used in recent studies and the one which has contributed the most to the current confusion in terminology; is based on the measurement of blue-emitting fluorescence in the organic solvent extracts of tissues[8]. This material is claimed to be derived from lipofuscin although there is absolutely no direct evidence to support this assertion. Indeed, studies using this method have generated much controversy because the results obtained by this methodology are often at odds with those obtained by microspectrofluorometry or by microscopy. We have previously provided various instances of such discrepancies[10,14]. We therefore suggest that the term

"lipofuscin" be only used for the material that specifically exhibits *in situ* the characteristics of lipofuscin, namely greenish-yellowish autofluorescence eminating from the lysosomal vacuome. In our opinion, microspectrofluorometry of the *in situ* granules is currently the most reliable method for the quantification of lipofuscin. We emphatically suggest that the blue-emitting fluorescent material should not be termed "lipofuscin or lipofuscin-related or lipofuscin-derived".

EVIDENCE SUPPORTING OXIDATIVE STRESS AS A FACTOR IN LIPO-FUSCINOGENESIS

There is a body of evidence, based on both *in vitro* and *in vivo* studies, which supports the concept that oxidative stress is a causal factor in lipofuscinogenesis.The *in vitro* evidence is based on studies on rat cardiac myocytes[15,16] and human glial cells[8].

CARDIAC MYOCYTES

Cardiac myocytes of neonatal rats can be kept in culture for about a month under the conditions used in our laboratory and when the cells are cultured directly on glass or on plastic. In one study, effects of 5 %, 20 %, and 40 % ambient oxygen on the amount of lipofuscin were monitored for 12 days of *in vitro* age by microspectrofluorometry (Fig. 1). After 7 days of in vitro age, cells exhibited considerable amounts of lipofuscin that were higher in cells kept in 20 % and 40 % oxygen than those kept in 5 % oxygen. More marked differences in the quantity of lipofuscin emerged at 12 days of age. As compared to those kept in 5 % oxygen, the amount of lipofuscin was 47 % and 250 % greater in cells placed in 20 % and 40 % oxygen atmosphere, respectively. Between 7 and 12 days of in vitro age, lipofuscin amount increased by 1 %, 14 %, and 89 % in cells kept under 5 %, 20 %, and 40 % oxygen, respectively. Results of this study clearly indicated that age-related increase in the amount of lipofuscin was directly related to the concentration of ambient oxygen.The mechanism by which hyperoxia induces lipofuscinogenesis may involve superoxide anion radical (O_2^-). A direct relationship between hyperoxia and increased mitochondrial O_2^- generation has been demonstrated[17]. It is widely believed that the effects of hyperoxia on cells are mediated by oxygen-derived free radicals (O_2^- and $HO\cdot$) and H_2O_2. It can thus be reasonably argued that since they are normally produced during cell metabolism O_2^- and H_2O_2 may also be involved in lipofuscinogenesis.

To further validate our model system, fluorescent characteristics and ultrastructural morphology of the lipofuscin-containing organelles was studied (Figs. 2,3). Using different barrrier filters the emission maxima of autofluorescence, exhibited by lipofuscin in the cardiac myocytes was found to range between 510 to 530 nm[15]. The corrected emission maxima was, however, around 570 nm. Electron microscopic examination of the myocytes was conducted at 3, 8, and 12 days of age in order to also elucidate the morphogenesis of lipofuscin. At 3 days of age, lipofuscin-related material exhibited a dense and finely or coarsely granular matrix with interspersed membranous lamellae. At 8 and 12 days of *in vitro* age, lipofuscin-containing organelles exhibited a progressively larger size and greater structural heterogeneity (Fig. 3).

In the current view, the most reactive, partially reduced, oxygen species in oxidative stress is the hydroxyl free radical ($HO\cdot$)[19]. We therefore investigated the effects of putative enhanced generation of $HO\cdot$ on lipofuscin formation. Transition metal (e.g., Fe and Cu) compounds apparently play a key role in the production of $HO\cdot$ by the reduction of H_2O_2 via an O_2^--driven Fenton reaction.

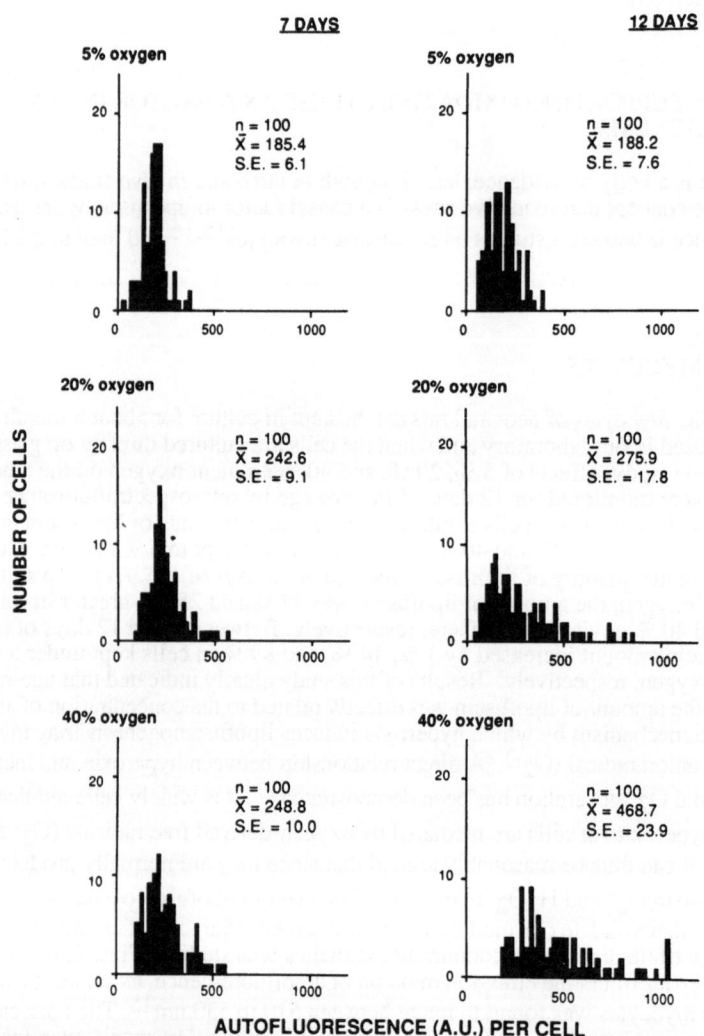

Fig 1. Computer-generated histograms of autofluorescence exhibit by cardiac myocytes kept in 5%, 20%, and 40% oxygen for 7- and 12 days. Abscissa: Autofluorescence/cell (arbitrary units). Ordinate: Number of cells. A total of 100 cells were examined in each group. (Reproduced from Ref. 15).

$$x -- Fe^{3+} + O_2^- ----> x -- Fe^{2+} + O_2$$
$$x -- Fe^{2+} + H_2O_2 ----> x - Fe^{3+} + OH^- + HO\cdot$$

Thus effects of 20 µM ferric iron, added to the culture medium, on lipofuscin levels were examined in cardiac myocytes kept under 20 % and 40 % ambient oxygen[16]. At both, 6 and 12 days of *in vitro* age, the amount of lipofuscin found in myocytes exposed to 20 µM ferric iron was markedly greater than in the controls. Again, the effect were more pronounced under 40 % than under 20 % oxygen. In a separate experiment, effects of various concentrations of desferrioxamine, which binds iron, on lipofuscin content were examined[16]. The amount of lipofuscin was found to be inversely related to the concentration of desferrioxamine. Interestingly, the protective effects of desferrioxamine against lipofuscin accumulation were more pronounced in cells kept under 40 % oxygen as compared to those kept under 20 % oxygen. Thus oxygen and iron seem to act synergistically to accelerate lipofuscin accumulation as would be expected if oxygen free radicals were playing a causal role in lipofuscinogenesis.

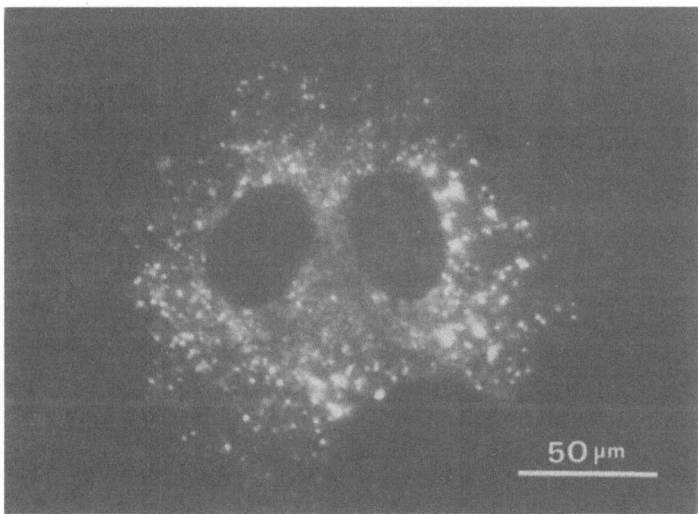

Fig. 2. Photomicrograph of cardiac myocytes from a culture kept in 40% oxygen, 55% nitrogen and 5% carbon dioxide for 12 days, using blue wave light. Cells contain numerous autofluorescent granules in the perinuclear area. (Reproduced from Ref. 15).

Electron microscopic examination and energy dispersive X-ray microanalysis of cells exposed to ferric iron[16] indicated that iron is sequestered in the lipofuscin-containing organelles (Fig. 4). Presumably, iron is taken up by endocytosis and then stored within secondary lysosomes where the acidic conditions would favor the loose binding of iron. It is therefore possible that lipofuscin formation, resulting from the reactions of reduced oxygen species may occur mostly in the lysosomal vacuome.

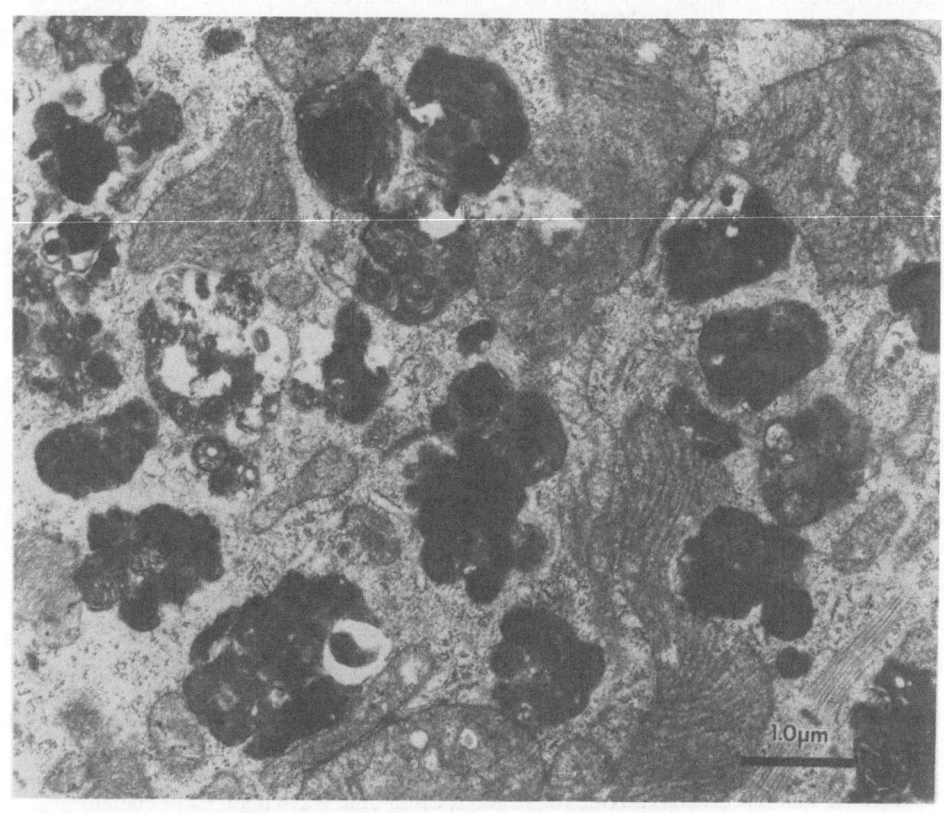

Fig 3 Electron micrograph of a cardiac myocyte after 8 days in culture in 20% oxygen.

To further test the involvement of lipid peroxidation, effects of vitamin E on lipofuscinogenesis were examined. Exposure of rat cardiac myocytes to vitamin E markedly decreased the rate of lipofuscin accumulation. The protective effects were more pronounced under 20% and 40 % oxygen atmosphere than under 5 % oxygen.

Ethanol, a reductant, also decreased the rate of lipofuscin accumulation in cardiac myocytes at a concentration of 3.1 mM, however, the rate of lipofuscinogenesis was accelerated in 12.5 mM ethanol. Both of these effects were more pronounced under 40 % ambient oxygen than under 20 % ambient oxygen.

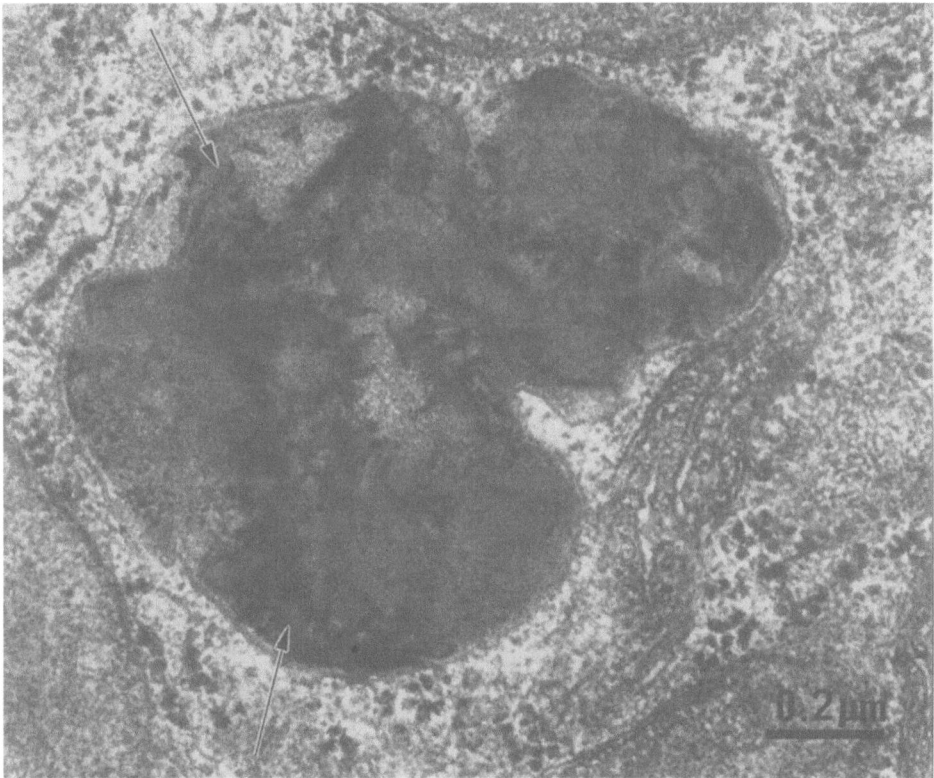

Fig 4. Electron micrograph of lipofuscin-containing organelle in an 8-day-old myocyte cultured in the medium containing 30 μM ferric chloride showing dense needle-shaped material, presumably iron (arrow). (Reproduced from Ref. 16).

GLIAL CELLS

Diploid human glial cells kept in the state of density-dependent inhibition of growth, *in vitro,* accumulate lipofuscin within the lysosomal vacuome[8]. Lipofuscin was found to accelerate in direct correspondence to oxygen tension. For example, after 12 weeks in culture, cells exposed to 40 % ambient oxygen contained about 4-fold greater amount of lipofuscin than those exposed to 5 % oxygen. Exposure of glial cells for 12 weeks to vitamin C/Fe, that putatively generates HO·, increased the amount of lipofuscin 2-fold as compared to controls (Fig. 5). DMSO, reduced glutathione, and vitamin E/Se decreased the rate of accumulation of lipofuscin[8].

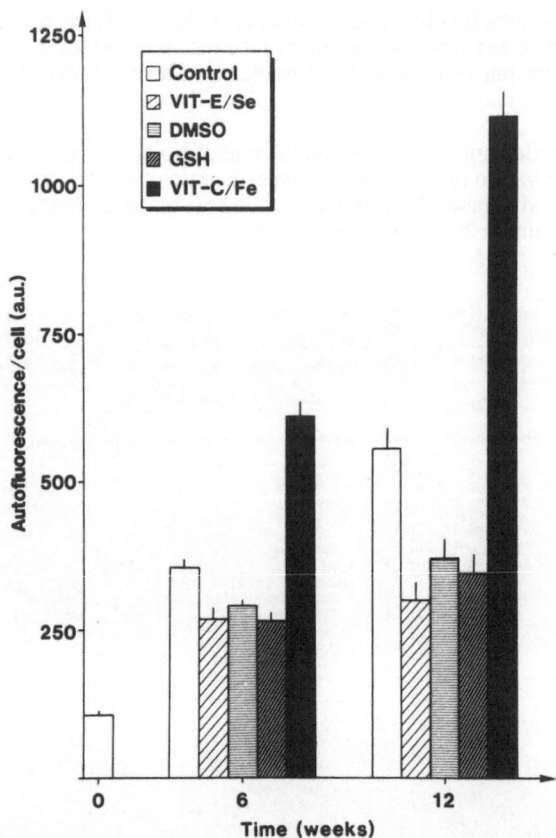

Fig. 5. Abscissa: time (weeks). Ordinate: median autofluorescence per cell for cells grown under the indicated conditions and subcultivated at different time following density-dependent inhibition of growth. Antioxidants such as DMSO, GSH and Vit-E/Se slow down the rate of lipofuscin formation, while pro-oxidants such as Vit-C/Fe increase the rate of lipofuscin in comparison to controls at both 6 and 12 weeks. (Reproduced from Ref. 8).

Results of these studies were thus consistent with those obtained in cultured rat cardiac myocytes. Altogether, our results stongly support a causal role for oxygen free radicals in lipofuscinogenesis. They do not , however, fully elucidate the underlying mechanisms. The role of O_2^- and H_2O_2 is indicated by studies employing various oxygen tensions; The role of HO· is suggested by the studies using ferric iron. And, the role of lipid peroxidation in lipofuscinogenesis is indicated by the ameliorative effects of vitamin E against lipofuscin accumulation. The process by which lipid peroxidation leads to lipofuscin formation is, however, a matter of controversy. Tappel and his coworkers have proposed a chemical model of lipid peroxidation leading to the formation of Schiff's base-containing fluorescent material, which is claimed to be responsible for the autofluorescence of lipofuscin[7]. The main weakness of this hypothesis is that while Schiff's base emits a blue fluorescence, lipofuscin, *in situ* emits yellowish or greenish fluorescence. Formation of fluorophores of such coloration cannot presently be reconciled with Tappel's model. However, this should not taken to mean that lipid peroxidation is not involved in lipofuscinogenesis. Quite clearly, we do not as yet understand the nature of the reactions that link oxidative damage with the formation of the autofluorescent, cross-linked lipoproteinous, lipofuscin material that is stored by cells in the lysosomal vacuome.

LIPOFUSCIN AS AN INDICATOR OF RATE OF AGING

The results of comparative studies on the rates of lipofuscin accumulation in animals of varying longevities and experimental studies on the factors influencing lipofuscin formation indicate that lipofuscin accretion is linked to the aging process. It is, of course, well known that lipofuscin accumulates progressively with age in a variety of cell types. However, more notaby, the rate of lipofuscin accumulation responds to variations in the rate of aging. The rate of lipofuscin accumulation in cardiac myocytes of dogs is approximately 5.5 time faster than in humans, which is, inversely, comparable to the difference in their life spans[20]. The same is valid for short-lived mammals such as mouse and rat as compared to long-lived species.

Experimental variations in the rate of aging have been found to be reflected in the rates of lipofuscin accumulation in mammals as well as insects. For example, when the life spans of the Turkish hamsters were prolonged by induced hibernation, lipofuscin accumulation in the heart and the brain was found to be slower in hibernators than in non-hibernators[21]. Similarly, the amount of lipofuscin-like granules in three different cell types, detected by electron microscopy, was found to be lower in houseflies, whose life span had been experimentally extended by decrease in the level of their physical activity[22,23]. There are also several other examples in the literature where tissues undergoing sub-normal level of activity were reported to also exhibit reduced amounts of lipofuscin (see Ref. 1).

Altogether, the aforementioned evidence suggests that lipofuscin accumulation is causally related to oxidative stress and is also linked to the aging process. This relationship supports the concept that the mechanism of lipofuscinogenesis and aging may involve in common the reactions of the active oxygen species. An increase in metabolic rate of the houseflies by elevation of ambient temperature has been found to enhance the rate of lipid peroxidation as measured by the exhalation of n-pentane and to shorten life span. Further knowledge of the chemical composition of lipofuscin and reactions involved in lipofuscinogenesis is, however, necessary to fully understand the relationship between lipofuscin and the aging process.

ACKNOWLEDGEMENTS

Research of the authors has been supported by grants from N.I.H. - N.I.A. and Swedish Medical Research Council (No. 4481).

REFERENCES

1. R.S Sohal, and L.S Wolfe, Lipofuscin: characteristics and significance. *Progr. Brain Res.*, 70:171-183 (1986).
2. R.S Sohal., ed., *Age Pigments*, Elsevier/North Holland, Amsterdam (1981).
3. H.Donato and R.S.Sohal, Lipofuscin, in *Handbook of Biochemistry in Aging*, J.Florini, ed., CRC Press, Cleveland, 221-227 (1981).

4. C.L.Dolman, and P.M.MacLeod, Lipofuscin and its relation to aging, in *Advances in Cellular Neurobiology*, Vol 2, S.Fedoroff, and L.Hertz, eds., Academic Press, New York pp. 205-247 (1981).

5. M.Wolman, Lipid pigments (chromolipids): their origin, nature and significance, in *Pathobiology Annual*, Vol 10, H.L.Joachim, ed., Raven Press, New York 253-267 (1980).

6. J.Miquel, J.Oro, K.G.Bensch, and J.E.Johnson Jr., Lipofuscin: fine structural and biochemical studies, in *Free Radicals in Biology*, Vol 1, W.A. Pryor, ed., Academic Press, New York 133-182 (1977).

7. A.L.Tappel, Lipid peroxidation and fluorescent molecular damage to membranes, in *Pathobiology of Cell Membranes*, Vol 1, B.Trump, and A.Arstila, eds., Academic Press, New York 145-172 (1975).

8. H.H.Thaw, U.T.Brunk, and V.P.Collins, Influence of oxygen tension, pro-oxidants and anti-oxidants on the formation of lipid peroxidation products (lipofuscin) in individual, cultivated human glial cells. *Mech. Ageing Dev.* 24: 211-223 (1984).

9. U. Brunk, and E.Cadenas, The potential intermediate role of lysosomes in oxygen free radical pathology, *APMIS* 96:3-13 (1988).

10. U.T.Brunk, and R.S. Sohal, in *Lipid Peroxidation.*, vol II, CRC Press, Boca Raton (in press).

11. M.Elleder, Chemical characterization of age pigments, in *Age Pigments*, R.S.Sohal, ed., Elsevier/North Holland, Amsterdam 204-241 (1981).

12. L.S. Wolfe, S.Gauthier, and H.D.Durham, Dolichols and phosphorylated dolichols in the neuronal lipofuscinosis, other lysosomal storage diseases and Alzheimers disease. Induction of autolysosomes in fibroblasts, in *Lipofuscin - 1987: State of the Art*, I.Zs.-Nagy, ed., Akademiai Kiado, Budapest and Elsevier, Amsterdam 389-411 (1988).

13. G.O. Ivy, and J.W.Gurd, A proteinase inhibitor model of lipofuscin formation, in *Lipofuscin- 1987: State of the Art*, I.Zs.-Nagy, ed., Akademiai Kiado, Budapest and Elsevier, Amsterdam 83-108 (1988).

14. R.S.Sohal, Quantification of lipofuscin: a critique of the current methodology, in *Advances in Age Pigment Research*, E.J. Totaro *et al.*, , eds., Pergamon Press, Oxford 85-91 (1987).

15. R.S.Sohal, M.R.Marzabadi, D.Galaris, and U.T.Brunk, Effect of ambient oxygen concentration on lipofuscin accumulation in cultured rat heart myocytes - a novel *in vitro* model of lipofuscinogenesis, *Free Rad. Biol. Med.* 6: 23-30 (1988).

16. R.M.Marzabadi, R.S.Sohal, and U.T.Brunk, Effect of ferric iron and desferrioxamine on lipofuscin accumulation in cultured rat heart myocytes. *Mech. Ageing Dev.*, 46:145-157 (1988).

17. J.F.Turrens, B.A.Freeman, J.G.Lewitt, and J.D.Crapo, The effect of hyperoxia on superoxide production by lung submitochondrial particles. *Arch. Biochem. Biophys.* 217: 401-410 (1982).

18. B.Halliwell, Free radicals, oxygen toxicity and aging, in *Age Pigments*, R.S.Sohal, (ed.), Elsevier, Amsterdam pp. 1-62 (1981).

19. B.Halliwell, and J.M.C.Gutteridge, The importance of free radicals and catalytic metals in human diseases, *Mol. Aspects Med.*, 8:89-193 (1985).

20. J. Munnel, and R. Getty, Rate of accumulation of cardiac lipofuscin in the aging canine. *J. Gerontol.* 23:154-158 (1968).

21. E.D. Papafrangos, and C.P. Lyman, Lipofuscin accumulation and hibernation in Turkish hamster, *Mesocricetus brandti, J. Gerontol.* 37:417-421 (1982).

22. R.S. Sohal, and J. Donato, Effect of experimental prolongation of life span on lipofuscin content and lysosomal enzyme activity in the brain of the housefly, *Musca domestica. J. Gerontol.* 34:489-496 (1979).

23. R.S. Sohal, Relationship between metabolic rate, lipofuscin accumulation and lysosomal enzyme activity during aging in the adult housefly, *Musca domestica. Exp. Gerontol.* 16:347-353 (1981).

24. R.S. Sohal, A. Muller, B Kolezko, and H Sies, Effect of age and ambient termperature on *n*-pentane production in adult housefly, *Musca domestica. Mech. Ageing Develop.* 29:317-326 (1985).

DISCUSSION

BERTONI-FREDDARI: Since in your houseflies the amounts of lipofuscin that accumulate in cells have been determined by electron microscopy, did you observe any mitochondrial changes attributable to oxidative stress?

SOHAL: We have seen some mitochondrial alterations, but nothing that we can say "diagnostic".

Zs.-NAGY: I have two comments; the first, on terminology. Although practically everyone is referring to SOD as an antioxidant, it should be realized that superoxide ion is a reductive agent, and therefore, SOD should be called "antireductant". So, in any calculation about the ratio between prooxidants and antioxidants, if SOD is included into the antioxidant category, this is obviously wrong.

SOHAL: All what I did was to look at the levels of SOD, glutathione and other antioxidants as well as prooxidants and find no correlations of their levels and maximum life span. I am not an expert in SOD, but what you say is the subject of wide controversy in the field.

Zs.-NAGY: Well, if the logic indicates that SOD is an antireductant, then superoxide ion radical can be the best scavenger of hydroxyl free radical. In fact, if you make in vitro spin trapping experiments, you will have clear proof that every type of ubiquinone has hydroxyl free radical scavenging effect, but the mediator of that reaction is superoxide ion radical. This means that we have to change our idea about the overall harmful effect of superoxide ion. The other comment I have is that at the end of your presentation, you strongly declared that the old idea about accumulation of damage in the aging process should be discarded, and I disagree with that. What you have said in your slide about physical training doesn't make sense for me, because if I do physical exercise, my muscle volume will really increase, and I will be then able to do more physical work. What you have shown, it may be valid perhaps for the postmitotic cells you have in the fly, but not valid for higher animals and mammals.

SOHAL: You may be right, but I did not present my work on cellular differentiation which forms the basis of my hypothesis that oxidative stress can regulate gene activity.

Zs.-NAGY: I do agree that hypotheses are important, but you should not make hypotheses in the air, because if you discard everything that has been done so far, and you propose something new without any background, it will collapse.

HARMAN: Dr. Sohal, you looked at lipofuscin accumulation as a function of aging. Was that an exponential function?

SOHAL: We found it to be somewhat linear rather than exponential.

HARMAN: I think that everybody would agree that aging is genetically programmed. The question is how as a program in growth and differentiation would be, or is a program in the sense that it is a random sort of thing which comes out that you would live a hundred years or so at the most? I will opt for the latter explanation.

SOHAL: I will opt for the first one because there is no difference in antioxidant levels, but there is a very significant difference in the rate of production.

HARMAN: There are a number of studies indicating that the mitochondrial efficiency goes down with age. Both biochemically and by electron microscopy, you will see mitochondrial alterations, and this is where the problem is. The inner mitochondrial membrane is one of the most complex membranes in the body, and some of the arguments that have been raised against the idea that free radical reactions are the major cause of aging have been simply ignoring the fact that at least most of the compounds that have been used do not get into the mitochondrial matrix and the mitochondrial chain.

SOHAL: I don't know whether we will ever have the ability to put things into the mitochondria and expand the maximum life span.

IVY: I would totally agree with the concept that aging is a programmed event, and that is the oxidative stress per se and not time which we will be concerned with. However, I think that it is possible that the program is going on at a rate determined by oxidative stress. But we cannot ignore that it is the production of that damage that is causing the manifestation of aging and the deterioration of the organism.

SOHAL: I agree with you. The point I am trying to make is that free radicals should not be thought as deleterious. They should be considered as playing a physiological role concerned with cellular differentiation and the maintenance of the differentiated state. I agree that there is damage, but I don't know how a damage that is supposedly redundant is driving the genetic program.

IVY: I think that random damage can occur at a given rate depending on the process causing it.

SOHAL: Age related changes are very reproducible, and how random damage can produce reproducible changes is difficult to explain.

HARMAN: Since in aging we frequently talk about the damage produced by free radical reactions, we should emphasize that probably the vast majority of these reactions are absolutely necessary to our function. Without it, we won't be here. But, because of the nature of this chemical reaction, certain amounts of free radicals are randomly leaking. As we sit down or stand up here, we are continuously bombarded by random free radicals from cosmic and terrestrial radiations. The greatest bulk of damaging free radical reactions, however, are those involved in normal function.

Zs.-NAGY: Dr. Sohal, your declaration that the new thing in your hypothesis is that free radicals are physiologically necessary is nothing new. Your hypothesis may sound as a new one, but is not new at all.

SOHAL: All what I am saying is on the basis of what we found experimentally. All that I am saying is that perhaps we should also look at the free radicals as physiologically important and regulated by the cells, rather than simply think in terms of damage.

KATZ: According to your theory, you are saying that what is really programmed is the total metabolic capacity and not what determines life span. So, what your hypothesis will predict is that people who exercise more will live a shorter life. It is any way in which you can test this hypothesis in mammals or warm animals rather than in flies?

SOHAL: I think that your question can be answered by saying that most of the energy that is expended by mammals is to maintain the basal metabolic rate. The energy expended by jogging, for example, is a very small fraction. Different organisms are built differently, and some are better models to discover basic biological principles than others. Perhaps flies, whose all somatic cells are postmitotic, are a better model than mammals.

PORTA: I have a comment and a question. The comment is about the well known problem of unwarranted extrapolations and biological generalizations between such different species as flies and mammals. The aging process is undoubtedly a universal phenomenon, and the fly model may be important for the understanding of certain mechanisms of aging, but I am disturbed by the generalizations and extrapolations you made in your presentation. For example, you mentioned that the oxidative potentials, as measured by the TBA reactive substances in vitro, increase with age. Although this may be perhaps the case in your flies, all the available data in mammals are clearly different. In man as well as in rodents, the oxidative potentials in blood and diverse solid tissues usually increase until maturity, but rather sharply decline during aging. Regarding the decrease in the levels of vitamin E with age that you mentioned and even showed in one of your slides, you should recognize that most of the data in mammals have rather consistently indicated that in practically all tissues, whether composed mainly by fixed or by renewable cells, the concentrations of vitamin E generally increase during the whole life span.

SOHAL: I emphasized vitamin E decline in flies because their diet did not contain this vitamin E, and therefore the tissue decline is an indication of consumption or elimination. All I can say is that flies live longest and consumed most amounts of oxygen if fed sucrose.

PORTA: Well, if this is all you can say, I still don't understand how you can make generalizations and propose a hypothesis with very limited and conflicting data. Now, my question is about the influence of ambient oxygen concentrations on lipopigment formation in cultured rat cardiac myocytes and human glial cells. We all agree, of course, that lipofuscin can be used as a marker of aging. This has been known for more than one hundred years, but I don't know why, based on the data you presented on cultured cells, we have to accept that the age-pigment can be used as a marker of oxidative stress. Do you realize that even the lowest concentration of 5% oxygen used in your in vitro studies is much higher than the physiologically normal concentrations of oxygen in the microenvironment of these cells in vivo?

SOHAL: All what we did was to increase the oxygen concentration as a mean to increase the generation of superoxide radical and hydroxyl radical.

KITANI: This point of excessive partial tension of oxygen as commonly used in vitro biochemical and pharmacological studies is very important. It has been found in Japan, for example, that if we use more physiological oxygen tension in vitro, the enzymatic activities of the cells considerably differ from those under excessive tensions.

HARMAN: I would like to say a few words in defense of exercise. It is true that if we exercise the production of free radicals increases not only in muscles but also in liver. While there is not good evidence that people who exercise moderately do live longer, there is, however, evidence that as they grow older they are in better health conditions; the skeleton muscles and joints are better preserved anatomically and functionally, so it would be a good idea to stop this discussion and go swimming in this beautiful place.

LIPOFUSCIN-LIKE SUBSTANCES ACCUMULATE RAPIDLY IN BRAIN, RETINA AND

INTERNAL ORGANS WITH CYSTEINE PROTEASE INHIBITION

G.O. Ivy, S. Kanai, M. Ohta, G. Smith, Y. Sato
M. Kobayashi and K. Kitani

University of Toronto, Scarborough, Ontario
Canada and Tokyo Metropolitan Institute of
Gerontology, Tokyo, Japan

SUMMARY

The protease inhibitor, leupeptin, has been previously shown to cause
an accumulation of lysosomally associated intracytoplasmic dense bodies
resembling lipofuscin when administered intraventricularly to the brains
of young rats (Ivy et al., Science, 226:985, 1984). These findings
support the idea that lipofuscin formation during normal aging involves
perturbed proteolytic degradation. In the current study, we delineate
more precisely the proteolytic mechanisms in question and examine the
effects of protease inhibition on other tissues and species.

Four sets of experiments were done. In the first, rats were administered leupeptin (L; an inhibitor of cysteine and some serine proteases),
E-64C (a cysteine protease inhibitor), chloroquine (C; a lysosomal enzyme
inhibitor), aprotinin (A; a serine protease inhibitor) or physiological
saline (S) into the lateral ventricle of the brain at a constant rate for
two weeks using an osmotic mini-pump. In the second set, beagle dogs were
administered L or S intraventricularly in a similar fashion. In the third
set, young rats received intraocular injections of L, E64C, C, A or S from
1 to 9 days at 24 hr intervals. In the fourth set, young rats and mice
were administered L, E-64C or S intraperitoneally via an osmotic mini-pump
for 2 or more weeks. The tissue of all animals was processed for light and
electron microscopy as well as for fluorescence analysis.

In rat brain cells (especially hippocampus and cerebellum), substances which by morphological, histochemical and fluorescence emission
criteria resemble lipofuscin, accumulated following L, C or E64C, but not
A or S treatment. A similar buildup of lipofuscin-like substance occured
in canine brain following L but not S infusion. In retinal pigment
epithelial cells, undigested rod outer segment discs accumulated after L,
C or E-64C and possibly A, but not after S injections. With increasing
survival times, the accumulated substances became morphologically more
similar to lipofuscin. Cells of liver (both hepatocytes and non-parenchymal cells) and kidney (mainly renal tubule cells) also accumulated
substances resembling lipofuscin in response to L but not S infusion;
other internal organs were differentially affected.

Lipofuscin and Ceroid Pigments
Edited by E. A. Porta
Plenum Press, New York, 1990

These results demonstrate 1) that specific inhibition of cysteine proteases is sufficient to cause accumulation of lipofuscin-like substances in brain and retina and 2) that several species and organ systems are similarly affected by protease inhibition. Taken together, these findings are consistent with the idea that defective or decreased proteolysis by lysosomal cysteine proteases is responsible for lipofuscin accumulation during normal aging.

KEY WORDS: Lipofuscin, protease inhibitors, proteolysis, lysosomes, brain, retina, internal organs

INTRODUCTION

Lipofuscin has long been perceived as a hallmark of cellular aging, both because it is relatively indiscriminate in regard to the type of tissue or organism it builds up in and because it accumulates linearly with age. These attributes demarcate lipofuscin as a potential key to the mechanism(s) underlying the aging process. Understanding the mechanisms of lipofuscin formation then, may ultimately unlock the mysteries of aging.

There are a number of ways in which a buildup of lipofuscin-like substances (LLS) may be brought about in young tissues (Porta et al., 1988). For example, vitamin E deficiency (Katz et al, 1984), excess heavy metals (Gutteridge, 1988; Pisanti et al, 1988), various diseases (Goebel, 1988, Koppang, 1988, Witkop et al., 1988), various toxins (Zuckerman and Geist, 1981) and inhibitors of lysosomal enzymes (Bhattacharyya et al, 1983; Ishikawa, et al, 1983; Ivy et al, 1984) all cause a buildup of substances in the cytoplasm of various cell types which closely resemble the lipofuscin that accumulates during the process of normal aging. Such treatments have in common that they overtax the metabolic machinery of the cell, leading to storage of cellular components which would normally be recycled through the lysosomal system. The various treatments may thus tell us much or nothing at all about what exactly taxes the cellular metabolic machinery during normal aging.

For about five years, we have been working on a "protease-inhibitor model of aging" which was initially based on the discovery that inhibition of lysosomal proteases by leupeptin causes a massive accumulation of LLS in brain cells of young rats (Ivy et al, 1984). The power of this model is that it also produces other manifestations of aging, at least in brain tissue. These include selective neuronal death, intraneuronal aggregation of neurofilaments, accumulation of abnormal tau molecules in dendrites and perikarya, and intracellular accumulation of ubiquitinated proteins (Ivy, 1987; 1988; Ivy et al., 1986; 1989).

If this model is to be validated as a means of studying lipofuscin formation and the aging process in general, however, it must be tested in other tissues and animals. This has been one of our major goals in the last few years and we now report on the effects of protease inhibition in dog brain, rat retina and rat and mouse internal organs. Our second major goal has been to more precisely define the classes of proteases responsible for causing the various effects observed. We have thus administered inhibitors of different classes of proteases to rat brain, retina and internal organs in order to compare their effects.

We present evidence here that specific inhibition of lysosomal cysteine (thiol) proteases is sufficient to cause accumulation of LLS in several cell types and we discuss mechanisms by which decreased activity of these proteases may cause lipofuscin deposition during normal aging.

MATERIALS AND METHODS

Drugs

Leupeptin, an inhibitor of both lysosomal and cytoplasmic cysteine proteases, and, to a lesser degree cytoplasmic serine proteases, was obtained from Boehringer Mannheim and Peptide Inst., Osaka. E-64C, a specific inhibitor of lysosomal and cytoplasmic cysteine proteases was a generous gift from Dr. K. Hanada of Taisho Pharmaceutical Co. Chloroquine, a general inhibitor of a lysosomal enzymes and aprotinin, a specific inhibitor of serine proteases, were obtained from Sigma. Leupeptin, aprotinin and chloroquine were dissolved in buffered physiological saline solution and E-64C was dissolved in saturated $NaHCO_3$. In each of the experiments described below, physiological saline was used as the control vehicle for leupeptin, aprotinin and chloroquine; 3.1% saline was used as the control for E-64C, to control for osmolarity.

Rat Brains

Sprague Dawley rats (40-60d) were administered leupeptin (5-60 mg/ml), chloroquine (40-60 mg/ml), aprotinin (40-60 mg/ml), E64C (40-50 mg/ml) or vehicle into the lateral ventricle of the brain using a mini osmotic pump (Alzet, Alza Corp, Palo Alto, CA). The pumps administered $0.5 \mu 1$ of drug or vehicle into the brain every hour at a constant rate for 2 weeks. With leupeptin, the time course was varied from 1-16 weeks, using pump replacement. At time of sacrifice, rats were perfused intracardially for light microscopy (saline followed by 4% paraformaldehyde in phosphate buffer) or electron microscopy (saline followed by 2.5% paraformaldehyde plus 2.5% glutaraldehyde in phosphate buffer). For light microscopy, brains were embedded in paraffin or were frozen sectioned; for electron microscopy, standard procedures were used for taking ultrathin sections. Paraffin sections were examined for presence of autofluorescent inclusions under UV light with an Olympus photomicroscope; emission spectra were obtained as described previously (Kitani et al 1988a). Other paraffin sections were stained with Periodic Acid Schiff (PAS), Nile Blue Sulfate (NBS) or Toluidine Blue (TB) for examination of the histochemical properties of induced LLS.

Dog Brains

Beagle dogs (8-10 mo) were infused for two weeks with leupeptin (0.6mg/kg/day) or saline into the lateral ventricle using a mini osmotic pump (Alza Corp). At time of sacrifice dogs were perfused intracardially with saline followed by 4% paraformaldehyde in phosphate buffer. The brains were removed and sections of various brain regions were placed in EM fixative for electron microscopy. The rest of the brain was embedded in parrafin, sectioned, and later examined for autofluorescence and for the presence of LLS, as described for rat brains.

Rat Retina

Sprague Dawley rats (60d) were injected intraocularly to the vitreous, one eye with $1 \mu 1$ leupeptin, E-64C, aprotinin or chloroquine (all at 200 mg/ml), and the other eye with vehicle. For leupeptin treated rats, one injection per eye per day was given for 1-9 days; for other drugs, one injection per eye per day was given for 3 days. Retinas from untreated rats (3-30 mo) were also used. Rats were anesthetised and eyes dissected between 10:00 and 11:30 AM to control for photoreceptor disc shedding. For light microscopy, tissue was fixed in 4% paraformaldehyde and for EM, the fixation was as for rat brain. Portions of retina just nasal to the optic nerve were dissected and embedded in either (1) glycol methacrylate,

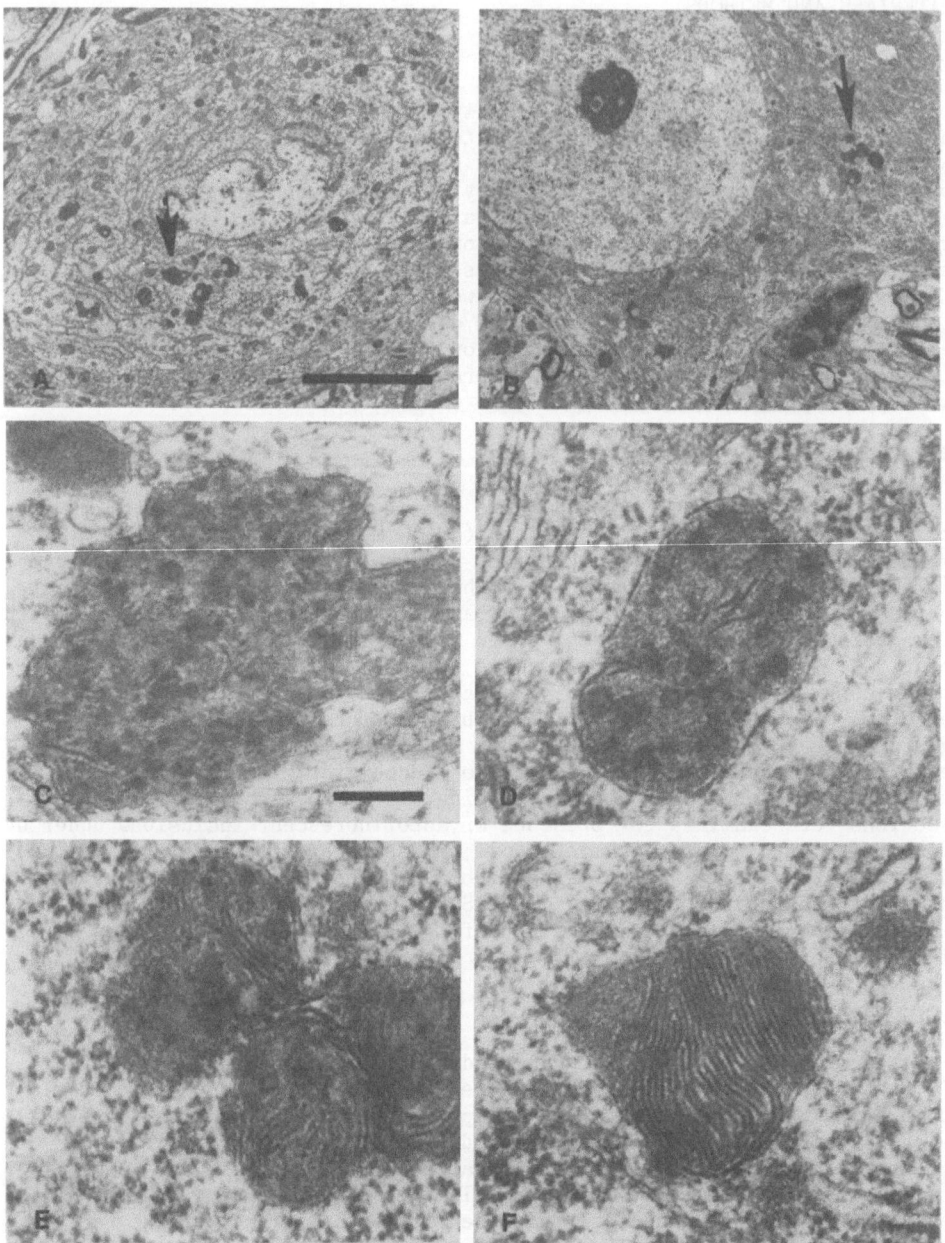

Fig. 1. Electron micrographs of lipofuscin-like substances induced by
intraventricular infusion of leupeptin or E-64C in polymorphic cells in
the hilar region of hippocampal dentate gyrus. A. Low magnification of
the secondary lysosomes (arrow) induced by leupepting (0.48 mg/day) B. Low
magnification of the secondary lysosomes (arrow) induced by E-64C. The
fine morphology of the induced secondary lysosomes is similarly heterogeneous with both treatments and includes fine granular substance with
denser granular and membranous inclusions (C, leupeptin; D, E-64C) as well
as fingerprint-like profiles (E, leupeptin, F, E-64C). Scale bar in A
applies also to B and equals 5 μm; bar in C applies to C-F and equals
0.2 μm.

for sectioning at 1 μm, (2) paraffin, for fluorescence microscopy or (3) epon, for EM analysis.

Rat and Mouse Internal Organs

Fisher 344 or Wistar rats and C57/BL mice (4wks) were infused with leupeptin, E-64C or saline, as described in Kitani et al (1990). Briefly, leupeptin was administered at the rate of from 1 to 50 mg/100g/day using a mini osmotic pump (Alza) implanted IP; treatment was generally for 2 weeks. At sacrifice, animals were perfused intracardially for light or electron microscopy and processed and examined as described for rat brains. Samples of liver, kidney, pancreas, intestine, heart and spleen were taken.

RESULTS

Rat Brains

Light Microscopy

At the LM level, various brain regions in leupeptin treated brain exhibited an intracellular (both neuronal and glial) accumulation of substances with histochemical properties resembling lipofuscin. As demonstrated previously, the substances were positive for PAS, NBS, TB and several other stains which also stain lipofuscin (Ivy et al, 1984). The major brain regions affected were hippocampal dentate gyrus and cerebellar Purkinje cells, though most of the large neurons of the brain showed heavy accumulation. The effect was found to be dose and time dependent (Ivy et al., 1984). The fluorescence spectrum of the induced substance under UV illumination was found to peak at 480 nm with a shoulder to 520-540 nm, similar to the lipofuscin of identically treated normal aged rat brains (Ivy and Gurd, 1988).

Treatment with E-64C induced an intracellular accumulation of substances with similar staining and fluorescence properties as found for leupeptin. The effect was not as dramatic as with leupeptin but was seen clearly in the dentate gyrus and in Purkinje cells.

Treatment with chloroquine gave results similar to E-64C at the light microscopic level, and treatment with aprotinin or with either vehicle showed no effect.

Electron Microscopy

We have previously demonstrated that leupeptin (at 0.5mg/day/2wks) causes a dramatic accumulation of substances resembling lipofuscin in dentate gyrus granule cells at the EM level (Ivy et al 1984). Large polymorphic neurons of the hilar region of hippocampal dentate gyrus also display a substantial accumulation of such substance (Fig. 1A). Like lipofuscin, the deposits are morphologically heterogeneous, with fine granular matrix being the most common. The matrix often contains small, denser aggregates and membrane-like inclusions (Fig. 1C), as well as fingerprint-like profiles (Fig. 1E), and is often associated with vacuoles and denser lysosomes. Similar results were obtained following E-64C treatment. Figure 1B shows an accumulation of lipofuscin-like substances in a polymorphic cell from an E-64C treated rat, and Figures 1D and 1F demonstrate that the morphological subtypes of the accumulated substances resemble those induced by leupeptin.

In cerebellar Purkinje cells of both leupeptin and E-64C treated rats, a notable buildup of dense bodies is evident. The fine morphology of these

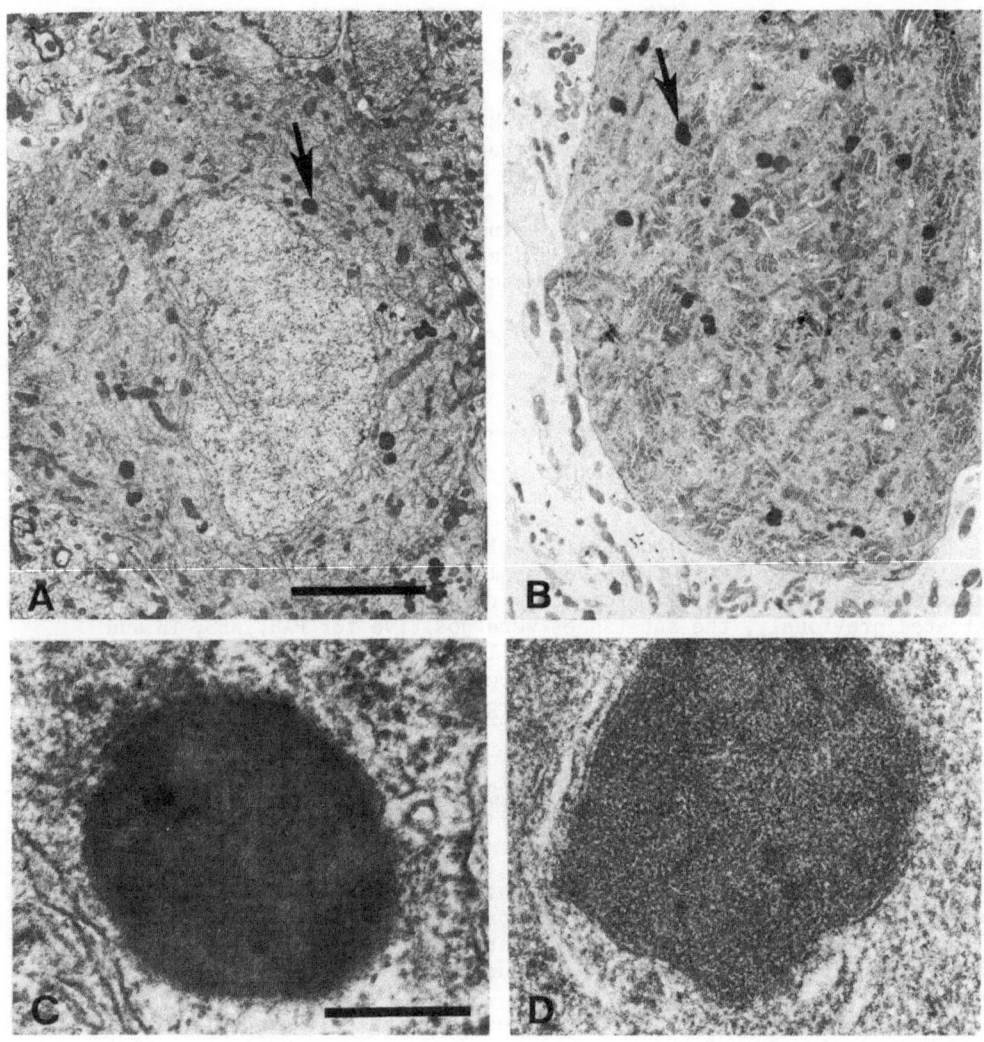

Fig. 2. Electron micrographs of lipofuscin-like bodies induced by
leupeptin in cerebellar Purkinje cells of rat (A and C) and dog (B and D)
brains. A and B show the relative density of these secondary lysosomes
while B and D show the similar fine morphologies. Scale bar in A applies
to B and equals 5 μm. Scale bar in C applies to D and equals 0.2 μm.

bodies is generally fine granular after either treatment (Fig. 2A and C)
and, interestingly, does not show as much morphological variability as do
the deposits found in polymorphic neurons of hippocampus.

Following chloroquine treatment, numerous dense bodies of complex
morphology were seen in dentate gyrus granule cells and polymorphic cells.
These included the morphologic types induced by leupeptin and E-64C.
However, large membranous whorls were at least as numerous and vacuoles
were quite common (Ivy et al., 1984). Treatment with aprotinin or with
either vehicle caused no buildup of LLS in dentate gyrus granule or
polymorphic cells or in Purkinje cells.

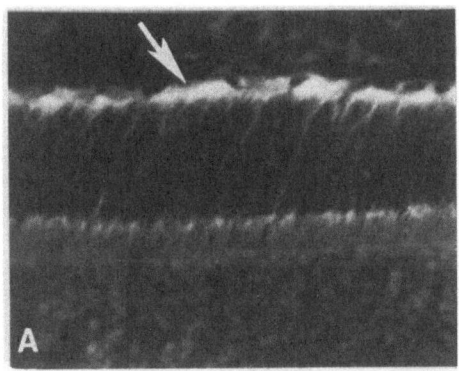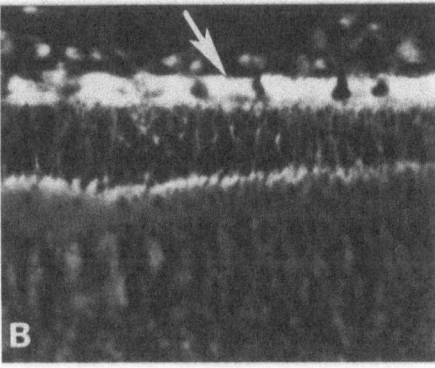

Fig. 3. Photomicrographs of autofluorescence in the retinal pigment epithelial cells (arrows) of (A) a saline treated retina and (B) a leupeptin treated retina (.2 mg/day, 2 days). Note the increased thickness and fluorescence of the cells in the leupeptin treated retina.

Dog Brains

Infusion of leupeptin into the lateral ventricle of dogs for two weeks produced an accumulation of LLS in dentate gyrus granule cells and in cerebellar Purkinje cells which was notable at both the LM and EM level. Figure 2B and D demonstrate the leupeptin-induced buildup of LLS in Purkinje cells of the dog, as compared to the rat (Fig. 2 A and C).

Retina

Time course of changes with leupeptin treatment

Following leupeptin, but not saline treatment, there was a buildup of bright fluorescent substances in the retinal pigment epithelium (RPE) (Fig. 3). A similar buildup of fluorescent substances was seen in aged retinas (not shown). Histological analysis revealed that numerous cytoplasmic inclusions in these cells were PAS, NBS and TB positive, as were the inclusions found in aged retina.

RPE cells in normal young or saline treated rat retinas contain dense deposits which are composed of mainly phagosomes which display the characteristic membrane bound vacuole containing membranous discs of the rod outer segments. These often appear to be in various stages of digestion, and as such are secondary lysosomes. At low magnification under the electron microscope, these deposits are scattered throughout the cytoplasm of the RPE cells (Fig. 4A). During aging of the retina, however, larger and denser aggregates gradually accumulate in the RPE cells (Fig. 4B) ; these are lipofuscin-like in morphology. A similar accumulation is seen after only 24 hours of leupeptin treatment (Fig. 4C). Progressive treatment with leupeptin was found to cause progressive accumulation of dense secondary lysosomes, with concomitant swelling of RPE cells, eventual degeneration of RPE and rod cells, as well as general retinal degeneration by 9 days of treatment (Smith, 1988). As well, the fine morphology of the secondary lysosomes changed with increased treatment time with leupeptin, as described below.

Fig. 5A demonstrates the morphological subtypes of secondary lyso-somes in RPE of a retina from an untreated 3 month old rat. Morphologies vary from apparently intact discs to discs in various stages of digestion and to roundish bodies of homogeneous substance with medium density.

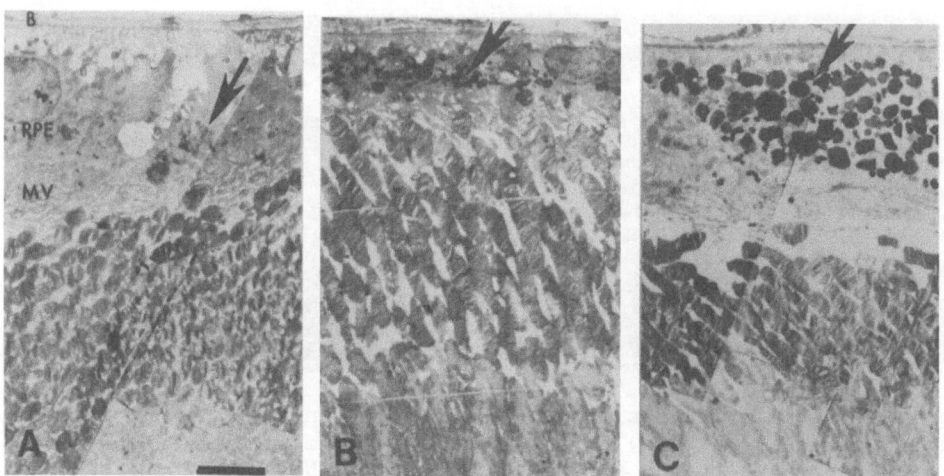

Fig. 4. Low magnification electrom micrographs of the retinas of a saline
treated 3 month old rat (A), a normal 26 month old rat (B) and a 3 month
old leupeptin treated rat (.2 mg/ 1/24 hr). Note the increased buildup of
secondary lysosomes in the RPE with aging (B, arrow) and the dramatic
buildup after leupeptin treatment (C, arrow). B, Bruchs membrane; RPE,
retinal pigment epithelium; MV, microvilli. Scale bar equals 5 μ m.

During normal aging, a proportion of the dense bodies present in the RPE
gradually turn smaller, denser and more numerous (Fig. 5B), although
secondary lysosomes with discs in various apparent stages of digestion can
also be seen (in rats which still have intact rod cells).

    Saline treatment of retina caused no obvious change in the number or
morphology of secondary lysosomes in RPE cells (Fig. 5C). However, after
1 day of leupeptin treatment an obvious increase in the number of
secondary lysosomes was seen (Fig. 4C). The fine morphology of these
bodies was similar to those found in normal young and saline treated
retinas, although many of the bodies appeared larger than normal, as
though several had fused together (Fig. 5D). After 4 doses of leupeptin
treatment in retina, a proportion of the secondary lysosomes were
condensed, and vacuoles were sometimes present (Fig. 5E). Leupeptin
treatment for 7 days caused further condensation of many of the secondary
lysosomes, such that their appearance was indistinguishable from that of
the bodies seen in RPE cells of aged retina (Fig. 5F and B respectively).
As well, secondary lysosomes with rod outer segments in various apparent
stages of digestion were present (Fig. 5F), and these were much more
numerous than those seen in the aged retina.

Changes induced by other protease inhibitors

    Following treatment with E-64C, more TB positive dense bodies were
present than in the normal or saline treated retina. These results were
confirmed at the EM level, where the fine morphology of the inclusions
was most similar to that of the normal phagosomes or of secondary
lysosomes, with the membranous structure of the outer segment discs
apparent (Fig. 6A). However, a small proportion of the bodies were
condensed, as seen in normal aging.

    LM examination of chloroquine treated retina revealed numerous TB
positive inclusions in the RPE cells. At the EM level, the inclusions
displayed many morphologies including normal phagosomes and new secondary

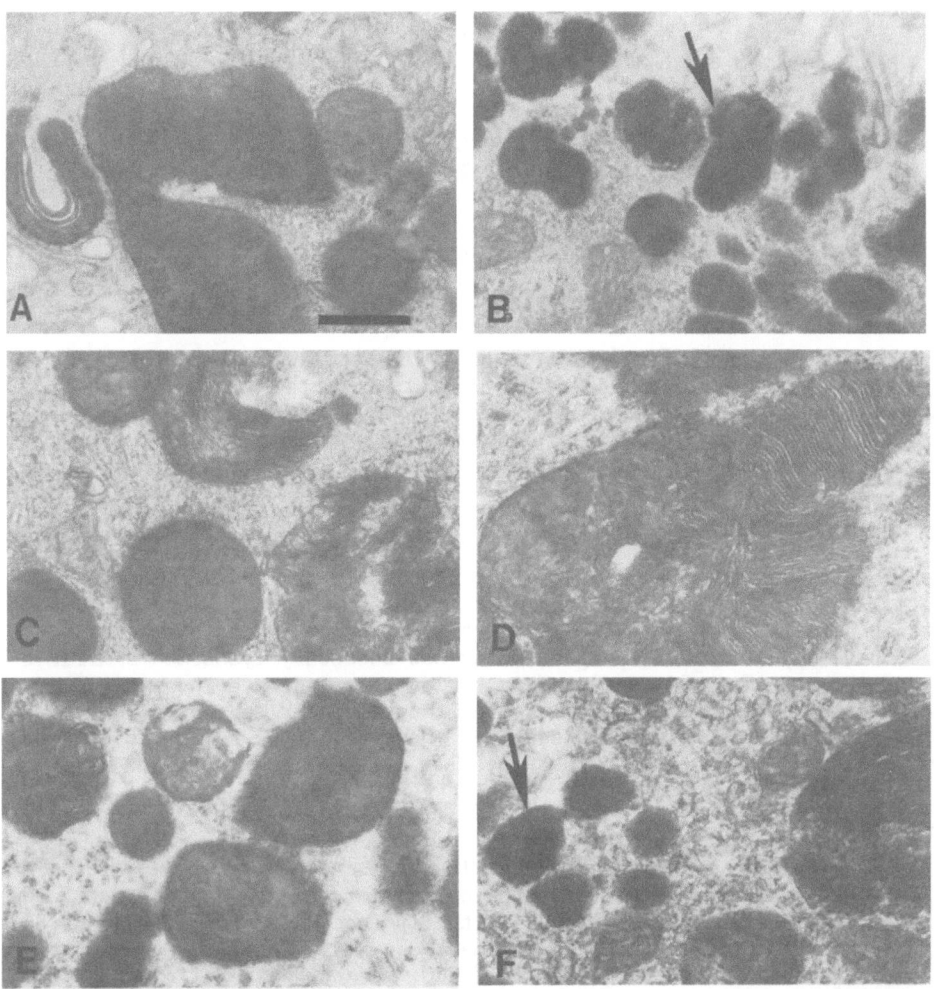

Fig. 5. Fine morphology of the secondary lysosomes in the retinal pigment epithelium of (A) a normal 3 month old rat and (B) a normal 26 month old rat as compared to retinas of 3 mo old rats injected with saline (C) or leupeptin at (D) .2 mg/ 1/day for 1 day, (E) .2 mg/ 1/day for 4 days and (F) .2 mg/ 1/day for 7 days. Note that in normal and control retinas, the secondary lysosomes contain what appears to be mainly rod outer segments in various stages of digestion. By 1 day of leupeptin treatment (D) the secondary lysosomes are enlarged and more irregular; by 4 days of treatment (E) many of them are condensing and losing fine structure and by 7 days of treatment (F) condensed bodies resembling the lipofuscin found in the aged retina (B) are common (arrows). Scale bar in A applies to all micrographs and equals 0.5 μ m.

lysosomes, phagosomes with cores being more dense than the periphery (Fig. 6B), lipid droplets and lipofuscin-like inclusions, such as are seen in the aged retina.

The results with aprotinin treatment were variable in that in two rats outer segment disc shedding occurred to the extent that no outer segments remained, and in one of these a detachment between the photo-receptors and the RPE had occurred. This latter rat showed a buildup of

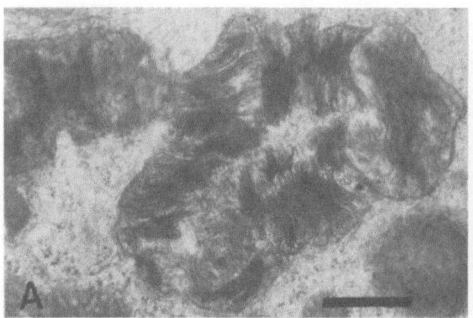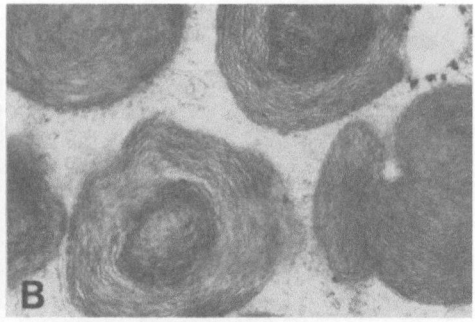

Fig. 6. Fine morphology of the secondary lysosomes in the retinal pigment epithelium of the retina of 3 month old rats injected with (A) E-64C (.2 mg/ 1/d for 3 days) and (B) chloroquine (.2 mg/ 1/d for 3 days). In both cases a dramatic accumulation of secondary lysosomes which appeared to contain undigested rod outer segment discs was evident. Scale bar equals 0.5 μm.

dense bodies in the space between the photoreceptors and RPE and the former rat showed a buildup in the RPE itself. One rat which had an intact sensory retina displayed some buildup of dense bodies in the RPE, and these appeared to be phagosomes or new secondary lysosomes similar to those shown in Fig. 5A and C. This buildup may indicate involvement of serine proteases in the initial degradation of outer segment disks prerequisite to lysosomal degradation. However, as we have only one animal showing this effect, these results must be replicated.

Internal Organs

Examination of tissue under UV fluorescence revealed a deposition of fluorescent granules with a lipofuscin-like morphology in hepatocytes and Kupffer cells of the liver treated with lipofuscin or E-64C, as described in detail in Kitani et al (this volume). In kidney cells of young rats treated with leupeptin (Fig. 7B) or E-64C, but not with saline (Fig. 7A), there was a similar accumulation of fluorescent granules in proximal convoluted tubules and, to a lesser extent, in the glomerulus. Fluorescent lipofuscin granules were also seen to accumulate preferentially in proximal tubules of kidneys of normal aged rats. With increasing doses of leupeptin, the granules became larger and the shape and deposition were very similar to that seen in aged rat kidneys. The fluorescent profiles of the granules were also small (not shown). Indeed, the only difference between the deposition in leupeptin treated and aged kidneys was that in the treated kidneys, PAS staining was a clear red color while in the aged kidney the lipofuscin stained brownish with the PAS stain. In EM examination, the deposits were granular with associated vacuoules, as seen in aged kidneys. Fluorescent granules were also observed in the pancreas, spleen and lung, however to a lesser extent than in the liver and kidney. In EM examination, the deposits in pancreas appeared to be abnormal zymogen granules with dense patchy and membranous inclusions, as seen in aged pancreas. EM analysis of spleen and lung has not been completed. Heart and intestine showed no clear deposition by light, fluorescence or electron microscopy.

DISCUSSION

Our results provide evidence that specific inhibition of lysosomal

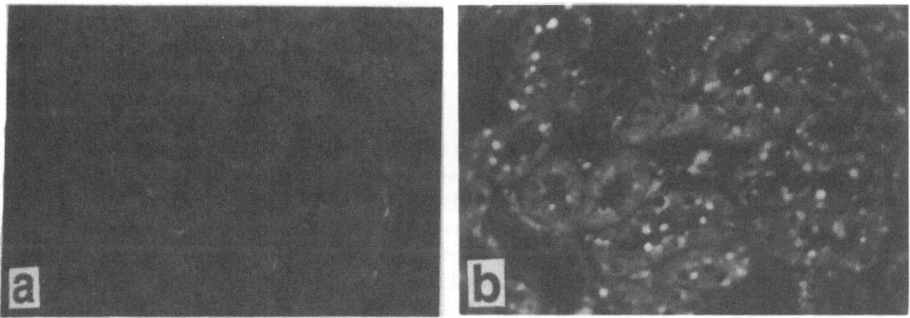

Fig. 7. Photomicrographs of autofluorescent inclusions in the kidneys of saline treated and (B) leupeptin treated (5 mg/100g/day for 2 weeks) rats. Note the dramatic increase in autofluorescent bodies after leupeptin treatment; such an increase is also seen in aged kidneys (X1000).

cysteine proteases causes a buildup of lipofuscin-like substances in several types of cells, tissues and animals. Leupeptin is a reversible inhibitor of lysosomal and cytoplasmic cysteine proteases, as well as of some cytoplasmic serine proteases (Toyo-oka et al., 1978; Barrett, 1980), while E64C specifi- cally inhibits only the cysteine proteases by binding to them irreversibly (Hanada et al., 1978; Hashida et al., 1980; Tamai et al., 1981). The fact that E64C causes a buildup of LLS in both brain and retina with morphologies similar or identical to the substance induced by leupeptin indicates that inhibition of serine proteases is probably not causing the LLS buildup in leupeptin treated animals. This idea is further supported by the fact that specific inhibition of serine proteases by aprotinin does not cause an intracellular buildup of LLS in brain. Thus, the enzymes responsible are either the cytoplasmic calcium activated neutral cysteine proteases (CANPs or calpains) or the lysosomal cysteine proteases (cathepsins B, H, L, S and others). However, the fact that inhibition of lysosomal proteases (and other lysosomal enzymes) by chloroquine induces substances with the same subset of CL morphologies as found with E64C and leupeptin strongly supports the idea that specific inhibition of lysosomal cysteine proteases causes the particular LLS buildup we have found in brain, retina and internal organs. This sugges- tion is consistent with the fact that the lysosomal thiol proteases are responsible for the majority of cellular protein turnover (Barrett, 1977) and that the LLS of leupeptin treated brains is primarily composed of protein (Ivy and Gurd, 1988). It is also consistent with evidence that the composition of LLS in cells of sheep afflicted with ceroid-lipofus- cinosis is primarily proteinaceous (Palmer et al., 1986).

The technical problem of identifying the precise composition of natural aged lipofuscin has been its relative indigestibility. This is thought to be due to polymerization of the constituent compounds resulting from oxidative damage over time (reviewed in Zs.-Nagy, 1988). Lipofuscin is, however, well known to be a heterogeneous mixture of proteins, lipids, carbohydrates, and metals (Elleder, 1981). In this regard, it is important to note that the LLS induced by E64C and leupeptin is also heterogeneous, containing both lipid vacuoles and carbohydrates. As pointed out by Ivy and Gurd (1988), it is not unlikely that failure to catabolize proteins may result in associated lipids and carbohydrate groups remaining together with the undigested proteins.

We find that while the morphologies of LLS induced by leupeptin or E64C in brain and internal organs may be generally described as granular with membranous inclusions and vacuoles, there is substantial variability among the organs and even among different brain cell types. For example, Purkinje cells contain almost exclusively fine granular LLS while the LLS in polymorphic cells of dentate gyrus has denser granular or membranous inclusions in a fine granular matrix. Hepatocytes contain fine granular LLS with membranous inclusions that often appear to be mitochondria in various stages of degradation. Kupffer cells contain LLS with perhaps the greatest variety of morphologies and inclusions while the LLS in RPE cells appears to be composed almost exclusively of membranes of rod outer segment discs in various stages of degradation. This variability in LLS morphology must surely be due to the different cellular constituents normally catabolized by the different cell types. In this light, we should not expect the lipofuscin of normal aged cells to be of identical composition among different cell types; this point is especially relevant to investigations into the fluorescent properties of lipofuscin.

The retina has proven to be an excellent system for studying LLS accumulation because of the ease of drug administration, the ease of identifying the composition of the majority of the induced LLS and the ease of following the changes in morphology of the LLS over time. Several of our findings reported here with leupeptin treatment were also seen by Katz and Shanker, (1989). We have found that with short treatment times (1 day) with leupeptin or E64C, most of the LLS in RPE cells is easily identifiable as rod outer segment discs. Over time (2-5 days of the treatment), the morphologies become more variable, as apparent partial digestion of the contents of the secondary lysosomes occurs. By 7 days of treatment, many of the LLS bodies have the dense compact appearance of lipofuscin from normal aged RPE cells. Thus, with residual time in the cell, induced LLS can change its morphology to become indistinguishable from lipofuscin. During the normal aging process, then, if lysosomal proteolysis becomes sluggish, causing cellular constituents to remain undigested in the secondary lysosome for a prolonged period of time, these constituents may aggregate and gradually become indistinguishable from lipofuscin.

The mechanisms underlying these changes are not known, but the best theory at present appears to be damage to the lysosomal constituents by free radicals. The free radical theory (reviewed in Harman, 1988) states that cellular constituents such as plasma membranes, mitochondria and endoplasmic reticulum are damaged by extremely reactive ions which are a natural by-product of oxidative metabolism. Such damage to the future substrates of lysosomal enzymes could cause them to become undigestible and thus to accumulate in secondary lysosomes and become lipofuscin. This may well occur. Our data, however, together with the facts that thiol groups are known to be especially susceptible to oxidative damage and that many oxidative processes are occurring in lysosomes, support the idea that direct damage to lysosomal cysteine proteases causes a buildup of their substrates, which then gradually become lipofuscin. As the substrates remain longer in the lysosome, they will also have a greater chance of being damaged by free radicals and of becoming increasingly indigestible. Several studies have demonstrated that oxygen radicals directly attack proteins (Chio and Tappel, 1969; Zs.-Nagy and Floyd, 1984) and cross link them (Zs.-Nagy and Nagy, 1980; Nagy and Zs. Nagy, 1984).

The ideas and data presented above are consistent with evidence that protein turnover becomes slower with age in a variety of species (Prasana and Lane, 1979; Resnick and Gershon, 1979; Sharma et al., 1979; Lavie et al., 1981; Resnick et al., 1981) and that an increasing proportion of certain enzymes exist in varying states of denaturation in old animals (reviewed in Resnick et al, 1985; Rothstein, 1985). These data have led Rothstein and colleagues to propose that the "dwell time" of enzymes in aged cells is longer than in young cells, predisposing them to subtle denaturation such as would occur if they were left at room temperature in a test tube overnight. Different classes of enzymes are more stable than others and so might not be damaged by the increased dwell time. We would add that oxidative processes might also be active during this time and that with the very sensitive thiol proteases this increased dwell time might be the critical factor in their demise. The thiol proteases, then may be the Achilles heel of the cell which causes the metabolic changes leading to lipofuscinogenesis and to other manifestations of aging (see references in Introduction).

In sum, while free radical damage to both the substrates of lysosomal enzymes and to the thiol proteases themselves may cause lipofuscinogenesis, we think that damage to the proteases is more likely to be responsible for most of the lipofuscin accumulation with age.

In conclusion, our studies demonstrate that inhibition of lysosomal thiol proteases causes a buildup of lipofuscin-like substances in a variety of tissues and across species. Our studies in retina, in particular, further demonstrate that the morphology of the LLS changes over time, becoming more condensed, like natural lipofuscin. Together with evidence presented above that decreased proteolysis occurs during aging, our results indicate that lipofuscinogenesis results from defective or decreased catabolism of cellular constituents by lysosomal thiol proteases. The crucial question now is precisely why do these proteases lose their vigor with age? The answer to this question may well contain the key to mechanisms underlying aging itself.

ACKNOWLEDGEMENTS

We are indebted to Dr. K. Hanada of Taisho Pharmaceutical Co. for his generous gift of E64C. The authors gratefully acknowledge funding of this research by the Natural Sciences and Engineering Council of Canada and the National Institute on Aging (to G.I.) as well as the Tokyo Metropolitan Institute of Gerontology (to K.K.). A portion of this work was performed with the assistance of the Exchange Program between the National Institute on Aging, USA, and the Tokyo Metropolitan Institute of Gerontology, Japan.

REFERENCES

Barrett, A. J., ed., 1977, "Proteinases in mammalian cells and tissues", North-Holland Publishing Co., New York.

Barrett, A. J., 1980, The many forms and functions of cellular proteinases, Fed. Proc. 39:9.

Bhachatarrya, T. K., Chatterjee, T. K., Ghosh, J. J., 1983, Effects of chloroquine on lysosomal enzymes, NADPH-induced lipid peroxidation, and antioxidant enzymes of rat retina. Biochem. Pharmachol. 32:2965.

Chio, K. S., and Tappel, A. L., 1969, Inactivation of ribonuclease and other enzymes by peroxidizing lipids and by malonaldehyde, Biochemistry, 8:2827.

Elleder, M., 1981, Chemical characterization of age pigments, in: "Age Pigments", R. S. Sohal, ed., Elsevier, Amsterdam.

Goebel, H. H., 1988, Ultrastructure of disease-related lipopigments, in: "Lipofuscin-1987: State of the Art", I. Zs.-Nagy, ed, Elsevier, Amsterdam.

Gutteridge, J. M. C., 1988, Damage to biological molecules by iron and copper complexes, in: "Lipofuscin-1987: State of the Art", I. Zs.-Nagy, ed., Elsevier, Amsterdam.

Hanada, K., Tamai, M., Morimoto, S. Adachi, T., Ohmura, S., Sawada, J. and
Tanaka, I., 1978, Inhibitory activities of E64 derivatives on papain,
Agric. Biol. Chem., 42:537.

Harman, D., 1988, Free radical theory of aging: current status, in:
"Lipofuscin-1987: State of the Art, I. Zs.-Nagy, ed., Elsevier,
Amsterdam.

Hashida, S., Totawari, T., Kominami, E. and Katunuma, N., 1980,
Inhibitions by E64 derivatives of rat liver cathepsin B and cathepsin L
in vitro and in vivo, J. Biochem.,88:1805.

Ishikawa, T., Furuno, K. and Kato, K., 1983, Ultrastructural studies of
autolysosomes in rat hepatocytes after leupeptin treatment, Exp. Cell
Res., 144:15.

Ivy, G.O., 1987, A proteinase inhibitor model of aging: Implications for
decreased neuronal plasticity, in: "Neural Plasticity, Learning and
Memory", N. W. Milgram, C. M. MacLeod and T. L. Petit, eds., Alan R.
Liss, New York.

Ivy, G.O., 1988, Decreased neural plasticity in aging and certain
pathological conditions: possible roles of protein turnover, in: "Neural
Plasticity: A Lifespan Approach", T. L. Petit and G. O. Ivy, eds., Alan
R. Liss, New York.

Ivy, G. O., Do, J. T., Baudry, M. and Lynch, G., 1986, Neurofilamentous
swellings in proximal axons of Purkinje cells induced by a thiol
proteinase inhibitor, Soc. Neurosci. Abs., 12:1507.

Ivy, G. O. and Gurd, J. W., 1988, A proteinase inhibitor model of
lipofuscin formation, in: "Lipofuscin-1987: State of the Art", I.
Zs.-Nagy, ed., Elsevier, Amsterdam.

Ivy, G. O., Kitani, K. and Ihara, Y., 1989, Anomalous accumulation of τ
and ubiquitin immunoreactivities in rat brain caused by protease
inhibition and by normal aging: a clue to PHF pathogenesis? Brain Res.,
498:360.

Ivy, G. O., Schottler, F., Baudry, M., Lynch, G., 1984, Inhibitors of
lysosomal enzymes: accumulation of lipofuscin-like dense bodies in the
brain, Science, 226:985.

Katz, M. L., Robison Jr., W.G., Herrmann, R.K., Groome, A.B. and Bieri,
J.G., 1984, Lipofuscin accumulation resulting from senescence and
vitamin E deficiency: spectral properties and tissue distribution, Mech.
Ageing Dev., 25:149.

Katz, M. L. and Shanker, M. J., 1989, Development of lipofuscin-like
fluorescence in the retinal pigment epithelium in response to protease
inhibitor treatment, Mech. Ageing Dev., 49:23.

Koppang, N., 1988, The English setter with ceroid-lipofuscinosis. A model
for the human juvenile type of ceroid-lipofuscinosis, in:
"Lipofuscin-1987: State of the Art", I. Zs.-Nagy, ed., Amsterdam,
Elsevier.

Lavie, L., Reznick, A. Z. and Gershon, D., 1981, Decreased protein and
puromycinylpeptide degradation in livers of senescent mice, Biochem. J.,
202:47.

Nagy, K. and Zs.-Nagy, I., 1984, Alterations in the molecular weight
distribution of proteins in rat brain synaptosomes during aging and
centrophenoxine treatment, Arch. Gerontol. Geriatr., 2:23.

Palmer, D. N., Barns, G., Husbands, D. R. and Jolly, R. D., 1986,
Ceroid-lipofuscinosis in sheep (II). The major component of the
lipopigment in liver, kidney, pancreas and brain is low molecular weight
protein, J. Biol. Chem., 261:1773.

Pisanti, F.A., Aloj-Totaro, E. and Cuomo, V., 1988, Effect of heavy metals
on the marine mycete Corollospora maritima, in: Lipofuscin-1987: State
of the Art", I. Zs. Nagy, ed., Elsevier, Amsterdam.

Porta, E. A., Mower, H. F., Moroye, M., Lee, Ch. and Palumbo, N. E., 1988,
Differential features between lipofuscin (age pigment) and various
experimentally produced "ceroid pigments", in: "Lipofuscin-1987: State
of the Art", I. Zs.-Nagy, ed., Elsevier, Amsterdam.

Prasanna, H. R. and Lane, R. S., 1979, Protein degradation in aged
   nematodes (Turbatrix aceti), Biochem. Biophys. Res. Commun. 86:552.
Reiss, U., Rothstein, M., 1974, Heat-labile isozymes of isocitrate lyase
   from aging Turbatrix aceti, Biochem. Biophys. Res. Commun., 61:1012.
Reznick, A. Z., Dovrat, A., Rosenfelder, L., Shpund, S. and Gershon, D.
   1985, Defective enzyme molecules in cells of aging animals are partially
   denatured, totally inactive, normal degradation intermediates, in:
   "Modification of proteins during aging", Alan R. Liss, New York.
Reznick, A. Z. and Gershon, D., 1979, The effect of age on the protein
   degradation system in the nematode, Turbatrix aceti, Mech. Ageing
   Dev., 11:403.
Reznick, A. Z., Lavie, L., Gershon, H. E. and Gershon, D., 1981, Age-
   associated accumulation of altered FDP aldolase B in mice, FEBS Lett.,
   128:221.
Rothstein, M., 1985, The alteration of enzymes in aging, in: "Modification
   of proteins during aging", Alan R Liss, New York.
Sharma, H. K., Prasanna, H. R., Lane, R. S. and Rothstein, M., 1979, The
   effect of age on enolase turnover in the free-living nematode, Turbatrix
   aceti, Arch. Biochem. Biophys., 194:275.
Smith, G., 1988, The effects of protease inhibition and aging in the rat
   retina, Master's Thesis, University of Toronto.
Tamai, M., Hanada, K., Adachi, T., Oguma, K., Kashiwagi, K. Omura, S. and
   Ohzeki, M., 1981, Papain inhibitions by optically active E64 analogues,
   J. Biochem., 90:255.
Toyo-oka, T., Shimizu, T. and Masaki, T., 1978 Inhibition of proteolytic
   activity of calcium activated neutral protease by leupeptin and
   antipain, Biochem. Biophys. Res. Commun. 82: 484.
Witkop, C. J., White, J. G., Townsend, D., Sedano, H. O., Cal, S. X.,
   Babcock, M., Krumwiede, M., Keenen, K., Love, J. E. and Wolfe, L. S.,
   1988, Ceroid storage disease in Hermanski-Pudlack syndrome: induction in
   animal models, in: "Lipofuscin-1987: State of the Art", I. Zs.-Nagy,
   ed., Elsevier, Amsterdam.
Zs.-Nagy, I., 1988, The theoretical background and cellular autoregulation
   of biological waste product formation, in: "Lipofuscin-1987: State of
   the Art", I. Zs.-Nagy, ed., Elsevier, Amsterdam.
Zs.-Nagy, I. and Floyd, R. A., 1984, Hydroxyl free radical reactions with
   amino acids and proteins studied by electron spin resonance spectroscopy
   and spin trapping, Biochim. Biophys. Acta., 790:238.
Zs.-Nagy,I. and Nagy, K., 1980, On the role of cross-linking of cellular
   proteins in aging, Mech. Aging Dev., 14:245.
Zuckerman, B. M. and Geist, M. A. (1981) Effect of nutrition and chemical
   agents on lipofuscin formation. In: Age Pigments. Sohal, R. S. (ed),
   Elsevier, Amsterdam, pp 283-302.

DISCUSSION

WITKOP: I want to congratulate you on your lovely presentation and exhaustive work.
My main comment is that we have observed the same sequences of ceroid formation in
the organs of patients with Hermansky-Pudlak syndrome. The pigment is first seen in
the proximal tubules of the kidney and in this organ is not associated with
macrophages, and we don't see too much pathology because the tubular epithelial cells
shed-off into the urine, and new epithelial cells grow. Did you look at the macrophages
of the lung?

IVY: Yes, but I did not find anything quite remarkable, probably because the drug was
infused into the peritoneal cavity.

SOHAL: At the end of your presentation, you said that the mechanism in leupeptin-
induced lipofuscin-like formation may be similar to that normally occurring in aging,
but you have not shown that this is what in fact occurs in aging. In order to strengthen
your case, you have to determine whether the proteinase activities go down, because if
they do not go down, you are wasting your time.

IVY: I was careful to say in my conclusion that our data is consistent with the hypothesis, and as I said in my introduction, there are numerous ways to cause a lipofuscin-like pigment build-up, but as far as tissue distribution of lesions, at least in our brain, our method more closely mimics those of aging. In regard to proteases, cathepsin B and H activities decrease with age. Also, it is known that cathepsin B activity is decreased in patients with progeria.

ELLEDER: I would like to know if in this model of proteinase inhibition you would be able to remove the accumulated pigment in tissue sections by enzymatic proteolysis at least during the early stages of its formation.

IVY: Although we have not done a proteolysis study in tissue sections, we have done this in the test tube using the pigment isolated by the method of Palmer et al., and the built-up substances are capable of being proteolysed.

PORTA: I think that Dr. Elleder's question is also important in relation to the problem of permanence and insolubility of the leupeptin-induced pigment in tissues. For example, if you kill the animal few days after treatment, will the pigment be there like in the case in age-pigment and mature ceroid pigments?

IVY: We have done time-course studies as you suggested, and although we have not performed quantitative analysis, we observed that the pigment did decrease with time. We don't know if eventually all the pigment will disappear or not.

JOLLY: When you isolated the pigment bodies and ran them on gel, did you get electrophoretic bands different from normal tissues?

IVY: We have normal protein bands and no evidence of abnormal proteins.

HALL: When you isolated the pigment and digested with proteolytic enzymes, what happened with the fluorophores? Did they still remain in the residue.

IVY: We did not do that experiment. We did all our studies before digestion.

HERVONEN: Did you ever treat very old animals with leupeptin? It would be of interest to see whether the pigment would be similar in 2-year old rats and mice.

IVY: We have not used old animals. I think it may be somewhat difficult because of the survival rate of old animals, but I would expect to have a larger proportion of young looking lipofuscin granules instead of the more condensed granules.

KITANI: I can answer in part Dr. Hervonen's question because in collaboration with Dr. Zs.-Nagy, we treated young, middle aged and old mice with leupeptin. The problem is that leupeptin is more toxic in old animals, so it is not easy to administer doses comparable to those given to young mice. Furthermore, mice are more resistant than rats to the effects of leupeptin in terms of developing lipofuscin-like pigments. Anyway, at least in the kidneys, the morphology and distribution of lipofuscin-like pigment in the proximal convoluted tubules of leupeptin-treated animals is very similar to what is found in old untreated animals. We are not neurophysiologists, but in our group we have recently looked at the dopamine receptors in the striatum in leupeptin-treated rats, and we have found a decrease in dopamine receptor binding number similar to what is found in old animals and man. Therefore, in terms of function, there are at least some similarities between leupeptin and aging effects.

PORTA: Dr. Kitani, since you mentioned the problem of toxicity of leupeptin in old animals, I wonder whether these toxic effects may increase the turnover of cellular membranes and introduce a pathological factor here that may not be necessarily present in the formation of lipofuscin in the normal process of aging. In fact, some years ago, Dr. Ivy has shown that leupeptin was seriously affecting to the point of killing the cerebellar Purkinje cells. So, the toxic effects of leupeptin in some cell populations is a factor that should be seriously considered in the interpretation of the results, and particularly in any attempts to find homologies with the uncomplicated aging process.

KITANI: I totally agree with that. Not only toxic effects, but also pharmacologic effects of leupeptin other than the protease inhibition should be always considered in comparing leupeptin data with the aging process.

ALHO: If I understood you correctly, Dr. Ivy, you found that in the brain there was an increased pigment accumulation in the dentate gyrus and in the Purkinje cells. Are these the only cells affected by leupeptin?

IVY: No, there are many other brain cells affected, especially the large neurons of the neocortex layers 3 and 5, some neurons in the striatum and thalamus, large neurons in the brain stem and in the hippocampus.

ELLEDER: I know that you have studied with Dr. Wolfe the increasing amounts of dolichols and probably the phosphorylated dolichols in the leupeptin-treated animals. Have you extended now these studies to monoglycosylated dolichols?

IVY: No, we haven't.

AGE PIGMENTS IN DIFFERENT POPULATIONS OF PERIPHERAL

NEURONS IN VIVO AND IN VITRO

Jari Koistinaho, Kaisa Hartikainen, Kimmo Hatanpää and Antti
Hervonen

Laboratory of Gerontology, Department of Public Health
University of Tampere, Tampere, Finland

SUMMARY

The distribution and ultrastructure of lipopigments in the rat
sympathetic, vagus and spinal ganglion neurons were studied in vivo and in
vitro using fluorescence and electron microscopy. Newborn, 3-6 mo and 24-30
mo-old male Wistar rats were used. In vivo, the age pigments in the
sympathetic neurons showed a tendency to form unipolar or bipolar caps,
whereas in the vagus and spinal ganglion neurons pigment granules were
packed in the peripheral area of the perikarya during aging. Ultrastructurally, lipid-like vacuoles and a rather homogeneous matrix were the
components shared by pigment bodies of all types of peripheral neurons.
However, pigment granules in sympathetic neurons frequently had a third,
osmiophilic component, which likely represents neuromelanin.In vitro, the
cytoplasmic area occupied by autofluorescent pigments was increased in
most of the neurons. Some neurons, however, showed the same amount of
lipopigments as in vivo. In electron microscopy, age pigment granules
typical of each type of neuron were found, and their number and intracellular distribution seemed to be comparable with those in vivo. In most of
the neurons cultured from all ages and of all types of ganglion, there
appeared to be accumulations of another, very homogeneous and large type of
pigment body. In some cases, they were structurally connected with classical
pigment bodies or they had a finger print-like substructure. Large homogeneous pigment bodies were also seen in surrounding satellite cells. All
these changes were most frequently seen in cultures of spinal ganglia from
old animals. It is concluded that although classical age pigments maintain
their characteristics in cultured peripheral neurons, there is, in addition, a
rapid accumulation of ceroid-like pigments, which may be caused by the
inability of the cultured neurons to cope with increased peroxidative damage.

Key words: Age pigments; in vivo; in vitro; peripheral ganglia;
autofluorescence; electron microscopy

INTRODUCTION

Age pigments accumulate in a different and selective manner in neuron populations of the central and peripheral nervous systems of man (Obersteiner, 1903; Friede, 1962; Braak, 1974; Braak, 1978; Mann et al., 1978; Dowson, 1982; Koistinaho et al., 1986a) and other vertebrates (Dayan, 1971; Ferrendelli et al., 1971; Samorajski et al., 1968; Brizzee et al., 1974; Nandy and Vijayan, 1979; Koistinaho, 1986; Koistinaho et al., 1989a). The accumulation rate may be related to the function and metabolic activity of neurons (Ferrendelli et al., 1971; Sohal, 1981; Roy et al., 1984) or to differencies in the activities of endogeneous and exogeneous antioxidants (Sohal et al., 1985; Meydani et al., 1986).

The fluorescence properties of the age pigments may be associated with the age and different neurotransmitters of the neurons (Hervonen et al., 1986; Koistinaho et al. 1986b). The ultrastructure of age pigment bodies seems to have a large range of variability, but it has also its own characteristics in distinct regions of the central and peripheral nervous systems (Samorajski et al., 1968; Hasan and Glees, 1976; Brizzee and Ordy, 1981; Koistinaho, 1986; Hervonen et al., 1986). In addition, cytoplasmic distribution of age pigment granules is altered differentially during aging in several neuron populations (Nandy, 1971; Hasan and Glees, 1976; Koistinaho, 1986).

In order to throw fresh light on the heterogeneity in the morphological properties of the age pigments of different neuron populations, this study seeks to determine whether this heterogeneity is also maintained in cultured peripheral neurons, when the neurons are isolated from their natural environment and target organs. Three neuron populations were compared: sympathetic neurons, which represent noradrenergic motor neurons, vagal neurons, which represent visceral sensory neurons, and the neurons of the dorsal root ganglion, which represent the primary sensory neurons.

MATERIAL AND METHODS

Newborn, 3-6 mo and 24-30 mo-old male Wistar rats were used. They were anaesthetized with ether inhalation and decapitated. Superior cervical ganglia (SCG), sensory vagal ganglia (VG) and thoracic and lumbar dorsal root ganglia (DRG) were dissected out using sterile instruments, and placed in separate plastic dishes (Nunc, Denmark) containing Dulbecco's Ca-Mg-free phosphate buffer (GIBCO, Grand Island, NY). Some ganglia were fixed with 3% glutaraldehyde for 4 hrs and processed for electron microscopy, or used as unfixed tissues and cut in cryostat at 10 um for fluorescence microscopy. The rest of the ganglia were desheathed and minced with microscissors. They were then treated with 0.05% trypsin (Sigma, St. Louis, MO) for 20 min et 37ºC, followed by incubation with 0.025% trypsin, 0.5% collagenase (type I, Sigma) for 40 min at the same temperature. They were partly dissociated by three series of 15 passages each through a Pasteur pipette. The dissociated cells and remaining ganglion tissues were placed in plastic culture dishes (Nunc, Denmark), coated with poly-D-lysine (Sigma), and maintained in Eagle's minimum essential medium (MEM) (GIBCO) with 3.7 NaHCO$_3$, 6 g/liter glucose, 0.1% gentamycin (Sigma), 10% (v/v) horse serum (GIBCO) and 5% fetal calf serum (GIBCO). Nerve growth factor (Sigma) was included at a final concentration of 30 nM. To attach explants onto the dish, they were covered with sterilized glass cover slides.

Cells were maintained at 37°C in humified 10% CO$_2$-90% air (v/v). After 5 days, 5-fluoro-2' deoxyuridine (15 ug/ml) and uridine (35 ug/ml) were added to inhibit proliferation of Schwann cells and fibroblasts. The cultures were maintained for 7-28 days.

The cultures were fixed with 3% glutaraldehyde for 30 min at room temperature. Native or toluidine blue stained cultures were viewed under Olympus Vanox T fluorescence microscope, and then processed for routine electron microscopy using osmium tetroxide post-fixation and lead citrate and uranyl acetate stainings. A Jeol 1200 X electron microscope was used.

RESULTS

Age pigments in vivo

The distribution of pigment granules in different types of peripheral neurons is shown in Figs 1-3. In sympathetic neurons pigment granules showed a tendency to form unipolar or bipolar caps at the age of 6 months; this was more obvious in animals over 24 months. In the vagus ganglion the pigment distributed diffusely throughout the cytoplasm at the age of 4-6 months, and evenly in the peripheral area of the perikarya in old animals. In the DRG pigment granules were packed even more peripherally in the neuronal somata than in the vagus ganglion; this was especially typical of the large neurons.

Electron microscopy revealed two components in pigment granules, which were shared by all types of peripheral neurons: ovel or round lipid vacuoles of a varying size and number, and a rather homogeneous matrix (Figs 4-9). However, pigment granules in the SCG frequently had a third, osmiophilic component, which was most typical of sympathetic neurons from old animals. In the VG pigment granules sometimes had a matrix composed of arrays of curved, slightly darker bands than the rest of the matrix (Figs 6-7), whereas the pigment granules in the DRG occasionally had a lamellar structure (Figs 8-9).

Age pigments in vitro

Age pigment autofluorescence in the cultured SCG, VG and DRG are shown in Figs 10-13. The cytoplasmic area occupied by age pigments and the intensity of pigment autofluorescence were increased in most of the cultured neurons both in explants and in dissociated cells. The alteration occured in all cultures from newborn, young adult and aged rats, being observed at 10-14 days in cultures from adult and aged animals and at 21 days in cultures

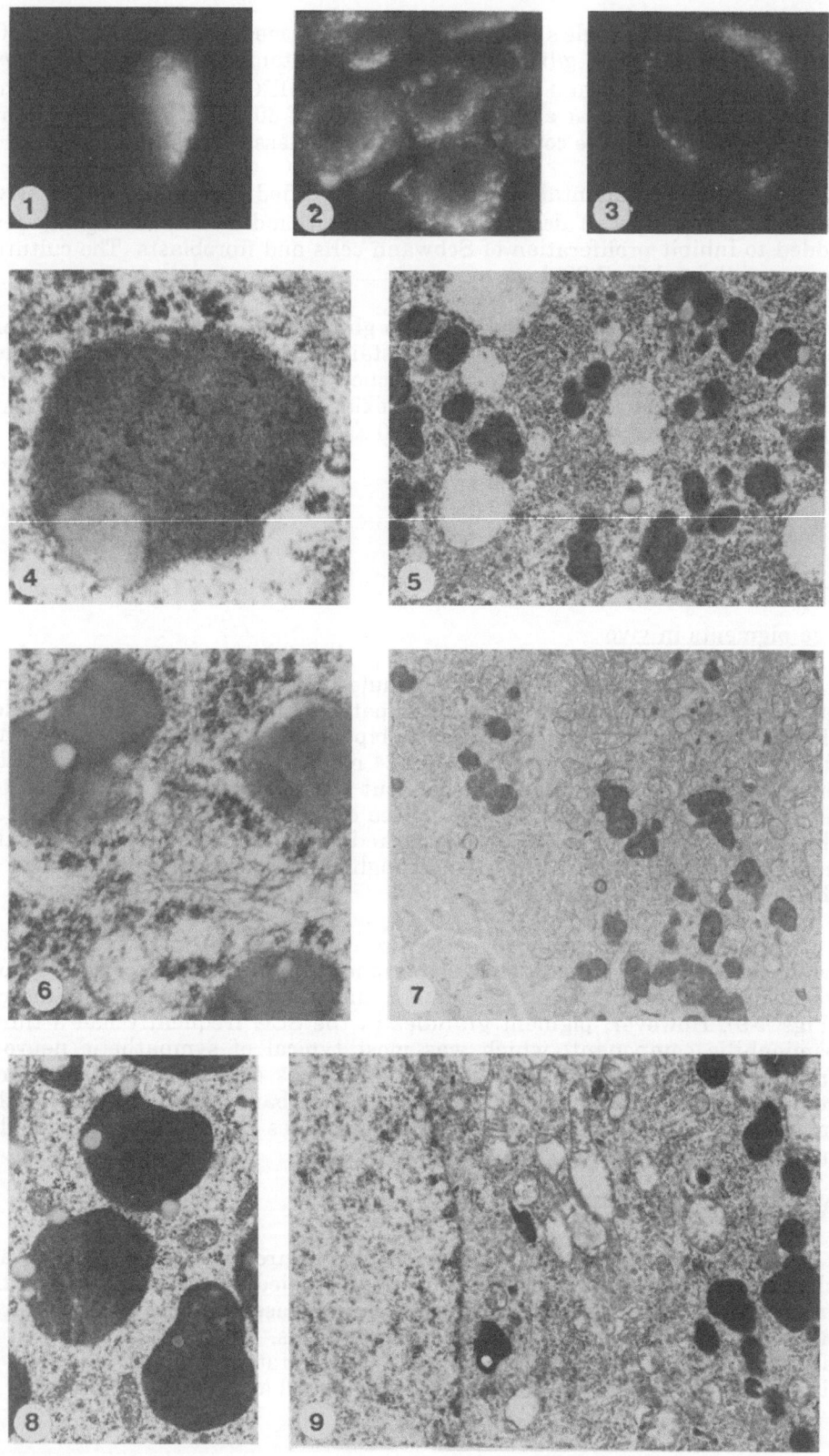

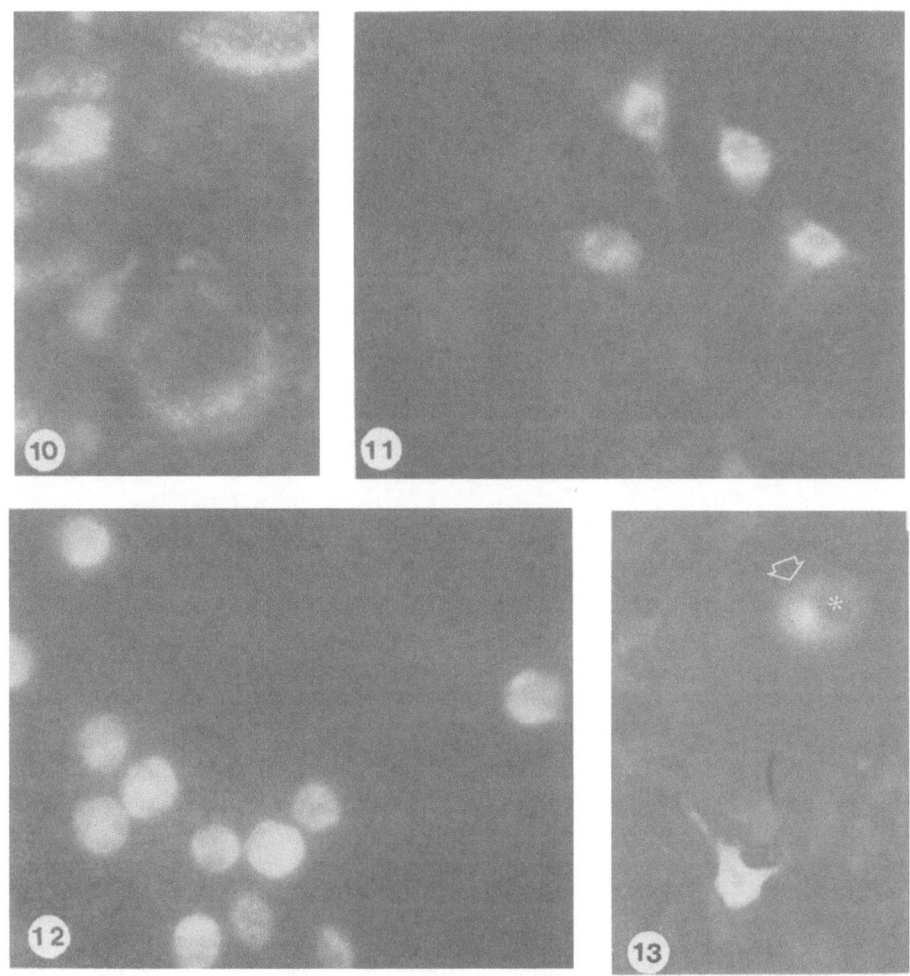

<u>Figures 1-13</u>

Figs 1-3 show the autofluorescence of age pigments in a sympathetic neuron (1), vagal neurons (2) and in a large DRG neuron. Note the different distributions of pigment granules in each neuron population. A 24-month-old rat. Figs 4 and 5 show the structure of pigment granules in the SCG. The clusters of highly osmiophilic material are characteristic of sympthetic neurons of aged rats (5). Figs 6 and 7 show the structure of pigment granules in the VG of 30-month-old rat. Pigment granules are smaller than in the SCG. In most of the granules the matrix is heterogeneous with arrays of curved bands. Figs 8 and 9 show the structure of pigment granules in the DRG of a 24-month-old rat. Most of the pigment granules are composed of two components: lipid-like vacuoles and homogeneous matrix. Figs 10-13 show the autofluorescence of pigment granules in cultured DRG explant (10), VG neurons (11) and SCG neurons (12-13). The distributions in the DRG and VG neurons resemble those in vivo conditions, while some heavily pigmentated SCG neurons show a perinuclear distribution of pigments. In Fig. 13 the arrow points to the neuron, in which pigment granules are still packed unipolarly.

Magnification: x400 (1-3,10), x58000 (4,6), x16000 (5,7,9), x30000 (8), x500 (10), x250 (11-13).

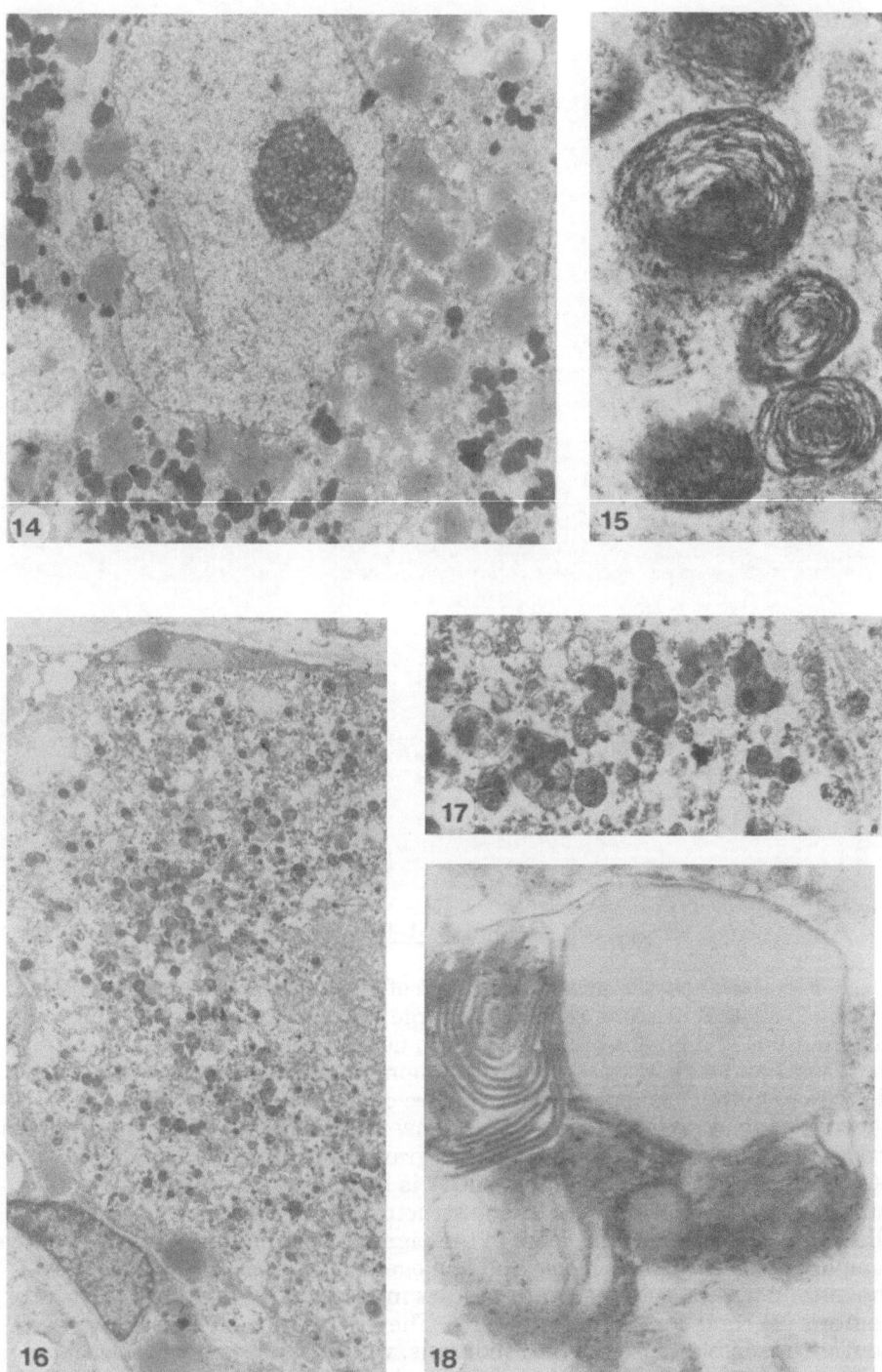

from newborn animals, respectively. Neurons, which did not seem to have increased amounts of pigments, showed similar cytoplasmic distributions of age pigments in all neuron populations as in vivo. In most neurons almost the entire cytoplasmic areas were occupied by age pigments in cultured neurons.

In electron microscopy, age pigment granules typical of each type of neuron were found, and their number appeared to be the same or only slightly increased compared to the neurons studied in vivo. Intracellular locations also resembled those seen in vivo conditions. However, in allmost all neurons of these ganglia there appeared to be accumulations of another, very homogeneous and large type of granule (Fig. 14). In some cases they formed heterogeneous inclusions structurally connected with classical pigment bodies, especially in the DRG. Homogeneous granules were seen in neurons cultured from newborn, young and aged rats, but they were more typical of the cultures from aged animals. In the DRG some inclusions consisted of classical pigment and homogeneous granules together with degenerated mitochondria or with bodies having a finger print-like substructure (Figs 16-18). Large homogeneous pigment bodies were also seen in satellite cells surrounding the neurons.

In some neurons in the SCG and DRG explants from old rats, round lamellar myelin bodies were found (Fig. 15).

DISCUSSION

The intraneuronal distributions and ultrastructures of age pigment granules in vivo were different in each population of aging peripheral neurons; this is consistent with earlier reports (Samorajski et al., 1968; Miquel et al., 1983; Boellaard and Schlote, 1986; Koistinaho, 1986; Koistinaho et al., 1989a). These differences may be associated with the presence of different neurotransmitters synthesized in the neurons, or with the different metabolic activities of these neuron populations. There is also some evidence that the third, osmiophilic component, which was seen only in the pigment granules of sympathetic neurons, represents neuromelanin, which accumulates probably due to the degradation of catecholamines. This occurs only in aminergic neurons (Graham, 1979; Barden, 1969; Hervonen et al., 1986; Koistinaho et al., 1986b).

―――――――――――――――――――― Figures 14-18

Electron micrographs from SCG neurons (Figs 14-15) and DRG neurons (Figs 16-18) cultured from 30-month-old rat. Note the accumulation of large, homogeneous inclusions in addition to the classical age pigment granules in SCG neurons. In Fig. 15 a group of myelin bodies. In DRG neurons the cytoplasm is sometimes occupied by heterogeneous pigment bodies composed of vacuoles, homogeneous or heterogeneous matrix and large, homogeneous inclusions. In Fig. 16 homogeneous inclusions are also seen in surrounding satellite cells.
Magnifications: x8250 (14,17), x29000 (15), x4100 (16), x58000 (18).

In culture, some of the neurons exhibited similar amounts and similar cytoplasmic distributions of age pigments as in the aging neurons in vivo, although most of the neurons of each population seemed to contain increased amounts of autofluorescent age pigments. The rate of degeneration and accumulation of age pigments may be more rapid in cultured than in noncultured cells (Nandy et al., 1978; Brunk and Collins, 1981; Marzabadi et al., 1988). Transplanted neurons from rat embryos or adults also show accelerated degeneration processes (Hoovler et al., 1985; Stenevi et al., 1976; Koistinaho et al., 1989b). When neurons are cultured from postnatal animals, they loose their axons and dendrites. During enzymatic and mechanical dissociation the neurons become to exposed other factors, which can disturb or damage the cells. As neurotoxin- and axotomy-induced neuronal damage has been associated with accelerated accumulation of age pigments in peripheral neurons (Hasan and Glees, 1976; Koistinaho and Hervonen, 1989a), it is possible that in the present study the culture procedure itself caused the increase of pigment bodies. On the other hand, culture conditions and the nutrients in the culture medium may not be optimal or comparative to those in vivo, which can also affect age pigment accumulation in cultured cells.

Electron microscopy revealed that classical age pigment granules were maintained in cultured neurons and had a intraneuronal location typical of each neuron population in vivo. Although there is always considerable loss of neurons during and shortly after plating the cells, the results suggest that pigmented neurons even from very old rats are not significantly more sensitive to the stress caused by the culture procedure than unpigmented or only slightly pigmented neurons from newborn or young adult rats.

Most of the cultured DRG, VG and SCG neurons were filled by large homogeneous inclusions, which were also frequently seen in surrounding satellite cells. Considering the rather low density of classical age pigment granules, it is likely that these inclusions emitted a substantial part of the pigment autofluorescence seen in cultured neurons under fluorescence microscope. Interestingly, similar structurally finger print or homogeneous bodies with a bright yellow autofluorescence have been described in the brain and muscles of ceroid lipofuscinosis patients and animals suffering from similar diseases (Haltia et al., 1973; Armstrong and Koppang, 1981; Oliver, 1981). Brightly fluorescent yellow lipopigments, which have been ascribed to be "ceroid" pigments, are also reported in acute vitamin E deficiency (Nandy, 1971; Hasan and Glees, 1976). Ceroid-like pigment bodies may reflect acute pathological changes in the environment or functions of the neurons in general.

The accumulation or "ceroid" pigments in cultured peripheral neurons was very rapid being observed at 10-14 days in cultures from adult and aged animals. It could be assumed that the appearance of this pigment could be the result of an imbalance between oxidative stress and exo- or endogeneous antioxidants. However, the studies of the effects of antioxidant deficiency do not support this hypothesis, because the response in pigment accumulation is usually observed after much longer period of time than ten days in mammalian tissues (Katz et al., 1978; Katz et al., 1984; Hermann et al., 1985; Koistinaho et al., 1989a). Instead, such a rapid accumulation of pigment bodies has been caused by inhibitors of lysosomal enzymes (Ivy et al., 1984; Constantopoulos et al., 1980; Katz and Shanker, 1989; Koistinaho and Hervonen, 1989b). Treatment with leupeptin, a protease inhibitor, induces lipofuscin-like inclusions in eight hours in brain neurons (Ivy et al., 1984) and in 30 hours in retinal pigment epithelium of rat (Katz and Shanker, 1989), both of the tissues representing cell types known to accumulate excessively age-related pigments. Although peroxidative damage might be increased and contribute the accumulation of "ceroid" pigments in cultured peripheral neurons, it is possible that altered activities of lysosomal enzymes also play a role in this phenomen.

REFERENCES

Armstrong, D. and Koppang, N., 1981, Ceroid lipofuscinosis, a model for aging, in: "Age pigments", R.S. Sohal, ed., Elsevier/North-Holland, New York.

Barden, H., 1969, The histochemical relationship of neuromelanin and lipofuscin. J. Neuropathol. Exp. Neurol. 38:419.

Boellaard, J.W. and Schlote, W., 1986, Ultrastructural heterogeneity of neuronal lipofuscin in the normal human cerebral cortex. Acta Neuropathol. (Berl) 71:285.

Braak, H., 1974, On pigment-loaded stellate cells within layer II and III of the human isocortex. A Golgi and pigmentoarchitectonic study. Cell Tiss. Res. 115:91.

Braak, H., 1978, On pigment architectonics of the human telencephalic cortex, in: "Architectonics of the cerebellar cortex", Brazier, Petsche, eds, Raven Press, New York.

Brizzee, K.R., Ordy, J.M., and Kaack, B., 1974, Early appearance and regional differences in intraneuronal and extraneuronal lipofuscin accumulation with age in the brain of a nonhuman primate (Macaca Mulatta). J. Geront. 29:366.

Brizzee, K.R. and Ordy, J.M., 1981, Age pigments, cell loss and functional implications in the brain, in: "Age pigments", R.S. Sohal, ed., Elsevier/North-Holland, New York.

Brunk U.T. and Collins, V.P., 1981, Lysosomes and age pigments in cultured cells, in: "Age pigments", R.S. Sohal, ed., Elsevier/North-Holland, New York.

Constantopoulos, G., Rees, S., Cragg, B.G., Barranger, J.A., and Brady, R.O., 1980, Experimental animal model for mucopolysaccharidosis: Suramin-induced glycosaminoglycan and sphingolipid accumulation in the rat. Proc. Natl. Acad. Sci. USA 77:3700.

Dayan, A.D., 1971, Comparative neuropathology of ageing, studies on the brains of 47 species of vertebrates. Brain, 94:31.

Dowson, J.H., 1982, Neuronal lipofuscin accumulation in ageing and Alzheimer dementia. A pathogenic mechanism? Br. J. Psychiat. 140:142.

Ferrendelli, J.A., Sedwick, W.G., Suntzeff, V., 1971, Regional energy metabolism and lipofuscin accumulation in mouse brain during aging. J. Neuropath. Exp. Neurol. 30:638.

Friede, R.L., 1962, Lipids and Lipofuscin. Topographic brain chemistry, Academic Press, New York.

Graham, D.G., 1978, Oxidative pathways for catecholamines in the genesis of neuromelanin and cytotoxic quinones. Molec. Pharmacol. 14:633.

Haltia, M., Rapola, J., and Santavuori, P., 1973, Infantile type of so-called Neuronal Ceroid-Lipofuscinosis. Histochemical and Electron microscopic studies. Acta Neuropathol. (Berl.) 26:157.

Hasan, M. and Glees, P., 1976, Lipofuscin in Neuronal Aging and Disease. Georg Thieme Publishers, Stuttgart.

Hermann, R.K., Robinson, W.G., Jr., Bieri, J.G. and Spitznas, M., 1985, Lipofuscin accumulation in extraocular muscle of rats deficient in vitamins E and A. Grafe's Arch. Exp. Ophtalmol. 223:272.

Hervonen, A., Koistinaho, J., Alho, H., Helen, P., Santer, R.S., and Rapoport, S.I., 1986, Age-related heterogeneity of the lipopigments in human sympathetic ganglion. Mech. Ageing Dev. 35:17.

Hoovler, D., Tang, Y., and Bernstein, J., 1985, Viability of mature superior cervical ganglia transplants in peripheral nerve of adult rat. Brain Res. 348:168.

Ivy, G.O., Wenzel, J., Baudry, M. and Lynch, G., 1984, Inhibitors of lysosomal enzymes: accumulation of lipofuscin-like dense bodies in the brain. Science 226:985.

Katz, M., Stone, W., and Dratz, E., 1978, Fluorescent pigment accumulation in retinal pigment epithelium of antioxidant-deficient rats. Invest. Ophtalmol. Vis. Sci. 17:1049.

Katz, M.L., Robinson, W.G., Herrman, R.L., Groome, A.B., and Bieri, J.G., 1984, Lipofuscin accumulation resulting from senescence and vitamin E deficiency: spectral properties and tissue distribution. Mech. Ageing Dev. 25:149.

Koistinaho, J., 1986, Difference in the age-related accumulation of lipopigments in the adrenergic and nonadrenergic peripheral neurons in the male rat. Gerontology 32:300.

Koistinaho, J., Sorvaniemi, M., Alho, H., and Hervonen, A., 1986a, Microspectrofluorometric quantitation of autofluorescent lipopigments in the human sympathetic ganglia. Mech. Ageing Dev. 37:79.

Koistinaho, J., Honkaniemi, J., and Hervonen, A., 1986b, The effect of bleaching on the lipopigments in the human sympathetic neurons. Mech. Ageing Dev. 37:91.

Koistinaho, J. and Hervonen, A., 1989a, Neuronal degeneration and lipopigment formation in rat sympathetic ganglion after treatment with high-dose guanethidine. Neurosci. Lett. 102:349.

Koistinaho, J. and Hervonen, A., 1989b, Suramin-induced accumulation of autofluorescent pigment bodies in peripheral ganglia and adrenal gland. Neurosci. Lett., in press.

Koistinaho, J., Alho, H., and Hervonen, A., 1989a, Effect of vitamin E and selenium supplement on the aging peripheral neurons of the male Spraque-Dawley rat. Mech. Ageing Dev., in press.

Koistinaho, J., Suhonen, J., and Hervonen, A., 1989b, Autoransplants of adrenergic neurons from 18 months old rats survive in adrenals. Neurology (suppl. 1) 39:349.

Mann, D.M., Yates, P.O., and Stamp, J.E., 1978, The relationship between lipofuscin pigment and ageing in the human nervous system. J. Neurol. Sci. 37:83.

Marzabadi, M.R., Sohal, R.S., and Brunk, U.T., 1988, Effect of ferric iron and desferrioxamine on lipofuscin accumulation in cultured rat heart myocytes. Mech. Ageing Dev. 46:145.

Meydani, M., Macauley, J.B., and Blumber, J.B., 1986, Influence of dietary vitamin E, selenium and age on regional distribution of alpha-tocopherol in the rat brain. Lipids 21:786.

Miquel, J., Johnson, J.E., and Cervos-Navarro, J., 1983, Comparison of CNS aging in human and experimental animals, in: "Brain aging: Neuropathology and neuropharmacology. Aging ser.", Cervos-Navarro, J. and Sarkander, S., eds, Raven Press, New York.

Nandy, K., 1971, Properties of neuronal lipofuscin pigment in mice. <u>Acta Neuropath. (Berl)</u> 19:25.

Nandy, K., Baste, C., and Schneider, F.H., 1978, Further studies on the effects of centrophenoxine on lipofuscin pigment in neuroblastoma cells in culture: an electron microscopic study. <u>Gerontology</u> 13:311.

Nandy, K. and Vijayan, V.K., 1979, Cerebellar cell populations, lipofuscin pigment and acetylcholinesterase activity, <u>in:</u> "Aging in nonhuman primates", Bowden, K., ed., Van Nostrand & Reinhold, New York.

Obersteiner, H., 1903, Uber das hellgelbe Pigment in den Nervenzellen und das Vorkommen weiterer feltähnlicher Körper in Centralnervensystem. <u>Archs. Neurol.</u> Inst., Univ. Wien 10:245.

Oliver, C., 1981, Lipofuscin and ceroid accumulation in experimental animals, <u>in:</u> "Age Pigments", R.S. Sohal, ed., Elsevier/North-Holland, New York.

Roy, D., Pathak, D.N., Singh, R., 1984, Effect of chlorpromazine on the activities of antioxidant enzymes and lipid peroxidation in the various regions of aging rat brain. <u>J. Neurochem.</u> 42:628.

Samorajski, T., Ordy, J.M., Dady-Reimer, P., 1968, Lipofuscin pigment accumulation in the nervous system of aging mice. <u>Anat. Rec.</u> 160:555.

Sohal, R.S., Allen, R.G., Farmer, K.J., Newton, R.K., and Toy, P.L., 1985, Effect of exogeneous antioxidants on the levels of endogeneous antioxidants, lipid-soluable fluorescent material and life span in the housefly, musca domestica. <u>Mech. Ageing Dev.</u> 31:329.

Steveni, U., Björklund, A., and Svendgaard, N., 1976, Transplantation of central and peripheral monoamine neurons to the adult rat brain: Techniques and conditions for survival. <u>Brain Res.</u> 114:1.

AGE-RELATED MORPHOMETRIC AND HISTOCHEMICAL FEATURES

OF RAT SYMPATHETIC NEURONS

Pia Jaatinen, Jari Koistinaho and Antti Hervonen

Laboratory of Gerontology, Department of Public Health
and Tampere Brain Research Center, University of Tampere
Box 607, SF-33101 Tampere (Finland)

SUMMARY

The correlation between catecholamine histofluorescence, tyrosine hydroxylase (TH) immunoreactivity and accumulation of age pigment was studied in the superior cervical ganglia (SCGL) of young (3 months) and old (28 months) rats. In the young animals there was a positive correlation between TH-immunoreactivity and catecholamine stores in most of the neurons. A pigment accumulation covering the profile area was found in approximately 10 per cent of the neurons in the young rats. In these neurons strong TH-immunoreactivity was associated with weak catecholamine histofluorescence.

In the ganglia of the old animals there were considerable differences between individual cells in TH-activities, catecholamine stores and amounts of age pigment. In addition, there was a marked uncoupling between TH-activity and catecholamine fluorescence in a number of neurons, i.e. there were neurons with strong TH-immunoreactivity but weak catecholamine histo- fluorescence and vice versa. The functional implications of this uncoupling are discussed in the article.

INTRODUCTION

Age-related changes in the histochemistry of the sympathetic nervous system have been characterized recently in several studies (for a review see Hervonen et al. 1986)[1]. In most sympathetic ganglia the norepinephrine (NE)

Key words: Sympathetic neuron; Tyrosine hydroxylase; Catecholamines; Lipopigments; Aging; Rat

Lipofuscin and Ceroid Pigments
Edited by E. A. Porta
Plenum Press, New York, 1990

content of the principal neurons decreases with age[2,3,4]. The reduction has been suggested to take place mainly in the non-granular catecholamine stores[5,6]. The SCGL of the rat, however, maintains its norepinephrine levels into old age4. The activities of tyrosine hydroxylase (the rate limiting enzyme of catecholamine synthesis) and choline acetyltransferase (ChAT, the enzyme synthesizing the main preganglionary transmitter acetylcholine) are significantly higher in the SCGL and adrenal medulla of old rats than in the young one[7,8]. Further evidence of an increased overall metabolic activity and norepinephrine turnover in SCGL of aged rats has been gained by estimation of glucose utilization and quantitation of formaldehyde-induced catecholamine fluorescence (FIF) after administration of reserpine and alpha-methyl-para-tyrosine[8,9].

The studies performed on the aging sympathetic ganglia so far mostly concern the ganglion as a whole. Little emphasis has been laid on the increasing heterogeneity of the neuron populations during aging. We decided therefore to correlate adrenergic markers with age pigment accumulation at the level of single neurons.

MATERIAL AND METHODS

6 young (3 months) and 6 old (28 months) male Wistar rats (Alko Research Laboratories, Helsinki, Finland) were used. The rats were housed under standard conditions with ad libitum access to food and water. They were killed under canbon dioxid anesthesia by decapitation. The SCGL were removed and frozen immediately in liquid nitrogen. Gatecholamines were demon- strated using the FIF-method[10] as follows: The samples were freeze-dried for 7 days at -40°C using phosphorous pentoxide as a water trap under a vacuum of 10^{-4} Torr. The specimens were then exposed to paraformaldehyde vapour for 60 minutes at +80°C, embedded in paraffin and sectioned serially at 10 μm. Every tenth section was photographed through an Olympus Vanox-T fluorescence microscope equipped with a special filter block for the detection of monoamines (filter block V, excitation light wavelength 395 to 415 nm, emission light 455 nm and up). The same filter combination was suitable for visualizing the autofluorescent pigments as well - the excitation maximum of lipopigment being at 420 nm and the emission maximum at 510-520 nm[11,12].

TH-immunoreactivity was demonstrated using a modified[13] peroxidase- antiperoxidase (PAP) technique of Sternberger[14] on the sections previously studied with fluorescence microscope. The dilution of the TH-antiserum (Eugene Tech. International) was 1:100 and the incubation time was 20 hours at +4°C. Goat anti-rabbit serum (1:40) and PAP (1:40) incubations were carried out for 30 minutes at room temperature. Diamino-benzidine was used as chromogen. In the control sections the specific antiserum was replaced by normal rabbit serum diluted 1:100. The immunostained sections were then photographed using visible light through the same microscope and with the same magnification (total magnification 287,5x) as in the fluorescence photographs. The neuron diameters, packing density and correlations between TH-activity, catecholamine fluorescence and amount of age pigment in each neuron were estimated from 2-3 representative sections (distance 100 μm or more) in the middle part of the ganglia.

Cell size was measured as the maximum diameter of the neurons at the plane of the section - only neurons showing large nuclear profiles were estimated. Neuronal packing density per unit area of 90 000 μm^2 was counted from as many unit areas as possible on each section. The numerical data were analyzed with Student's two-way t-test.

RESULTS

In the SCGL of the young rats (3 months) the overall appearance of FIF was rather homogeneous, i.e. most of the neurons exhibited a moderate fluorescence intensity (Fig. 1A). A large proportion of neurons exhibited a barely visible yellow pigment fluorescence, and there was also a population of neurons (mostly small ones) that apparently contained no autofluorescent age pigment (Fig. 3B). A fine granular pigment accumulation covering the whole profile area was seen in 10-15% of the neurons. These pigmented neurons had a weak catecholamine fluorescence, but in most cases their TH-activity was strong (Fig. 7A,B). With the exception of this minor neuron population, there was a positive correlation between TH-activity and catecholamine stores in most of the neurons of the young rats (Fig. 3A,B).

The neurons in the ganglia of the old rats were characterized by increased heterogeneity (28 months).[1] There were large groups of practically FIF-negative neurons (Fig. 2A) and, on the other hand, numerous neurons with extremely strong catecholamine fluorescence (Fig. 2A,6). There was no distinct difference between the two age-groups in the intensity of TH-immunostaining. A peculiar finding in the old rats was the uncoupling between TH-activity and catecholamine histo- fluorescence (Fig. 4A,B). There were neurons with strong TH-immunostaining but weak FIF; large neurons (diameter $\geq$ 34 μm) predominated in this subpopulation. Additionally, there existed neurons with weak immunostaining but intense FIF; this subpopulation consisted mainly of small neurons (diameter $\leq$ 21 μm). An accumulation of age pigments was seen in more than 90 per cent of the neurons - either diffusely distributed throughout the cytoplasm or clustered as uni- or bipolar caps (Fig. 6). No distinct general correlation was detected between the amount of age pigment and the intensity of FIF in individual neurons. Only neurons extremely packed with lipopigment showed loss of both TH and FIF (Fig. 8A,B).

The average maximum diameter of the neurons was 25.7 $\pm$ 1.28 μm (mean $\pm$ SD) in the 3-month-old rats and 29.3 $\pm$ 1.47 μm (mean $\pm$ SD) in the 28-month-old rats; the difference was statistically significant at p < 0.001. The neuronal packing density (expressed as number of neurons with large nuclear profile / 90 000 μm^2) was 42.4 $\pm$ 1.4 (mean $\pm$ SD) in the young rats and 26.0 $\pm$ 2.5 (mean $\pm$ SD) in the old rats; p < 0.001. The distributions of the neuronal diameters in the two age groups are shown in Table I.

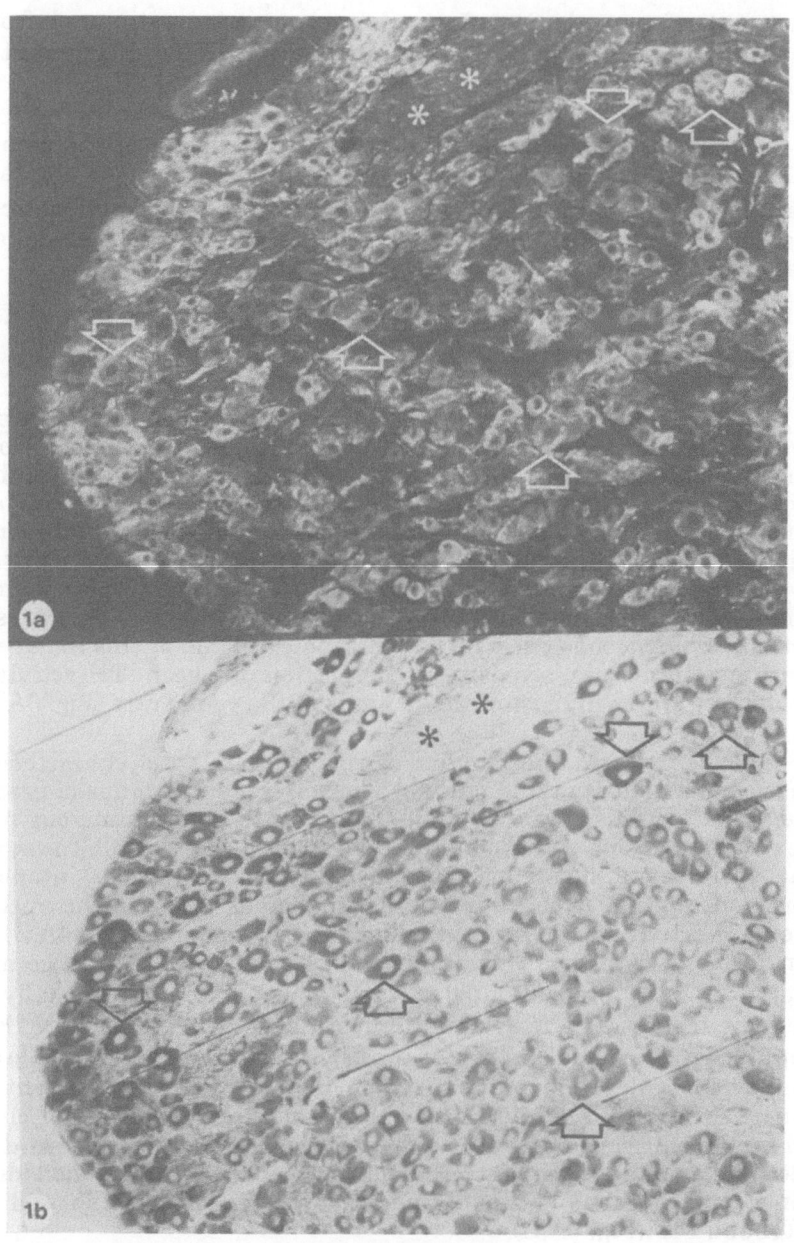

<u>Figures 1A and B</u>

A doublestained section of a superior cervical ganglion of a young rat. Fig. 1B fluorescence micrograph of a paraformaldehyde gas treated freeze-dried section. The noradrenergic neurons emit blue fluorescence typical of norepinephrine. Finely granular yellow spots representing lipofuscin granules can be detected only using larger magnification and colour prints (Fig. 5-8).

Fig. 1B shows the same view after subsequent immunohistochemical demonstration of tyrosine hydroxylase (TH). The intensity of formaldehyde induced catecholamine fluorescence and the TH immunoreactivity correlate positively in the majority of the neurons. The arrows point to a few of such cells and cell groups. The asterisk marks a nerve bundle. X 180.

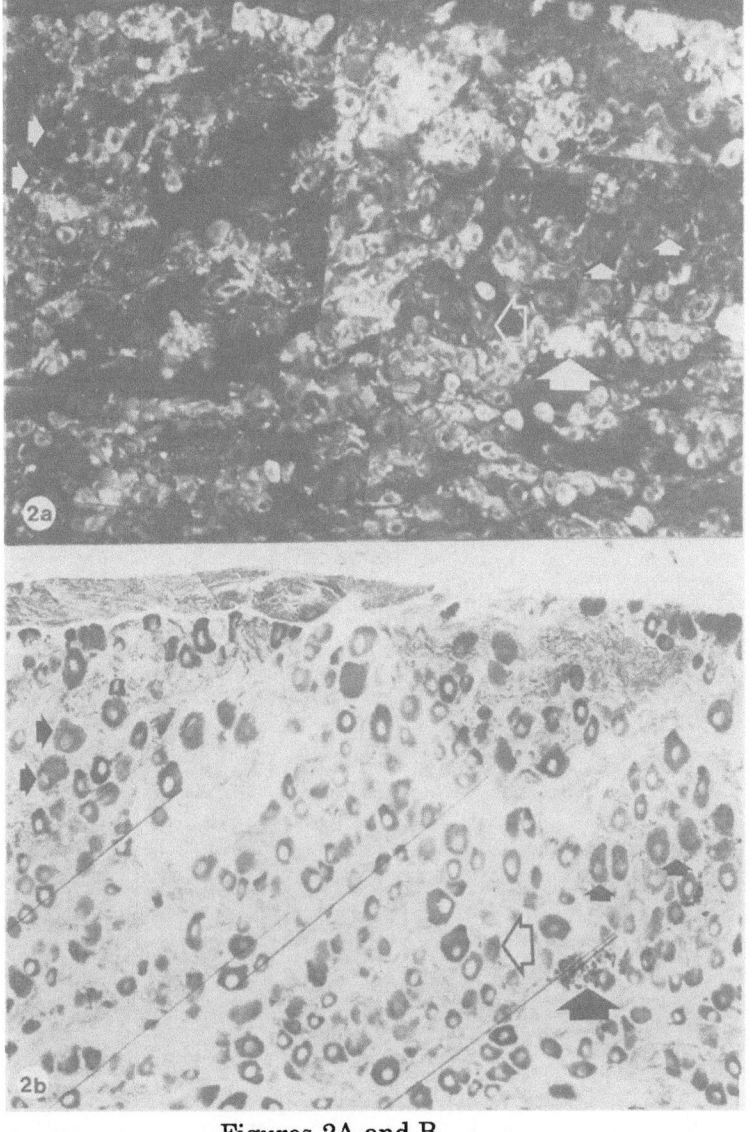

Figures 2A and B

Identical pair of micrographs as in Fig. 1 but describing the superior cervical ganglion of a 32-month-old rat. The fluorescence micrograph shows the accumulation of large collections of yellow lipopigment fluorescence in the majority of the cells. This can only be judged from the colour prints (Figs. 5-8).

There is a wide heterogeneity in the intensity of the formaldehyde induced fluorescence as well as in the TH immunoreactivity of the aged neurons. The small solid arrows point to neurons which show moderate to strong TH immunoreactivity but barely any catecholamine histofluorescence. The large outline arrow points to a group of cells showing intense lipopigment fluorescence, TH immunoreactivity but no catecholamine histofluorescence. Large groups of neurons in the middle of the figure show a positive correlation between catecholamine fluorescence and TH immunoreactivity. These cells are mostly small neurons devoid of lipopigments. X 180.

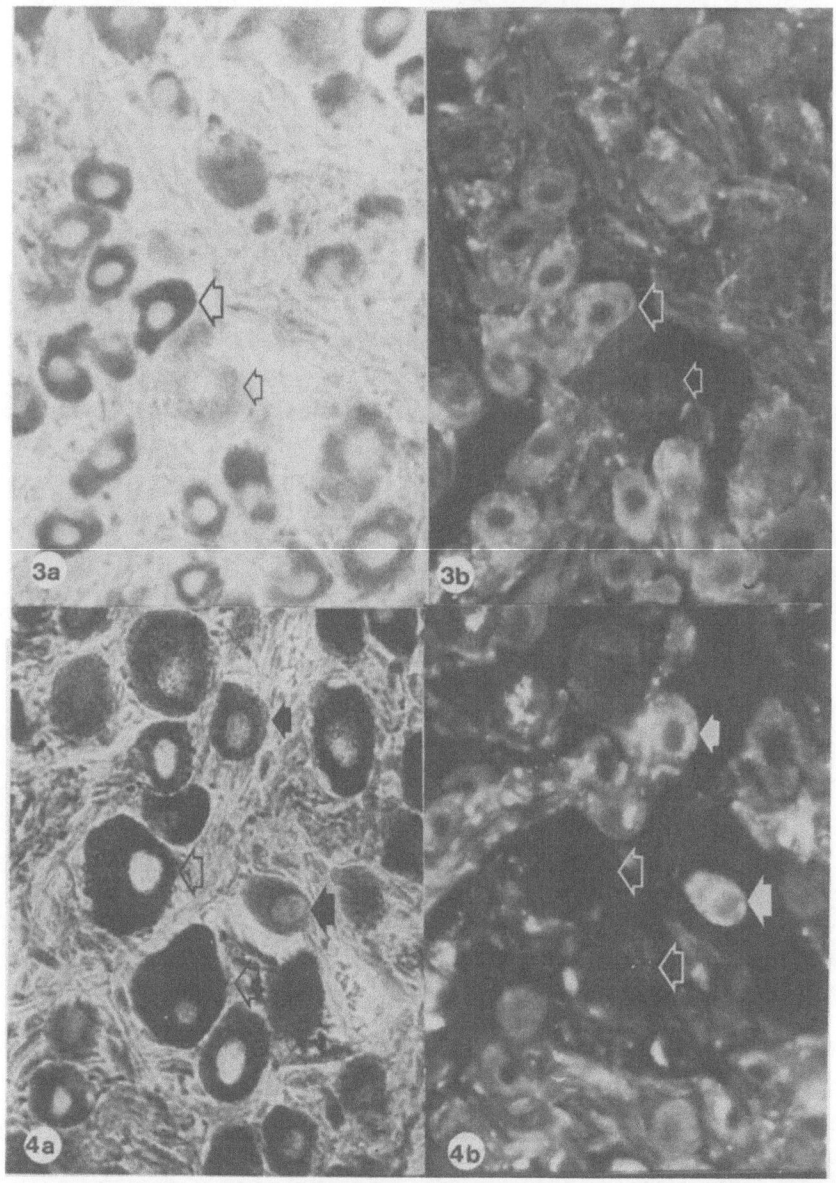

Figures 3A and B

A larger magnification of the same field in a ganglion of a young rat. The arrows point to a pair of neurons showing a good example of the correlation of TH and catecholamine histofluorescence in younger neurons. X400.

Figures 4A and B

A pair of micrographs comparable to Fig. 3 from a 32-month-old rat. The outline arrows point to neurons with strong TH but no catecholamine histofluorescence. The solid arrows point to neurons showing the opposite relationship between the two markers: strong catecholamine histofluorescence but moderate to weak TH immunoreactivity. X 400.

DISCUSSION

There has been comparatively little research into the morphometric changes occurring in sympathetic ganglia during aging, and the results that have been reported are somewhat controversial. Baker and Santer[15] found an increased size of sympathetic neurons in aged Wistar rats. Järvi et al., by contrast, found no age-related changes in the diameter of human sympathetic neurons[16]. Neuronal packing density decreases significantly in the sympathetic ganglia of aging Wistar rats[15] but no changes have been observed in the ganglia of man[16] or the Fischer-344 rat[8]. In the present study the diameter of the neurons was found to increase significantly during aging (25.7 ± 1.28 µm in the 3-month-old rats vs. 29.3 ± 1.47 µm in the 28-month-old rats). At the same time the neuronal packing density decreased by 39% from 42.4 ± 1.4 to 26.0 ± 2.5 per 90 000 µm^2. The somewhat controversial results of morphometric studies presumably reflect real differences between species or strains, although methodological differences may have contributed to the controversy observed.

The increase in neuronal size during aging has been explained by reference to the passive expansion of neurons in response to decreased pressure from surrounding tissues[15]. Both nerve cell death and changes in the amount or composition of interstitial elements may contribute to the decreased pressure within the ganglion. Markov and Riaschikov[17] suggested that the large neurons occurring in late reproductive age might result from swelling of the neuronal cytoplasm as a consequence of exhaustion of functional resources. Our own findings do not support the idea of swelling as a sign of degeneration since most of the large neurons in the aged rats showed low FIF and strong TH-IR, suggesting a high functional state in this neuronal subpopulation. The large neurons also contained only relatively small amounts of age pigments.

Alho et al. showed a significant increase in SCGL neuronal diameter in rats subjected to mild physical exercise throughout their lives[18]. They suggested that the increase in cell size was due to the prolonged and repeated stimulation. The functional activity of SCGL neurons has been shown to increase during aging[7,8,9] and thus increased stimulation of the ganglion cells might explain the age-related expansion, or part of it, as well. Another possible explanation is an increased supply of nerve growth factor (NGF) from target organs to the remaining nerve cells, as the number of neurons apparently decreases during aging[19].

An age-related decrease has been reported to occur in the density of peripheral sympathetic networks in various organs of the rat[20-24]. Of the organs innervated by the SCGL, a decreased sympathetic innervation has been observed in the heart[21] and in major cephalic arteries[20]. The density of the innervation of the iris, however, seems to maintain a level comparable to that of young animals in 24-month-old rats[25]. These results may indicate differences in the rate of cellular death between subpopulations of neurons projecting to different target organs. Smolen[26] has reported morphological and biochemical differences between SCGL neurons innervating the submandibular gland (SMG) and the iris. The sympathetic neurons projecting to the SMG were significantly larger than those projecting to the

Figure 5

A fluorescence micrograph demonstrating the variations in the fluorescence intensity and cell size in the young superior cervical ganglion. Only scattered yellow lipopigment granules are visible. X 440.

Figure 6

A fluorescence micrograph demonstrating the variations of fluorescence intensity and lipopigment accumulation in the old superior cervical ganglion neurons. Note the extremely brightly fluorescent cells with only few pigment granules (outline arrows) and the neurons with larger accumulations of lipopigments (solid arrows) and weaker catecholamine histofluorescence. X 440.

Figures 7A and B

A pair of micrographs illustrating the good general correlation of TH immunoreactivity and catecholamine histofluorescence in the young ganglia (To be compared with Fig. 8). X 400.

Figures 8A and B

A pair of micrographs showing the correlation between lipopigments TH and catecholamines in the old sympathetic neurons. The arrows point to cells devoid of catecholamine fluorescence and containing clusters of age pigment granules. The small neurons showing intense catecholamine fluorescence are devoid of age pigments. X 400.

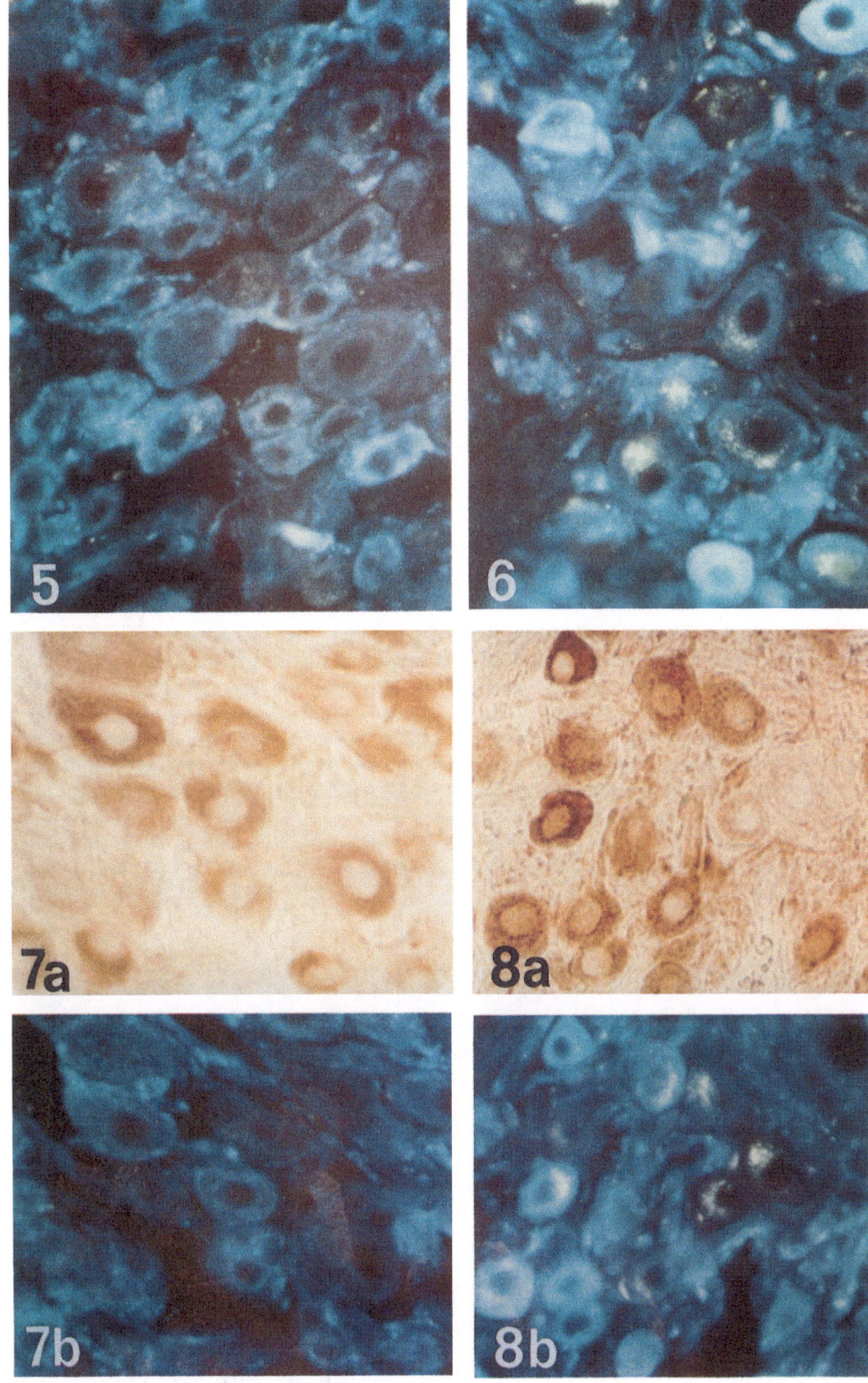

iris, and the rate of NE-turnover was much higher in the SMG than in the iris. Morphological and functional differences between neuronal subpopulations were also apparent in the present study. The subpopulation with high NE-turnover (intense TH-immunostaining in connection with low cytoplasmic NE-stores) consisted mainly of large neurons ($\geq$ 34 µm in diameter). This is in accordance with the findings of Alho et al.[18] and Smolen[26]. The population with low to medium TH-IR but relatively high FIF-intensity consisted of small and medium-sized neurons, which may include e.g. the neurons projecting to the iris.

Differences in the rate of lipopigment accumulation have been reported between different brain regions (for a review see 11) as well as between different autonomic ganglia[11,27]. The regional selectivity in the rate of pigment accumulation has been explained by differences in the levels of metabolic activity[28-31] and antioxidant enzymes[32].

In the present study more than 90 per cent of the SCGL neurons in the 28-month-old rats were found to contain lipopigments. The level of lipopigment autofluorescence (LAF) varied significantly from neuron to neuron, which probably reflects differences between neurons in the oxidative stress to which they have been exposed during their life time. The level of presynaptic stimulation determines to a great extent the functional activity of the ganglion cells, and a proper functioning of the presynaptic inhibitory modulator systems may be vital in protecting the neurons from excessive stimulation[16].

The intraneuronal distribution of the lipopigments was different in the two age groups. In the 3-month-old rats the pigments were diffusely distributed throughout the cytoplasm, but in the older group the granules often formed uni- or bipolar caps. A similar condensation of pigment granules has previously been observed in the central nervous system[33,34] and in peripheral adrenergic neurons, but not in nonadrenergic peripheral neurons[11]. The significance of the condensation is not clear, but it has been suggested by Glees and Gopinath[34] that the localization of the pigment may be related to a transfer of the pigment into satellite cells located close to the pigmented areas of the neurons.

A moderate accumulation of age pigments did not generally seem to affect the neuronal catecholamine stores or TH-IR, which implies a rather inert character of the pigment. However, a few extremely pigmented neurons were seen that showed no TH-IR and no FIF. Thus, an extreme filling of the cytoplasm with age pigment can be regarded as a sign of cellular death.

Increased heterogeneity was the most prominent feature of the neurons in the old animals - concerning both the amount of lipopigments and the intensity of FIF and TH-IR. The originally fairly homogeneous neuron population - with moderate FIF and TH-IR and weak or no LAF - had turned into several populations with different histochemical features (Fig. 2) by the age of 28 months. Further experimental studies are needed to determine the influence of e.g. target organs and presynaptic input on the fate of individual ganglion cells during aging.

REFERENCES

1. A. Hervonen, M. Partanen, P. Helén, J. Koistinaho, H. Alho, D. M.
    Baker, J. E. Johnson and R. M. Santer, The sympathetic neuron
    as a model of neuronal aging. In: "Neurohistochemistry: Modern
    Methods and Applications", P. Panula, H. Päivärinta and S.
    Soinila, eds., Alan R. Liss, Inc., New York (1986).
2. A. Hervonen, A. Vaalasti, M. Partanen, L. Kanerva and H. Hervonen,
    Effects of aging on the histochemically demonstrable
    catecholamines and acetyl-cholinesterase of human sympathetic
    ganglia, J. Neurocytol. 7:11-23 (1978).
3. R. M. Santer, Fluorescence histochemical evidence for decreased
    noradrenaline synthesis in sympathetic neurons of aged rats,
    Neurosci. Lett., 15:177-180 (1979).
4. M. Partanen, A. Hervonen and S. I. Rapoport, Microspectro-
    fluorometric quantitation of histochemically demonstrable
    catecholamines in peripheral and brain
    catecholamine-containing neurons in male Fischer-344 rats at
    different ages. In: "The Aging Brain: Cellular and Molecular
    Mechanisms of Aging in the Nervous System", E. Giacobini, G.
    Filogamo, G. Giacobini and A. Vernadakis, eds., Raven Press,
    New York (1982).
5. O. Eränkö, Light and electron microscopic histochemical evidence of
    granular and nongranular storage of catecholamines in the
    sympathetic ganglion of the rat, Histochem. J. 4:213-224 (1972).
6. M.Partanen, Aging and Neuroendocrine Charasteristics of the
    Hypogastric (Main Pelvic) Ganglion of the Male Rat. Acta
    Universitatis Tamperensis ser. A vol. 112, Publications of the
    Faculty of Medicine 23, Tampere (1980).
7. D. J. Reis, R. A. Ross and T. H. Joh, Changes in the activity and
    amounts of enzymes synthesizing catecholamines and
    acetylcholine in brain, adrenal medulla and sympathetic ganglia
    of aged rat and mouse, Brain Res., 136:465-474 (1977).
8. M. Partanen, S. I. Rapoport and E. D. London, Glucose utilization in
    sympathetic ganglia of male Fischer-344 rats at different ages, J.
    Aut. Nerv. Syst., 5:391-398 (1982).
9. M. Partanen, S. B. Waller, E. D. London and A. Hervonen, Indices of
    neurotransmitter synthesis and release in aging sympathetic
    nervous system, Neurobiol. Aging, 6:227-232 (1985).
10. O. Eränkö, The practical histochemical demonstration of catecholamines
    by formaldehyde induced fluorescence, J. Roy. Microsc. Soc.,
    87:259-276 (1967).
11. Difference in the age-related accumulation of lipopigments in the
    adrenergic and non-adrenergic peripheral neurons in the male
    rat, Gerontology, 32:300-307 (1986).
12. A. Hervonen, J. Koistinaho, H. Alho, P. Helén, R. M. Santer and S. I.
    Rapoport, Age-related heterogeneity of lipopigments in human
    sympathetic ganglia, Mech. Age. Dev., 35:17-29 (1986).
13. A.Hervonen, V. M. Pickel, T. H. Reis, I. Linnoila, L. Kanerva and R. J.
    Miller, Immunohistochemical localization of the catechol-
    amine-synthesizing enzymes and neuropeptides in the
    catecholamine storing cells of human fetal sympathetic nervous
    system, Adv. Biochem. Psychopharmacol., 25:3373-3378 (1980).
14. L. A. Sternberger, "Immunocytochemistry", Prentice-Hall, Englewood
    Cliffs, NJ (1974).

15. D. M. Baker and R. M. Santer, Morphometric studies on pre- and paravertebral sympathetic neurons in the rat: changes with age, Mech. Age. Dev., 42:139-145 (1988).

16. R. Järvi, P. Helén, M. Peltohuikko, S. I. Rapoport and A. Hervonen, Age-related changes of encephalinergic innervation of human sympathetic neurons, Mech. Age. Dev., 42:139-145 (1988).

17. V. N. Markov and S. N. Riashchikov, Effect of different fixation methods and partial chemical sympathetic denervation on cell size and 3H-leucine incorporation by rat sympathetic neurocytes, Tsitologiia, 29:54-58 (1987).

18. H. Alho, J. Koistinaho, V. Kovanen, H. Suominen and A. Hervonen, Effect of prolonged physical training on the histochemically demonstrable catecholamines in the sympathetic neurons, the adrenal glands and extra-adrenal catecholamine storing cells of the rat, J. Aut. Nerv. Syst., 10:181-191 (1984).

19. R. Levi-Montalchini and B. Booker, Excessive growth of the sympathetic ganglia evoked by a protein isolated from mouse salivary glands, Proc. Natl. Acad. Sci., 46:373-391 (1960).

20. R. M. Santer, Fluorescence histochemical observations on the adrenergic innervation of the cardiovascular system in the aged rat, Brain Res. Bull., 9:667-672 (1982).

21. M. R. McLean, P. Bursztyn Goldberg and J. Roberts, An ultra-structural study of the effects of age on sympathetic innervation and atrial tissue in the rat, J. Molec. Cell. Cardiol., 15:75-92 (1983).

22. F. Amenta and M. C. Mione, Age-related changes in the noradrenergic innervation of the coronary arteries in old rats: a fluorescent histochemical study, J. Aut. Nerv. Syst., 22:247-251 (1988).

23. S. Y. Felten, D. L. Bellinger, T. J. Collier, P. D. Coleman and D. L. Felten, Decreased sympathetic innervation of spleen in aged Fischer 344 rats, Neurobiol. Aging, 8:159-165 (1987).

24. D. M. Baker and R. M. Santer, A quantitative study of the effects of age on the noradrenergic innervation of Auerbach's plexus in the rat, Mech. Age. Dev., 42:147-158 (1988).

25. T. Cowen and R. M. Santer, The effect of age on the ability of noradrenergic nerves to regrow after injury by the noradrenergic neurotoxin, DSP-4, J. Aut. Nerv. Syst. (submitted).

26. A. J. Smolen, "Heterogeneity of target specific populations of sympathetic neurons in the rat superior cervical sympathetic ganglion", Presented at the 18th annual meeting of the Society for neuroscience, Toronto, Ontario (1988).

27. J. Koistinaho, M. Sorvaniemi, H. Alho and A. Hervonen, Micro-spectrofluorometric quantitation of autofluorescent lipopigments in the human sympathetic ganglia, Mech. Age. Dev., 37:17-29 (1986).

28. R. L. Friede, "Lipids and Lipofuscin. Topographic Brain Chemistry", Academic Press, New York (1962).

29. J. A. Ferrendelli, W. G. Sedgwick and V. Suntzeff, Regional energy metabolism and lipofuscin accumulation in mouse brain during aging, J. Neuropathol. Exp. Neurol., 30:638-649 (1971).

30. H. E. Hirsch, Enzyme levels of individual neurons in relation to lipofuscin content, J. Histochem. Cyhtochem., 18:268-270 (1970).

31. R. S. Sohal, ed., "Age Pigments", Elsevier, Amsterdam (1981).

32. D. Roy, D. N. Pathak and R. Singh, Effect of chlorpromazine on the activities of antioxidant enzymes and lipid peroxidation in the various regions of aging rat brain, J. Neurochem., 42:628-633 (1984).

33. R. Whiteford and R. Getty, Distribution of lipofuscin in the canine and porcine brain as related to aging, J. Geront., 27:31-44 (1966).
34. P. Glees and G. Gopinath, Age changes in the centrally and peripherally located sensory neurons in rat, Z. Zellforsch., 141:285-298 (1973).

DISCUSSION

WITKOP: Thank you for your lovely presentation. I have just one cautionary advice. The same phenomenon that you see in brain cells also occurs in other cells that accumulate ceroid. For example, in the renal epithelial cells, there is initially a bright yellow fluorescence and the ultrastructure of the pigment is similar to the early ultrastructure that you have shown. Later on, it gets very dark and displays orange fluorescence, but there is no neuromelanin in these renal cells. The same phenomenon occurs in several other tissues such as gut, lungs, macrophages, etc. The changes in the autofluorescence are associated with changes in the ultrastructure.

HERVONEN: We are well aware of that, and actually all these aspects have been discovered early in this century, like many other things in the area of pigments. However, when we compare these adrenergic and nonadrenergic neurons, only the adrenergics show the change from yellow to orange. The presence of catecholamines is the prerequisite for the neuromelanin production. You find neuromelanin only in the catecholaminergic neurons also in the brain. We have studied different types of neurons, and we don't see that happening in the nonadrenergic neurons.

ELLEDER: I think that your evidence for neuromelanin is somewhat indirect, because if you have proven that there was an increase in autofluorescence after bleaching with hydrogen peroxide, this is a more or less generalized phenomenon that can be seen in many lipofuscin-storing cells.

HERVONEN: Due to the lack of time I did not go into details, but we have tried most of the histochemical reactions specific for neuromelanin, and all indicated the presence of neuromelanin.

IVY: You mentioned in the form of a hypothesis that you think that you saw some intermediate products, probably neurotoxic. Could you expand on this?

HERVONEN: There is some evidence from the works of Graham and coworkers in the late 70's where they studied the process of melanization. Before ending-up with melanin, you have quinones which in some form, tested in vitro, caused the death of the neurons. They concluded, therefore, that these intermediate products may be responsible for the dysfunction of the cells.

GOEBEL: When you talked about the removal of the submandibular gland, what you actually did was to cut-off the very terminal of all the innervating axons. So, how do you know that is really the removal of the gland and not a rather nonspecific or multifactorial retrograde degeneration of the innervating neurons that gives you this kind of atrophy in the superior cervical ganglia?

HERVONEN: That's a good question, because I have have a good answer. This situation is only in the mouse submandibular gland, but not in the rat. So, we have used the rat as a kind of control. If you remove some submandibular gland from the rat, there is some slight decrease in the fluorescent intensity, but nothing happens to the neurons. They regenerate their axons in 3 - 4 weeks, and then the ganglia is completely normal, while in the mouse, the size of the neurons remains smaller, and the intensity of the catecholamine histofluorescence remains diminished.

GOEBEL: The other question is about the detrimental effects of melanization. Could it not be the other way around? Is it not possible that because the atrophy may reduce the lysosomal enzymatic functions, this may end up in the production of these peculiar lipopigments?

HERVONEN: I am not strongly inclined for either view. It is just that the accumulation of this material with the wash-off of other functional parameters suggests that this could be the way.

IVY: I am not so convinced by the differences in your results in mice and rats. Is there any evidence that there is a different dependence on NGF in one versus the other? Maybe the neurons of rats just don't need it so much.

HERVONEN: In vitro experiments indicate that both species need about similar amounts. In fact, this has been used as a bioassay for the presence of NGF in both species. Nothing is known about the NGF dependence of aging adrenergic neurons.

MORPHOLOGICAL, PHYSIOLOGICAL AND BIOCHEMICAL ALTERATIONS IN LIVERS OF

RODENTS INDUCED BY PROTEASE INHIBITORS: A COMPARISON WITH OLD LIVERS

Kenichi Kitani, Minoru Ohta, Setsuko Kanai, Munetaka Nokubo, Yuko Sato, Koichiro Otsubo[1] and Gwen O. Ivy[2]

First Laboratory of Clinical Physiology and [1]Department of Clinical Pathology, Tokyo Metropolitan Institute of Gerontology, 35-2 Sakaecho, Itabashi-ku, Tokyo-173, Japan; [2]Life Science Division, Univ. of Toronto, Scarborough College, 1265 Military Trail, Scarborough, Ontario, Canada

SUMMARY

A "Protease inhibitor model of aging" has been proposed primarily based on observations on brain tissues exposed to a thiol protease inhibitor, leupeptin (Ivy et al., 1984a). In order to validate this model in terms of a mechanism of cellular aging, as well as of lipofuscin formation in particular, attempts have been made to induce lipofuscin in hepatocytes in young rodent (rat and mouse) livers by continuous i.p. infusion of two different thiol protease inhibitors, leupeptin and E-64C. With doses of leupeptin higher than 1.0 mg/100g/day for 2 wks, a fine granular lipo-fuscin-like deposition with distinct yellowish-green fluorescence was induced in young rat hepatocytes. The deposition became greater in degree with increasing leupeptin doses. In Kupffer cells and other endothelial cells, fluorescent granules were also induced. In contrast to rat livers, lipofuscin-like pigments induced in hepatocytes in mice were much less, even with a higher dose (20 mg/100 g/day). E-64C also induced the accumu-lation of lipofuscin-like pigments at a dose of 5 mg/100 g/day, their characteristics being very similar to those induced by leupeptin, but the accumulation being smaller in degree. The fluorescence of leupeptin induced lipopigments was yellowish-green having a peak around 520 nm in emission profile, closely resembling that observed in old rat livers. The hepatobiliary transport functions such as biliary transport maximum (Tm) for sulfobromophthalain and the biliary recovery of iv injected ouabain which are known to decline with age tended to decline in young (6-wk-old) rats administerd with leupeptin at a dose of 5 mg/100 g/day for 2 wks. On the other hand, dolichol concentration in leupeptin treated livers was not increased in comparison to control livers, whereas in old rat livers, the dolichol concentration was more than 2 times greater than in young livers. A clear-dose-dependent deposition of ceroid-lipofuscin induced in young rodent livers by protease inhibitors strongly suggests that the "Protease inhibitor model" is generally valid not only for the brain but for other tissues such as the liver, and for two different thiol protease inhibitors.

KEY WORDS: Ceroid-lipofuscin, hepatocyte, nonparenchymal cells, rodent livers, leupeptin, E-64C

INTRODUCTION

Although many hypotheses for cellular mechanism(s) of aging have been proposed in the past, there is no universally accepted theory for aging up to now. Some theories have been invalidated on an experimental basis, while others albeit attractive, are difficult to directly prove in the experimental systems presently available. The "Protease inhibitor model of aging (Ivy et al., 1984a)" is unique, since it provides not only a theoretical background for at least some important aspects of the cellular aging process but also an experimental system to further validate the theory. Using rat brains exposed to leupeptin, a thiol protease inhibitor, they have demonstrated a considerable accumulation of ceroid-lipofuscin (C-L) closely resembling lipofuscin found in tissues of old animals. Further, subsequent studies demonstrated certain other biochemical and morphological alterations in leupeptin treated brain cells which can also be found in physiological and pathological aging processes (Ivy, 1987; Ivy et al., 1984b, 1986, 1988, 1989; Wolfe et al., 1987). So far, however, experimental evidence for this thesis has been limited to the rat brain and to a single protease inhibitor, leupeptin. If this hypothesis is generally valid, some of above findings should be reproducible in other tissues and with other protease inhibitors as well. The present study aimed to induce first of all C-L in hepatocytes in young rats and mice using two different thiol protease inhibitors, leupeptin and E-64C. Hepatocytes have been chosen, because this cell type has been best characterized in the past in terms of the aging process by morphological, physiological and biochemical approaches (Kitani, 1978, 1982, 1986), so that once C-L is produced in hepatocytes, many parameters that have been reported to change during aging can be examined. So far, however, attempts to induce C-L in hepatocytes by leupeptin are mostly without success (Ivy et al., 1986; Porta et al.,1988). Only Witkop et al. (1988) reported their preliminary results on leupeptin induced C-L in hepatocytes in a hereditarily enzyme defficient mouse strain. E-64 is a group of protease inhibitors exclusively acting on thiol proteases (Hanada et al., 1978; Hashida et al., 1980; Tamai et al., 1981). In contrast to leupeptin, which is a reversible inhibitor for proteases, E-64 molecules are irreversibly bound to proteases, leading to their inactivation (Hashida et al., 1980). Although many (somewhat) different E-64 species are available, E-64C is the only species which can be dissolved in water in high concentration (Hanada et al., 1978). This is required for the use of the osmotic minipump for which water should be used as a solvent.

We introduce in this report, the first general success in the induction of massive accumulation of C-L in hepatocytes in young rats by leupeptin and E-64C. Furthermore, we also report some characteristics of leupeptin treated livers based on our preliminary studies using morphological, physiological and biochemical methodologies in comparison to aged livers.

MATERIAL AND METHODS

Animals

Specific pathogen free (SPF) male Fischer-344 rats (Charles River Japan, Atsugi), Wistar derived rats and C57/BL mice (Shizuoke Jikken Dobutsu, Hamamatsu) were obtained at the age of 4 wks. Animals treated with protease inhibitors or saline solution were maintained in the animal facility of the institute under clean conventional conditions (Kitani et al., 1978a). In most studies the infusion was started at the age of 4 wks, and at the age of 6 wks, various studies were performed. Aged animals of the same strain and sex were also originally obtained from the same sources as described above and raised in the institute's aging farm in SPF condition. Husbandry conditions, life spans and pathologies in the later period of their lives have been reported elsewhere (Nokubo, 1985).

Protease Inhibitor Treatments

Leupeptin (Ac-L-Leu-L-Leu-L-Arginal·1/2 H_2SO_4·H_2O) was obtained from commercial source (Peptide Inst. Osaka). E-64C was a generous gift from Dr. K. Hanada of Taisho Pharmaceutical Company. Leupeptin and E-64C were dissolved in a saline solution and saturated $NaHCO_3$ solution respectively. Continuous infusion of these solutions was performed by the use of an osmotic minipump (Alzet, Alza, Palo Alto, CA) by simply implanting a minipump containing a solution in intraperitoneal cavity under light ether anesthesia. The duration of the treatment was two wks, unless specifically stated. Different doses of leupeptin ranging from 1 mg/100 g/day to 50 mg/100 g/day were used. In most experiments, a priming dose of 5 mg/100 g/day was given on the day of the start of infusion. The dose of E-64C was 4-5 mg/100 g/day. Since the body weight of young animals was increasing rapidly, the dose was determined on the basis of the estimated body weight at the middle of the infusion period. Control animals were infused with an isovolumetric saline solution.

Morphological Observations

Liver tissues were prepared from these animals for various morphological examinations. For light microscopic studies, rodents were perfused intracardially with saline followed by buffered paraformaldehyde solution for 30 min. Livers were then removed, embedded in paraffin, sectioned at 10 μm and stained for HE, PAS and toluidine blue. Sections without staining were used for fluorescence micrography as well as microfluorometry for emission profiles of C-L. The details of the procedure were reported previously (Kitani et al., 1988a). For electron microscopic studies, rodents were perfused and livers were processed as described previously (Ivy et al., 1987a).

Physiological Studies

Ouabain excretion. In different groups of young Fischer rats treated with 1 or 5 mg/100 g/day of leupeptin for 2 wks, the biliary excretion of iv injected ouabain was examined. This material is a neutral cardiac glycoside with a steroidal structure. Animals were anesthetized with pentobarbital and 5- or 10-min bile samples were collected for 60 min after an iv injection of ouabain (0.1 mg/100 g) together with [^{3}H]ouabain (NEN, Boston MA). From the specific radioactivity and flow rate of a bile sample, the biliary recovery of iv injected ouabain was calculated. Animals were 6 wk old when the ouabain study was performed. The efficient biliary excretion without biotransformation and its age-dependent decrease in rats have been reported by the authors (Kitani et al., 1978b, 1988b; Sato et al., 1987).

Sulfobromophthalein (BSP) Tm (transport maximum) studies. Wistar derived rats treated with 1 or 5 mg/100 g/day dose of leupeptin were studied at different time intervals (7 to 14 days) after the start of an ip infusion of leupeptin. The experimental procedures have been reported elsewhere in detail (Kanai et al., 1985; Kitani et al., 1978a, 1981). In brief, animals were anesthetized with pentobarbital and the bile was collected every 10 min under the continuous iv infusion of BSP solution at a rate approximately 1.5 times higher than the estimated Tm value. From the BSP concentration and bile flow rate, the biliary excretion rate of BSP was calculated. The BSP Tm was calculated as the average of the three highest excretion rates during the experiment. The age-dependent decrease in BSP Tm value has been reported for 5 different rat strains of both sexes (Kanai et al., 1985; Kitani et al., 1978a, 1981).

Dolichol accumulation. Since one of the authors (G.I.) previously
found that in brains of rats treated with leupeptin, as well as of old rats
dolichol concentrations were much higher than that of non-treated young rat
brains (Ivy et al., 1984b; Wolfe et al., 1987), dolichol concentrations in
liver tissues were determined in control (saline treated) and leupeptin-
treated young, and non-treated old rats.

A sample of pig pancreatic dolichols was a generous gift from Dr. K.
Yamada of Eisai Co., Ltd. This sample was purified by a silica gel chromatography with an elution solvent of cyclohexane-chloroform (1:1). The
ratio of different dolichols in the purified sample was determined by mass
spectrometry. Rats were sacrificed by decapitation and livers were
excised, frozen by liq. N_2 and stored at –80°C until use. Dolichol determination was carried out essentially by the method as described by Adair
and Keller (1985) with minor modifications. Briefly, liver was homogenized
in 2 volumes of distilled water with a Polytron homogenizer. To 1 ml of
the homogenate in a screw capped test tube was added 1 ml of 1N HCl. The
mixture was allowed to stand at room temperature for 45 min, then heated at
70°C for 45 min. After incubation, 1 ml of 60% KOH was added to the mixture and an alkaline saponification was performed in a boiling water bath
for 1 hr. The reaction mixture was cooled and extracted three times with 3
ml hexane. The extracts were pooled and washed with 5% acetic acid. The
hexane layer was removed and evaporated to dryness. The residue was dissolved in 2 ml of methanol and applied to a Sep-Pak C18 column. After
washing the column with 10 ml of methanol, dolichols were eluted with 5 ml
acetone. The acetone fraction was taken to dryness and redissolved in 500
ul ethanol. Fifty µl of this preparation was injected for HPLC analysis.
The HPLC system consisted of a JASCO 880 solvent delivery system, a
Rheodyne model 7125 sample injector, a Shimadzu SPD-6A uv spectrometric
detector, and a Tosoh TSK-Gel ODS 80TM reversed-phase column (4.6 mm i.d. x
15 cm). All runs were performed isocratically with a solvent system consisting of ethanol–methanol–isopropanol (90:5:5) at 0.8 ml/min. The quantitation of dolichols was performed by comparing the peak hights with those
of the standard pig pancreatic dolichols. The recovery of dolichol added
to liver homogenates was practically complete (100.6 $\pm$ 4.0%, n=6).

RESULTS

Figure 1-b shows a typical example of PAS stained liver sections
prepared from rats treated with leupeptin after 24-h fasting. In liver
tissues treated with leupeptin, hepatocytes tended to be generally greater
in size in comparison with saline treated controls (Fig. 1-a). Further the
irregularity in hepatocyte size was also observed. Inside hepatocytes,
fine granular structures (usually round in shape) stained densely with PAS
were frequently observed in leupeptin treated livers. In non-parenchymal
cells, PAS positive structures usually were stained more densely than
particles in hepatocytes. These granules were also toluidine blue positive. Further, these structures fluoresced a greenish-yellow fluorescence
under a fluoresence microscope with excitation by a mercury lamp (360 nm).

In Fig. 2, examples of fluorescence micrograms are shown. The deposition of fluorescent granules in hepatocytes and Kupffer cells was clearly
dose dependent in the dose range from 1 to 5 mg/100 g/day. With higher
dosages (above 5 mg/100 g/day), the dose dependency was also observed but
the dose effect was not striking as was observed with lower dose studies.
In general, C-L deposition in leupeptin-treated young rat livers at the
dose of 5 mg/100 g/day or higher (Fig.2-d,e) was generally greater than
that found in old rat livers (Fig. 2-b). In leupeptin-treated livers,

almost all hepatocytes contained uniformly fluorescent granules. In contrast, in old livers some hepatocytes contained many lipofuscin granules while some adjacent cells contained little. Uneven intracellular distribution of lipofuscin was also noted in old hepatocyte. In contrast to rat livers, in mouse livers, the deposition of fluorescent materials was much smaller in degree. Even in animals treated with 20 mg/100 g/day for 2 wks, the depostion was very minor, although there was a distinct deposition (Fig. 2-h) in comparison to control livers (Fig. 2-g). Interestingly, there was practically no fluorescent material in Kupffer cells and other non-parenchymal cells in mouse liver given this high dose (20 mg/100 g/day, 2 wks). In rats given E-64C, the accumulation of fluorescent C-L was also induced with a dose higher than 3 mg/100 g/day (Fig. 2-f). However, the deposition was generally milder in comparion to livers of rats given a comparable dose of leupeptin.

Figure 3 shows electron micrographs of hepatocyte (Fig. 3-a,b) and non-parenchymal cells (Fig. 3-c) in rats given leupeptin and in hepatocytes of old rats (Fig. 3-d). Granular electron-dense inclusions were abundantly seen in hepatocytes as well as in Kupffer cells in livers of leupeptin-treated young rats. These structures corresponded to fluorescent materials seen in fluorescence pictures (Fig. 2).

Figure 4 shows examples of fluorescence emission profiles of the C-L in livers from leupeptin-treated (Fig. 4-b) and aged animals (Fig. 4-c) in comparison to young control animals (Fig. 4-a). The emission profiles of C-L in hepatocytes was in the range from 480 to 600 nm with a peak around 520 to 540 nm in both hepatocytes and Kupffer cells. These emission patterns were quite similar to those obtained from old livers (Fig. 4-c). Relative intensities of the peak of fluorescence profiles in leupeptin-treated livers were 3 to 4 times greater than those of control livers and were comparable to those of old livers.

Figure 5-a shows results of the ouabain excretion studies. In rats treated with a low dose (1 mg/100 g/day) for 2 wks, the excretion was somewhat greater than in non-treated rats, although the difference was not statistically significant. With a higher dose of leupeptin, the ouabain-excretion for the first 10-min period tended to be lower than the control

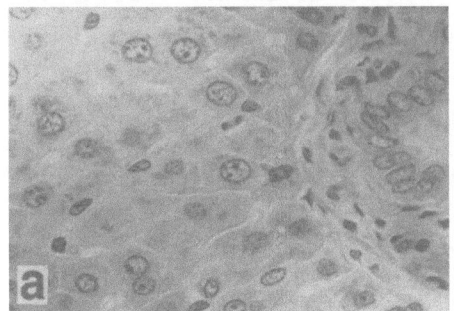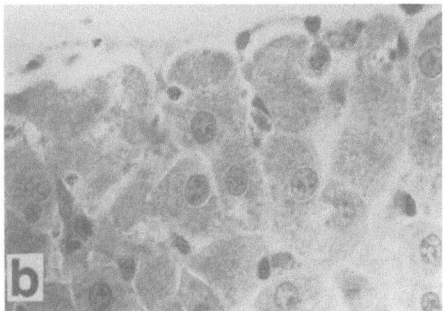

Fig. 1. Light micrographs of a young control liver (1-a), young, leupeptin-treated (20 mg/100 g/day for 2 wks) liver (1-b). PAS staining after 24 hr fasting, X1000. Irregularities in hepatocyte size are more pronounced in a leupeptin-treatd liver as observed in old livers in comparison to young control livers. Hepatocyte size is generally greater in leupeptin-treated liver (1-b) than young control liver (1-a). PAS positive fine granules are abundantly observed in livers from leupeptin- treated rats.

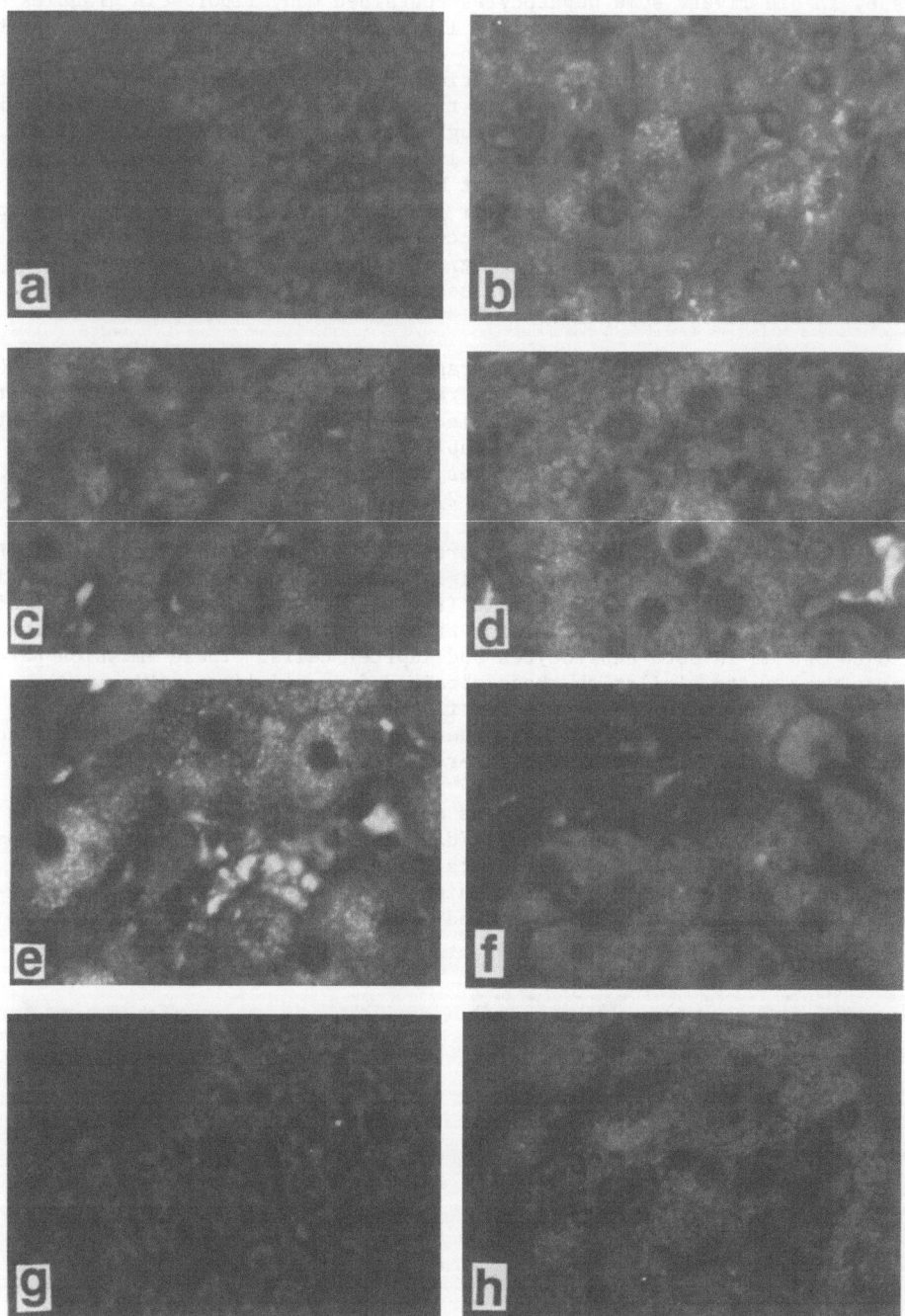

Fig. 2. Examples of fluorescence micrographs without staining.
a, saline-treated (2 wks) young rat liver; b, non-treated
old rat liver, c, leupeptin-treated (1 mg/100 g/day, 2 wks)
rat liver; d, leupeptin-treated (5 mg/100 g/day, 2 wks) rat
liver; e, leupeptin-treated (20 mg/100 g/day, 2 wk) rat
liver; f, E-64C-treated (4 mg/100 g/day, 2 wks) rat liver;
g, saline-treated (2 wks) young mouse liver; h, leupeptin-
treated (20 mg/100 g/day, 2 wks) mouse liver.

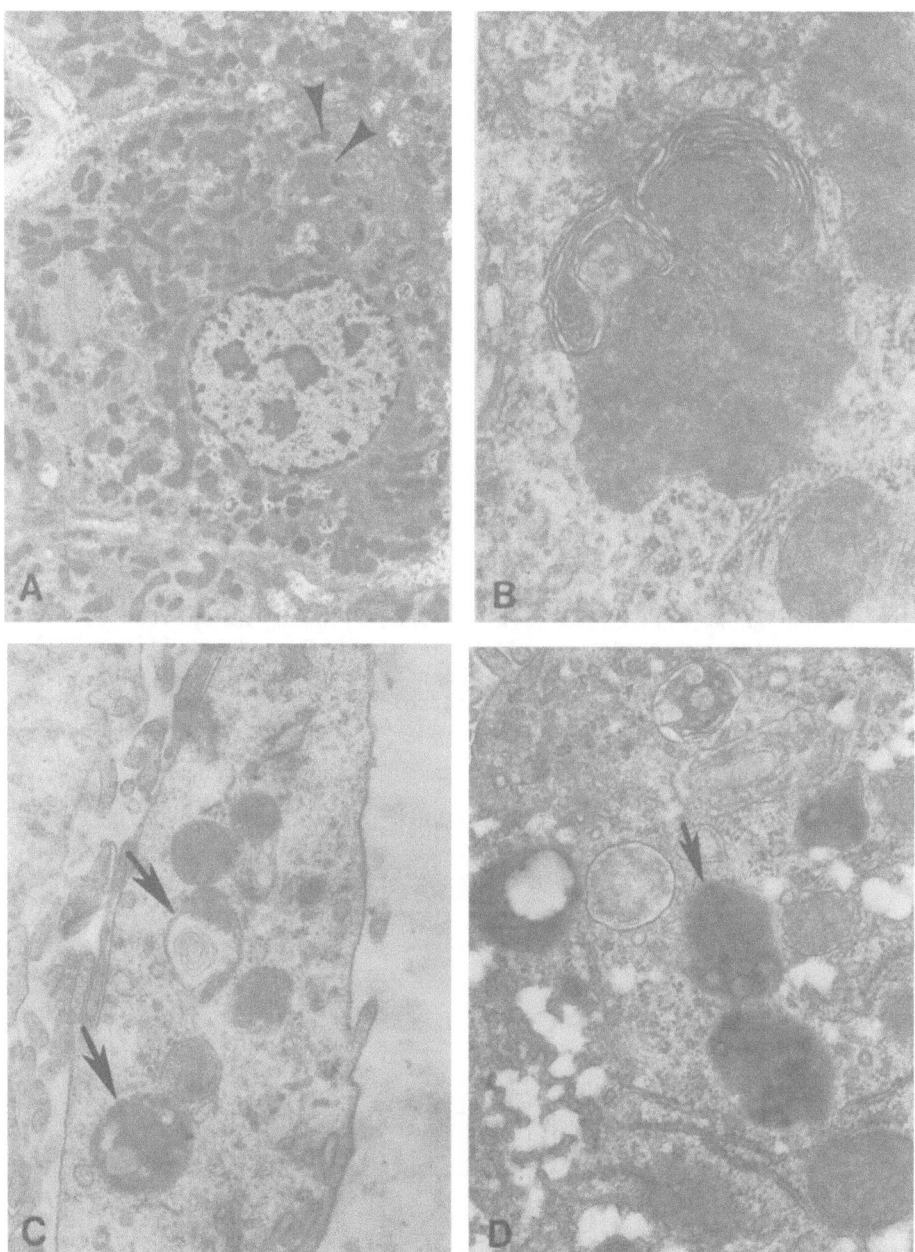

Fig. 3. Electron micrographs of a hepatocyte (a, x3500; b, x33,000)
and a Kupffer cell (c, x22,000) of leupeptin treated rats
and a hepatocyte of an old rat (27mo) (d, x22,000). Electron dense inclusion bodies of heterogeneous morphology
(arrows) are obviously seen in leupeptin-treated liver
cells.

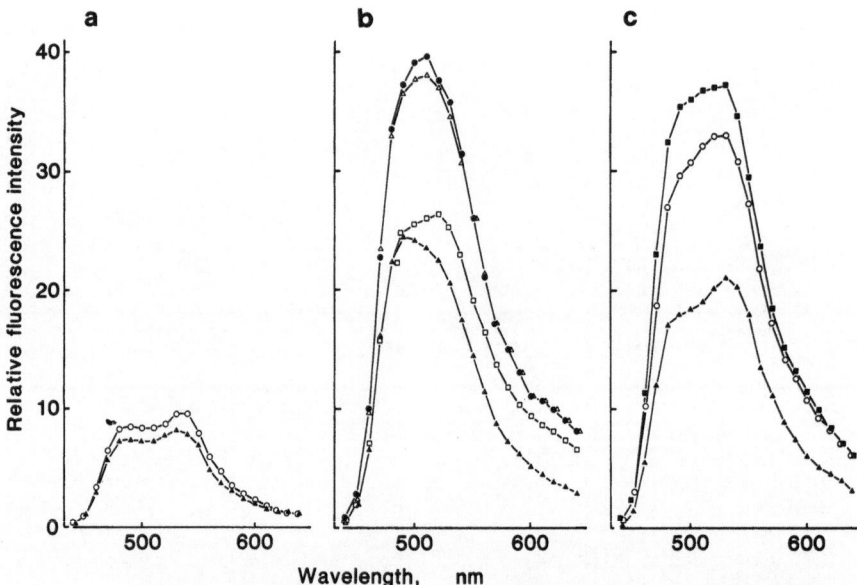

Fig. 4. Examples of emission profiles obtained by microfluorometry. Profiles were obtained by using an excitation wave
length of 405 nm, dichroic mirror V, and cut filter Y475.
Technical details have been reported previously (Kitani et
al., 1988a). a, young control hepatocytes; b, leupeptin-
treated hepatocytes (20 mg/100 g/day for 2 wks); c, old
non-treated hepatocytes.

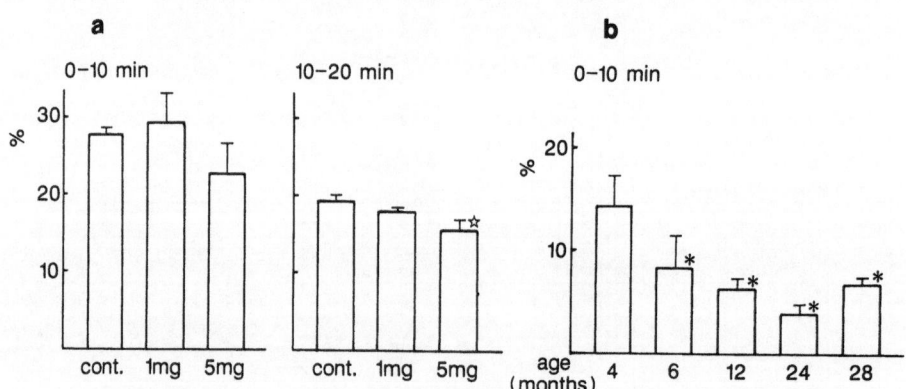

Fig. 5. Summary of ouabain excretion studies in leupeptin-treated
young rats (5-a, 1 or 5 mg/100 g/day, 2 wks) and aging rats
(5-b) of F-344. All values were expressed as the percent of
the administered dose recovered in bile during the period
indicated. The data on aging rats (5-b) was reported
previosly (Sato et al., 1987). ☆, *, Significantly different
from the control value and the value in 4-month-old rats respectively (P<0.05, ANOVA + Scheffè's test).

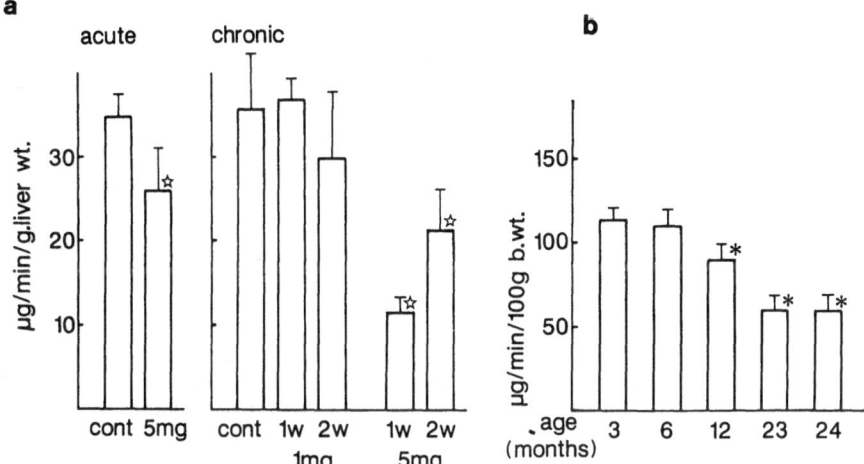

Fig. 6. Summary of BSP Tm studies in leupeptin-treated young rats
(6-a) and aging rats (6-b) (Wistar derived). Fig. 6-a: Acute
experiments were done 2h after a single ip injection of
leupeptin (5 mg/100 g, left panel). Chronic studies were
done after a continuous i.p. infusion of leupeptin (1 or 5
mg/100 g/day for 1 or 2 wks, right panel). The data in Fig.
6-b were reported previously (Kitani et al., 1978a).
☆, *, Significantly different from the control value and the
value in 3-month-old rats respectively (P<0.05, ANOVA +
Scheffè's test).

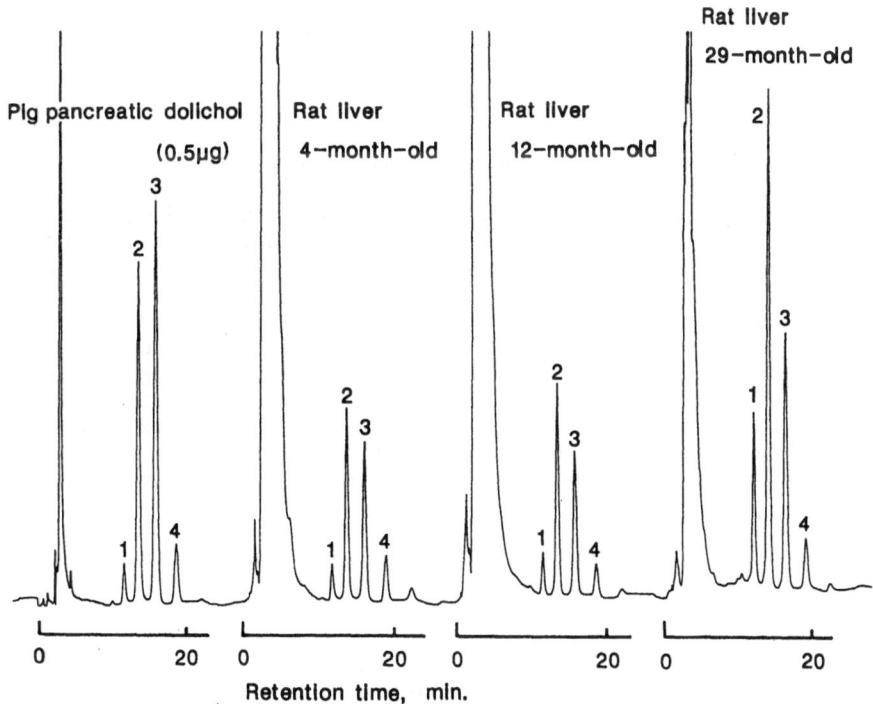

Fig. 7. Representative chromatograms of dolichols of pig pancreas and
of livers obtained from rats of different ages. Peaks 1, 2,
3, 4 correspond to D-17, D-18, D-19 and D-20 respectively.

value, although the difference was not statistically significant. The recovery from 5 to 10 min was, however, significantly lower (data not shown). The difference was statistically significant for the second 10-min recovery. Fig. 5-b shows the first 10-min recovery of ouabain in rats of different ages reported in our previous study using the same rat strain and sex (Sato et al., 1987). An age-dependent decline of this parameter is obvious. The second 10-min recovery value also similarly declined with age (Sato et al., 1987).

Figure 6 shows a part of our study on BSP Tm in young rats given leupeptin as well as our previous results on aging rats (Kitani et al., 1978a). The pretreatment with leupeptin at a dose of 1.0 mg/100 g/day for 2 wks did not significantly affect the BSP Tm in comparison to the control value, although the 2-wk value tended to be lower. A higher dose (5 mg/100 g/day) for 1 wk did decrease the BSP Tm considerably; however, the 2-wk treatment yielded a value greater than the 1-wk treatment value approaching the control value. Further, the BSP Tm was significantly lower 2 hr after a single ip dose of 5 mg/100 g. Fig. 6-b shows a summary of our previous study on the same rat strain (Kitani et al., 1978a). An age-dependent decline in BSP Tm is obvious.

Figure 7 shows examples of profiles of HPLC of extracts of livers from rats of different ages. A dominant peak is seen for D-18 dolichol. Several other peaks (D-16, 17, 19, 20, 21) were also observed. Most peaks tended to be higher in old livers. A relative increase for the D-18 peak was also seen. Dolichol concentration (µg/g) calculated as the sum of D-17, D-18, D-19 and D-20 was 30.7 ± 4.3, 32.2 ± 4.7, and 65.2 ± 8.2 for 4-, 12- and 29-month-old rat livers respectively (each, n=4), showing a clear age dependency. In contrast, dolichol concentration in livers treated with leupeptin (10 mg/100 g/day, 2 wks) examined at the age of 6 wks was not significantly different (25.2 ± 1.4 µg/g, n=5) from saline-treated age-matched control values (24.3 ± 1.0 µg/g, n=4, P>0.05).

DISCUSSION

Ceroid-lipofuscin Formation by Protease Inhibitors

The "Protease inhibitor model of aging" (Ivy et al., 1984a) predicts the decline of proteolysis in cells during aging based on the experiments on the brain exposed to a thiol protease inhibitor, leupeptin. If this hypothesis is correct, this phenomenon could be reproduced in other cell types including hepatocytes. As discussed earlier in the introduction, attempts to produce C-L in hepatocytes by leupeptin have gained only a very limited success. One of the authors (Ivy et al., 1986), using the dose of 0.5 mg/day for 2 successive days found electron-dense bodies in Kupffer cells but nothing abnormal in hepatocytes. From these results, however, she speculated that a higher dose of leupeptin should be able to produce morphological changes in hepatocytes similar to those produced in brain cells. Porta et al. (1988) also failed in inducing C-L in hepatocytes by leupeptin (0.3 mg/100 g/day for 10 days). The doses used in these studies were apparently much smaller than those used in the present study. Indeed, C-L deposition produced by leupeptin at a dose of 1.0 mg/100 g/day for 2 wks was rather mild and was hardly recognizable by a light microscopic examination on stained sections. However, the fluorescence was clearly visible under UV illumination. Further, with increasing doses, the formation of fine granular electron-dense fluorescent structures are clearly observed in hepatocytes under the electron microscope. The fluorescence characteristics of these C-L bodies were very similar to those observed for lipofuscin in hepatocytes of old rats (Fig. 3).

The characteristic features of emission profiles as examined by a microfluorometry procedure are subject to a number of artifactual pitfalls (Douson et al., 1982; Eldred et al., 1982). Among them, a decreasing efficiency of photomultipliers with increasing wavelength may deform an actual emission profile (Eldred, 1982). Although we did not make any correction for figures presented in Fig. 3 for this factor, we have recently been able to examine emission profiles for these C-L by another new instrument with a full correction for this factor. We found, however, emission patterns are essentially similar to what we presented in Fig. 3 even with this new instrument. Since the characteristics of lipofuscin in aged livers as examined by the same system were very similar, we can conclude that these two different fluorescent pigments produced by leupeptin and aging have very similar fluorescence characteristics. With other methods, such as PAS and toluidine blue staining and electron microscopy, there remains little doubt that we have succeeded in the induction of C-L in young rat hepatocytes by leupeptin and to a milder extent by another thiol protease inhibitor, E-64C. The milder effect of E-64C may be due to a somewhat lower inhibitory effect of E-64C on cathepsin B in comparison to leupeptin (Tanaka et al., 1981), or to a shorter duration of inhibitory action, when administered via the i.p. route (Kominami et al., 1980). From these observations, we can conclude that leupeptin, or other thiol protease inhibitors, if given in sufficient amounts, can yield C-L formation in young rat hepatocytes which is very similar to lipofuscin in old hepatocytes.

Intracellular Distribution and Effects of Protease Inhibitors

Previous studies have shown that leupeptin (Dennis and Aronson, 1985) as well as E-64 (Hashida et al., 1982), when given i.p., rapidly penetrate into hepatocytes, presumably by permeation rather than pinocytosis. Further, both of these inhibitors initially distribute in the cytosol fraction mostly in free form, subsequently accumulating in the mithochondrial-lysosomal fraction largely in protein-bound form (Hashida et al, 1982; Dennis and Aronson, 1985). The concentration of protein bound E-64 in this fraction was found to change reciprocally with cathepsin B activitiy (Hashida et al., 1982). Thus, it is highly likely that these enzyme inhibitors directly inhibited lysosomal thiol protease activities, slowing down the proteolysis finally leading to the formation of C-L in these cells, as suggested previously (Ivy et al., 1984) for brain cells, although the possibility of the additional inhibition of cytosolic calcium activated neutral protease (CANP) activity cannot be excluded in a rigorous sense.

Both inhibitors are known to inhibit cathepsin B but the effect on cathespin L is stronger for E-64 (Kominani et al., 1980). Intraperitoneal injection of leupeptin (Kominami et al., 1980) as well as E-64 (Hashida et al., 1982) at a dose of 0.5 mg/100 g was reported to be sufficient to maximally inhibit cathepsin B activity (by 70-90 %) in rat liver but the effect was very transient and the activity returned to the pretreatment level in 24 hrs (Kominami et al., 1980). Acid phosphatase positive autolysosomes induced by a single ip injection of 2.0 mg/100 g of leupeptin disappeared completely in 24 hrs (Ishikawa et al., 1983). The rapid recovery of enzyme activity and morphologic alterations is explained by a rather rapid and efficient excretion of leupeptin into the bile as well as rapid inactivation in the cytosolic fraction as reported by Dennis and Aronson (1985). This rapid clearance of the drug from the liver, coupled with the high proteolytic enzyme activities in the liver, may explain why much higher doses of inhibitors were needed to produce C-L in hepatocytes than previously found for brain cells (Ivy et al., 1984a). It is also apparent that doses used in previous liver studies (Ivy et al., 1986; Porta et al., 1988) are smaller than needed for the formation of C-L in rat hepatocytes.

It is interesting that in the present study as well as in a previous one (Ivy et al., 1986), C-L were induced in Kupffer cells much earlier and to a greater extent than in hepatocytes in rats. In acute liver injuries by toxic substances, the deposition of fluorescent material (ceroid) in Kupffer cells or other phagocytic cells are usually considered to be owing to the phagocytosis of these cell types of necrotic debris coming from hepatocyte necrosis and are regarded as evidence for the latter. While this possibility can not be totally excluded in our study, the sole accumulation of C-L in Kupffer cells with the lack of fluorescent materials in hepatocytes in livers treated with lower doses provides an alternative and more plausible explanation for the C-L accumulation in Kupffer cells. Lysosomal proteases are abundant in Kupffer cells (Brouwer et al., 1988) and may also be subject to inhibition by leupeptin treatment. Protein turnover may be more rapid in Kupffer cells than in hepatocytes. Once protein degradation is interrupted in cells having more rapid turnover, undigested proteins may accumulate more rapidly. Alternatively it is also possible that proteases in Kupffer cells were more affected by leupeptin (e.g. due to higher accumulation of leupeptin in Kupffer cells). In any case, it seems most likely that the accumulation of C-L in Kupffer cells is also the result of the direct effect of leupeptin. The very rapid and efficient inhibition of protein degradation by leupeptin in cultured macrophages has been previously reported (Dean, 1979). The surprisingly lower susceptibility of hepatocytes, and Kupffer cells in particular in mice to leupeptin treatment was a rather unexpected finding in view of the very small dose (0.35 ug/day for 10 days) successfully used by Witkop et al. (1988) on ep/ep mice, even if the latter strain was an enzyme defficient one. This may be related to possible higher proteolytic enzyme activities in livers of this particular mouse strain that we used than in rats, or to other unknown causes. This remains to be explored in the future.

Functional Alterations in Leupeptin-Treated Livers

Rat hepatocyte functions are known to be very resistant to the aging process. Microsomal as well as cytosolic enzyme activities are known to be very well preserved in livers of female rats and of certain mouse strains (Fujita et al., 1982, 1986; Kitani 1988). A drastic decline in many enzyme activities in the microsomal mono-oxygenase system as well as in cytosolic enzymes such as glutathione S-transferase, as observed in male rat livers, was shown to be due to the feminization of male rat liver and not to aging per se (Fujita et al., 1985; Kamataki et al., 1985; Kitani, 1988). In contrast, hepatocyte surface membrane functions such as ouabain uptake appear to be progressively decreased with age in rats (Ohta et al.,1988). The parameters examined in the present study such as BSP Tm (Kitani et al., 1978, 1981; Kanai et al., 1985) and ouabain excretion (Kitani et al. 1978, 1988b; Sato et al., 1987) were consistently shown to progressively decrease with age.

Liver functions as assessed by BSP Tm and the biliary excretion of ouabain tended to be lower in rats treated with leupeptin. Although the exact mechanism(s) for the age-dependent decline in surface membrane functions of hepatocytes remains unknown, we suggested that the possible decline in the protein mobility of surface membranes as revealed by fluorescence recovery after photobleaching (Zs.-Nagy et al., 1986; Kitani et al., 1988b) may be at least a partial factor, since we repeatedly observed the decline in lateral mobility of surface membrane proteins almost in a linear fashion with age (Zs.-Nagy et al., 1986, Kitani et al., 1988b). In this regard, we also examined the membrane protein mobility of leupeptin-treated livers. Our preliminary results have shown a clear tendency of the rise of lateral diffusion constant of surface membrane proteins in leupeptin-treated hepatocytes in rats as well as in mice, in contrast to a decline of this factor in aging livers (Zs.-Nagy, et al.

unpublished observation). At the same time, however, leupeptin treatment caused an increase in the immobile fraction of surface membrane proteins. Thus, the overall influence of leupeptin treatment on surface membrane functions appears to be regulated by the balance of these two opposing factors. Our present results showing some increase in ouabain excretion with a low dose and a decrease with a higher dose appear to fit such an interpretation. The mechanisms underlying the BSP Tm decline with age is more complicated. It may partly be due to the decline in cytosolic enzyme activity of glutathione S-transferase and partly to a possible alteration of physicochemical characteristics of the canalicular membrane. There might be some other unknown factor (e.g. a decline in energy for transport). The results of the present study clearly showed the decline in BSP Tm after 1-wk treatment of leupeptin at a dose of 5 mg/100 g/day. However, a single injection of leupeptin also caused a clear decline in BSP Tm. Thus, it is possible that the decline in BSP Tm was caused by a direct interaction of leupeptin with the excretion of BSP (and/or an acute effect on membrane), since an efficient biliary excretion of leupeptin is known to occur (Dennis and Aronson, 1985). A higher BSP Tm after 2 wks of treatment than that after 1-wk treatment suggests that the capacity of the liver to compensate for a suppressive intervention with liver function is efficient. Such an adaptive capability of the liver, however, may complicate the interpretation of the results after treatment with protease inhibitors. In sum, the results of the present study have shown that leupeptin treatment causes functional alterations of the liver similar to those found during aging. It remains to be established whether these changes in leupeptin-treated livers and aged livers share the common underlying mechanisms.

The failure in demonstration of an increase in dolichol concentration in leupeptin-treated livers was unexpected and is difficult to explain in light of the theory that dolichols are an essential component of C-L (Wolfe et al., 1987, 1988). Previous studies (Ivy et al., 1984b; Wolfe et al., 1987) have shown that in aged brains as well as leupeptin-treated young brains, the dolichol concentration was significantly increased in comparison to young non-treated brains. Our results on livers are at variance in this regard, since the dolichol concentration was almost identical for control and leupeptin-treated young livers, while in old livers it was more than 2-fold higher. The dolichol concentration in the liver in young rats found in our study is one order of magnitude lower than reported values in human liver (Wolfe et al., 1987) but agrees well with values previously reported for rat liver (Keller et al., 1985; Keller and Nellis, 1986; Yamada et al., 1986). Further, the increase in dolichol concentration in old livers as observed in the present study also agrees with a previous study (Keller and Nellis, 1986) which showed a significant increase with age in dolichol concentration in rat liver, although the latter study investigated only up to 14 wks of age in the rat. Wolfe et al. (1988) reported that in livers of patients with neuronal ceroid lipofuscinosis, the dolichol (and dolichyl phosphate) content was not increased, but they attributed this observation to the relative paucity of ceroid accumulation in the liver in these patients. Since C-L accumulation was generally greater in livers treated with high doses of leupeptin than in old livers in the present study, we are unable to explain the lack of increase in dolichol concentration in leupeptin treated livers. Our results are, however, very preliminary in nature. Furthermore, we have been unable to quantitate dolichyl phosphate in the liver, because a suitable internal standard was not available. Thus a careful back-up study is necessary to draw a definite conclusion in this regard. However, if these results are confirmed in future studies, it may suggest that the accumulation of C-L can occur without an increase in dolichol concentration.

Similarities and Differences Between Leupeptin-treated and Aged Livers

Despite the lack of evidence for an increase in dolichol concentration

by leupeptin treatment, the morphological evidence obtained is sufficient
for support of the "Protease inhibitor model of aging" in hepatocyte.
Thus, the next step should be a comparison of functional and biochemical
changes between leupeptin-treated and aged livers as attempted in the
present study. It should be kept in mind, however, that hepatocytes have a
tremendous reserve capacity and further can respond in many ways to many
different stimulations. Ichihara and coworkers have shown that the activity of Hb-hydrolase, an acid protease, was enhanced 6 fold after leupeptin
(but not E-64) treatment (Tanaka et al., 1979, 1981). There may be some
other enzymes induced by treatment of protease inhibitors. Further,
leupeptin was reported to increase activities of cathepsins B,L,D and acid
phosphatase in the cytosol fraction of livers possibly by releasing lysosomal enzymes into this fraction by means of the labilization of lysosomal
membranes (Kominami et al., 1981). Enzymatically less effective but immunologically intact enzyme forms were reported to appear for aldolase and
superoxide dismutase in old animal livers (Gershon and Gershon, 1973; Russ
et al., 1977). A similar enzyme modification was also reported for aldolase in the cytosol fraction of leupeptin-treated liver (Kominami et al.,
1981). Interestingly, however, the enzyme modification in old livers was
reported to be prevented by the addition of leupeptin to samples (Petell
and Lebherz, 1979), while an in vivo leupeptin administration induced such
a modification in young livers, presumably by increasing the activities of
cathepsins in the cytosol fraction (Kominami et al., 1980, 1981). Furthermore, while in old hepatocytes both protein synthesis and degradation may
be declining in harmony with each other, available evidence suggests that
leupeptin does not decrease protein synthesis (Tanaka et al., 1979; Libby
and Goldberg, 1978; Seglen et al., 1979). The use of very young (immature)
rats in the present study may also create another potential pitfall, since
many functional parameters are changing rapidly during this developmental
period. Thus, the overall functional consequences of leupeptin treatment
in young rat liver may be very complex and may not be identical to those
occurring in old livers. The results presented here are only the first
step in our attempt to explore the functional alterations in leupeptin
treated livers. Thus, although we have obtained some results showing similarities in functional alterations between leupeptin-treated livers and
aged liver, we do not immediately claim that the underlying mechanisms are
the same, since leupeptin may directly modify membrane functions other than
by the slowing-down of protein turnover.

Future Potentialities and Conclusions

Despite these complexities as discussed above, a clear demonstration
of lipopigment accumulation in hepatocytes by protease inhibitor administration suggests that important alterations simulating cellular aging were
induced by these inhibitors in hepatocytes as previously shown in brain
cells. Further, as is elaborated in another paper in this volume (Ivy et
al., 1990) protease inhibitors can induce C-L in other internal organs such
as pancreas, kidney, spleen and lung, as well as retina. These results
suggest that the "Protease inhibitor model of aging", originally proposed
on the basis of observations in brain cells, is generally valid for most
cell types. Further investigations to explore the similarities and differences between leupeptin-treated and aged organs may provide new information
about the essential processes of cellular aging. Further, studies comparing different organs treated with protease inhibitors may also elucidate
aging processes common to all organs and specific to each organ. Demonstration of C-L deposition in other varieties of organs by leupeptin treatment indicates that such experimental approaches are now possible. The

model, therefore, provides not only a theoretical basis for the cellular mechanisms of aging but a wide variety of experimental approaches to further elucidate mechanisms of cellur aging.

ACKNOWLEDGMENTS

This study was partly supported by grants in aid for the research project "Pharmacodynamics of the Brain" (1987-1990) of the Tokyo Metropolitan Insitute of Gerontology. The authors gratefully acknowledge the generous supply of E-64C by Dr. K. Hamada and pig pancreatic dolichol samples by Dr. K. Yamada. The skillful secretarial work by Mrs. T. Ohara is also gratefully acknowledged.

REFERENCES

Adair, W. L., and Keller, K., 1985, Isolation and assay of dolichol and dolichyl phosphate, in: "Methods in Enzymology", J. H. Law and H. C. Rilling, eds., Acad. Press, San Diego, London.

Aoyagi, T., Miyata, S., Nanbo, M., Kojima, F., Matsuzaki, M., Ishizuka, M., Takeuchi, T., and Umezawa, H., 1969, Biological activities of leupeptins. J. Antibiot. 22:558.

Brouwer, A., DeLeeuw, A. M., Barelds, K. J., Knook, D. L., 1988, Aging of sinusoidal liver cells, in:"Aging in Liver and Gastrointestinal Tract" L. Bianchi, P. Holt, O. F. W. James, R. N. Butler, eds., MTP Press, Lancaster.

Dean, R. T., 1979, Macrophage protein turnover. Evidence for lysosomal participation in basal proteolysis. Biochem. J. 180:339.

Dennis, P. A., and Aronson JR., N. A., 1985, Metabolism of [^{3}H]leupeptin by rat liver, Arch. Biochem. Biopys. 240:768.

Dowson, J. H., 1982, The evaluation of autofluorescence emission spectra derived from neuronal pigment, J. Microscopy 128:261.

Eldred, G. E., Miller, G. V., Stark, W. S., Feeney-Burns, L., 1982, Lipofuscin: Resolution of discrepant fluorescence data, Science 216, 757.

Fujita, S., Uesugi, T., Kitagawa, H., Suzuki, T., and Kitani, K., 1982, Hepatic microsomal monooxygenase and azoreductase activities in aging Fischer-344 rats. Importance of sex difference for aging study, in: "Liver and Aging-1982, Liver and Drugs", K. Kitani, ed., Elsevier/ North Holland, Amsterdam.

Fujita, S., Kitagawa, H., Chiba, M., Suzuki, T., Ohta, M., and Kitani, K., 1985, Age and sex-associated differences in the relative abundance of multiple species of cytochrome P-450 system, Biochem. Pharmacol. 34:1861.

Fujita, S., Chiba, M., Suzuki, T., and Kitani, K., 1986, Effect of senescence on the hepatic metabolism of drugs affecting the central nervous system in rats and mice, in: "Liver and Aging-1986, Liver and Brain", K. Kitani, ed., Elsevier Science Publishers, Amsterdam.

Gershon, H., and Gershon, D., 1973, Inactive enzyme molecules in aging mice: Liver aldolase, Proc. Natl. Acad. Sci. USA, 70:909.

Hanada, K., Tamai, M., Morimoto, S., Adachi, T., Ohmura, S., Sawada, J., and Tanaka, I., 1978, Inhibitory activities of E-64 derivatives on papain, Agric. Biol. Chem. 42:537.

Hashida, S., Towatari, T., Kominami, E., and Katunuma, N., 1980, Inhibitions by E-64 derivatives of rat liver cathepsin B and cathepsin L in vitro and in vivo, J. Biochem. 88:1805.

Hashida, S., Kominami, E., and Katunuma, N., 1982, Inhibitions of cathepsin B and cathepsin L by E-64 in vivo II. Incorporation of [^{3}H] E-64 into rat liver lysosomes in vivo. J. Biochem. 91:1373.

Ishikawa, T., Furuno, K., and Kato, K., 1983, Ultrastructural studies of autolysosomes in rat hepatocytes after leupeptin treatment, Exp. Cell Res. 144:15.

Ivy, G. O., 1987, Decreased neural plasticity in aging and certain pathological conditions: possible roles of protein-turnover, in: "Neural Plasticity: A Lifespan Approach", T. Petit and G. Ivy, eds., Liss, New York.

Ivy, G. O., Schottler, F., Wenzel, J., Baudry, M., Lynch, G., 1984a, Inhibitors of lysosomal enzymes: Accumulation of lipofuscin-like dense bodies in the brain, Science 985.

Ivy, G. O., Wolfe, L. S., Houston, K., Baudry, M., Lynch, G., 1984b, Lysosomal enzyme inhibitors cause the accumulation of ceroid lipofuscin and dolichols in rat brain, Neurosci. Abstr. 10:885.

Ivy, G. O., Schottler, F., Baudry, M., Lynch, G., 1986, Leupeptin causes several manifestations of aging in brain and liver of young rats, in: "Liver and Aging-1986, Liver and Brain," K. Kitani, ed., Elsevier Science Publishers, Amsterdam.

Ivy, G. O., and Gurd, J. W., 1988, A Proteinase inhibitor model of lipofuscin formation, in: "Lipofuscin-1987, State of the Art," I. Zs.-Nagy, ed., Excerpta Medica, Amsterdam.

Ivy, G. O., Kitani, K., and Ihara, Y., 1989, Anomalous accumulation of and ubiquitin immunoreactivities in rat brain caused by protease inhibition and by normal aging: a clue to PHF pathogenesis?, Brain Res. (in press).

Ivy, G. O., Kanai, S., Ohta, M., Smith, G., Sato, Y., Kobayashi, M., and Kitani, K., 1990, "Lipofuscin-like substances accumulate rapidly in brain, retina and internal organs with cystein protease inhibition", in: "Lipofuscin and Ceroid Pigments-1989, State of the Art", E. A. Porta ed., Prenum Press, New York.

Kamataki, T., Maeda, K., Shimada, M., Kitani, K., Nagai, T., and Kato, R., 1985, Age-related alteration in the activities of drug-metabolizing enzymes and contents of sex-specific forms of cytochrome P-450 in liver microsomes from male and female rats, J. Pharmacol. Exp. Ther. 233:222.

Kanai, S., Kitani, K., Fujita, S., Kitagawa, H., 1985, The hepatic handling of sulfobromophthalein in aging Fischer-344 rats: In vivo and in vitro studies. Arch. Gerontol. Geriatr. 4:73.

Keller, R. K., Fuller, M. S., Rottler, G.D., Connelly, L. W., 1985, Extraction of dolichyl phosphate and its quantitation by straight-phase high-performance liquid chromatography, Anal. Biochem. 147:166.

Keller, R. K., and Nellis, S. W., 1986, Quantitation of dolichyl phosphate and dolichol in major organs of the rat as a function of age, Lipids 21:353.

Kitani, K. (ed), 1978, "Liver and Aging-1978", Elsevier/North-Holland, Amsterdam.

Kitani, K. (ed), 1982, "Liver and Aging:-1982, Liver and Drugs", Elsevier Biomedical, Amsterdam.

Kitani, K. (ed), 1986, "Liver and Aging-1986, Liver and Brain", Elsevier Science Publishers, Amsterdam.

Kitani, K., 1988, Drugs and the aging liver, Life Chem. Rep. 6:143.

Kitani, K., Kanai, S., Miura, R., 1978a. Hepatic metabolism of sulfobromophthalein (BSP) and indocyanine green (ICG) in aging rats, in: "Liver and Aging-1978", K. Kitani ed., Elsevier/North-Holland, Amsterdam.

Kitani, K., Kanai, S., Miura, Morita, R., and Kasahara, M., 1978b, The effect of aging on the biliary excretion of ouabain in the rat, Exp. Gerontol. 3:9.

Kitani, K., Zurcher, C., van Bezooijen, C. F. A., 1981, The effect of aging on the hepatic metabolism of sulfobromophthalein in BN/Bi female and WAG/Rij male and female rats, Mech. Ageing Dev. 7:381.

Kitani, K., Nokubo, M., Ohta, M., and Zs.-Nagy, I., 1988a, Characterization of peroxide-induced autofluorescence of hepatocyte plasma membrane proteins in relation to age pigments in rats, in: "Lipofuscin-1987: State of the Art" I. Zs.-Nagy ed., Excerpta Medica, Amsterdam.

Kitani, K., Zs.-Nagy, I., Kanai, S., and Ohta, M., 1988b, Correlation between the biliary excretion of ouabain and the lateral mobility of hepatocyte plasma membrane proteins in the rat - The effects of age and spironolactone pretreatment, Hepatology 8:125.

Kominami, E., Hashida, S., and Katunuma, N., 1980, Inhibitions of degradation of rat liver aldolase and lactic dehydrogenase by N- N-(L-3-trans-carboxyoxirane-2-carbonyl)-L-leucyl agmatine or leupeptin in vivo, Biochem. Biophys. Res. Commun. 93:713.

Kominami, E., Hashida, S., and Katsunuma, N., 1981, Proteolytic modification of rat liver fructose 1,6-bisphosphate aldolase by administration of leupeptin in vivo, Biophys. Biochim. Acta 659:378.

Libby, P., and Goldberg, A. L., 1978, Leupeptin, a protease inhibitor, decreases protein degradation in normal and diseased muscles, Science 199:534.

Nokubo, M., 1985, Physical-chemical and biochemical differences in liver plasma membranes in aging F-344 rats. J. Gerontol. 40:409.

Ohta, M., Kanai, S., Sato, Y., and Kitani, K., 1988, Age-dependent decrease in the hepatic uptake and biliary excretion of ouabain in rats, Biochem. Pharmacol. 37:935.

Petell, J. K., and Lebherz, H. G., 1979, Properties and metabolism of fructose diphosphate aldolase in livers of "old" and "young" mice, J. Biol. Chem., 254:8179.

Porta, E. A., Mower, H. F., Moroye, M., Lee, Ch, and Palumbo, N. E., 1988, Differential features between lipofuscin (age pigment) and various experimentally produced "ceroid pigments", in: "Lipofuscin-1987, State of the Art," I. Zs.-Nagy, ed., Excerpta Medica, New York.

Russ, U., Lavie, S., Jacobus, S., Dresnick, J., Gershon, H., and Gershon, D., 1977, Studies on altered enzyme molecules from livers of aging animals, in: "Liver and Ageing," D. Platt, ed., Schattauer Verlag, Stuttgart.

Sato, Y., Kanai, S., and Kitani, K., 1987, Biliary excretion of ouabain in aging male and female F-344 rats, Arch. Gerontol. Geriatr. 6:141.

Seglen, O., Grinde, B., and Solheim, A. E., 1979, Inhibition of the lysosomal pathway of protein degradation in isolated rat hepatocytes by ammonia, methylamine, chloroquine and leupeptin, Eur. J. Biochem. 95:215.

Tamai, M., Hanada, K., Adachi, T., Oguma, K., Kashiwagi, K., Omura, S., and Ohzeki, M., 1981, Papain inhibitions by optically active E-64 analogs, J.Biochem. 90:255.

Tanaka, K., Ikegaki, N., and Ichihara, A., 1979, Induction of hemoglobin-hydrolase activity by the thiol-protease inhibitors leupeptin and antipain in adult rat liver cells in primary culture, Biochem. Biophys. Res. Commun. 91:102.

Tanaka, K., Ikegami, N., and Ichihara, A., 1981, Effects of leupeptin and pepstatin on protein turnover in adult rat hepatocytes in primary culture. Arch. Biochem. Biophys. 208:296.

Witkop, C. J., White, J. G., Townsend, D., Sedano, H. O., Cal, S. X., Babcock, M., Krumwiede, M., Keenan, K., Love, J. E. and Wolfe, L.S., 1988, Ceroid storage disease in Hermansky-Pudlack syndrome: induction in animal models, in: "Lipofuscin-1987, State of the Art," I. Zs.-Nagy, ed., Excerpta Medica, Amsterdam.

Wolfe, L. S., Ivy, G. O., and Witkop, C. J., 1987, Dolichols, lysosomal membrane turnover and relationships to the accumulation of ceroid and lipofuscin in inherited disease, Alzheimer's disease and aging, Chim. Scripta 27:79.

Wolfe, L. S., Gauthier, S., and Durham, H. D., 1988, "Dolichols and phosphorylated dolichols in the neuronal ceroid lipofuscinosis, other lysosomal storage diseases and Alzheimer disease. Induction of auto-lysosomes in fibroblasts", in: "Lipofuscin-1987, State of the Art", I. Zs.-Nagy, ed., Excerpta Medica, Amsterdam.

Yamada, K., Abe, S., Suzuki, T., Katayama, K., Sato, T., 1986, A high-performance liquid chromatographic method for the determination of dolichyl phosphates in tissues. Anal. Biochem. 156:380.

Zs.-Nagy, I., Kitani, K., Ohta, M., Imahori, K., 1986, Age-dependent decrease of the lateral diffusion constant of proteins in the plasma membrane of hepatocytes as revealed by fluorescence recovery after photobleaching in tissue smears, Arch. Gerontol. Geriatr. 5:131.

DISCUSSION

GOEBEL: This a wonderful model to produce lipopigment, but I am a little bit skeptical about considering this a good model of accelerating aging. My first question is if the leupeptin remains with the pigment that accrues in the cells, or is this just a very temporary effect? The other question is what happens if you stop the treatment, does the pigment stay there for the rest of the life of the animal, or does it go away?

KITANI: There is good evidence that leupeptin enters very rapidly into the cells and at first is found in the cytosol and then in particle fractions. Although we haven't done concentrations studies, leupeptin is metabolized and excreted rather rapidly. That's the reason for infusing constantly leupeptin by means of osmotic minipumps. In regard to the lipofuscin-like pigment, we know that it remains in the cell for some time, although we don't know yet for how long. Part of the pigment, at least in hepatocytes, is removed.

GOEBEL: Is the accumulation of the pigment itself detrimental to the cell metabolism?

KITANI: We don't know whether the pigment is detrimental or not. In fact, this was one of the reasons for doing this experiment. We did find some functional alterations in hepatocytes, but we could not establish by certain a cause-effect relation between lipopigment accumulation and function.

HERVONEN: In presenting one of your slides, you said that the autofluorescent characteristics of the leupeptin-induced pigment was similar to that of age-pigment, but in the previous slide, where you showed the spectra of both pigments, it was clear that they are not similar.

KITANI: We have seen some differences between age-pigment and leupeptin-induced pigment in terms of emission profiles, but also we can see many differences between individual measurements in the same preparations from aged livers and leupeptin-treated hepatocytes. Some profiles look very much identical in both cases. But you are correct in pointing out that in the aging hepatocytes the shoulder of the emission spectrum is shifted toward higher wavelengths.

GOEBEL: You showed a slide with fluorescent data, and in the saline-treated controls there was still some fluorescence. How do you explain this?

KITANI: Well, every tissue has some autofluorescence, not as high as that emitted by lipofuscin-like pigments, but always present.

HARMAN: One question on the difference between rats and mice. Is the protein turnover in hepatocytes about the same, and is there a difference in the lysosomal concentrations in these cells between rats and mice?

KITANI: Although I cannot precisely answer this question, there is at least a marked difference in terms of leupeptin susceptibility between rats and mice. The exact cause is presently unknown, but it is important to recognize that there are clear quantitative differences in susceptibility between species.

THE LACK OF AGE-PIGMENTS AND THE ALTERATIONS IN INTRACELLULAR MONOVALENT
ELECTROLYTES IN SPONTANEOUSLY HYPERTENSIVE, STROKE-PRONE (SHRsp) RATS AS
REVEALED BY ELECTRON MICROSCOPY AND X-RAY MICROANALYSIS

I. Zs.-Nagy, V. Zs.-Nagy, T. Casoli and Gy. Lustyik[+]

Verzár International Laboratory for Experimental Gerontolo-
gy (VILEG), Italian Section, Department of Gerontological
Research, INRCA, Via Birarelli 8, 60121 Ancona/Italy and
[+]VILEG, Hungarian Section, University Medical School, 4012
Debrecen/Hungary

SUMMARY

Male, spontaneously hypertensive, stroke-prone (SHRsp) rats established by Okamoto et al. (1974) were studied. About 80 % of the males of
this strain have a particularly short life span (33-41 weeks); they display a considerable hypertension (above 220 mmHg) and a tendency for plurifocal brain strokes. Hypertension and strokes can be provoked in an
accelerated and synchronized fashion by supplementing 1 % NaCl into their
drinking water. Symptoms of the appearance of brain strokes can be judged
from characteristic signs of motor disorders, and can be established also
by pathohistology. Since hypertension and arteriosclerosis are frequently
involved in aging, the question we intended to answer was whether these
animals may represent a model of the normal aging process or not. Two
approaches are described: (1) Accumulation of lipofuscin granules in
their brain, liver and myocardium was followed by transmission electron
microscopy before and after the appearance of strokes. It has been established that these tissues do not show any typical accumulation of lipofuscin granules, although submicroscopic signs of an enhanced damage of
cell organelles (especially of mitochondria in liver and brain cells, but
not in myocardium) were encountered. (2) The intracellular monovalent
composition in the brain and liver was measured by using bulk-specimen
X-ray microanalysis. The intracellular Na-content (mEq/kg water) was
significantly higher (170-200 %) in both the brain and liver cells,
whereas the K-content increased only moderately (118-130 %). The results
suggest that although the SHRsp rats do not represent a direct model for
the normal aging process from the point of view of lipofuscin accumulation, the shifts of the monovalent electrolyte contents in the brain and
liver cells observed already in the youngest ages, are similar to those
observed in aged normal rats. The theoretical consequences of such a conclusion are discussed.

KEY WORDS

spontaneously hypertensive stroke-prone (SHRsp) rats, hypertension, lipofuscin, brain, liver, myocardium, mitochondrial damage, X-ray microanalysis, intracellular monovalent electrolytes

INTRODUCTION

It is an important task of the experimental biology and medicine to identify animal models of human diseases, in order to facilitate the clarification of the underlying pathological phenomena and to test the efficiency of certain drugs in preventing or curing the respective disease. One of the recent developments in this field is represented by the selection of spontaneously hypertensive rats (SHR) from the inbred Wistar rat strain of the Kyoto University (Okamoto, 1962; Okamoto and Aoki, 1963). Further studies and careful selective work led to the establishment of a stroke-prone version of the SHR rats (Okamoto et al., 1974). These letter rats are termed now in the literature as SHRsp (= SHRSP or SHR-sp) rats (Tomita et al., 1978, 1979; Shibota et al., 1978; Nagaoka et al., 1989; Nagai et al., 1989; Suno et al., 1989).

Numerous blood and tissue parameters have been studied so far in the SHRsp rats: cerebrovascular permeability (Shibota et al., 1978), antioxidative potential (Tomita et al., 1978), lipid peroxides and related enzyme activities (Tomita et al., 1979), pharmacological effects of Idebenone on various parameters (Nagaoka et al., 1989; Nagai et al., 1989; Suno et al., 1989). However, to best of our knowledge, no studies have been undertaken to clarify whether the male SHRsp rats display any similarity to the normal aging process or not. This question seemed to be all the more justified, since hypertension and arteriosclerosis are very frequently present in aged humans or animals.

Since December 1987 our group is keeping an SHRsp strain (kind gift of the TAKEDA Research Laboratories, Osaka, Japan). We intend to clarify some general and special characteristics of these animals. The present studies were aimed to reveal whether some known, basic phenomena of normal aging take place in the male SHRsp rats. Namely, two approaches have been realized: (1) The accumulation of lipofuscin granules in their brain, liver and myocardium was followed by transmission electron microscopy before and after the appearance of strokes. This approach was based on the general knowledge according to which the lipofuscin, since the discovery of Hannover (1842), has been considered as the most characteristic marker of aging (see for ref.: Kikugawa, 1988, Eldred and Katz, 1988). (2) The intracellular monovalent composition in the brain and liver was measured by using bulk-specimen X-ray microanalysis. This approach was justified because a considerable age-dependent increase of the intracellular potassium content was measured in the brain cortical cells, the hepatocytes and myocardium (see for details: Zs.-Nagy, 1983, 1988a; Lustyik and Zs.-Nagy, 1985, 1988), and according to the membrane hypothesis of aging (MHA), the membrane alterations underlying these changes of the electrolyte composition are of crucial importance in the cell maturation and aging (Zs.-Nagy, 1978, 1986, 1987, 1988b, 1989; Zs.-Nagy et al., 1988).

MATERIALS AND METHODS

Our SHRsp colony differs from that of the original one of Osaka only in one aspect. Namely, the animals were kept in Osaka under specific pathogen free (SPF) conditions, whereas in our animal house they are raised under conventional conditions. The first 6 couples of 2 month of age after having arrived by air from Osaka to Ancona, were placed in a separate room where the first new generation was also born. After having reached another 3 month period after birth, the animals were brought (1 cage per week) into the normal animal house. Apparently they did not get any infective disease from the conventional Wistar rats, and did not show any unusual phenomena either immediately or afterwards.

The male SHRsp rats have a particularly short medium life span (33-
41 weeks). They develop a considerable hypertension (Table 1) by the age
of 2-3 months, and according to Okamoto et al. (1974), about 80 % of them
has a strong tendency for plurifocal brain strokes over the age of 100
days. Since the female SHRsp rats show similar phenomena less frequently
and only at later ages (Okamoto et al., 1974), all the experiments were
carried out on males.

The blood pressure was monitored without anesthesy of the rats by an
instrument (EIKI, Budapest, Hungary) measuring systolic blood pressure on
the tail by a properly designed system using an optical detecting method
of blood flux converted into blood pressure units. The animals are kept
in a temeperature-controlled cage at 36-37°C during the measurement. They
must be trained or habituated for this measurement for about 2 weeks.

Transmission electron microscopy

These experiments were carried out on 2 age groups of male SHRsp
rats: Group 1 (Young group) was 5 weeks old. These animals did not show
yet hypertension, neither strokes were encountered. Group 2 (Adult group)
was 5.5 months old, the level of hypertension was above 240 mmHg, and
plurifocal strokes were present in the brain of each of them (2-11 visible smaller or larger loci in the brain sectioned in about 0.5 mm
slices or in the subarachnoideal space). Each group consisted of 5 rats.

The animals were killed by decapitation. Small tissue blocks of
brain parietal cortex (from the apparently healthy regions), liver and
myocardium were dissected and fixed on 3 % glutaraldehyde in Sörensen
phosphate buffer at pH 7.35 for 22 hrs at 4°C. This was followed by washing (3 changes) in the same buffer for 1 hr. Postfixation was carried out
in 1 % OsO$_4$ in the same buffer for 1 hr. Dehydration was carried out in
alcohol series, completed by propylene oxide, and embedding in Durcupan
ACM. Sections were made on an LKB Ultrotome III ultramicrotome, contrasting with uranylacetate (40 min) and lead-citrate (5 min). Micrographs
were taken by a JEOL JEM 100B microscope at 60 or 80 kV.

Bulk specimen X-ray microanalysis

These experiments were carried out on the brain (and in part on the
liver) of 3 age groups of male SHRsp rats: Group 1 (Young-1 group) consisted of 1 month old rats, still without hypertension and strokes. Group
2 (Young-2 group) was 2 months old, the level of hypertension was above
220 mmHg, several visible smaller or larger signs of strokes were present
in the brain of these animals. Group 3 (relatively healthy, long survivor
adult group) was 13 months old, with a moderate level of hypertension,
and without apparent signs of strokes. These males will be commented in
the description of the results. Each group consisted of 3-5 rats.

Small pieces (1x1x2 mm) of the brain parietal cortex and the liver
were processed for energy-dispersive X-ray microanalysis by the freeze-
fracture freeze-drying (FFFD) technique described in detail elsewhere
(Zs.-Nagy et al., 1977; Zs.-Nagy 1983, 1988a). Here we list only the main
steps of the procedure with some practical information.

(a) Samples were deep frozen in isopentane cooled to its melting
point by liquid nitrogen. This procedure must be carried out immediately
after the removal of the tissue sample. The frozen samples can be maintained for a long time under liquid nitrogen without any risk.

(b) Fracture of the frozen samples by cooled scissors, still in the
frozen state, in order to explore intact intracellular compartments. This

step is realized immediately before the next procedure. The freshly broken surface remains recognizable even after the freeze-drying procedure on the basis of its white color, whereas the other surfaces cut during the excision are contaminated by blood.

(c) Freeze-drying of the samples in order to exclude the possibility of the redistribution of the light elements between the intra- and extracellular spaces. This process was carried out in vacuum evaporators of JEOL JEE-4B or JEE-4X types, each equipped with a JEOL EE-ACE freeze-drying attachment. The specimen temperature is kept around $-80^{\circ}C$, whereas the "cold finger" is at liquid nitrogen temperature. Endvacuum of 10^{-5} Torr can be achieved, and experience shows the necessary time for drying out a given amount of tissue pieces (5-6 hrs). At last the specimen temperature is raised to about $25^{\circ}C$. After this procedure, the FFFD specimens can be maintained in vacuum of about 10^{-3} Torr, in order to avoid any rehydration of them from the humidity present in air. The FFFD specimens are mounted on a properly designed specimen holder immediately before use by means of a suitable conductive glue. Care is taken to reduce the necessary time for keeping them at atmospheric pressure to the absolutely necessary minimum.

(d) Scanning electron microscopy of the bulk specimens was carried out without using any coating layer, at 10 kV accelerating voltage in secondary electron image mode. As we have shown it elsewhere (Zs.-Nagy et al., 1977; Zs.-Nagy, 1983), no considerable electrostatic charging of the FFFD specimens comes into being. The two main compartments, the nucleus and the cytoplasm can easily be recognized in this type of preparations. X-ray microanalysis of the nuclei and cytoplasm was carried out by using a JEOL JSM 35C model equipped with an EDAX 711 type analyzer (EDAX System F), or a PHILIPS 512 scanning microscope equipped with an EDAX 9100 system. Although the 2 systems apply various parameters as regards incident beam current, effective beam current, working distance, takeoff angle, etc., the results obtained can be compared, since the proper use of standards in the two systems really eliminates the differences in the analytical parameters (Zs.-Nagy, 1983, 1988a, Zs.-Nagy and Casoli, 1989).

(e) Computer handling of the spectra results in the necessary information for the use of the mass fraction method of Hall et al. (1973) extended for bulk specimens (Zs.-Nagy and Pieri, 1976; Zs.-Nagy et al., 1977; Zs.-Nagy, 1983, 1988a, Zs.-Nagy and Casoli, 1989). This involves a background subtraction, calculation of the net peak integrals and the necessary "white count number" in the properly selected range of background. In possession of the peak-to-background ratios obtained under identical conditions on properly selected bulk standard crystals (Zs.-Nagy and Pieri 1976; Zs.-Nagy et al., 1977; Zs.-Nagy and Casoli, 1989), this procedure gives relative elemental concentrations in the dry mass of the FFFD specimen.

(f) Calculation of the monovalent electrolyte concentrations for the intranuclear and intracellular water. For this purpose it is possible to measure the intracellular water content in identical tissues by using a specially designed cryo-technique (Zs.-Nagy et al., 1982; Lustyik and Zs.-Nagy, 1985, 1988). Once the age dependence of the intracellular water content is established for a given strain, one can extrapolate these values for other age groups, too, without any great risk. The nucleus contains always 2-3 weight % more water than the cytoplasm (Horváth et al., 1984; Lustyik and Zs.-Nagy, 1985, 1988), and this fact can be used also for estimating the real intracytoplasmic water content of the cells. The final goal of this method is to obtain the intracellular monovalent ion concentrations in mEq/kg water units. Details of this technique can be found elsewhere (Zs.-Nagy, 1983, 1988a; Zs.-Nagy and Casoli, 1989).

Statistical evaluation of the data

For each analyzed cell nucleus and cytoplasm elemental concentrations of Na, K and Cl were calculated. Since the data obtained in the same type of cells belong to a normal distribution, it is possible to pool together the data from various animals of the same age (or experimental conditions). Properly chosen methods of the "t" test (Snedecor and Cochran, 1980) can be applied to reveal the significance of the differences in the parameters obtained. The evaluation was carried out by suitable computer programs.

RESULTS

The blood pressure levels

The SHRsp rats of 1 month of age show only 130–150 mmHg blood pressure which is quite similar to that we observe in normotensive, conventional CFY males of 2-3 months of age. The blood-pressure of the male SHRsp rats increases very sharply afterwards: we demonstrate the blood pressure levels by showing the data obtained in two randomly selected groups of 3-month-old males: they were monitored for a 2-weeks period (one measurement per week). These data are summarized in Table 1. It should be stressed that the results of the very first measurements of the blood pressure display a broad scatter: these results were disregarded, since they are considered as training or habituation data until the animals learned to support the procedure. Blood pressure values were regularly collected after at least 2 weeks of training. The rats included in the measurement of blood pressure (Table 1) received drinking water supplemented by 0.9 % NaCl. It is important to stress that the rats should be kept at 37° before and during the measurements, otherwise the blood vessels of the tail remain closed, compromising the measurement.

Repeated measurements of the blood pressure under standardized conditions prove unanimously that the hypertension of the SHRsp males is very considerable and persists in time (Table 1). It should be stressed that for technical reasons it was not possible to measure the blood pressure of each animal which was used for the various studies, nevertheless, the data of Table 1 prove that one can expect very similar tendencies in most of the males of this strain, although some individual differences may certainly exist (Okamoto et al., 1974).

The onset of hypertension is followed by the appearance of strokes within about 2-3 weeks. Some changes in the animal's behavior can show unanimously the presence of strokes. For example, the animals show some characteristic, abnormal leg movements, their hairs become irregular, etc. In some animals the strokes are so extensive that they die in a short time, however, in most of the cases, they survive the usually small strokes, but have again and again new strokes. It should be noted, however, that the wide scatter of the medium survival of the males derives from two factors: (1) Some rats die very early due to strokes; and (2) one can observe several males of the SHRsp strain which have only a moderate hypertension, and do not display any sign of strokes. The presence of these exceptional males was observed also by Okamoto et al. (1974). These male rats survive much longer than the average mentioned above (more than 1 year), therefore, we called them for simplicity as long-survivors). We do not know the exact frequency of occurrence of such males in our SHRsp strain, since our colony has been established only 2 years ago, nevertheless, these animals can be the best controls for numerous observations, since they are of the same genetic origin. As a matter of fact, we analyzed some of these animals in the present experiments, too.

Table 1, Blood pressure values (mmHg) in 2 randomly selected groups of 3-month-old male SHRsp rats during an observation period of 2 weeks subsequent to the habituation of the rats (Means of 5 animals $\pm$ S.E.M.)

Time	Group A	Group B
2nd week	261 $\pm$ 11	254 $\pm$ 13
3rd week	276 $\pm$ 14	274 $\pm$ 6
4th week	270 $\pm$ 13	276 $\pm$ 13

Note: Numbers of weeks are counted from the very first measurement. The differences within the groups or between them are statistically not significant.

Transmission electron microscopic results

Detailed studies of hundreds of sections from the brain cortex, the liver and the myocardium of SHRsp rats have been carried out. These investigations revealed that the submicroscopic structure of the studied cells in Group 1 (young) rats was very similar to that any normotensive Wistar or CFY rats. They display a completely normal ultrastructure of the studied cells, and do not show any conspicuous appearance of lipofuscin or lipofuscin-like granules in these tissues.

The Group 2 (adult) rats studied have had hypertension and strokes before killing them, as shown by the stroke-symptoms, the behavior of the rats, by the macroscopic alterations and by the identified cerebrovascular lesions (bleeding spots) in the brains of them. Rats with the same symptoms of strokes usually would die within 2-4 weeks, even if the NaCl supplementation to the drinking water is suspended. It involves that these rats were very near to the end of their life span, i.e., can be considered as relatively old rats. In spite of this situation, practically no lipofuscin accumulation could be detected in the brain cells, hepatocytes and myocardium of these animals.

Some fine structural alterations could, however, be observed in the brain and liver cells. These alterations concerned mostly the mitochondria: unusual, strange structural formations can be seen as demonstrated in Figures 1 and 2 for the liver. Some mitochondria show irregular extensions of their external membranes, some parts of their matrix seems to be empty, the formation of multilamellar structures can be observed, etc. With some modifications, alterations of the mitochondria were found also in the brain cell cytoplasm (Figures 3 and 4): the dominant feature in the brain is the appearance of large, swollen mitochondria. However, it is striking, that the mitochondria of the myocardium remained mostly intact from a structural point of view (Figure 5).

The lysosome-like structures demonstrated in Figure 2 deserve a particular interest. They are membrane bound, spheric formations containing a fine granular matrix (which is identical in its appearance with the mitochondrial matrix), and an other type of granular substance. These structures are almost certainly of auotophagosomal character. Detailed enzyme histochemical analysis will be necessary to reveal the real nature of these granules.

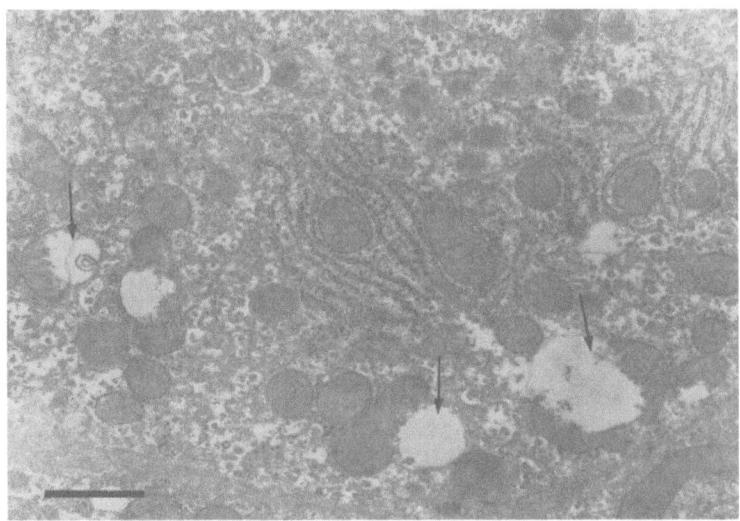

Figure 1. Low power electron micrographic detail of a hepatocyte from a 5.5 months old SHRsp rat. The general ultrastructural appearance of the cell is quite intact, however, some mitochondria display irregularities (arrows). lipofuscin granules are completely missing from the cell. Bar in all figures = 1 μm.

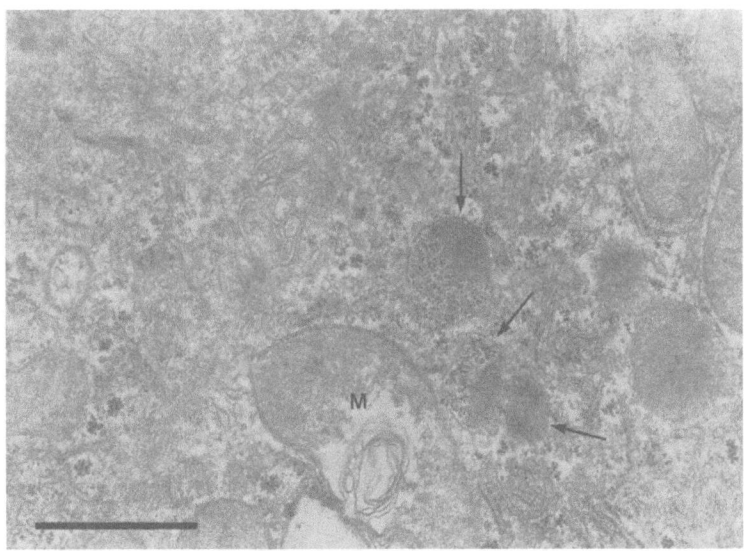

Figure 2. Higher power electron micrographic detail of a hepatocyte from a 5.5 months old SHRsp rat. The mitochondrion (M) shows serious signs of damage, whereas relatively large, lysosome-like bodies indicated by arrows appear in its vicinity.

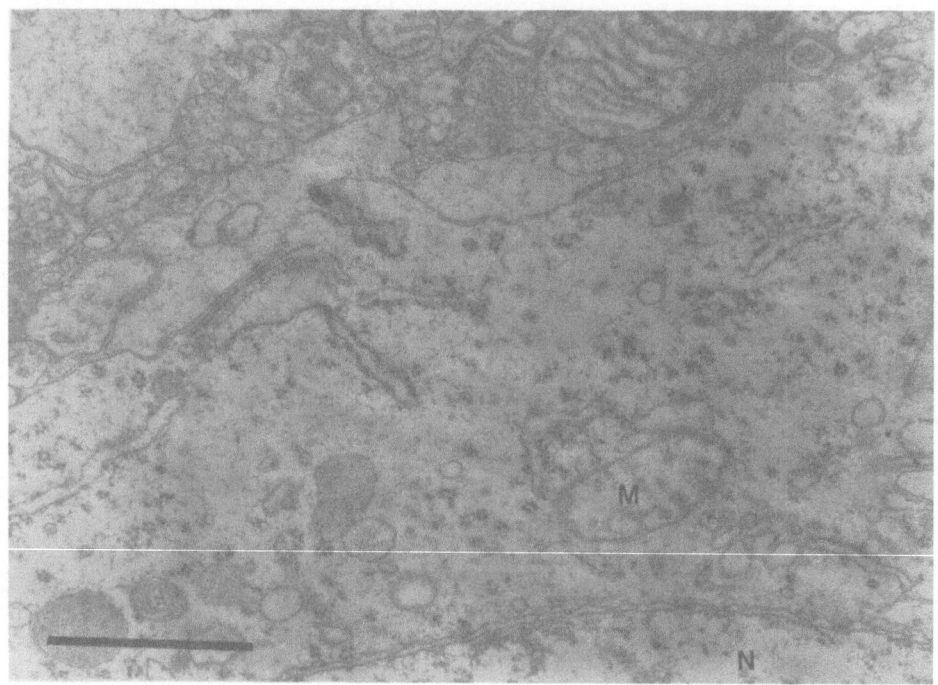

Figure 3. An overview of a neuronal perikaryon from the parietal cortex of a 5.5 months old SHRsp rat. The overall ultrastructural features can be seen without any sign of lipofuscin accumulation. N - nucleus; M - mitochondrion.

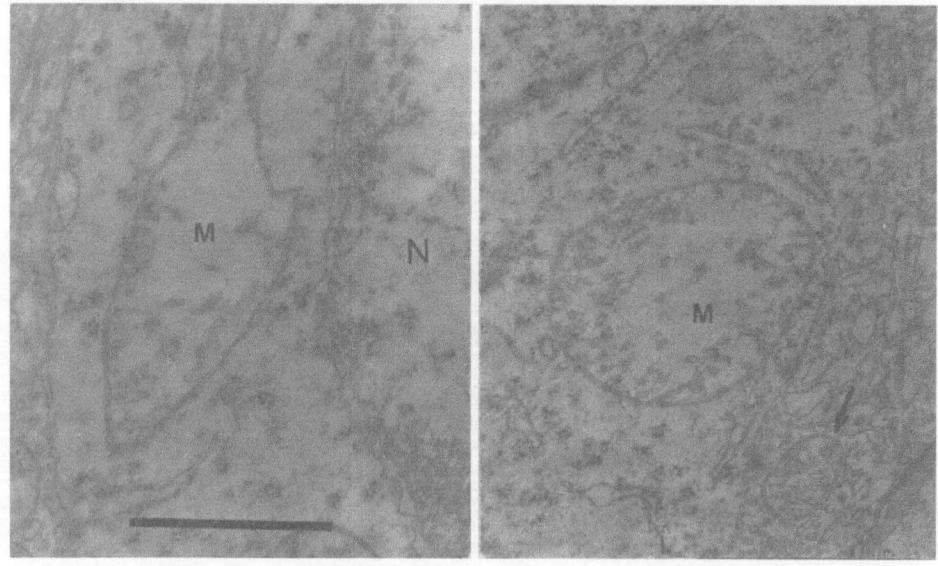

Figure 4. Details of 2 neuronal perikarya from the parietal cortex of a 5.5 months old SHRsp rat. Heavily damaged, swollen mitochondria (M) can be seen. N - nucleus. Note a mitochondrion of normal size in the right lower corner of the figure (arrow).

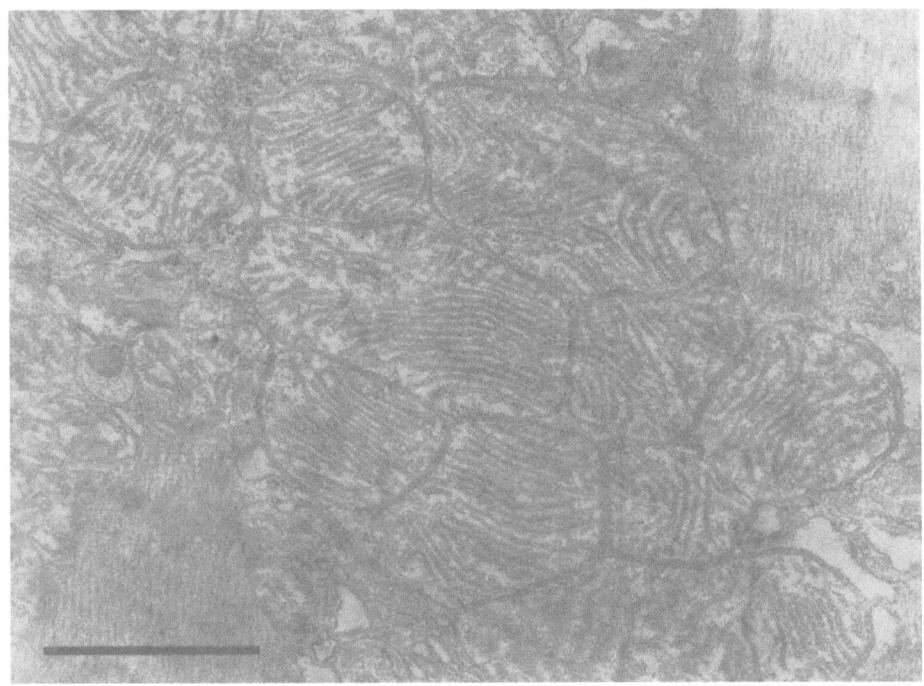

Figure 5. Higher power electron micrographic detail of the myocardium from a 5.5 months old SHRsp rat. The mitochondria show no apparent signs of structural damage.

The intracellular monovalent electrolytes

The results obtained by this technique are summarized in Table 2. For comparison, we reported the Na, Cl, and K-concentrations observed in the brain cells of normotensive CFY rats (Lustyik and Zs.-Nagy, 1985). The results obtained in the brain cells of SHRsp rats show unanimously that these male rats have such a high Na-content in their nerve cells already at the age of 1-2 months, as the normotensive rats at the age of 25 months (Table 2). The K-content of the nerve cell cytoplasm is also higher in the SHRsp rats than in the normotensive controls of the same age, nevertheless, this increase is of much lower proportion. In other words, the intracellular Na/K ratio increased significantly in the SHRsp rats, mostly because of the almost 100 % increase of the Na-content. The total monovalent ion content of the SHRsp rat brain cells is also very significantly higher than in the controls of the same age.

The significance of the increased Na- and K-contents is underlined by the results observed in the long-survivors of the SHRsp strain (Table 2). These latter rats were already 13 months old, but had only moderate hypertension and did not have strokes. The Na- and K-contents correspond practically to those of the normotensive rats. It should also be stressed, that one can encounter also in the young age (around 1 month) some male SHRsp rats with almost normal Na- and K-contents. These may represent those individuals which will become the long-survivors. It is important that the 2-month-old SHRsp rats having had already strokes, display very high Na- and K-concentrations in their neuronal cytoplasm.

In the liver of the same normal rats the intracytoplasmic Na- and K-contents varied between 16-26 and 159-164 mEq/kg water, respectively,

Table 2. Intracytoplasmic water and monovalent ion concentrations in the brain and liver cells of male CFY and SHRsp rats of various ages as revealed by FFFD bulk specimen X-ray microanalysis. Mean values of $\underline{n}$ measurements in each cell type $\pm$ S.E.M.

Cell type	Na^+ mEq/kg	Cl^- mEq/kg	K^+ mEq/kg	Total mEq/kg	Age (months)
BRAIN (CFY): Water content:	78.0 $\pm$ 0.4 % in $\underline{n}$ = 110 measurements				
$\underline{n}$ = 61	49.0 $\pm$ 3.0	32.0 $\pm$ 2.0	147.0 $\pm$ 4.0	223.0 $\pm$ 5.0	1.0
BRAIN (CFY): Water content:	76.3 $\pm$ 0.6 % in $\underline{n}$ = 76 measurements				
$\underline{n}$ = 31	77.0 $\pm$ 4.0	38.0 $\pm$ 2.0	151.0 $\pm$ 5.0	266.0 $\pm$ 7.0	11.0
BRAIN (CFY): Water content:	71.9 $\pm$ 0.5 % in $\underline{n}$ = 160 measurements				
$\underline{n}$ = 60	92.0 $\pm$ 4.0	47.0 $\pm$ 2.0	211.0 $\pm$ 5.0	350.0 $\pm$ 6.0	25.0
BRAIN (SHRsp): Water content:	78.0 % (taken from other measurements)				
$\underline{n}$ = 50	83.7 $\pm$ 5.2	37.7 $\pm$ 1.3	173.9 $\pm$ 2.8	295.3 $\pm$ 3.9	1.0
BRAIN (SHRsp): Water content:	78.0 % (taken from other measurements)				
$\underline{n}$ = 25	93.6 $\pm$ 4.0	56.6 $\pm$ 3.7	197.0 $\pm$ 5.0	347.2 $\pm$ 4.6	2.0
BRAIN (SHRsp): Water content:	76.3 % (taken from other measurements)				
$\underline{n}$ = 25	68.2 $\pm$ 9.9	34.5 $\pm$ 2.1	171.6 $\pm$ 4.5	274.3 $\pm$ 5.2	13.0
LIVER (SHRsp): Water content:	70.0 % (taken from other measurements)				
$\underline{n}$ = 25	51.1 $\pm$ 5.4	40.8 $\pm$ 2.7	225.7 $\pm$ 4.3	317.6 $\pm$ 3.3	2.0

Note: Parameters were calculated individually for each spectrum. Control data belonging to CFY male rats have been published before (Lustyik and Zs.-Nagy, 1985). The water contents of the SHRsp rats are based on measurements carried out on another group of the same strain. The experiments were realized by means of an EDAX System F except the 2-month-old SHRsp rat brain and liver for which an EDAX 9100 system was used.

during their life span (data not shown in Table 2). Since the present X-ray microanalytic studies can be regarded for the liver only of preliminary character, we wish only to point out to the fact that the liver cells of a 2-month-old SHRsp rat (having had strokes) showed an increase of about 100 % of intracellular Na-concentration, and also the K-content was much higher than in the hepatocytes of the normotensive CFY rats of the highest age studied (Table 2) (Lustyik and Zs.-Nagy, 1985).

DISCUSSION

It should be stressed that to the best of our knowledge, the present studies represent the first attempt to investigate SHRsp rats from the points of view of a current aging theory. Therefore, the results obtained can be interpreted only in tentative terms. The possibilities of interpretation outlined below, in spite of their speculative character which we are aware of, may be useful for the design of new experiments, and detailed further studies may answer the basic question, whether the short survival of the SHRsp rats is due to an accelerated aging process, or it is attributable to a specific pathological process.

Our first approach revealed that the brain, liver and myocardium of the SHRsp rats does not accumulate lipofuscin granules even after the occurrence of strokes. This fact may indicate that the overall conditions for the elimination of the waste products are not compromised in these animals. As a matter of fact, the intracellular water contents are still rather high in the neurons (Table 2) permitting the maintenance of a proper cycling of the replacement and decomposition of the damaged cellular components, as it is outlined by the MHA (Zs.-Nagy, 1988b). It seems to be worth while to check the effect of leupeptin on the SHRsp rats, in order to reveal the role of the lysosomal thiol-proteases in the overall elimination of the waste products (Ivy et al., 1984, 1986; Ivy, 1987; Ivy and Gurd, 1988; Wolfe et al., 1988).

The observed structural alterations of the mitochondria, however, demonstrate that the cellular energy supply is probably not perfect. It would be interesting to know the extent of the eventual energy shortage of these cells, or whether this is due to a lower expression of the relevant genes, i.e., certain enzymes are missing at least in part or not. Obviously, it is also possible that not only enzymes of the energy productive mechanisms are missing in the SHRsp rats, but also some effectors of specific functions in the mitochondria and the cell membrane (e.g., the "Na-pump" enzyme, or others).

The possibility of a genetically determined energy shortage may also be relevant from the point of view of the homeostatic regulation of the cells. It is well known that the intracellular Na-content is kept low as compared to the extracellular level of this ion by means of an energy-requiring, active pumping mechanism. It is interesting in this aspect that normal aging is accompanied by an increase of the Na-content of the brain cells, and the young SHRsp male rats showed an elevated Na-content in their brain and liver cells (Table 2). Since this shift in the ionic composition represents an increase of the Na/K ratio, it may bring the resting potential of neurons and neuroeffector cells (e.g., smooth muscle of blood vessel wall) nearer to the discharge-threshold level. In other words, the statistical probability of non-desired action potentials will increase in the SHRsp rats. Such a mechanism may underline the development of the hypertension and the strokes, too. The relevance of the increased Na-content of the cells in this respect is also demonstrated by the case of the long-survivors of this strain: they do not show this increase in Na-content at all. Although our present studies on the ionic composition is only of preliminary character, the data obtained so far suggest the importance of a deeper analysis of these parameters.

Another conclusion can also be drawn from these experiments: the shifts in Na- and K-contents of the SHRsp rats resemble the tendencies we observed in normal, aging rats. This indicates that the alteration of the homeostatic parameters of the intracellular space may have a very considerable influence on the whole regulation procedure of the cell functions. Such a phenomenon is consistent with the idea that alterations in the cell membrane structure and functions may play the leading role in the age-dependent deterioration of the neuronal and other functions, as predicted by the MHA (Zs.-Nagy, 1978, 1986, 1987, 1988b, 1989; Zs.-Nagy et al., 1988). On the other hand, the frequent occurrence of hypertension and strokes in the elderly may also be correlated with the same type of shifts in the intracellular ionic composition as seen in the SHRsp rats.

REFERENCES

Eldred, G.E., and Katz, M.L., 1988, Possible mechanism for lipofuscinogenesis in the retinal pigment epithelium and other tissues, in: "Lipofuscin - 1987: State of the Art", p.: 185, I. Zs.-Nagy, ed., Akadémiai Kiadó, Budapest, and Elsevier Science Publishers, Amsterdam.

Hall, T.A., Clarke-Anderson, H., and Appleton, T., 1973, The use of thin specimens for X-ray microanalysis in biology. J. Microsc., 99: 177.

Hannover, A., 1842, Mikroskopische Undersögelser af Nervensystemet. Kgl. Danske Videns Kabernes Selskabs Naturv. og. Math. Afh. Copenhagen, 10: 1, (in Danish).

Horváth, I., Zs.-Nagy, I., Lustyik, Gy., and Váradi, Gy. 1983, The effect of primycin on the intracellular monovalent ion and water contents of rat hepatocytes as revealed by energy dispersive X-ray microanalysis and interference microscopy. Tissue and Cell, 15: 515.

Ivy, G.O., 1987, A proteinase inhibitor model of aging: implications for decreased neuronal plasticity, in: "Neuroplasticity, Learning and Memory", p.: 125, N.W. Milgram, and C.M. MacLeod, eds., Alan R. Liss Inc., New York.

Ivy, G.O., and Gurd, J.W., 1988. A proteinase inhibitor model of lipofuscin formation, in: "Lipofuscin - 1987: State of the Art", p.: 83, I. Zs.-Nagy, ed., Akadémiai Kiadó, Budapest, and Elsevier Science Publishers, Amsterdam.

Ivy, G.O., Schottler, F., Wenzel, J., Baudry, M., and Lynch, G., 1984, Inhibitors of lysosomal enzymes: accumulation of lipofuscin-like dense bodies in the brain. Science, 226: 985.

Ivy, G.O., Schottler, F., Baudry, M., and Lynch, G., 1984, Leupeptin causes several manifestations of aging in brain and liver of young rats, in: "Liver and Aging - 1986, Liver and Brain", p.: 389, K. Kitani, ed., Elsevier Science Publishers, Amsterdam, New York, Oxford.

Kikugawa, K., 1988, Involvement of lipid oxidation products in the formation of lipofuscin, in: "Lipofuscin - 1987: State of the Art", p.: 51, I. Zs.-Nagy, ed., Akadémiai Kiadó, Budapest, and Elsevier Science Publishers, Amsterdam.

Lustyik, Gy., and Zs.-Nagy, I., 1985, Alterations of the intracellular water and ion concentrations in brain and liver cells during aging as revealed by energy dispersive X-ray microanalysis of bulk specimens. Scanning Electron Microscopy, 1985/I: 323.

Lustyik, Gy., and Zs.-Nagy, I., 1988, Age-dependent dehydration of post-mitotic cells as measured by X-ray microanalysis of bulk specimens. Scanning Microsc., 2: 289.

Nagai, Y., Yoshida, K., Narumi, S., Tanayama, S., and Nagaoka, A., 1989, Brain distribution of idebenone and its effect on local cerebral glucose utilization in rats. Arch. Gerontol. Geriatr., 8: 257.

Nagaoka, A., Kakihana, M., and Fujiwara, K., 1989, Effects of idebenone on neurological deficits following cerebrovascular lesions in stroke-prone spontaneously hypertensive rats. Arch. Gerontol. Geriatr., 8: 203.

Okamoto, K., 1962, Experimental studies on the relationship between endocrine organs and hypertension. Folia Endocr. Jap., 38: 782, (in Japanese).

Okamoto, K., and Aoki, K., 1963, Development of a strain of spontaneously hypertensive rats. Jap. Circ. J., 27: 282.

Okamoto, K., Yamori, Y., and Nagaoka, A., 1974, Establishment of the stroke-prone spontaneously hypertensive rat (SHR). Circulation Res. Suppl. I. 34-35: I-143.

Shibota, M., Shino, A., and Nagaoka, A., 1978, Cerebrovascular permeability in stroke-prone spontaneously hypertensive rats. Exp. Mol. Pathol., 28: 330.

Snedecor, G.W., and Cochran, W.G., 1980, "Statistical Methods", 7th Edition, pp: 39-82, Iowa State University Press, Ames, Iowa.

Suno, M., Shibota, M., and Nagaoka, A., 1989, Effects of idebenone on lipid peroxidation and hemolysis in erythrocytes of stroke-prone spontaneously hypertensive rats. Arch. Gerontol. Geriatr., 8: 307.

Tomita. I., Sano, M., Serizawa, S., Ohta, K., and Katou, M., 1978, Studies on the antioxidative potential of tissues and blood in spontaneously hypertensive rats (SHR), stroke-prone SHR (SHRSP) and normotensive Wistar Kyoto rat (WKR), Jpn. Heart J., 19: 671.

Tomita. I., Sano, M., Serizawa, S., Ohta, K., and Katou, M., 1979, Fluctuation of lipid peroxides and related enzyme activities at time of stroke in stroke-prone spontaneously hypertensive rats. Stroke, 10: 323.

Wolfe, L.S., Gauthier, S., and Durham ,H.D., 1988, Dolichols and phosphorylated dolichols in the neuronal ceroid lipofuscinoses, other lysosomal stroage diseases and Alzheimer disease. Induction of autolysosomes in fibroblasts, in: "Lipofuscin - 1987: State of the Art", p.: 389, I. Zs.-Nagy, ed., Akadémiai Kiadó, Budapest, and Elsevier Science Publishers, Amsterdam.

Zs.-Nagy, I., 1978, A membrane hypothesis of aging. J. theor. Biol., 75: 189.

Zs.-Nagy, I., 1983, Energy dispersive X-ray microanalysis of biological bulk specimens: a review on the method and its application to experimental gerontology and cancer research. Scanning Electron Microscopy, 1983/III: 1255.

Zs.-Nagy, I., 1986, Common mechanisms of cellular aging in brain and liver in the light of the membrane hypothesis of aging, in: "Liver and Aging - 1986, Liver and Brain", p.: 373, K. Kitani, ed., Elsevier Science Publishers, Amsterdam, New York, Oxford.

Zs.-Nagy, I., 1987, An attempt to answer the questions of theoretical gerontology on the basis of the membrane hypothesis of aging. Advances in the Biosciences, 64: 393.

Zs.-Nagy, I., 1988a, The bulk specimen X-ray microanalysis of freeze-fractured, freeze-dried tissues in gerontological research. Scanning Microsc., 2: 301.

Zs.-Nagy, I., 1988b, The theoretical background and cellular autoregulation of biological waste product formation, in: "Lipofuscin - 1987: State of the Art", p.: 23, I. Zs.-Nagy, ed., Akadémiai Kiadó, Budapest, and Elsevier Science Publishers, Amsterdam.

Zs.-Nagy, I.,1989, Functional consequences of free radical damage to cell membranes. in: "CRC Handbook of Free Radicals and Antioxidants in Biomedicine", Vol. I., p.: 199, J. Miquel, A.T. Quintanilha and H. Weber, eds., CRC Press, Inc., Boca Raton, Florida.

Zs.-Nagy, I., and Casoli, T., 1989, A review on the extension of Hall's method of quantification to bulk specimen X-ray microanalysis. <u>Scanning Microsc.</u>, (in press)

Zs.-Nagy, I., and Pieri, C., 1976, Choice of standards for quantitative X-ray microanalysis of biological bulk specimens. <u>EDAX EDITOR</u>, 6:28.

Zs.-Nagy, I., Pieri, C., Giuli, C., Bertoni-Freddari, C., and Zs.-Nagy, V., 1977, Energy dispersive X-ray microanalysis of the electrolytes in biological bulk specimen. I. Specimen preparation, beam penetration and quantitative analysis. <u>J. Ultrastruct. Res.</u>, 58: 22.

Zs.-Nagy, I., Lustyik, Gy., and Bertoni-Freddari, C., 1982, Intracellular water and dry mass content as measured in bulk specimens by energy-dispersive X-ray microanalysis. <u>Tissue and Cell</u>, 14: 47.

Zs.-Nagy, I., Cutler, R.G., and Semsei, I., 1988, Dysdifferentiation hypothesis of aging and cancer: a comparison with the membrane hypothesis of aging. <u>Ann. N.Y. Acad. Sci.</u>, 521: 215.

DISCUSSION

ALHO: One question on this method of x-ray analysis. Do you feel that it makes a difference if the sodium content is intra or extracellular? Can you say whether the sodium is intra or extracellular?

Zs.-NAGY: Yes, we can determine that. We know the size of the cell, so if you have a neuron which has for instance 50 microns in diameter, and you know that the penetration is not more than 5 microns, you can put your analyzing beam inside the cell, and not at the periphery. This is one point, and the other is that these methodological aspects have been thoroughly discussed during the last ten years and now the method is well established and widely accepted. We can actually measure intracellular ion contents in cells not smaller than 10 microns in diameter.

ALHO: Can you determine the distribution of these ions in the cells?

Zs.-NAGY: We cannot say anything about distribution because we obtain an intracellular average.

BRUNK: You showed some micrographs with distorted mitochondria, and I would like to suggest that these may be artifacts due to improper hypotonic perfusion. What total and effective osmotic pressure have you utilized?

Zs.-NAGY: We used the classical solutions and buffers with the same osmolarity which is used for normal tissues. Your suggestion that the swelling may be due to the use of improper osmolarity may be true, except that the same solutions did not produce any distortion or swelling in normal rats.

HERVONEN: In your last slide you presented your very interesting hypothesis on aging changes, but I am disturbed about some statements contained in it. It seems to me that these statements are not really accurate because many of the major enzyme activities, especially in the rat brain, remain the same. There are not signs of decrease in the cholinergic system; the acetylcholinesterase remains practically unchanged.

Zs.-NAGY: One thing is to measure the overall tissue activity of an enzyme and another is to say that the gene expression of the enzyme activity has not changed.

HERVONEN: At least, what we can measure does not change with aging, and on this basis, we apparently disagree.

GOEBEL: I want to know something about the spontaneously-hypertensive stroke-proned model in rats. If I am not mistaken, you get brain edema in hypertension possibly due to an impaired blood-brain barrier. To which extent this may affect your electrolytes inside the brain, including your intracellular potassium? Another point is that you mentioned in your slides the words "brain cells". Do you mean neurons, astrocytes, oligodendrocytes? If you have hypertension, the edema largely takes place in the white matter and also results in chronic astrocytosis.

Zs.-NAGY: Yes, structural signs of brain edema are present in these animals. We were trying to analyze by transmission electron microscopy and x-ray analysis the neurons of certain size, because sometimes it is not possible to distinguish, particularly by x-ray analysis between glial cells and neurons. I do not exclude that all these phenomena may contribute to the ionic composition, but we also measured the water content, and there was not too much difference with normal neurons.

INCOMPLETE PROTEOLYSIS MAY CONTRIBUTE TO LIPOFUSCIN

ACCUMULATION IN THE RETINAL PIGMENT EPITHELIUM

Martin L. Katz

Department of Ophthalmology
University of Missouri School of Medicine
Columbia, MO 65212

SUMMARY

A substantial amount of experimental evidence indicates that the bulk of retinal pigment epithelial (RPE) lipofuscin is derived from components of the photoreceptor outer segments. Antioxidant nutrient deficiency promotes RPE lipofuscin deposition, suggesting that oxidation of outer segment components is involved in the formation of this pigment. However, the fluorescent products of *in vitro* oxidation of retinal constituents differ from the fluorophores associated with lipofuscin. Thus, the role of nonenzymatic oxidation in lipofuscin formation is probably indirect; it may inhibit proteolysis, which in turn allows nonoxidative amino acid modifications to accumulate. Direct inhibition of proteolysis by the RPE *in vivo* results in a rapid appearance of lipofuscin-like fluorescence in the undegraded proteins. This suggests that protein components of the outer segments are probably the major precursors of RPE lipofuscin. The modifications undergone by these proteins that are responsible for their accumulation in the RPE and for lipofuscin fluorescence remain to be determined.

INTRODUCTION

Despite extensive research, the chemical nature and molecular origins of lipofuscin pigments have yet to be determined. Evidence suggests that these autofluorescent pigments may be products of nonenzymatic oxidative degradation of cellular constituents. For example, dietary deficiencies in a variety of antioxidants have been shown to result in the deposition of a lipofuscin-like material in many cell types (Katz, et al., 1978, 1982; Oliver, 1981). However, attempts to generate lipofuscin fluorophores by in vitro oxidation of tissue preparations have been unsuccessful (Eldred and Katz, 1988, 1989). Thus, the formation of lipofuscin inclusions in cells appears to be a more complex process than that involved in the oxidation of tissue constitutents alone. Additional experiments will therefore be necessary to develop an understanding of the process of lipofuscin formation and of the chemical composition of this pigment.

The retinal pigment epithelium (RPE) of the eye is a convenient tissue in which to study lipofuscin formation and composition. This tissue consists of a single layer of cells separating the photoreceptor cells of the neural retina from their primary blood supply in the choroid (Fig. 1). The RPE can easily be isolated from adjacent tissues, both anatomically and surgically, and it can therefore be analyzed in relatively pure

---

Key Words: aging, retina, photoreceptors, protein turnover, oxidation, lipofuscin

Lipofuscin and Ceroid Pigments
Edited by E. A. Porta
Plenum Press, New York, 1990

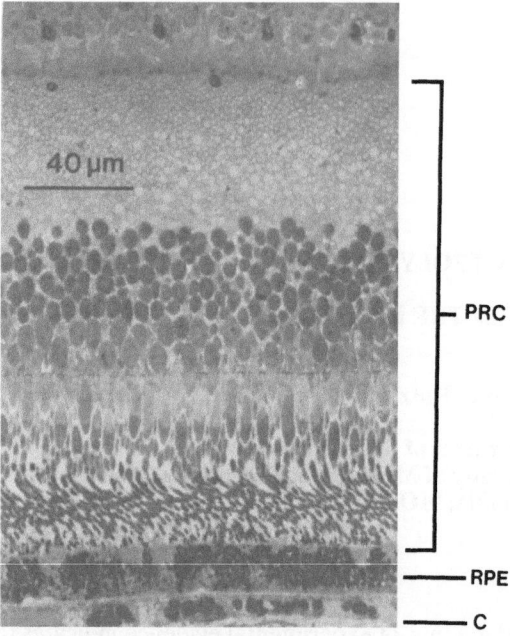

Fig. 1. Light micrograph of a cross-section of the outer region of a monkey retina. The photoreceptor cells (PRC) lie adjacent to the RPE, which is composed of a continuous sheet of cells just above a bed of large capillaries in the choroid (C). Micrograph kindly provided by L. Feeney-Burns.

form. During the aging process, there is a progressive increase in the lipofuscin content of the RPE (Wing, et al., 1978; Katz and Robison, 1984). Investigations into the factors that regulate the rate of RPE lipofuscin deposition are leading to a better understanding of the origins of this pigment.

PHOTORECEPTOR OUTER SEGMENTS CONTAIN LIPOFUSCIN PRECURSORS

Vertebrate photoreceptor cells each have a specialized region (the outer segment) which contains a high concentration of membrane-bound visual pigment proteins (Fig. 2). In rod photoreceptors, the outer segment consists of a cylindrical extension of the plasma membrane that encases a stack of flattened membrane sacs called discs. The outer segment disc membranes undergo a continuous process of renewal. New discs are synthesized and added at the bases of the outer segments, and the older discs are displaced toward the outer segment tips, which are adjacent to the the RPE. Periodically, packets of the oldest discs are shed from the tips of the outer segments and are phagocytosed by the RPE, which apparently degrades them to their molecular components. The rate of outer segment turnover is very high; approximately 10 to 15% of the photoreceptor outer segment mass is phagocytosed and degraded by the RPE every day (Young, 1967, 1971).

Accumulating evidence suggests that the bulk of RPE lipofuscin is derived from components of the phagocytosed photoreceptor outer segments. Ultrastructural examination of the RPE indicates that these cells contain phagosomes with intact outer segment discs, lipofuscin granules, and inclusions with morphological appearances that are intermediate between those of phagosomes and lipofuscin. The presence of these intermediate inclusions suggests that lipofuscin may be derived from constituents of the phagosomes. Support for this possibility comes from the observation that removal of the source of phagosomes reduces lipofuscin accumulation in the RPE. A hereditary defect in the RCS rat strain results in specific degeneration of the retinal photoreceptor cells shortly after they develop. Thus, for most of their lives, RCS rats produce no

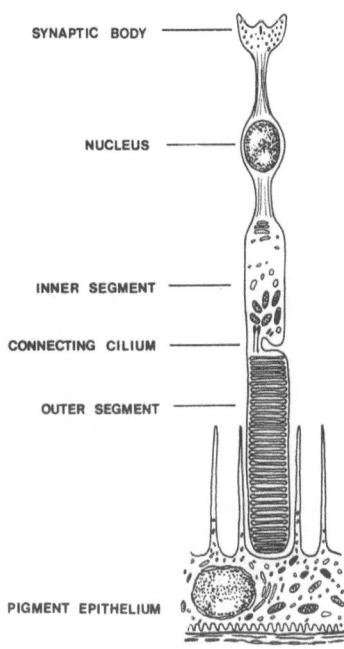

SYNAPTIC BODY

NUCLEUS

INNER SEGMENT

CONNECTING CILIUM

OUTER SEGMENT

PIGMENT EPITHELIUM

Fig. 2. Diagram illustrating the structure of a mammalian rod photoreceptor cell and its relationship to the RPE.

outer segments for phagocytosis by the RPE. In these animals, the age-related increase in RPE lipofuscin content is much less than that observed in rats with normal retinas (Katz, et al., 1986). This observation supports the hypothesis that phagocytosed photoreceptor outer segments are the major precursors for RPE lipofuscin granule contents.

Additional evidence for this hypothesis was obtained from investigations into the effects of bright light exposure on the retina (Katz and Eldred, 1989a,b). Albino rats reared under bright cyclic light were found to suffer a specific loss of photoreceptor cells from the retina. The light-induced destruction of the photoreceptor cells was accompanied by a siginificant reduction in the age-related accumulation of lipofuscin in the RPE. Again, in the absence of a source of photoreceptor outer segments, RPE lipofuscin deposition is reduced.

A more direct indication that the outer segments contain precursors for RPE lipofuscin was obtained upon examination of the retinas of RCS rats at a time when the photoreceptor cells were udergoing degeneration. The degenerating outer segments, which apparently cannot be phagocytosed by the RPE, were found to develop a lipofuscin-like autofluorescence. Thus, at least the fluorophores in RPE lipofuscin are very likely to develop from constituents of the outer segments. Further evidence for this comes from the recent finding by Boulton, et al. (1989) that feeding cultured RPE isolated photoreceptor outer segments results in the deposition of lipofuscin-like autofluorescent granules in these cells (Fig. 3).

In summary, an overwhelming amount of evidence suggests that the bulk of the RPE lipofuscin granule contents are derived from the photoreceptor outer segments. Given this knowledge, it should be possible to identify the molecular precursors of RPE lipofuscin among the outer segment constituents, and to determine the mechanisms by which these precursors are converted into autofluorescent RPE inclusions. Progress is currently being made toward these goals.

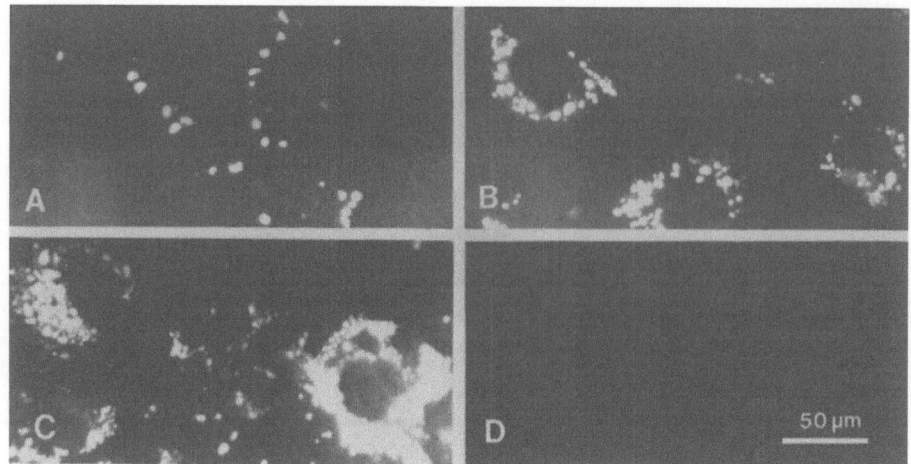

Fig. 3. Fluorescence micrographs illustrating the effect of feeding rod outer segments to cultured RPE cells. Bovine rod outer segments were fed daily to cultured human RPE cells for 2 weeks (**A**), 1 month (**B**), and 3 months (**C**). A control culture had the media changed daily for 3 months, but was not exposed to rod outer segments (**D**). The development of lipofuscin-like fluorescence in the RPE was clearly dependent on the presence of rod outer segments. [Figure adapted from Boulton, et al. (1989) with permission].

POSSIBLE MOLECULAR PRECURSORS FOR LIPOFUSCIN FORMATION

The photoreceptor outer segments, from which RPE lipofuscin is apparently derived, are composed primarily of protein and lipid, and also contain a number of inorganic compounds and smaller organic molecules, including amino acids, nucleotides, and large amounts of vitamin A. Which of these constituents are involved in the formation of RPE lipofuscin granules? According to the most widely recognized theory, the process of lipofuscin formation is initiated by the oxidative degradation of cellular lipids, particularly polyunsaturated fatty acids (Donato, 1981). Among the products of nonenzymatic oxidation of polyunsaturated fatty acids are a number of aldehydes, including malonaldehyde. It has been proposed that malonaldehyde formed by this process crosslinks amino groups of cellular constitutents, such as proteins, forming products that are responsible for the autofluorescence of lipofuscin granules (Donato, 1981). If this theory is correct, the photoreceptor outer segments should be particularly good precursors for lipofuscin formation. The outer segment disc membranes are quite rich in the highly unsaturated fatty acid docosahexaenoate, a 22-carbon fatty acid with 6 double bonds (Fliesler and Anderson, 1983). The more double bonds a fatty acid has, the more susceptible it is to oxidation (Holman and Elmer, 1947). Thus the outer segments have abundant substrates for aldehyde formation. Amino groups with which these aldehydes could react are also present in large amounts in the outer segments. The outer segment membranes contain approximately 50% protein, most of which is the visual pigment (Fliesler and Anderson, 1983). Visual pigment proteins are rich in lysine and arginine residues (Applebury and Hargrave, 1986), which would be capable of reacting with aldehydes generated by membrane lipid oxidation. In addition, 30 to 40 mol% of outer segment membrane lipids are phosphatidylethanolamine, which also has an amino group that can react with oxidation-generated aldehydes. Based on their molecular composition, therefore, the outer segments would be expected to be good candidates for lipofuscin fluorophore formation via the proposed oxidative pathway.

Results of dietary experiments are consistent with the mechanism described above for the formation of RPE lipofuscin fluorophores from the constituents of photoreceptor outer segments. Animals fed diets deficient in antioxidant nutrients accumulate autofluorescent lipofuscin-like pigments in the RPE at substantially faster rates than do animals fed diets containing adequate levels of these nutrients (Katz, et al., 1978, 1982, 1986; Robison, et al., 1979, 1980). Recent experiments designed to examine the effects of nonenzymatic oxidation on the generation of fluorophores from retinal components *in vitro*, however, do not support the hypothesis that products of oxidation are directly responsible for the generation of lipofuscin fluorophores. A number of fluorescent products were formed by incubating retinal preparations under oxidizing conditions, but these compounds differed substantially from lipofuscin-derived fluorophores in both their chromatographic and fluorescence properties (Eldred and Katz, 1989). Malonaldehyde-crosslinked amines are blue-emitting fluorophores, while lipofuscin fluorophores are yellow-emitting. Thus, it is likely that the theory that the lipofuscin granule contents are primarily malonaldehyde crosslinking products is probably incorrect, or at least is in need of substantial modification.

OUTER SEGMENT PROTEINS MAY BE LIPOFUSCIN PRECURSORS

Are there any clues that may suggest alternative mechanisms for the formation of RPE lipofuscin granule contents? A number of years ago Ivy and colleagues (1984) discovered that inhibition of cellular proteases led to the rapid accumulation of lipofuscin-like inclusions in cells of the brain. This finding suggested that proteins may be the major precursors of lipofuscin granule contents. Recently, a similar effect of protease inhibition on the retina has been observed (Katz and Shanker, 1989). A single intraocular injection of the protease inhibitor leupeptin was found to substantially block degradation of phagocytosed outer segments by the RPE. The phagosomes rapidly developed a lipofuscin-like autofluorescence; within one day of leupeptin injection, RPE autofluorescence was increased an average of 80%. Once formed, the lipofuscin-like inclusions did not appear to be degraded by the RPE; one week after a single leupeptin treatment, the RPE had approximately twice the autofluorescence of untreated control eyes. The appearance of a lipofuscin-like fluorescence in the undegraded phagosomes is analogous to the development of a similar fluorescence in the undegraded outer segment debris that accumulates adjacent to the RPE in RCS rats, and probably occurs via a similar mechanism. Based on the effects of leupeptin, it is likely that outer segment proteins are the major precursors for RPE lipofuscin granules. The mechanism for lipofuscin formation must involve modification of these proteins.

Additional support for these conclusions comes from the finding that RPE lipofuscin content is highly dependent on vitamin A. Animals fed diets deficient in the forms of vitamin A that can be used for visual pigment synthesis have substantially reduced amounts of RPE lipofuscin (Robison, et al., 1980; Katz, et al., 1986). The autofluorescence of the degenerating photoreceptor cells in the RCS rats is also reduced by vitamin A deficiency (Katz, et al., 1987). Retinal levels of the visual pigment rhodopsin and of the protein component of the pigment (opsin) decline in response to vitamin A deficiency (Carter-Dawson, et al., 1979). The inverse correlation between retinal opsin levels and RPE lipofuscin content or lipofuscin-like fluorescence in degenerating photoreceptors indicates that the opsins are likely to be major precursors for RPE lipofuscin.

What type of modification, other than malonaldehyde crosslinking, might the opsin protein undergo to account for the development of fluorescence and the inability of the RPE to degrade the modified protein? At present, the nature of the adducts to opsin or other outer segment proteins that are responsible for the development of fluorescence are unknown. Numerous enzymatic and nonenzymatic covalent modifications of proteins, including opsin, have been characterized (Wold, 1981; Stadtman, 1988; Aton, 1984; Fliesler, et al., 1984; Applebury and Hargrave, 1986). Some of these modifications, such as nonenzymatic glycation, are accompanied by the development of fluorescence (Pongor, et al., 1984). Recently, a number of fluorophores have been found that appear to be covalently bound to proteins from normal brains (Fig. 4). Similar protein-bound fluorophores may be responsible, at least in part, for the autofluorescence of RPE lipofuscin granules. Determination of the

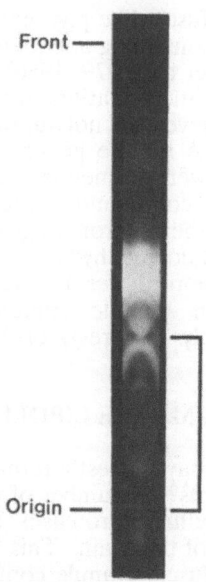

Fig. 4. Thin layer chromatogram of HCl digest of crude protein preparation from normal human brain. The unstained chromatogram was photographed under ultraviolet light. The area of the chromatogram indicated by a bracket contains a number of fluorophores that were present in the digests of brain proteins but not in digests of purified proteins obtained from commerical sources. The fluorophores with greater mobility were generated both from brain proteins and from purified protein standards. A number of fluorophores that were apparently covalently bound to protein were released by acid digestion.

structures of these fluorescent adducts will give some indication of the mechanisms by which they are formed.

Not only might adducts to proteins be responsible for the development of fluorescence, but covalent modification of certain amino acids may also block proteolysis, and thus account for the accumulation of lipofuscin during senescence. In ubiquitin-mediated proteolysis, for example, modification of specific lysine residues on the substrate protein can block proteolytic degradation of that protein (Gregori, et al., 1985). Likewise, nonenzymatic glycosylation also impairs proteolytic degradation of proteins (Vlassara, et al., 1986). The formation of fluorescent and proteolysis-inhibiting adducts apparently competes with normal protein turnover. If protein turnover is inhibited, as in the experiments with leupeptin, amino acid modifications apparently accumulate to the point at which they themselves permanently prevent degradation of modified proteins. Thus, the lipofuscin content of a cell probably reflects the balance between the rates of adduct formation and of protein breakdown. Changes in this balance will affect the rate of lipofuscin accumulation.

If the above scenario is correct, it is possible that the effect of antioxidant deficiency on lipofuscin accumulation is due primarily to a direct inhibition of proteolysis, rather than to oxidation-mediated modification of the proteins from which lipofuscin is largely derived. This would explain why the fluorophores associated with lipofuscin differ from those obtained upon *in vitro* oxidation of tissue components. Further experimentation is required to characterize the interactions between proteolysis rate and protein adduct formation in determining the lipofuscin content of the RPE and other cell types.

ACKNOWLEDGMENTS

The work described in this chapter was supported by National Institutes of Health grant EY-06458, and by grants from the Children's Brain Diseases Foundation and Research to Prevent Blindness, Inc.

REFERENCES

Applebury, M. L., and Hargrave, P. A., 1986, Molecular biology of the visual pigments, Vision Res., 26:1881.

Aton, B. R., Litman, B. J., and Jackson, M. L., 1984, Isolation and identification of phosphorylated species of rhodopsin, Biochem., 23:1737.

Boulton, M., McKechnie, N. M., Breda, J., Bayly, M. and Marshall, J., 1989, The formation of autofluorescent granules in cultured human RPE, Invest. Ophthalmol. Vis. Sci., 30:82.

Carter-Dawson, L., Kuwabara, T., O'Brien, P. J., and Bieri, J. G., 1979, Structural and biochemical changes in vitamin A-deficient rat retinas. Invest. Ophthalmol. Vis. Sci., 18:437.

Donato, H. Jr., 1981, Lipid peroxidation, cross-linking reactions, and aging, in: "Age Pigments," R. S. Sohal, ed., Elsevier/North-Holland, Amsterdam.

Eldred, G. E., and Katz, M. L., 1988, Possible mechanism for lipofuscinogenesis in the retinal pigment epithelium and other tissues, in: "Lipofuscin - 1987: State of the Art," I.Zs. - Nagy, ed., Elsevier Science Publishers, Amsterdam.

Eldred, G. E., and Katz, M. L., 1989, The autofluorescent products of lipid peroxidation may not be lipofuscin-like, Free Radical Biol. Med., (in press).

Fliesler, S. J., and Anderson, R. E., 1983, Chemistry and metabolism of lipids in the vertebrate retina, Prog. Lipid Res., 22:79.

Fliesler, S. J., Tabor, G. A., and Hollyfield, J. G., 1984, Glycoprotein synthesis in the human retina: localization of the lipid intermediate pathway, Exp. Eye Res., 39:153.

Gregori, L., Marriott, D., West, C. M. and Chan, V., 1985, Specific recognition of calmodulin from Dictyostelium discoidium by the ATP, ubiquitin-dependent degradative pathway, J. Biol. Chem., 260:5232.

Holman, R. T., and Elmer, O. C., 1947, The rate of oxidation of unsaturated fatty acids and esters, J. Am. Oil Chem. Soc., 24:127.

Ivy, G. O., Schottler, F., Wenzel, J., Baudry, M., and Lynch, G., 1984, Inhibitors of lysosomal enzymes: accumulation of lipofuscin-like dense bodies in the brain, Science, 226:985.

Katz, M. L., Drea, C. M., Eldred, G. E., Hess, H. H., and Robison, W. G., Jr., 1986, Influence of early photoreceptor degeneration on lipofuscin in the retinal pigment epithelium, Exp. Eye Res., 43:561.

Katz, M. L., Drea, C. M., and Robison, W. G., Jr., 1986, Relationship between dietary retinol and lipofuscin in the retinal pigment epithelium, Mech. Age. Dev., 35:291.

Katz, M. L., and Eldred, G. E., 1989, Failure of vitamin E to protect the retina against damage resulting from bright cyclic light exposure, Invest. Ophthalmol. Vis. Sci., 30:29.

Katz, M. L., and Eldred, G. E., 1989, Retinal light damage reduces autofluorescent pigment deposition in the retinal pigment epithelium, Invest. Ophthalmol. Vis. Sci., 30:37.

Katz M. L., Eldred, G. E., and Robison, W. G., Jr., 1987, Lipofuscin autofluorescence: evidence for vitamin A involvement in the retina, Mech. Age. Dev., 39:81.

Katz, M. L., Parker, K. R., Handelman, G. J., Bramel, T. L., and Dratz, E. A., 1982, Effects of antioxidant nutrient deficiency on the retina and retinal pigment epithelium of albino rats: a light and electron microscopic study, Exp. Eye Res., 34:339.

Katz, M. L., and Robison, W. G., 1984, Age-related changes in the retinal pigment epithelium of pigmented rats, Exp. Eye Res., 38:137.

Katz, M. L., and Shanker, M. J., 1989, Development of lipofuscin-like fluorescence in the retinal pigment epithelium in response to protease inhibitor treatment, Mech. Age. Dev., (in press).

Katz, M. L., Stone, W. L., and Dratz, E. A., 1978, Fluorescent pigment accumulation in retinal pigment epithelium of antioxidant-deficient rats, Invest. Ophthalmol. Vis. Sci., 17:1049.

Oliver, C., 1981, Lipofuscin and ceroid accumulation in experimental animals, in: "Age Pigments," R. S. Sohal, ed., Elsevier/North-Holland Biomedical Press, Amsterdam.

Pongor, S., Ulrich, P. C., Benesath, F. A., and Cerami, A., 1984, Aging of proteins: isolation and identification of a fluorescent chromophore from the reaction of polypeptides with glucose, Proc. Nat. Acad. Sci. U.S.A., 81:2684.

Robison, W. G., Kuwabara, T., and Bieri, J. G., 1979, Vitamin E deficiency and the retina: photoreceptor and pigment epithelial changes, Invest. Ophthalmol. Vis. Sci., 18:683.

Robison, W. G., Kuwabara, T. and Bieri, J. G., 1980, Deficiencies of vitamins E and A in the rat. Retinal damage and lipofuscin accumulation, Invest. Ophthalmol. Vis. Sci., 19:1030.

Stadtman, E. R., 1988, Protein modification in aging, J. Gerontol., 43:B112.

Vlassara, H., Brownlee, M., and Cerami, A., 1986, Nonenzymatic glycosylation: role in pathogenesis of diabetic complications. Clin. Chem., 32:B37.

Wing, G. L., Blanchard, G. C., and Wieter, J. J., 1978, The topography and age relationship of lipofuscin concentration in the retinal pigment epithelium, Invest. Ophthalmol. Vis. Sci., 17:601.

Wold, F., 1981, In vivo chemical modification of proteins (post-translational modification), Ann. Rev. Biochem., 50:783.

Young, R. W., 1967, The renewal of photoreceptor cell outer segments, J. Cell Biol. 33:61.

DISCUSSION

BRUNK: What happens to the retinal pigment epithelium (RPE) when it is loaded with lipofuscin? Does it retain its capacity to degrade still the outer segment?

KATZ: We examined that question by looking at the RPE in animals of different ages. In old animals, we still found phagocytosis of the shedded outer segments, but at a rate lower than in young animals. Whether this relative decrease was due to the presence of increased amounts of lipofuscin or to lower rate of synthesis of the outer segments, I don't know.

IVY: I just have a couple of comments. In your experiments, you presented evidence that after only one injection of leupeptin and one week survival, you have actually increased autofluorescence, and the conclusion you might have drawn is that the lipofuscin-like substances are not degraded. Earlier, Dr. Kitani mentioned that leupeptin is probably cleared very quickly from the liver. Indeed, leupeptin is known to be a reversible inhibitor. I also presented evidence that lipofuscin-like deposits induced by leupeptin can go away with time. I think that we need to be very cautious about interpreting all the data that seems to lead us to the conclusion that that the pigment is not degradable. Another set of experiments I did when I was in California was to implant the leupeptin-filled minipumps for two weeks and then removed the pumps and measured the activity of calpain, another cytoplasmic proteinase also inhibited by leupeptin. We found that 5 weeks after removal of the pump, the activity of calpain was still depressed. So, we don't know if leupeptin was still sticking around in the cell. In your model, the leupeptin may be hanging in the vitreous. We don't know yet enough about this drug clearance rate.

HERVONEN: Can you tell us the excitation and emission characteristics of the pigment that accumulated in the RPE?

KATZ: The excitation maxima is about 390 nm, and the emission cut-off is 510 nm, so in the photographs I presented, all emission was about 510 nm.

PORTA: I noticed that the diet used for the control animals was supplemented with 250 mg/kg of vitamin E, and this is obviously about 10 times higher than the normal requirements for rats. Do you have any particular reason for using such high levels of supplementation?

KATZ: The basal diet is high in unsaturated fat, about 10%, and therefore, we used that level of vitamin E in all our studies to properly supplement the animals.

PORTA: Yes, it is well known that dietary unsaturated fats increase the vitamin E requirements, but in terms of levels of fat, your diet is not too different from other basal diets used to test the effects of vitamin E deficiency and supplementation. My question was not intended to suggest that your level of supplementation was wrong, but on the contrary, I thought it could be of great interest to emphasize the need for high levels of supplementation in order to prevent the accumulation of ceroid pigment in the control animals. In our studies on the effects of vitamin E deficiency on ceroid formation in mice and rats, we found, for example, that we could not totally prevent the accumulation of ceroid in the control supplemented animals unless the levels of tocopherol were elevated to more than 20 mg per 100 g of diet.

GOEBEL: You mentioned several times the term "aging" in the formation of the lipopigment in the RPE. What you have in fact shown was that the pigment accumulation was related to the phagocytosis of the outer segments, but this is certainly not the case in the formation of lipopigments in neurons, hepatocytes or muscle fibers in which only autophagocytosis is taking place. So, the pigment you see in the RPE is probably quite different from the aging pigment or lipofuscin.

KATZ: The pigment that accumulates slowly over the years in the RPE of rats, mice and humans has at least the same autofluorescent spectral features of age-pigment. The reason why we think that this is a good model is precisely because we can clearly identify the precursors of the pigment; something not usually possible in other cells where autophagy only takes place. In relation to heterophagocytosis versus autophagocytosis, I don't think it makes a big difference whether the phagosomes come originally from outside or inside. Once they are inside the cell, the degradative processes are practically the same.

PALMER: I presume that the pigment that accumulates in the RPE is mainly rhodopsin. Is that correct?

KATZ: Well, at least rhodopsin is a major protein component of the outer segment, although not the only protein. Yes, presumably, a substantial fraction is rhodopsin. In our Department of Ophthalmology, Dr. Feeney-Burns has done some chemical labelling of age pigment in the human RPE using 3 antibodies which have been raised against rhodopsin. Unfortunately, all these antibodies are against also hydrophilic regions. I don't know if you are familiar with the structure of this rhodopsin protein molecule which has 7 hydrophobic regions and short connecting hydrophilic regions, plus the carboxy amino terminals which are hydrophilic and stick outside the membrane. Of the 3 antibodies against such hydrophilic regions, none label the lipofuscin granules. So, if the material that accumulates in aging is rhodopsin-derived, it may have either lost the hydrophilic components or they have been masked by covalent modifications, probably by attachments with fluorophores.

PALMER: Yes, this may well be a possibility, because if it aggregates in a different conformation, it may easily bury the antigenicity, even without modifying the protein at all.

KATZ: What we are planning to do is to isolate the peptides from the human age pigment and to sequence the individual peptides to see if rhodopsin is there by homology.

KITANI: I understand that you have been able to extract by chloroform an orange fluorescent material. I wonder if you can extract a similar fluorescent material from other tissues, because it may be that the orange fluorescent material is quite specific for RPE.

KATZ: First, let me tell you that the orange fluorescent component is only one of the fluorophores identified in RPE. Like in other tissues, lipofuscin has many other yellow and blue fluorophores common to all of them. To answer your question about extraction from other tissues, we don't have the technics to do that as we do have for the RPE where there is a uniform monolayer of cells easy to use for extraction of the abundant pigment accumulating there.

Zs.-NAGY: I think it is a vain effort to look for a homogeneous composition. These pigment substances "a priori" cannot be homogeneous in different tissues for various obvious reasons. One is that the cell recognition is very much sensitive in certain cases of proteins against conformational alterations, and less sensitive in other cases. In other cases of proteins, there must be numerous conformational changes until they are finally recognized and go into the phagocytic system. For me, it is impossible to characterize the lipofuscin as a chemical entity.

KATZ: We will never know if we don't look at the problem. We are not claiming that lipofuscin is homogeneous, but the general mechanisms in the genesis of the pigment could be the same. The real question is what could be causing these mechanisms. A mechanism could be covalent modification due, for instance, to protein oxidation; some modifications may involve attachment of fluorescent molecules with specific amino acids. These fluorophores could be products of oxidation or something else. The adducts to specific amino acids in the protein could block protein degradation whatever the proteinase is. I don't care too much about which proteinase in particular, because there are some 15 different mechanisms for protein degradation, and all have to recognize specific amino acid sequences. If you modify the amino acid, you can block the protein degradation, and it will accumulate in lysosomes. This is a general theory that can be tested in this model system.

Zs.-NAGY: I agree with you now.

ELLEDER: I don't think that we need to worry too much about the heterogeneity in the composition of the pigment. In other storage diseases, for example, that are precisely defined by missing enzymes, you can also have mixtures of various secondary components besides the primary and most important one. In Niemann-Pick disease, for instance, you have lots of other components such as cholesterol and glycolipids. In lipofuscin, there might be one very specific precursor and then thousand other secondary components accumulating in lysosomes. I think that the experimental model of Dr. Katz may let us determine the most important component.

CEROID PIGMENTS IN NUTRITIONAL AND ENVIRONMENTAL PATHOLOGY

PROTECTION FACTORS AGAINST FREE

RADICAL-INDUCED CEROIDOGENESIS

E. Aloj Totaro, L. Lucadamo and F.A. Pisanti

Stazione Zoologica, Napoli, Italy, and
Dip. Ecologia, Universita´ della Calabria
Arcavacata di Rende (CS), Italy

SUMMARY

The most important products of the combustion process are SO_2, NO_x, CO_2 and the heavy metals. When these substances come into contact with the biotic components of the ecosystems they produce an oxidative damage by means of a free radical mechanism. One of the significant natural sources of these oxides and metals are the volcanic emissions that contribute, either locally or more diffusely, to enrich the atmosphere with these substances. The area of Campi Flegrei (Naples, Italy) is an experimental model fit for studying the contemporary effect of the aforesaid oxidative agents, because it is characterized by a continous fumarolic activity, particularly in the area of the widest crater (Solfatara). We have made so two experiments utilizing rats and earthworms (Octolasium complanatum) to evaluate the following aspects in phylogenetically very different organisms: 1.the combined effect of the atmospheric pollutants, 2.the effect of the only heavy metals (Cu, Ni, Mn), 3.the protection action played by reduced glutathione in rats. The reduced glutathione being either a substrate of the glutathione proxidase or an oxyradicals scavenger, is one of the main protection agents against the above stress. Because many papers suggest that the mentioned atmospheric pollutants damage both animal and vegetable organisms by their oxidative properties, the reduced glutathione seems to be able to counteract efficaciously the damaging activity studied in terms of age pigments production.

INTRODUCTION

Many industrial and energy producing processes develop large amounts of waste, that are subsequently discharged haphazardly into the environment. The oxides of sulphur,

KEY WORDS: free radicals, heavy metals, rats, earthworms, volcanic emissions, glutathione

carbon and nitrogen are among the major contributors to atmospheric pollution (Ottar, 1983; Bettlheim and Littler, 1979; Brimblecombe and Stedman, 1982), while the heavy metals are released both into the surrounding water basins and into the air (Forstner and Wittmann, 1981; Nriagu, 1988; Galloway et al., 1982). All of these substances enter the troposphere after a possible chemical change, then fall back to earth in an area close to or far from the production site in function of orographic and metereological factors and contribute to the damage done on the ecosystem (ERL, 1983; Sposito and Page, 1984; Fisher, 1984). It must be underlined that there are natural sources of these substances: volcanic discharges, forest fires and bacterial metabolic process all come to mind (Adams et al., 1981; Carbonelle, 1981; Gravenhorst, 1985).

It has been known for some time that heavy metals damage living matter by the induction of free radical formation (Gutteridge, 1988). Recent studies have reported that SO2 and NOx also may produce negative effects by similar oxidative phenomena.

Our studies have been carried out in the Campi Flegrei area of Naples: this is a large volcanic region (about 12 Km in diameter) characterized by intense fumarolic which comes mainly from its largest crater (Solfatara) (Le Bronec et al., 1988). This location, in particular its east slope, is ideal for the evaluation of the impact that the gases discharged have on the integrity of the biological systems present.

Our aim was twofold: 1)to analyse the damaging action of the oxidizing components of the environmental pollutant operating either separately or contemporaneously, 2)to observe possible protective effects of reduced glutathione against the environmental oxidizing agents.

MATERIALS AND METHODS

We chose two completely different organisms for our study: Wistar rats and Octolasium complanatum worms. Animals from distinct evolutionary levels were chosen in order to evaluate responses of different genotypes and phenotypes to the induced oxidative stresses. The Wistar rats only were subjected to treatment with reduced glutathione.

Rats

165 Wistar rats of both sexes aged between 4 and 5 months, fed ad libitum, were used. Animals were divided in three groups. Group A (n=15) were used as controls: these were subdivided into a further 3 groups (A1-A3) each containing 5 animals, and kept in our laboratory rat house at a constant temperature of 20´C and humidity 70%. Group B animals (n=90), subdivided into 18 subgroups containing 5 individuals (B1-B18), were left exposed for 30 (B2, B5, B8, B11, B14, B17) or 60 (B3, B6, B9, B12, B15, B18) days to the fumarolic gas in Campi Flegrei, at 300 mt from the Solfatara crater. The rats of subgroups B7-B18 were allowed to drink water contaminated respectively (B7-B9 and B13-B15) with 80 micrograms/lt Cu and (B10-B12 and B16-B18) with 80 micrograms/lt Mn. Subgroups (B4-B6 and B13-B18) received 30 mg/kg b.wt. of reduced glutathione intramuscularly (i.m.)

Table 1. Experimental schedule - Rats

Group	environmental conditions	GSH treatment	day of sacrifice
A1	Lab.rat house	no	0
A2	temp.20´C	no	30
A3	humidity 70%	no	60
B1	volcanic gas	no	0
B2		no	30
B3		no	60
B4		yes	0
B5		yes	30
B6		yes	60
B7	volcanic gas	no	0
B8	+ Cu	no	30
B9	(80 microgr./l	no	60
B10	in drinking	yes	0
B11	water)	yes	30
B12		yes	60
B13	volcanic gas	no	0
B14	+ Mn	no	30
B15	(80 microgr./l	no	60
B16	in drinking	yes	0
B17	water)	yes	30
B18		yes	60
C1	Cu	no	0
C2	(80 microgr./l	no	30
C3	in drinking water)	no	60
C4	Mn	no	0
C5	(80 microgr./l	no	30
C6	in drinking water)	no	60
C7	Cu	yes	0
C8	(80 microgr./l	yes	30
C9	in drinking water)	yes	60
C10	Mn	yes	0
C11	(80 microgr./l	yes	30
C12	in drinking water)	yes	60

every 4 days for the duration of the experiment. For this period mean daily temperature and humidity were measured. The remaining 60 rats (group C) were divided in 12 subgroups (C1-C12) and kept next to the controls (group A). Subgroups C1-C3 and C7-C9 were allowed to drink water containing 80 micrograms/lt of copper; C4-C6 and C10-C12 had drinking water contaminated with 80 micrograms/lt of manganese. The subgroups C7-C9 and C10-C12 were also given 30 mg/kg b.wt. reduced glutathione i.m. every four days for either 30 (C8-C11) or 60 (C10-C12) days.

Table 2. Experimental schedule - Worms

Group	treatment	day of sacrifice
D1	Laboratory, sprayed daily	0
D2	with distilled water	15
D3		30
E1	volcanic gas	0
E2	sprayed daily	15
E3	with distilled water	30
E4	volcanic gas	0
E5	sprayed daily	15
E6	with Cu (80 microgr./l)	30
E7	volcanic gas	0
E8	sprayed daily	15
E9	with Mn (80 microgr./l)	30
F1	Laboratory, sprayed daily	0
F2	with Cu (80 microgr./l)	15
F3		30
F4	Laboratory, sprayed daily	0
F5	with Mn (80 microgr./l)	15
F6		30

Animals were killed on day 0 (A1, B1, B4, B7, B10, B13, B16, C1, C4, C7, C10), 30 (A2, B2, B5, B8, B11, B14, B17, C2, C8, C5, C11) or 60 (A3, B3, B6, B9, B12, B15, B18, C3, C6, C9, C12) (Table 1).

Worms

190 Octolasium complanatum worms were used and kept in plastic boxes (15x20x50 cm) containing "mull" humus. Thirty worms were used as controls (group D) and divided in 3 subgroups each containing 10 organisms (D1-D3) and left in our laboratory (20´C and 70% humidity). Ninety group E worms (subdivided in E1-E9, n=10) were situated in the same zone as group B rats (i. e. in the Campi Flegrei area). The subgroups E4, E5, E6 were watered with 80 micrograms/lt Cu and E7, E8, E9 with 80 micrograms/lt Mn. D1, D2, D3 and E1, E2, E3 subgroups were sprinkled with distilled water once daily. The remaining 60 worms, also kept in our lab, were divided into further 6 subgroups (n=10 per subgroup): F1-F3 were sprayed every day with an aqueous solution of Cu 80 micrograms/lt while F4-F6 were watered with 80 micrograms/lt Mn. Humus pH was measured every day and corrected, if necessary, with calcium carbonate. Worms were killed on day 0 (D1, E1, E4, E7, F1, F4), 15 (D2, E2, E5, E8, F2, F5) and 30 (D3, E3, E6, E9, F3, F6) (Table 2).
On the day of sacrifice, the animals were killed, the cerebral ganglia removed from the worms and the spinal ganglia removed from the rats. Biochemical tests assayes the TBA reactivity (Placer et al., 1966) and tissue lipo-

Table 3. Mean weekly temperature and humidity at
300 m from the Solfatara crater

Week	mean temperature (°C)	mean humidity %
1	20	64
2	21	74
3	20	59
4	22	60
5	20	73
6	20	68

fuscin-associated fluorescence (Kruk and Enesco, 1981). Morphological alterations were also observed using standard TEM techniques.

RESULTS

Mean weekly temperature and humidity of the Solfatara area are shown in table 3. A slight to moderate increase of TBA reactivity and lipofuscin fluorimetric values during the 0-30 and 30-60 day periods can be noted in rats of group B (B1-B3) when compared to control rat (group A) values (tables 4, 5). One can also see from these tables that these increases are not present in rats administered the reduced glutathione (subgroups B4-B6). TEM observations reveal the presence of phagosomes and an increased number of lipofuscin granules in the nerve tissue of these animals only after 60 days of exposure to the Solfatara gas (fig. 1).
Worms show a greater sensitivity than the rats: from day 15 onwards (subgroups E2 and E3) both TBA reactivity and lipofuscin production are significantly higher than the control values (group D) (tables 4, 5). At the end of the experiment one worm was found dead but non in a state of advanced decomposition.
The presence of Cu and Mn induce in the rat ganglia (subgroups C1 and C6) a rapid increase of both TBA reactivity and of fluorimetric values. From the data obtained, it seems that Cu and Mn are equally effective in alterating the above parameters. Even in this case, reduced glutathione treatment (subgroups C7-C12) attenuates the damage caused by the heavy metals, but levels do not return the values shown in the control animals.
The situation is different for the worms: Cu (subgroup F1-F3) is notably more efficient than Mn (subgroup F4 and F6) in producing alterations (tables 6, 7). Furthemore, Cu-induced damage is evident from day 15 while that induced by Mn is only noticiable after 30 days. Even in this part of the experiment, dead but intact worms were found in Cu-treated (n=2) and Mn-treated (n=1) groups.
TEM observations, both for rats and worms, show that the heavy metals are responsible for an increase in lipofuscin granule size and number (figs. 2, 4).
The results of exposure of rats and worms both to heavy metals and fumarolic emissions show clearly the presence of a synergy (tables 8, 9; fig. 3). In case of worms, however,

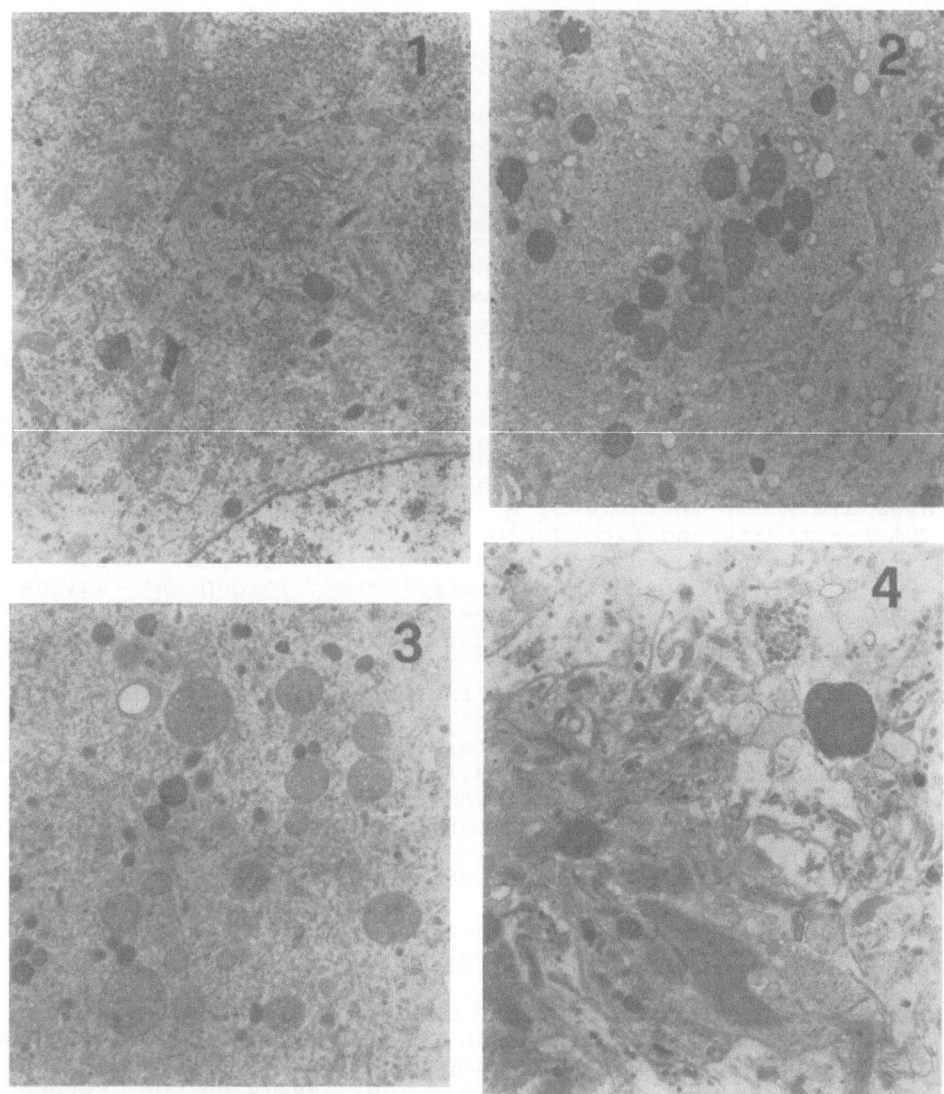

Fig. 1. Spinal ganglia nerve cell. Rat exposed 60 days to volcanic gas (10,000 x).

Fig. 2. Spinal ganglia nerve cell. Rat exposed 60 days to Cu treatment (6,000 x).

Fig. 3. Spinal ganglia nerve cell. Rat exposed 60 days to both volcanic gas and Cu treatment (8,000 x).

Fig. 4. Octolasium complanatum cerebral ganglia. Animal exposed 30 days to both volcanic gas and Cu treatment (13,000 x).

Table 4. TBA reactivity (a) in volcanic gas exposed animals.

Rats (group)	day 0	p	day 30	p	day 60	p
control (A)	43.4(0.4)	–	49.1(0.6)	–	56.1(0.2)	–
exposed (B1-B3)	44.1(0.2)	NS	55.2(0.1)	0.1	63.1(0.2)	0.1
exposed and GSH treated (B4-B6)	43.9(0.2)	NS	50.1(0.3)	NS	55.8(0.3)	NS

Worms (group)	day 0	p	day 15	p	day 30	p
control (D)	4.1(0.3)	–	4.2(0.3)	–	4.2(0.2)	–
exposed (E1-E3)	4.1(0.4)	NS	5.3(0.2)	0.5	6.2(0.1)	0.1

(a) nanomoles of malondialdehyde per g of tissue. Mean(+/-SD).
p = significancy (Student's t test) of the difference between previous and control value in the same day; NS= not significant.

Table 5. Lipofuscin-associated fluorescence (a) in volcanic gas exposed animals.

Rats (group)	day 0	p	day 30	p	day 60	p
control (A)	50.2(2.1)	–	61.1(1.1)	–	74.2(1.5)	–
exposed (B1-B3)	51.2(1.1)	NS	66.2(2.2)	0.5	79.3(2.0)	0.5
exposed and GSH treated (B4-B6)	51.5(2.0)	NS	62.5(1.0)	NS	73.8(1.4)	NS

Worms (group)	day 0	p	day 15	p	day 30	p
control (D)	2.1(0.1)	–	2.2(0.2)	–	2.1(0.1)	–
exposed (E1-E3)	2.2(0.1)	NS	2.8(0.2)	NS	3.4(0.3)	0.1

(a) units of fluorescence per 0.2 g of tissue. Mean(+/-SD).
p = significancy (Student's t test) of the difference between previous and control value in the same day; NS= not significant.

Table 6. TBA reactivity (a) in animals treated with heavy
metals.

Rats (group)	day 0	p	day 30	p	day 60	p
control (A)	43.4(0.4)	–	49.1(0.6)	–	56.1(0.2)	–
Cu (C1-C3)	42.8(0.1)	NS	61.2(0.7)	0.01	75.2(0.4)	0.01
Cu+GSH (C7-C9)	42.2(0.4)	NS	54.3(0.4)	0.1	60.9(0.2)	0.1
Mn (C4-C6)	43.5(0.2)	NS	58.7(0.4)	0.01	74.3(0.2)	0.01
Mn+GSH (C10-C12)	42.7(0.4)	NS	55.5(0.1)	0.1	61.7(0.3)	0.1

Worms (group)	day 0	p	day 15	p	day 30	p
control (D)	4.1(0.3)	–	4.2(0.3)	–	4.2(0.2)	–
Cu (F1-F3)	4.3(0.1)	NS	7.1(0.2)	0.01	9.2(0.1)	0.001
Mn (F4-F6)	4.2(0.2)	NS	4.3(0.3)	NS	5.8(0.2)	0.1

(a) nanomoles of malondialdehyde per g of tissue.
Mean(+/-SD).
p = significancy (Student's t test) of the difference
between previous and control value in the same day; NS= not
significant.

Table 7. Lipofuscin-associated fluorescence (a) in animals
treated with heavy metals.

Rats (group)	day 0	p	day 30	p	day 60	p
control (A)	50.2(2.1)	–	61.1(1.1)	–	74.2(1.5)	–
Cu (C1-C3)	52.3(1.8)	NS	75.2(2.4)	0.01	89.3(1.4)	0.01
Cu+GSH (C7-C9)	53.1(0.3)	NS	66.1(0.3)	0.1	79.1(0.5)	0.1
Mn (C4-C6)	51.4(2.2)	NS	73.5(3.1)	0.05	88.7(0.1)	0.01
Mn+GSH (C10-C12)	52.2(0.4)	NS	65.4(0.5)	0.1	78.3(0.4)	0.1

Worms (group)	day 0	p	day 15	p	day 30	p
control (D)	2.1(0.1)	–	2.2(0.2)	–	2.1(0.1)	–
Cu (F1-F3)	2.2(0.3)	NS	3.5(0.2)	0.1	4.7(0.3)	0.001
Mn (F4-F6)	2.1(0.4)	NS	2.3(0.3)	NS	3.2(0.1)	0.1

(a) units of fluorescence per 0.2 g of tissue. Mean(+/-SD).
p = significancy (Student's t test) of the difference
between previous and control value in the same day; NS= not
significant.

Table 8. TBA reactivity (a) in animals exposed to both volcanic gas and heavy metals.

Rats (group)	day 0	p	day 30	p	day 60	p
control (A)	43.4(0.4)	–	49.1(0.6)	–	56.1(0.2)	–
gas+Cu (B7-B9)	43.2(0.3)	NS	77.2(0.3)	0.001	92.5(0.3)	0.001
gas+Mn (B10-B12)	44.0(0.1)	NS	74.1(0.1)	0.005	92.1(0.4)	0.001
gas+Cu+GSH (B13-B15)	41.8(0.5)	NS	59.2(0.7)	0.01	70.8(0.2)	0.01
gas+Mn+GSH (B16-B18)	42.4(0.2)	NS	58.5(0.3)	0.01	68.7(0.7)	0.01

Worms (group)	day 0	p	day 15	p	day 30	p
control (D)	4.1(0.3)	–	4.2(0.3)	–	4.2(0.2)	–
gas+Cu (E4-E6)	4.3(0.1)	NS	11.1(0.5)	0.005	17.3(0.6)	0.001
gas+Mn (E7-E9)	4.2(0.2)	NS	5.6(0.3)	0.1	8.4(0.3)	0.05

(a) nanomoles of malondialdehyde per g of tissue. Mean(+/-SD).
p = significancy (Student´s t test) of the difference between previous and control value in the same day; NS= not significant.

Table 9. Lipofuscin-associated fluorescence (a) in animals exposed to both volcanic gas and heavy metals.

Rats (group)	day 0	p	day 30	p	day 60	p
control (A)	50.2(2.1)	–	61.1(1.1)	–	74.2(1.5)	–
gas+Cu (B7-B9)	51.4(1.7)	NS	89.3(1.2)	0.001	103(1.4)	0.001
gas+Mn (B10-B12)	49.0(2.4)	NS	87.2(1.3)	0.001	101(1.0)	0.001
gas+Cu+GSH (B13-B15)	52.1(1.3)	NS	72.3(1.7)	0.01	85.8(2.2)	0.01
gas+Mn+GSH (B16-B18)	51.8(2.2)	NS	70.8(1.3)	0.01	86.1(1.7)	0.01

Worms (group)	day 0	p	day 15	p	day 30	p
control (D)	2.1(0.1)	–	2.2(0.2)	–	2.1(0.1)	–
gas+Cu (E4-E6)	2.3(0.4)	NS	5.8(0.3)	0.001	9.0(0.4)	0.001
gas+Mn (E7-E9)	2.1(0.8)	NS	3.0(0.2)	0.1	4.7(0.3)	0.01

(a) units of fluorescence per 0.2 g of tissue. Mean(+/-SD).
p = significancy (Student´s t test) of the difference between previous and control value in the same day; NS= not significant.

manganese produces an additive effect and only in the animals sacrified at 30 day. Also in the boxes of gas and metals exposed worms were found 3 (for Cu) and 1 (for Mn) dead at the end of the treatment. The supply of reduced glutathione lowers both TBA reactivity and fluorimetric levels induced by the contemporary exposition to the pollutant.

DISCUSSION

One may postulate that the oxidizing components of atmospheric pollution act primarily on the external and respiratory surfaces of an organism. SO2 and NOx-induced damage of internal organs and tissue may be ascribed to the following. A certain quantity of the damaging species do not react at the external or respiratory surfaces of the organism and enter in the circulation where:
1) they reach, unmodified, distant tissues where they interact with the local cellular components, or,
2) they react with micro- or macromolecules contained in the body fluids generating relatively instable substances. These substances, and above all those that have a medium-high grade of stability, can migrate to other tissues and here attack and alter the morpho-functional integrity of area affected, in a direct or indirect manner.
SO2, once inside the organism, can be converted into SO3. radical within cells. Various authors suggest that this radical produces most of the damage (Halliwel and Gutteridge, 1985), while others think that SO3., in the presence of SO3-- and O2, induces the formation of the highly dangerous hydroxil radical (Sandaman and Gonzales, 1989). On the other hand, NO2 possesses a radical-like structure and can start peroxidative processes in the presence of organic compounds containing polyunsaturated bonds (Menzel, 1977). Furthermore, both NO2 and NO can react with intracellular hydrogen peroxide, producing further hydroxil radicals (Halliwel and Gutteridge, 1985).
Table 3 shows that environmental conditions of the exposed organisms fell within the norm for the whole of the experimental period. Exposure to the fumarolic gas, for 60 days, stimulates an increase in free radical damage of rat spinal ganglia most probably by mechanisms similar to those described above. Moreover, reduced glutathione administration eliminates this increase, suggesting that the damage produced is in fact due to oxidative phenomena.
Worms appear to be more sensitive to atmospheric pollution even though they live below the ground. This is hardly surprising considering that these anellides produce galleries and therefore are responsible for the airation of the superficial layers of the ground: therefore worms are also in direct contact with the oxidative components contained in the air. Furthermore, worms are known to be particularly sensitive to free radical-generating agents such as UV radiation and biocidal preparations (Ruppel and Laughlin, 1977; Karnak and Hamelink, 1982). This further justifies the appearance of oxidative damage in the exposed worms with respect to their control counterpart.
The data obtained, clearly show that the heavy metal treatment (in both of the experimental organisms) is efficient in producing an increase in the TBA reactivity and

lipofuscin fluorimetric indices. As already stated, Cu and Mn were both equally effective when given to rats. Throughout the experimental period these animals did not undergo behavioural or motor alterations. Even in this case, reduced glutathione administration was responsible for an attenuation, both at 30 and at 60 days, of the toxicity caused by the heavy metal treatment. However, with the doses of reduced glutathione used, levels did not return completely to those shown by control animals.

In worms, it was noticed that copper induces rapid and marked oxidative-type damage. On the other hand, manganese produced only slight, and, with respect to the controls, not very significant alterations. This difference could be due to:
1) a slower uptake of manganese,
2) the presence of more efficient Mn chelation, transport and accumulation mechanisms in the worms,
3) a different ability to catalize the formation of free radicals by Cu and Mn.

In both animals used the coexposition to gas pollutants and heavy metals generates a synergy, in fact, the effect produced is greater than the sum of the single effects. The TBA and fluorimetric levels in rats are much higher than would be expected from an additive effect alone, anyway, as in the treatment with only metals, the total damaging effect of Cu and Mn (contemporaneous to the exposition to fumarolic gases) is, once again, similar. In case of worms only Cu produced a synergy and this reflects the results obtained with metals where the copper was a damaging agent more efficient and rapid than manganese.

It is not clear how the synergy develops, probably the contemporary activities of these oxidant agents so alter the integrity of structures and metabolic functions that severe cell disorders are generated that, in turn, can increase the level of oxiradicals produced i.e. the deterioration of membranes containing cellular redox systems can provoke an increse of the physiological leaking of electrons that, after having reacted with oxygen, produce a "surplus" of superoxide.

Glutathione treatment shows that this antioxidant molecule reduces the synergistic damage by two-thirds.

In all three worm experiments, individuals were found dead. Due to the normal environmental conditions (pH, temperature and humidity) one may presume that their death was partially, if not totally, ascribable to the toxicity of the pollution encountered.

CONCLUSIONS

Exposure of organisms to the oxidative components, contained in a polluted environment, accelerates free radical-mediated cellular damage. Our data suggest that:
1) SO2 and NOx are responsible for an oxidative associated toxicity, similar to that induced by heavy metals;
2) the development of cellular damage depends on
a)the duration of exposure
b)the concentration of the pollutant
c)the type of pollutant and its chemical properties
d)the organisms ability to absorb the pollutant, and,
e)the prevention and/or repair mechanism present in the organism;

3) the administration of an antioxidant (e.g. reduced glutathione) protects the organism from the citotoxic effects of both dangerous species contained in volcanic gases and heavy metals. This point confirms the first and the importance of antioxidants in antiradical mechanisms;
4) because of the possible development of a synergy, the presence of such environmental pollutants as heavy metals in a volcanic area represents a situation of risk for the organisms living in that zone. The supply of an SH-group containing substance can reduces the free radical induced-damage also in the case of synergy.

REFERENCES

Adams, D. F., Farwell, S. O., Pack, M. R., and Robinson, E., 1981, Biogenic sulphur gas emissions from soils in Eastern and Southeastern United States, J. Air Poll. Contr. Assoc., 31: 1083-1089.

Bettlheim, J. and Littler, A., 1979, Historic trends in sulphur oxide emissions in Europe since 1865. CEGB Report PL-GS/E/1/79.

Brimblecombe, P. and Stedman, D. H., 1982, Historical evidence for a dramatic increase in the nitrate component of acid rain. Nature, 298:123.

Carbonelle, J., 1981, Preliminary results on contribuition of Mt Etna and Mt Stromboli to atmospheric SO2 and CO2. 2nd Meeting of WP 4, Aussois, France, April 28-29, Project COST 61a bis.

E.R.L., 1983, Acid rain: a review of the phenomenon in the EEC and Europe. Final report prepared for the Commission of the European Communities.

Fisher, B. C. A., 1984, The long range transport of air pollutants-some thoughts on the state of modelling, Atm. Env., 18: 553-562.

Forstner, U. and Wittmann, G. T. W., 1981, "Metal pollution in the acquatic environment", Springer Verlag, New York.

Gaddie, R. E. and Douglas, D. E., 1975, "Earthworms for ecology and profit", Bookworm Publishing Company, Ontario.

Galloway, J. N., Thornton, J. D., Norton, S. A., Volchok, M. L. and McLean, R.A.N, 1982, Trace metals in atmospheric deposition: a review and assessment, Atm. Env. 10: 1677-1700.

Garrison, S. and Page, A.L., 1984, Cycling of the metal ions in the soil environment, in "Circulation of the metals in the environment", A.L. Page, ed., Marcel Dekker Inc., New York.

Gravenhorst, G., 1985, Natural NOx emissions in the FRG, Meeting of WP 4 and 5, COST 611, Bilthoven, 23-25 September.

Gutteridge, J. M. C., 1988, Oxygen radicals, transition metals and aging, Advances in Biosciences, 64: 1-22.

Halliwell, B. and Gutteridge, J. M. C., 1985, "Free radicals in biology and medicine", Clarendon Press, Oxford.

Karnak, R. E. and Hamelink, J. L., 1982, A standardised method for determining the acute toxicity of chemicals to the earthworms, Ecotoxicol. Environ., 6: 216-222.

Kruk, P. and Enesco, H. E., 1981, Alfa tocopherol reduces

fluorescent age pigments levels in heart and brain of young mice, Experientia, 37: 1301-1304.

Le Bronec, J., Robe, H. C, and Allard, P., 1988, Distribuition and sources of acid gases in the atmosphere of Campi Flegrei Caldera Italy, in press.

Nriagu, J.O., A silent epidemic of environmental metal poisoning, 1988, Env. Poll., 50: 139-161.

Ottar, B., 1983, Air pollution emissions and ambient concentrations, in: "Proceedings CEC Symposium on Acid Deposition" H. Ott ed., Stangl, Karlsruhe.

Placer, Z. A., Cushmann, L. L. and Johnson, B. C., 1966, Estimation of product of lipid peroxidation (MDA) in biochemical systems, Anal. Biochem., 16: 359-364.

Ruppel, R. F. and Laughlin, C. W., 1977, Toxicity of some pesticides to earthworms, J. Kans. Entomol. Soc., 56: 113-118.

Sandaman, G. and Gonzales, H. G., 1989, Peroxidative process induced in bean lives by fumigation with sulphur dioxide, Env. Poll., 56: 145-154.

DISCUSSION

PORTA: Thank you Dr. Totaro for your very interesting presentation. Environmental pollution is an increasing worldwide problem affecting not only humans, but the whole ecosystem, and studies like yours in different animal species are in my opinion very important and timely. I am particularly interested in your data on rats indicating that the administration of reduced glutathione decreased the levels of TBA-reactive substances as well as lipopigment formation. Since there is some concern about the effectiveness of administering glutathione in order to build up this tripeptide in tissues, could you please tell us in more detail what dosage and route of administration did you use?

TOTARO: GSH was administered intramuscularly to rats at the dosage of 30 mg/kg body weight every 4 days.

IVY: What part of the nervous system was studied in rats?

TOTARO: We have only studied in rats the nerve cells of the spinal ganglia because of the cellular homogeneity in these ganglia.

KITANI: I just want to make a comment. You have shown that GSH really works in rats, although it is known that the elimination of administered GSH is very rapid. Therefore, even if it is effective, it would be convenient to have a special device to keep GSH in the body. Although such a device is not generally available, Dr. Inouye from Kumamoto University has developed a special delivery device for rats where SOD or GSH is attached to a special form of protein and, therefore, kept intravascularly for more prolonged periods and delivered to places where these substances are needed. It would be interesting to use this method in your future studies.

TOTARO: Thank you for your comments and suggestions. The delivery of antioxidant substances is indeed very important and a matter of great concern in our laboratory. In the next presentation by my colleague, Dr. Pisanti, he will present data on the administration of cythiolone, a homocysteine derivative, that gradually liberates SH and may help to maintain the levels of GSH in the body.

CEROIDOGENIC EFFECT OF IONIZING RADIATION

F. A. Pisanti, L. Lucadamo and E. Aloj Totaro

Stazione Zoologica, Napoli, Italy, and
Dip. Ecologia, Universita´ della Calabria
Arcavacata di Rende (CS), Italy

SUMMARY

The effect of the exposure of Torpedo marmorata to a single dose of 400 cGy of ionizing radiation and doses of 200 cGy/week for 4 weeks has been studied. The parameters measured were the TBA reactivity and lipofuscin production, a fluorescent pigment that is thought to derive from the reaction of malonaldehyde, a peroxide-degradation product, with free amino-groups. Acute irradiation was found to generate an increase of peroxidative damage in some tissues. This effect was inhibited by acetyl homocysteine-thiolactone, a drug that increases the activity of tissue superoxide dismutase. Chronic irradiation produces a severe increase in the generation of lipofuscin.

INTRODUCTION

Ionizing radiations damage biological systems essentially by producing free radicals and inducing molecular alterations (1). One of the most important parameter characterizing ionizing radiations is the linear energy transfer (LET) that represents the capability of the radiation to transfer its energy content to the biological structures passing through them. We can distinguish radiations at low, intermediate and high LET, with the last ones being the most damaging and the first ones the least. It´s known that the presence of oxygen in the tissues increases the level of alterations induced by radiations, but not in the same way for the three types of radiation. In fact the greatest damage occurs with low LET radiations and the least with high LET radiations. The latter probably are so powerful that they do not need the presence of oxygen to damage the organisms by means free-radical mechanisms.

In aerobes, free radicals induce the peroxidation of important components of biological structures such as

KEY WORDS: Torpedo, ionizing radiations, free radicals, lipofuscin, peroxidation, citiolone, TBA reactivity

Lipofuscin and Ceroid Pigments
Edited by E. A. Porta
Plenum Press, New York, 1990

lipids, proteins, and nucleic acids. One of the degradation products of lipoperoxides is malonaldehyde, which reacts with the free amine groups of proteins, so producing fluorescent pigments, known as age pigments or lipofuscin (2). Thus, lipofuscin and lipoperoxides production can be a marker of ionizing radiation-induced damage (3). However, while the formation of lipofuscin is a somewhat slow process, the production of lipoperoxides is fairly fast and can serve as an index of acute damage.

Little is known about the effect of radiations in marine organisms. It has been suggested that exposure to low linear energy transfer (LET) radiations modifies the structure of lipid membranes and their metabolism, which, in turn, leads to ion leakage and deranged osmotic equilibria (4, 5). Radiosensitivity experiments have shown that marine organisms are more resistant than mammals to the action of ionizing radiations. In fact, their LD50 (about 700 cGy) is twice or three times greater than that of mammals (6).

The metabolic characteristics involved in age pigment formation have been well documented in Torpedo marmorata, a batoid selachian (7, 8). Here we report the effects in Torpedo m. of ionizing radiations on lipoperoxides and lipofuscin production, and the radioprotective effect of a compound derived from homocysteine, citiolone (n-acetyl-homocysteine-thiolactone). This drug is a moderate free radical scavenger and it also stimulates superoxide dismutase (SOD) activity (9).

MATERIALS AND METHODS

The Torpedo marmorata studied were caught in the Gulf of Naples, they were of both sexes, weighed 350-450 grams and were about 1 year old (10). The animals were kept in tanks with circulating seawater at the Naples Zoological Station and fed, during the experiment, as described previously (11). The animals were divided in five groups of seven. The group A was panirradiated with a single dose of 400 cGy. Group B was given 8 mg/kg body weight of citiolone (Roussel Maestretti, Milano) per day for five days and then treated as group A. Group C was irradiated with 200 cGy once a week for four weeks. Group D was given citiolone (8 mg/kg b.w. per day) for five days and then irradiated as group C; citiolone was continued, at the same dose, throughout the four weeks of irradiation. Finally the controls (group E) were neither treated nor irradiated.

During irradiation, the torpedos were kept in tanks small enough to impede movement containing 10 cm of seawater, 65 cm from a low LET radiation (60Co) source (average yield: 23.6 R/min). Irradiation times were: 31min 44sec for 400 cGy and 15min 52sec for 200 cGy. The 400 cGy-irradiated animals (groups A and B) were killed 2 days after the irradiation; this experiment served to assess acute damage induced by radiations. The animals of groups C and D, which received a total dose of 800 cGy, were killed 10 days after the last irradiation to evaluate the chronic effects of the treatment. The animals were anaesthesized with MS 222 (Sandoz) 0.015% in seawater for 15min, and then samples of liver, forebrain, electric lobe, heart and electric muscle were removed for biochemical assays. The whole body was fixed using the perfusion technique (12) and the remaining part of the electric lobe was prepared for transmission electron microscopy. Ultrathin sections (LKB Ultratome III) were observed with a Philips 400 electron microscope.

The samples removed for the biochemical analyses were assayed to evaluate the level of peroxides measuring the TBA reactivity (13) and lipofuscin (fluorimetric assay, according Kruk and Enesco, 1981). Student´s "t" test was used for statistical analyses.

RESULTS

In preliminary trials some animals died a few days after having been irradiated with 400 cGy. This dose seems low, particularly because marine animals are considered more resistent to radiations than, e.g., mammals (5, 6). The low resistance of Torpedo m. could be because Torpedo electric lobe lacks SOD, which is a superoxide radical scavenger (8). Mammals that have received very high levels of radiations die within 24-48 hours from a central nervous system syndrome; the death seen in Torpedo could perhaps be attributed to their lack of antioxidants defenses.

Table 1 shows the TBA reactivity in the animals exposed to 400 cGy (groups A and B) compared with control levels. Some of the tissues of group A animals (not pretreated with citiolone) showed high levels of peroxidation, demonstrating that some tissues are more susceptible then others. Group B animals (pretreated with citiolone) had significantly lower MDA levels than group A.

There were no significant differences in lipofuscin levels between group A, B and controls (Table 2). Electron microscope observations of electric lobe confirm this result. Group A animals also showed numerous swollen mitochondria and vacuoles. Such degenerative endocellular processes were much less evident in the animals of group B, suggesting that citiolone exerts a radioprotective effect also at subcellular level.

Table 1. TBA reactivity in Torpedo m. exposed to ionizing radiations (400 cGy).

tissues	animals				
	control	p	group A	p	group B
electric lobe	2.16	<0.01	4.90	<0.01	4.08
forebrain	6.35	NS	6.53	NS	6.30
heart	2.08	<0.01	3.24	<0.01	2.36
liver	17.25	<0.05	24.22	<0.001	16.60
electric muscle	3.37	NS	3.85	NS	3.60

Results are expressed as micromoles of malonaldheyde/g protein. Group B: animals pretreated with citiolone (8 mg/kg b.w./day/5 days). Column p shows significancy (Student´s t test) of the difference between previous and next column (control/group A; group A/group B). NS: non significant.

Table 2. Lipofuscin production in Torpedo m. exposed to
ionizing radiations (400 cGy).

tissues	animals				
	control	p	group A	p	group B
electric lobe	3.95	NS	4.35	NS	3.97
forebrain	2.91	NS	2.98	NS	2.80
heart	2.01	NS	2.40	NS	2.10
liver	2.43	NS	2.88	NS	2.52
electric muscle	1.59	NS	1.61	NS	1.40

Results expressed in terms of fluorescence units/200 mg of
tissue. Group B: animals pretreated with citiolone (8 mg/kg
b.w./day/5 days). Column p shows significancy (Student's t
test) of the difference between previous and next column
(control/group A; group A/group B). NS: non significant.

Table 3 shows the TBA reactivity measured in Torpedo m.
exposed to 200 cGy for four weeks (group C and D). In the
group C animals the radioinduced production of TBA
reactivity is significantly lower than in the controls, but
it is not statistically different from that found in animals
of group D, irradiated with a 800 cGy dose (200 cGy/week for
4 weeks) and treated with citiolone. As shown in table 4,
lipofuscin levels were much higher in groups C and D than in
controls. Also in this case, citiolone did not affect
lipofuscin production. In fact, the levels found in group D
were not statistically different from those of group C. Once
again, biochemical data are confirmed by electron microscope
observations of numerous lipofuscin granules.

DISCUSSION
The results obtained with animals exposed to a single
dose of 400 cGy and killed two days later (group A) indicate
that the various tissues studied have different degrees of
radiosensitivity. In fact, with respect to controls, in the
radioexposed animals, the peroxidation marker (MDA
production) was increased by 126% in the electric lobe, by
55% in the heart and by 40% in the liver. Therefore these
tissues are more radiosensitive than the electric muscle and
the forebrain when the presence of malonaldehyde is not
significantly modified by exposure to radiations. It is
conceivable that the severe peroxidative radio-induced
damage in the electric lobe, could be related to the
deficiency of superoxide dismutase reported in this cerebral
area (8, 15). The radiosensitivity of liver and heart is
less easy to explain; however, it may be related to the high
concentrations of oxygen necessary for the intense aerobic
metabolism that occurs in these organs.
It was interesting to observe that citiolone produced a
significant reduction of peroxidation of tissues found to be
radiosensitive (liver, electric lobe, heart) but it did not
modify the extent of damage of "radioresistant" tissues
(electric muscle and forebrain). There are two possible
explanations of this:

Table 3. TBA reactivity in Torpedo m. exposed to
ionizing radiations (200 cGy/week/4 weeks).

tissues	animals				
	control	p	group C	p	group D
electric lobe	2.16	<0.001	1.40	NS	1.00
forebrain	6.35	<0.001	1.00	NS	0.90
heart	2.08	<0.001	1.03	NS	0.81
liver	17.25	<0.001	8.10	NS	7.20
electric muscle	3.37	<0.001	0.40	NS	0.32

Results are expressed as micromoles of malonaldheyde/g
protein. Group D: animals treated with citiolone (8 mg/kg
b.w./day). Column p shows significancy (Student's t test) of
the difference between previous and next column
(control/group C; group C/group D). NS: non significant.

Table 4. Lipofuscin production in Torpedo m. exposed to
ionizing radiations (200 cGy/week/4 weeks).

tissues	animals				
	control	p	group C	p	group D
electric lobe	3.95	<0.001	29.1	NS	26.0
forebrain	2.91	<0.001	23.2	NS	21.1
heart	2.01	<0.001	15.1	NS	11.9
liver	2.43	<0.001	11.2	NS	9.1
electric muscle	1.59	<0.001	22.0	NS	19.3

Results expressed in terms of fluorescence units/200 mg of
tissue. Group D: animals pretreated with citiolone (8 mg/kg
b.w./day). Column p shows significancy (Student's t test) of
the difference between previous and next column
(control/group C; group C/group D). NS: non significant.

1)the radioresistance of electric muscle and forebrain is
not due to antioxidant activity; or 2)the antioxidant system
in these tissues affords protection against ionizing
radiations. Of course, one possibility does not exclude the
other, and they can coexist.

The exposure of animals to a dose of 800 cGy given over
four weeks (group C) produced effects completely different
from those obtained with acute exposure. This confirms that
subdivision of dose triggers mechanisms of recovery of
cellular functionality (16). Many in vitro experiments
clearly show that the level of survived cells is higher when
a dose is subdivided than when a single dose of radiations
is given. In fact, in Torpedo there was a significant
increase of lipofuscin production and a reduction of TBA
reactivity. These data suggest that chronic induction of

peroxidative damage can be so severe as to give rise to a massive increase of cell turnover that causes, on one hand, the removal of vast damaged areas by autophagocytosis and, on the other, lipofuscin production.

The electron microscope studies revealed numerous lipofuscin granules in electric lobe cytoplasm, which indicates rearrangement of the cell architecture induced by widespread damage. Cythiolone treatment (group D) did not modify significantly the tissue reaction to chronic irradiation. On the contrary, animals pretreated with the drug and then exposed to only one dose of 400 cGy (group B) were protected from the effects of radiation as shown by inhibition of both lipofuscin and peroxides production. In addition, in these animals, the radioinduced degenerative subcellular processes (appearance of numerous vacuoles and swollen mitochondria) was clearly less than in the animals not treated with citiolone (group A).

In conclusion, chronic irradiation produces in Torpedo marmorata an increase of lipofuscinogenesis probably resulting from rearrangement of the damaged structures. Acute irradiation induces an increase of peroxidative damage only in tissues with greater radiosensitivity. Cythiolone (N-acetyl-homocysteine-thiolactone) exerts a radioprotective effect, which can be related to the free radical scavenger activity of the molecule (9) to the superoxide dismutase-activating effect of the drug (17), or both.

REFERENCES

1. H. B. Bielshi and J. M. Gebicki, Application of radiation chemistry to biology, in: "Free radicals in biology", W. A. Pryor, ed., vol. 3, Academic Press, New York (1977).
2. H. Donato, Lipid peroxidation, cross-linking reactions, and aging, in: "Age pigments", R. J. Sohal, ed., Elsevier N.H., Amsterdam (1981).
3. H. J. M. Hansen, Changes in the lipid fraction of Eel gills after ionizing irradiation in vivo and a shift fresh to sea water, Radiat. Res. 62: 216 (1973).
4. P. N. Srivastava and C. Tachi, Effect of irradiation on the excretion of radiosodium from the gills of goldfish Carassius auratus, Radiat. Res. 23: 222 (1964).
5. IAEA, Effects of ionizing radiation on acquatic organisms and ecosystems. International Atomic Energy Agency, Wien (1976)
6. H. Mitani, H. Etoh, and N. Egami, Resistance of a cultured fish cell line (CAF-NMI) to gamma-irradiation, Radiat. Res. 89: 334 (1982).
7. E. Aloj Totaro and F. A. Pisanti, Stereological analysis of lipofuscin in the CNS of Torpedo m.:correlation with superoxide dismutase distribution, Experientia 41: 1047 (1985).
8. E. Aloj Totaro and F. A. Pisanti, Age pigments and superoxide dismutase activity in the CNS of Torpedo m., Neuroscience letters 45: 341 (1984).
9. E. Aloj Totaro, F. A. Pisanti, and E. Liberatori, Possible interelation of acetyl-homocysteine-thiolactone in mechanisms of lipofuscinogenesis, Res. Comm. Chem. Pathol. Pharmacol. 47: 415 (1985).

10. E. Aloj totaro, F. A Pisanti, P. Russo, and P. Brunetti, Evaluation of aging parameters in Torpedo marmorata, Annls. Soc. Zool. Belg. 115: 203 (1985).
11. E. Aloj Totaro, Effect of centrofenoxine on the variation of some behavioural pattern in Topredo m., Acta Neurol. 34: 332 (1979).
12. E. Aloj Totaro and F. A. Pisanti, Preliminary observations at the electron microscope on the presence of neuronal lipofuscin in Torpedo m., Acta Neurol. 34: 322 (1979).
13. Z. A. Placer, L. L. Cushman, and B. C. Johnson, Estimation of product of lipid peroxidation (malonyl dialdehyde) in biochemical systems, Anal. Biochem., 16: 359 (1966).
14. P. Kruk and H. E. Enesco, Alfa-tocopherol reduces fluorescent age pigment levels in heart and brain of young mice, Experientia 37: 1301 (1981).
15. F. A. Pisanti, S. Frascatore, E. Aloj Totaro, and E. Vuttariello, Superoxide dismutase in the central nervous system of Torpedo marmorata, Arch. Gerontol. Geriatr. 2: 343 (1983).
16. J. E. Coeggle, "Effetti biologici della radiazioni ionizzanti", Minerva Medica, Torino (1982).
17. F. A. Pisanti, S. Frascatore, E. Vuttariello, and A. Grillo, The influence of acetyl-homocysteine--thiolactone on erithrocyte superoxide dismutase activity, Biochem. Med. Metab. Biol. 37: 265 (1987).

DISCUSSION

KATZ: It seems that the results of your chronic treatment would appear rather inconsistent with an oxidative mechanism, because you showed that at 200 cGy dose level, the malonaldehyde levels were lowered in the irradiated animals, and yet lipofuscin levels were higher.

PISANTI: Yes, this was a rather unexpected finding, but the lipopigment did increase.

Zs.-NAGY: One explanation could be that the TBA-reaction was done some time after irradiation, and probably the levels of reactive substances were no longer present at the moment they were measured.

KATZ: Yes, I am not questioning the effect of radiation on lipopigment formation. I simply think that at least the initial effects of malonaldehyde on proteins are fluorescent Schiff bases which are unstable and, therefore, most probably unrelated to the stable fluorophores of lipopigment formed by radiation damage or, in fact by any other kind of damage.

ICHIKAWA: You mentioned that cythiolone protects against radiation damage. What are the mechanisms of this protection? Is the drug really stimulating SOD or is it acting more as a general free radical scavenger?

PISANTI: Probably, both effects of cythiolone are playing a role. As a free radical scavenger, cythiolone effect appears to be only moderate. Apparently, the main effect could be through the induction of SOD.

ELLEDER: My question is more hypothetical. What is your view about the observation that during radiation there is an increase in the number of residual bodies which accumulate lipopigments? Do you think that the explanation is that there might be a hyperactive reaction of the intact lysosomal system scavenging damaged cell membranes, or is it caused by radiation hitting the lysosomes and making them unable to handle the normal cellular turnover?

PISANTI: Radiobiologists are not yet in agreement about this point, but one favoured hypothesis is that radiation may inhibit lysosomal enzymes.

BRUNK: I would like to make a comment on the effect of radiation on lysosomes. There is one cell very sensitive to radiation and this is the secretory cell of salivary glands. These cells contain secretory granules with very high content of zinc, manganese and iron. If you give the animals drugs that produce the release of the secretory granules, you could decrease their radiosensitivity down to almost what is normal in non-dividing cells. Our interpretation to these findings, which we published a couple of years ago, is that lysosomes which phagocytize these granules become very rich in these metals. So, when the cells are hit by radiation, free radicals are formed everywhere in the cell. It is then possible that superoxide ion radicals reduce trivalent iron to divalent iron within lysosomes, and this would lead to special sensitivity of surrounding cell membranes, because they are exposed to a metal catalyzed peroxidation and fragmentation. It seems that lysosomes are especially prone to radiation damage because in most cells, they are rich in heavy metals.

ELLEDER: I don't doubt that radiation hits everything in the cytoplasm, but for the formation of lipopigment, it seems that the crucial event is the inhibition of lysosomal enzymes.

EFFECT OF LIFELONG SELENIUM AND VITAMIN E DEFICIENCY OR
SUPPLEMENTATION ON PIGMENT ACCUMULATION IN RAT PERIPHERAL
TISSUES

H. Alho[1], J. Koistinaho[2], H.M. Laaksonen[2]
and A. Hervonen[2]

Department of Biomedical Sciences[1] and
Department of Public Health[2]
University of Tampere
POB 607, 33101 Tampere, Finland

SUMMARY

The accumulation of lipopigments during aging in several
peripheral organs and in the nervous system is considered
to be related to the peroxidation of unsaturated fatty
acids. In this study the effect of lifelong (until to 18
months) dietary antioxidants selenium and vitamin-E on
pigment accumulation in some peripheral tissues was
estimated using fluorescence and electron microscopy. In the
vitamin E deficiency group, there was increased pigment
accumulationin all peripheral tissues studied except the
hypogastric ganglion, where no change was observed. The
vitamin E supplementation degreased the pigment accumulation
in older animals in some of the tissues studied. At the
electron microscopical level the accumulated pigment in the
adrenal cortex showed a lipofuscin-like structure. Lifelong
selenium supplementation or deficiency did not significantly
alter pigment accumulation in any of the tissues studied. It
is possible that in many organs dietary selenium may not
play a critical role in lipofuscin formation.

INTRODUCTION

The attempt to understand the phenomenon of aging and
the accumulation of age pigments various hypotheses has
generated (see Sohal, 1981). The best known of these and the
most widely debated is the free radical hypothesis. Aging
tissues defend themselves against lipid peroxidation by exo-
and endogenous antioxidants (Harman, 1981; Sohal et al.,
1985). Considering the biological role of vitamin E as an
important antioxidants and preventive factor for lipid
peroxidation, the possible influence of this vitamin on
aging and pigment accumulation has been widely studied. Many

KEY WORDS: Lipofuscin, diet, vitamin E, selenium, peripheral
           tissues, fluorescence, electron microscopy

authors have shown that vitamin E deficiency increases the accumulation of age pigment and shortens the life span of mammals (Einarson et al., 1960; Harman, 1968; Desai et al., 1975; Sohal et. al., 1985; Sarter et al., 1987; see also Sohal, 1981). Additionally, some authors have shown that high doses of vitamin may decrease synthesis and the accumulation of age pigment (Blackett & Hall, 1981; Elmadfa et al., 1986). Clinical therapy with antioxidants such as tocopherol (vitamin E) and metyhionine has been reported to slow down degenerative processes (Siakotos et al., 1974) but this therapy has only been of limited benefit (Santavuori & Moren, 1977).

Selenium is attributed to function that is necessary for the synthesis of selenium-dependent glutathione peroxidase, an enzyme that catalyzes the oxidation reduction reaction between glutathione and peroxide (Noughi et al., 1973). Nutritional deficiency of selenium is accompanied by a decrease in the activity of glutathione peroxidase in several tissues (Combs & Combs, 1984). Dietary selenium supplementation increases the tissue glutathione peroxidase activity (Meyer et al.,1983) and there is some evidence that selenium plays an antioxidative role in lipid peroxidation related diseases (Harman, 1981; Westermark et al., 1982). There has been only limited research on the effect of dietary selenium on the accumulation of age pigment. The objectives of our experiments were to determine the effects of life long dietary supplementation and deficiency of selenium and vitamin E on pigment accumulation in rat tissues.

MATERIAL AND METHODS

The rats used in this study were fed on designed diets from weaning to sacrifice. Fifteen adult Sprague-Dawley female rats were fed from the day of weaning on special pellets (Astra, Sweden, R0-food) containing specially low concentrations of selenium and vitamin E. According to the manufacturer's analysis, the pellets contained 3 ug/kg of seleniun and 0.1 mg/kg of vitamin E. Selenium and vitamin E supplementation and deficiency was regulated by administering different concentrations of sodium selenite (Sigma) and alpha-tocopherol (Merck) into drinking water (Se: 90.1 mg/ml - 0 mg/ml; vit.-E: 207.2 ug/ml - 1.08 ug/ml). The consumption of these antioxidants per body weight was kept constant by regulating the drinking water concentration of the antioxidants at different ages of the rats. The daily allowance of selenium and vitamin E was estimated and three different doses were used (Table 1.). Two ug of selenium and 1.2 mg of alpha-tocopherol in diet / kg (body weight) were considered normal dietary values (Nutrient Requirements for Laboratory Animals, 1978).

The mother rats as well as their litter (at the age of three weeks) were divided into five groups and fed with different combinations of selenium and alpha-tocopherol diets (see Table 2).

Table 1. Estimated daily consumption of selenium and
alpha-tocopherol per kg/day

	Selenium	Alpha-tocopherol
Dose		
High	10ug	23mg
Normal	2ug	1.2mg
Low	0.2ug	0.12mg

Selenium and alpha-tocopherol were dissolved in water
(0 - 90.1 mg /ml and 0.67 - 207 ug/ml respectively)

Table 2. The different dietary groups

I	II	III	IV	V
Se_nE_n	Se_nE_l	Se_nE_h	Se_lE_n	Se_hE_n

n = normal, l = low and h = high dose
Se =selenium and E = alpha-tocopherol

The control animals were fed on normal, common
laboratory rat food from the same maufacturer (Astra, R3).
This food contains 200 ug/kg of selenium and 50 mg /kg of
vitamin E. The control group drank normal water.

At the age of 8 months eight rats from each group were
sacrificed. Because of high mortality in group II (low
vitamin E), all remaining 6 animals in this group were
sacrificed. The other animals were kept alive and sacrificed
at the age of 18 months. The animals were anesthetized with
sodium pentobarbital (25 mg/kg, i.p.). Their adrenal glands,
heart, prostate, testicles and hypogastric ganglia were
dissected. The hearts, testicles and adrenals were cleaned
and weighed. Some of the specimens were rapidly frozen in
liquid nitrogen and some fixed with immersion for electron
microscopy.

For fluorescence microscopy the samples were eihter
freeze-dried under vacuum for 8 days (Hervonen et al., 1986)
and embedded in vacuo in paraffin or cut frozen with a
cryostat. The autofluorescence of the pigment was studied
under a stabilized HBO high pressure mercury lamp with a
Nikon microphot-FXA microscope equipped with filter
combinations B-3A and B-2B suitable for pigment
autofluorescence. The amount of pigment was estimated from
three samples of each specimen by calculating the number of
pigment granules from colour slides (three from each sample)
in a standard size area.

For electron microscopy the adrenal glands were
processed with a standard method. Briefly, the samples were

fixed with 2.5% glutaraldehyde by immersion in a phosphate buffer for 3 h, followed by a postfixation with 1% osmium tetroxide at +4 ^{O}C for 1 hour. After dehydration and embedding ultrathin sections were stained with uranyl acetate and lead citrate and the sections were studied with a Jeol 1200X microscope. The properties of the pigment were estimated from standardized photomicrographs.

Selenium plasma levels were studied at the age of 6 and 15 months in all animals except group II (low vitamin E), which was studied at the age of 4 and 6 months using the method described by Behne et al., 1972. Serum tocopherol concentration was studied only at the age 6 months by the method described by Vatassery et al. (1982).

RESULTS

Body weight and mortality. In group II (low vitamin E) gain of body weight (BW) was significantly lower than in the other groups (BW mean at the age of 8 months 345g). Group IV (low selenium) also showed less gain (BW mean at the age of 8 months 355g) than the others (BW mean at 8 mo, 525g). Mortality was highest in group II (at the age of 8 mo, mortality 65.2%) and also rather high in group IV (8 mo, mortality 18.5%); the other groups did not differ from the control (mortality 7.8%) animals. Because of the high mortality in group II, the remaining animals were sacrificed at the age of 8 months.

Organ weight. Organ weights were evaluated by calculating ratio between the organ and body weight (BW). Different diets had different effects on organ/BW ratios (Table 3); the heart ratio was lowest in the low selenium group (IV) and highest in the high selenium group (V), while the testis ratio was highest in group I (Se_nE_n) and lowest in group II (Se_nE_l).

Table 3. Organ weight/body weight ratio

Group	I	II	III	IV	V
HW/BW	2.6±0.2	2.7±0.1	2.6±0.1	2.1±0.2*	2.8±0.1
TE/BW	3.8±0.3	2.1±0.1*	3.6±0.2	3.5±0.2	3.6±0.3

* p< 0.01 compared to group I. HW= heart, TW= testis and BW= body weigt (g), values x10^{-3}

Plasma concentration of seleniun and alpha-tocopherol. At the age of 6 months all groups except group V (high selenium) had significantly lower plasma selenium levels than the control animals (Fig. 1). The lowest selenium concentration (6 mo, 11 ng/ml) was found in group IV (low selenium) and the highest (6 mo, 235 ng/ml) in group V (high selenium). The control group (animals fed on normal rat food) showed higher selenium concentrations (6 mo, 289 ng/ml) than all others. The alpha-tocopherol concentration was measured only at the age of 6 months. The plasma levels were lower in all groups except in group III (high vit.-E) compared to controls (Fig. 2).

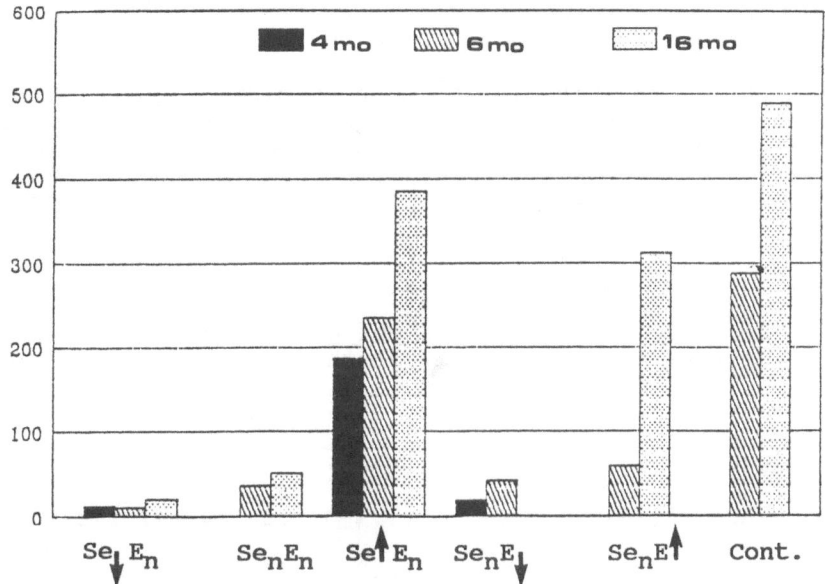

Fig.1. Serum selenium concentration (ng/ml) in different dietary groups.

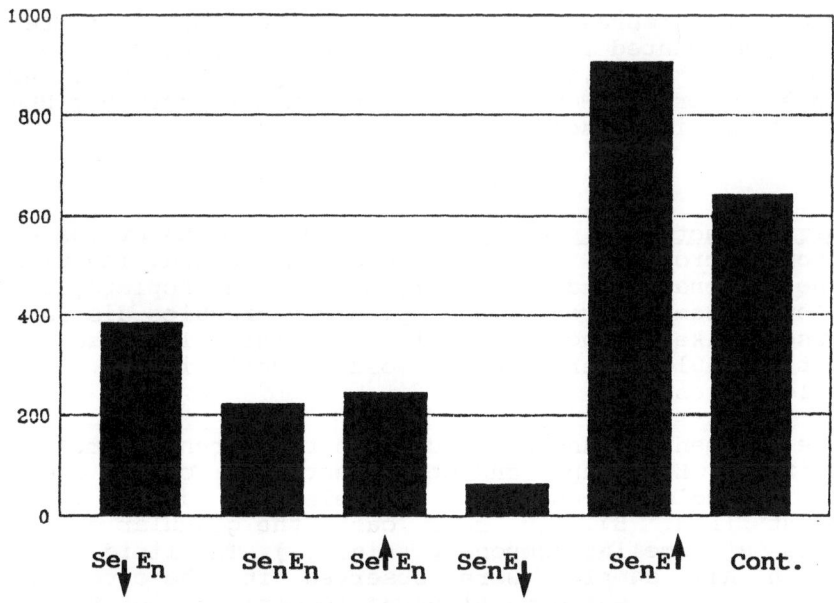

Fig. 2. Serum tocopherol concentration (ug/100ml) in different dietary groups at the age of 6 months.

Pigment accumulation. At the age of 8 months pigment accumulation was increased of the vitamin E deficiency group (II) in all tissues studied, except the hypogastric ganglion, where no change was observed (Table 4, Fig.3). Selenium supplementation or deficiency and vitamin E supplementation did not significantly alter pigment accumulation in any tissues studied (Table 4).

At the age of 18 months in all dietary groups normal age dependent pigment accumulation was observed in all the tissues studied . Between the different dietary groups pigment accumulation was increased in vitamin E deficiency group (II). In vitamin E supplementation group (III) the pigment accumulation was decreased compared to controls in all tissues studied except in the hypogastric ganglia where no difference was observed. Also at the age of 18 months selenium supplementation or deficiency had no effects on pigment accumulation.

Table 4. Amount of pigment grains in different tissues at the age of 8 months

Group	I	II	III	IV	V
Tissue					
AC	23.5±2	33.2±4*	21.1±3	25.1±4	22.8±3
HGG	15.2±2	17.1±3	16.7±1	16.4±3	15.1±1
HM	7.5±1	8.6±1*	7.1±1	7.8±2	7.3±1
PR	37.2±4	40.2±3*	35.6±3	38.1±3	36.3±2

The values represent the mean $\pm$ SD from three different animals counted from three 1.4×10^3 um^2 area of each specimen. * p< 0.01 compared to group I or control. AC= adrenal cortex; HGG= hypogastric ganglia, neurons; HM= heart cardiac muscle; PR= prostate, epithelial cells

Ultrastructural studies. The pigment was evaluated in electron microscopy only in the adrenal gland. The pigment in the adrenal gland was mainly lipid like droplets, but in the low vitamin group many pigment granules displayed a lipofuscin-like structure. (Fig. 4). Large lipofuscin-like pigment granules were found most commonly in the cells of zona glomelurosa.

The pigment granules found in the adrenal cortex from low vitamin E group consist essentially of an electron ludent vacuolar component and electron dense granular component (Fig. 5). In some cases the granular component exhibits a lamellar component (Fig. 5.). The lipid droplets found in all samples were observed in the cytoplasm of parencymal cells and were electron ludent; some of them were incorporated into mitochondria.

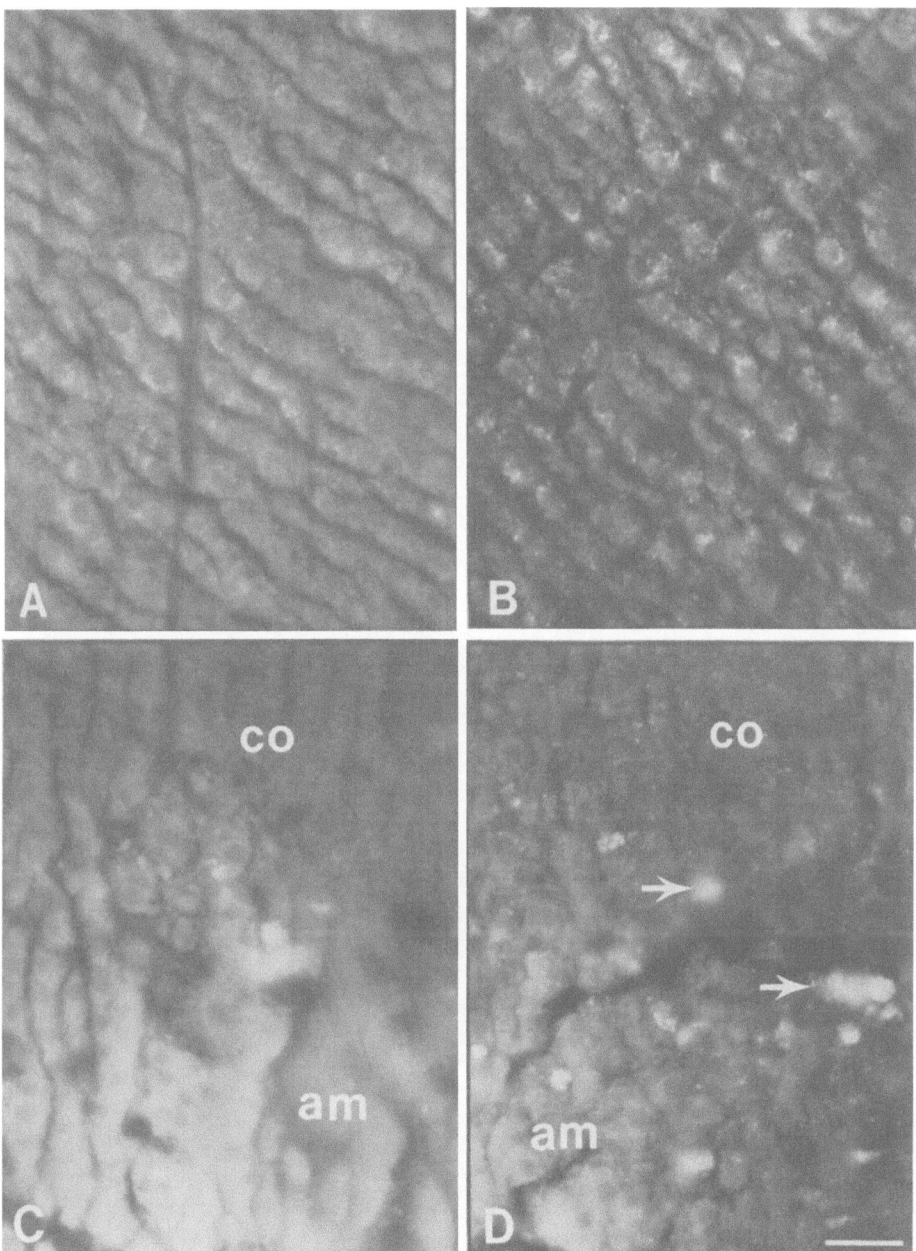

Fig. 3. Fluorescence micrographs of pigment autofluorescence in adrenal gland in different dietary groups (8 month-old animals). A + C = control rat; B + D = low vitamin E animal (group II); A and B represent identical areas in cortex and C and D represent areas between cortex (co) and medulla (am). Note the increased number of pigment granules in low vitamin E samples and large pigment deposits (arrows). Bar represents 20 um.

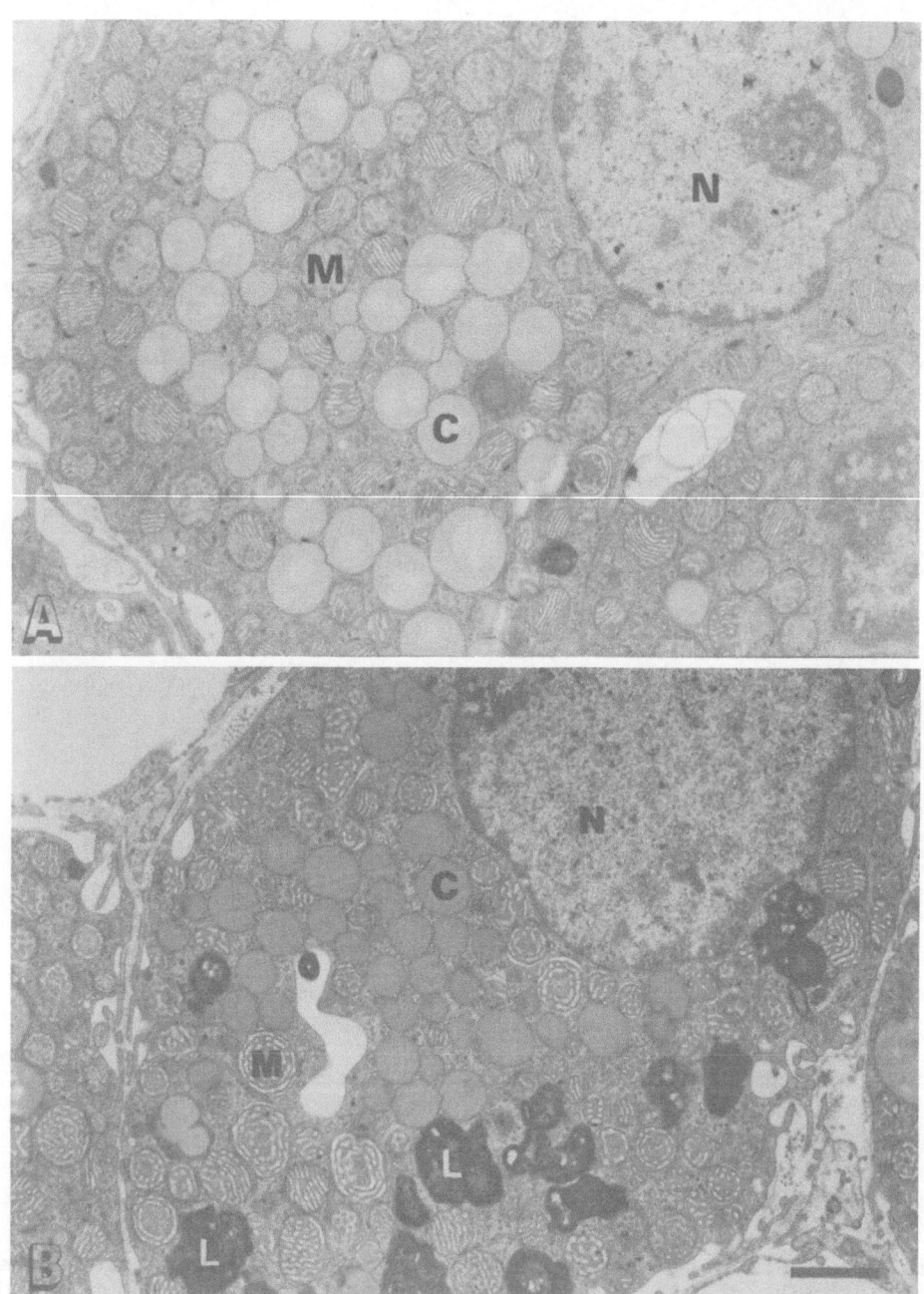

Fig. 4. Electron micrographs of pigment in rat adrenal cortex. A = control sample; B = low vitamin E sample. N= nucleus, m = mitochondrion, c = lipid droplet, l = lipofuscin. Note the increased number of lipofuscin in low vitamin E sample. Bar = 1.5 um.

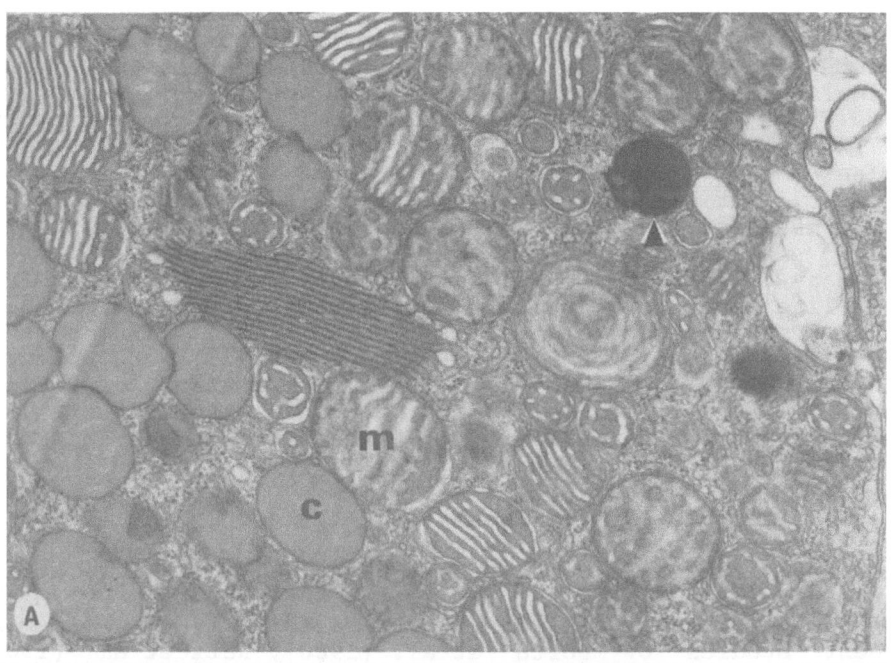

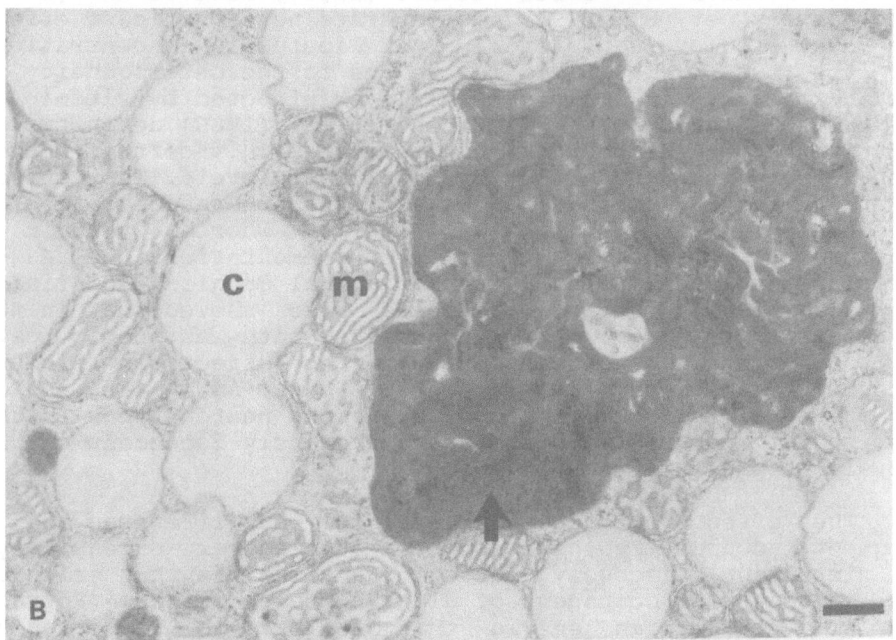

Fig. 5. Electron micrographs of pigment in rat adrenal
cortex. A, in a control sample the pigment consists mainly
of lipid droplets (c) and lysosomal deposits (arrow head)
and in low vitamin E sample (B) of large lipofuscin deposits
with lamellar components (arrow). Bar = 500 nm.

DISCUSSION

In the planning of this experiment it was hoped that the selenium and vitamin E serum concentrations of the experimental animals of high dietary selenium and tocopherol groups would exhibit higher serum levels than the normal animals (animals fed on normal rat food). This was not the case. The high selenium and vitamin E serum concentrations on normal amimals illustrates the high concentrations of selenium and tocopherol normally added in regular rat food. However, our dietary groups differed clearly from each other in terms of serum selenium and vitamin E concentrations.

Many authors have studied the effects of vitamin E on the life span and on lipofuscin formation (Tappel et al., 1973; Csallany et al., 1977; Porta et al., 1980; Katz et al., 1984; Elmadfa et al., 1986; Davies et al., 1987; Sarter et al., 1987; see also Sohal,1981). The general observation is that vitamin E deficiency decreases the life span and increases the accumulation of lipofuscin in many tissues. Our studies confirms these findings.

The nervous system seems to be less sensitive to the regulation of lipofuscin accumulation as an effect of dietary vitamin E. Some authors have claimed that vitamin E deficiency leads to the formation of age pigment in the central nervous system (Einarson & Telfold, 1960) while more recent studies by Porta et al. (1980), Katz et al. (1984) and Davies et al. (1987) demonstrate no significant effect of dietary levels of vitamin E on lipofuscin concentrations in the rat brain. This may be due to the heterogeneity of brain tissue, with only some cells be affected by vitamin E. A previous study by Koistinaho et al. (1989) demonstrates that in the peripheral nervous system only certain neuron populations are affected by vitamin E levels. They showed that dietary vitamin E had no effect on the accumulation of lipopigments in autonomic ganglia, whereas vitamin E deficiency increased and supplementation decreased lipofuscin accumulation in lumbar spinal ganglia (Koistinaho et al., 1989). Our own studies also showed that in the hypogastric ganglia, which contains both autonomic cell types, dietary vitamin E had no effect on pigment accumulation. Thus, it is possible that, as in the case of the central nervous system, different neuron types in the peripheral nervous system are differently dependent on the levels of plasma vitamin E.

The effects of dietary selenium on life span and on pigment accumulation have not been studied very extensively. An important discovery for aging studies was that selenium is an essential component of glutathione peroxidase (Rotruck et al., 1973), an enzyme that catalyzes the reduction of hydroperoxides and thus protects the cells from oxidative damage. A few studies have examined the effects of dietary selenium on the glutathione peroxidase system and on the accumulation of lipopigments. Behne and Wolters (1983) investigated the effects of dietary selenium on the distribution of glutathione peroxidase activity in several rat tissues. They found the highest level of glutathione peroxidase activity in the liver and erythrocytes and the lowest in the brain, adrenals and thymus. Tappel et al. (1973) included selenite in a diet containing high levels of several antioxidants. In these experiments, high levels of antioxidants were foun to reduce the amount of lipofuscin. However, the experimental protocol used did not allow the effects of individual antioxidants to be separately identified. In our experiments selenium deficiency increased mortality and gain of body weight. However, neither selenium deficiency nor supplementation had any effect on the accumulation of pigments in the adrenals, hypogastric ganglia or other tissues studied. This is in agreement with the previous studies by Koistinaho et al. (1989) who did not find any effect of dietary selenium on the accumulation rate of lipopigments in spinal or sympathetic ganglia. Although the plasma level of glutathione peroxidase correlates with the dietary level of selenium (Combs & Combs, 1984). Since the tissue activity of this enzyme varies (Behme & Wolters, 1983) it is possible that it does not play a critical role in lipopigment formation in all tissues; pigment accumulation may vary from tissue to tissue, and may not be seen for example in nervous system. Further studies should be carried out to identify the effects of selenium supplementation and deficiency at various stages during the life span to determine whether or not some critical period or selenium concentration exists where permanent cell damage results from such treatment.

REFERENCES

Blackett, A.D. and Hall, D.A., 1981, Tissue vitamin E levels and lipofuscin accumulation with age in the mouse, J Gerontol., 36:529.

Behne, D. and Jurgensen, H., 1978, Determination of trace elements in human blood serum and in the standard reference material bovine liver by instrumental neutron neutron activation analysis, J. Radioanal. Chem., 42:447.

Behne, D. and Wolters, W., 1983, Distribution of selenium and glutathione peroxidase in the rat, J. Nutrition, 113:456.

Csallany, A.S., Ayaz, K.L. and Su, L.C., 1977, Effect of dietary vitamin E and aging on tissue lipofuscin pigment concentration in mice, J. Nutr., 107,1792.

Combs, F.G. Jr. and Combs, S.B.,The nutritional biochemistry of selenium, Ann. Rev. Nutr., 4:257.

Davies, I., Davidson, Y. and Fotherinham, P., 1987, The effect of vitamin E deficiency on the introduction of age pigment in various tissues of the mouse, Exp. Gerontol., 22:127.

Desai, I.D., Fletcher, B.L. and Tappel, A., 1975, Fluorescent pigments from uterus of vitamin E deficient rats,Lipids, 10:307.

Einarson, L. and Telfold, I.R., 1960, Effect of vitamin E deficiency on the central nervous system in various laboratory animals,Biol. Skr. Dan. Vird. Selsk.,11:1.

Elmadfa, I., Both-Bendenbender, N., Sierakowski, B. and Steinhagen-Tiessen, E., 1986, Bedeutung von Vitamin E in Alter, Z. Gerontol., 19,206.

Harman, D., 1968, Free radical theory of aging: effect of free radical inhibitors on the life span of male LAF1 mice, <u>Gertontologist</u>,8:13.

Harman, D., 1981, The Aging Process,<u> Proc. Natl. Acad.Sci., USA</u>,78:7124.

Hervonen, A., Koistinaho, J., Alho, H., Helen, P., Santer, R.M. and Rapoport, S., 1986, Age related heterogeneity of lipopigments in human sympathetic ganglia, <u>Mech. Aeging Dev.</u>, 35:17.

Katz, M.L., Robinson, W.G., Herrmann, R.K., Groome, A.B. and Bieri J.G., 1984, Lipofuscin accumulation resulting from senescence and vitamin E deficiency: spectral properties and tissue distribution, <u>Mech. Age. Dev.</u>, 25:149.

Koistinaho, J., Alho, H. and Hervonen, A., 1989, Effect of vitamin E and selenium supplement on the aging peripheral neurons of the male Sparague-Dawley rat, <u>Mech. Age. Dev.</u>, in press.

Meyer, S.H., Ewan, R.C. and Beitz D.C., 1983, Effect of selenium on the subcellular distribution of glutathione peroxidase in rat liver, epidymal fat pad and seminal vesicle, <u>J. Nutrition</u>, 113:394.

Noughi, T., Cantor, A.H. and Milton, L., 1973, Mode of action of selenium and vitamin E in prevention of exudative diasthesis in chicks, <u>J Nutr.</u>, 103:1502.

Nutrient requirements of laboratory animals, 1978, National Research Council, third ed., vol 10, National Academy of Sciences, Washington D.C.

Porta, E.A., Nitta, R.T., Kia, K., Joun, N.S. and Nguyen, L., 1980, Effects of the type of dietary fat at two levels of vitamin E in Wistar male rats during development and aging, <u>Mech. Age. Dev.</u>, 13:319.

Rotruck, J.T., Pope, A.L., Ganther, H.E., Swanson, A.B., Hafeman, D.G. and Hoekstra, W.G., 1973, Selenium: Biochemical role as a component of glutathione peroxidase, <u>Science</u>, 179:558.

Santavuori, P. and Moren, P., 1977, Studies on the effects of selenium administration to neuronal ceroid lipofuscinosis of Spielmeyer-Sjögren type, <u>Neuropädiatrie</u>, 8:333.

Sarter, M. and VanDer Linde, A., 1987, Vitamin E deprivation in rats: Some behavioral and histochemical observations,<u>Neurobiol. Aging</u>, 8:297.

Siakotos, A.N., Koppang, N., Youmans, B.S. ansd Bucana,C., 1974, Blood and tissue levels of aplha-tocopherol in a disorder of lipid peroxidation, Battens disease, <u>Am. J. Clin: Nutr.</u>, 27:1152

Sohal, R.S., 1981, Age Pigments, Elsevier Biomedical Press, Amsterdam.

Sohal, R.S., Allen, R.G., Farmer, K.J., Newton, R.K. and Toy, P.L., 1985, Effect of exogenous antioxidants on the levels of endogenous antioxidants, lipid-soluble fluorescent material and life span in the housefly, musca domestica, <u>Mech. Ageing Dev.</u>, 31:329.

Tappel, A.L., Fletcher, B. and Deamer, D., 1973, Effect of antioxidants and nutrients on lipid peroxidation fluorescent products and ageing parameters in the mouse, <u>J. Gerontol.</u>, 28:415.

Vatassery, G.T., Krezowski, A.M. and Eckfelt, J.H.,1982, Vitamin E concentrations in human blood plasma and platelets, <u>Am. J. Clin. Nutr.</u>, 37:1020.

Westermark, T., Santavuori, P., Marklund, S., Pohja, P. and Salmi, A., 1982, Studies on the effects of selenium administration to neuronal ceroid lipofuscinosis patients with special reference to reduced glutathione, <u>in:</u> Ceroid-lipofuscinosis (Batten's disease), D. Armstrong, N. Koppang and J.A. Rider, eds., Elsevier Biomecical Press, Amsterdam.

DISCUSSION

ICHIKAWA: Since it has been shown that prolonged vitamin E deficiency is associated with underdevelopment of the nervous system, do you have any data on the weights of brains?

ALHO: We did not measure the weights of brains, because we freeze the brain as soon as the animals are killed.

PORTA: Your results are very interesting, and at least some of them appear to confirm what Dr. Katz and coworkers have previously shown about differences in tissue distribution between vitamin E defiency-induced ceroid and lipofuscin in aging rats (Mech. Age. Dev., 25:149, 1984). It is, however, quite intriguing that in your vitamin E deficient rats you found ceroid in the cells of the lumbar ganglia, but not in the other ganglia. Do you have any explanation for this difference?

ALHO: As you know, there are also differences in the distribution of vitamin E deficiency-induced ceroid pigment in different parts of the brain. Although hypothetically, this may be due to differences in the turnover of cells, or in their metabolic activity, it is not possible at this time to identify the real cause.

PORTA: Another interesting aspect of your results is the lack of effect of selenium deficiency on ceroid formation because this finding appears to be in line with the demonstration that the dietary supplementation of selenium to vitamin E deficient diets does not prevent the accumulation of ceroid in diverse tissues of rats (Christensen et al., Acta Pharm. Tox., 15:181, 1958; Porta et al., Lab. Invest., 18:283, 1968) and pigs (Grant and Thafvelin, Nord. Veter. 10:657, 1958). It seems logical to assume that the deficiency of vitamin E results in ceroid formation because the structural lipids become more susceptible to free radical attack and may undergo lipid peroxidation and subsequent co-polymerization with proteins. However, since there is evidence that either selenium deficiency or vitamin E deficiency increase the lipid peroxidative potentials of rat tissues to almost the same extent (Porta et al., J. Nutr., 107:1852, 1977), and yet the single deficiency of selenium is not associated with ceroid formation, it is difficult to assign an exclusive or fundamental role of lipid peroxidation in ceroidogenesis.

PALMER: The lower part of the south island of New Zealand is deficient in selenium, and although the levels of activity of glutathione peroxidase in the inhabitants is directly proportional to the soil content of selenium, this has no apparent untoward effects on the people, although it does affect animals. There is also an article by Beutler (Blood, 54:1, 1979) on glutathione peroxidase inherited disease of high incidence in the Mediterranean basin, and again it appears that nothing is wrong with these people. It seems, therefore, that on these bases, the antioxidant role of glutathione peroxidase might be questioned. Do you have any comment on this?

KATZ: I think I can respond to that. There is a large area in China where the soil is also selenium deficient, and many people in that region suffer from a cardiomyopathy called Keshan's Disease, which can be completely cured by selenium supplementation. Therefore, selenium is a required essential nutrient in man. I have a couple of questions. First, you said that vitamin E was added to the drinking water. However, vitamin E is not water soluble. So, how do you solubilize it?

ALHO: First, we solubilize in oil, and then it is emulsified in the water.

KATZ: Did you measure tissue levels of vitamin E to show that the animals were really receiving the vitamin.

ALHO: Yes, we measured serum levels of vitamin E.

KATZ: We did find an effect of selenium deficiency on lipopigment in experiments where we combined both selenium and vitamin E deficiency in a group of animals. In your study, you did not include such a group; so, it is possible that vitamin E can substitute for selenium when just vitamin E is present in the diet, but if you removed both, you may see an effect.

PORTA: Since Dr. Katz has raised the possibility that selenium deficiency might have an increasing effect on ceroid formation induced by vitamin E deficiency, I would like to make some comments and clarifications. First, to my knowledge and in agreement with Dr. Alho's results, there is no evidence that the prolonged single dietary deficiency of selenium per se induces ceroid formation in any mammalian tissue, including RPE. Second, in my opinion the question whether selenium deficiency may increase ceroid formation in tissues of vitamin E deficient animals, has not been yet resolved. The suggestion of Dr. Katz that in order to demonstrate such promoting effect it is necessary to include a group of rats deficient in both selenium and vitamin E, may appear logical if comparisons are then made between such a group and another deficient in vitamin E, but supplemented with selenium and not with a group supplemented with both nutrients, as has been done by Dr. Katz et al. in the case of ceroid in the RPE of rats (Exp. Eye Res., 34:339, 1982). However, it is important to recognize that if a diet is truly deficient in both selenium and vitamin E, it would be almost impossible to do in rats or mice a prolonged experiment as that of Dr. Alho, particularly if the feeding is started at the weaning period, because these animals will then invariably die in a matter of 2 months.

LIPOPIGMENTS IN VETERINARY PATHOLOGY: PATHOGENESIS AND TERMINOLOGY

R.D. Jolly, and R.R. Dalefield

Department of Veterinary Pathology and Public Health
Massey University
Palmerston North, New Zealand

SUMMARY

The lipopigments are a heterogenous group of pigments whose pathogenesis and terminology is confused. Whereas there is epidemiological and observational evidence that ceroid is derived from degeneration and peroxidation of unsaturated lipid, the assumption that all so-called lipopigments are similarly formed, is questioned. In particular, recent studies have distanced the pathogenesis of the pigment found in the ceroid-lipofuscinoses from that perceived for ceroid. The importance of protein rather than lipid in the pathogenesis of the pigment of ceroid-lipofuscinosis and of age pigment from the equine thyroid is noted. In the former the essential feature is storage of the DCCD binding protein subunit c of mitochondrial ATP synthase. There is a need for more analytical studies on isolated pigments which are generally more soluble than anticipated by the literature.

It is proposed that the term ceroid be limited to a family of pathological pigments where lipid degeneration and peroxidation is implied from obervational and/or epidemiological factors. The term age pigment is unequivocal and preferred for age related pigment not obviously complicated by other factors. The terms lipofuscin and lipopigment retain a usefulness as generic terms, particularly where the nature of the pigment is uncertain. The term ceroid-lipofuscinosis for the inherited storage diseases of children and animals is misleading. The term "proteolipid proteinosis" has been suggested to define this group of diseases but this is perhaps premature until their full pathogenesis is known.

INTRODUCTION

This paper concerns the terminology and pathogenesis of the lipopigments ceroid, age pigment and those of the inherited storage diseases known as the ceroid-lipofuscinoses. It is neither a definitive study nor a review of such pigments but draws on experience gained from studying a collective of cases in veterinary pathology. Some presently held concepts are questioned and the need to rationalise terminology is stressed.

KEY WORDS: Lipopigment, lipofuscin, ceroid, ceroid-lipofuscinosis, age pigment, thyroid

Lipopigments are commonly seen in a variety of diseases and as a function of age. They are described as poorly soluble yellow/brown cytosomes that fluoresce under ultraviolet light, stain for lipid in paraffin sections, are periodic acid Schiff (PAS) positive and have a tendency to be acid fast (Pearse 1985). These pigments are sometimes subdivided into age pigment and ceroid on the basis that age pigment is a normal age related phenomenon and ceroid is a pathological pigment (Porta and Hartcroft, 1969; Gedik and Totivić, 1983; Porta 1988a).

There has been a general assumption that peroxidation of lipid is a central and unifying factor in the pathogenesis of these various lipopigments. There is epidemiological evidence that this may be so for some pigments, but the hypothesis inherent in the assumption should not be taken as axiomatic in regard to the biogenesis of all so-called lipopigments. Recently there has been a questioning of the nature of the presumed fluorophore (Eldred, 1987; Eldred and Katz, 1988; Porta, 1988b) and a growing interest in the concept that protein, rather than lipid, may be the primary species involved in the genesis of some pigments (Ivy et al., 1984; Palmer et al., 1986b, 1988, 1989; Jolly et al., 1987; Davies, 1988; Dalefield et al., 1988).

CEROID

The term ceroid is a generic one for a family of lipopigments with the staining and tinctorial properties listed above that accumulate as part of a pathological process. It is used here in a more narrow sense to describe a lipopigment with these characteristics but one that clearly derives from lipid. It is typically found within macrophages but pigmentogenesis may begin prior to phagocytosis (heterophagy). For such early pigmentary material the term preceroid or interceroid has been used (Porta and Hartcroft 1969). Further maturation of pigment occurs within macrophages. In brain necroses, Schröder (1980) has described four chronological stages of pigment development on the basis of staining characteristics. Maturation in macrophages of the cerebral cortex occurred more rapidly than that in white matter. A rapidly maturing form known as hemoceroid occurs selectively in haemorrhagic brain tissue necrosis (Schröder and Reinartz 1980). Ceroid may occur extracellularly and as such maybe noted in atheroma.

The association of ceroid with degenerating lipid, diets high in unsaturated fatty acids, vitamin E deficiency and acceleration by the presence of haemoglobin (iron), implies that autooxidation plays an important role in pigmentogenesis (Porta, 1988a,b). This is supported by knowledge concerning in vitro oxidation of lipids and pigmentary changes in lipids injected or inhaled in animals (Thompson, 1969). Further experimental support for the role of unsaturated fatty acids comes from the elegant in vitro experiments involving mouse macrophages. Ceroid accumulated in those that phagocytosed artificial lipoproteins consisting of cholesteryl esters or acylglycerols of polyunsaturated fats complexed with bovine serum albumin or various polyamino acids. It did not form when monounsaturated lipid artificial lipoproteins were used (Carpenter et al., 1988).

In veterinary pathology a disease known as nutritional steatitis (paniculitis) or yellow fat disease occurs in carnivores fed diets usually derived from the fish industry, that are high in polyunsaturated fat and deficient in antioxidants (Yager and Scott, 1985; Brooks et al., 1985). In acute disease there is necrosis of adipocytes which show refractile insoluble material that stains variably with PAS and luxol fast blue, intensely with Sudan black and which fluoresces when excited by UV light

(Figure 1). This should be regarded as preceroid or interceroid (Porta and Hartcroft, 1969). Only one term is needed and preceroid is the more descriptive. Within a few days there is leucocytic infiltration with phagocytosis of preceroid by neutrophils and macrophages. At that intracellular stage it can be regarded as an early stage of ceroid formation.

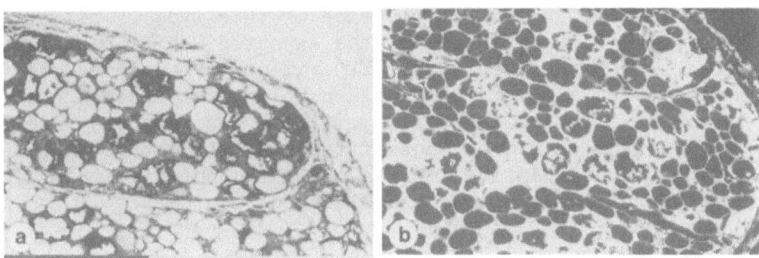

Figure 1. Preceroid in acute nutritional steatitis in a ferret (a) Sudan black, (b) Fluorescence microscopy.

Another syndrome resulting from similar dietary imbalances as well as malabsorption syndromes and pancreatitis, is the brown pigmentation of smooth muscle particularly that from the small intestine (Gedik and Tetovic, 1983; Barker and Van Dreumel, 1985; Porta, 1988a). One such outbreak concerned a kennel of dogs fed a high fat diet supplemented with corn and cotton seed oils (Cordes and Mosher, 1966). Plasma vitamin E levels were low. The epidemiological associations implied peroxidation of lipid and as such this pathological "lipofuscin" might well have been named ceroid. However its occurrence in long lived cells, the fact that prepigment precursors would be expected to enter the lysosomal system by autophagy rather than heterophagy, and a strong Schmorl's iron reduction stain, indicate it had features more in common with age pigment. It is cases such as this that lend credence to the views that lipofuscins in general, and by implication age pigment in particular, are derived from oxidised lipid.

In cats and dogs, multifocal lipid granulomas of liver and associated lymph node may show varying degrees of ceroid formation within macrophages (Figure 2). They are believed to derive from phagocytosis of lipid in fatty cysts consequent to fatty degeneration (Pritchard et al., 1983). In paraffin sections the macrophages appear partially vacuolated but insoluble contents may be yellow/brown, stain with PAS and Sudan black stains and have a yellow autofluorescence.

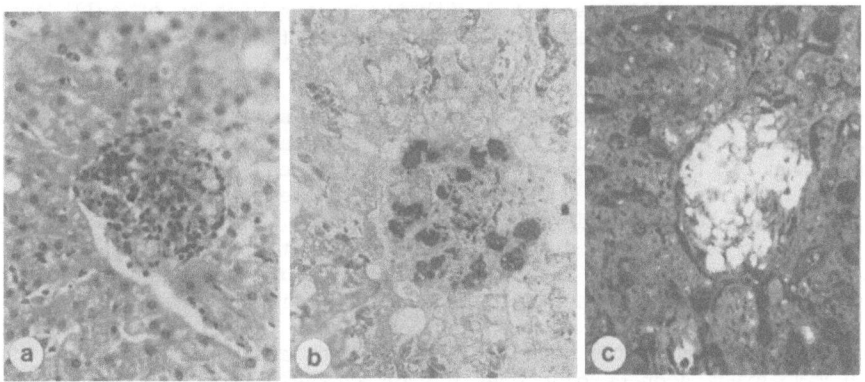

Figure 2. Ceroid in lipid granulomas in the liver of a cat (a) Haematoxylin and eosin, (b) Sudan black, (c) Fluorescence microscopy.

Ceroid may also be found within perivascular macrophages of the central nervous system in old animals or those with neuropathies. Such pigment may stain with luxol fast blue and it is implied that it derives from degenerating myelin.

Despite the strong implication of lipid peroxidation in the cases referred to above, the yellow fluorescence of this type of pigment in situ is at odds with the blue emissions of the expected Schiff bases formed by malonaldehyde cross linked proteins. Experimental Vitamin E deficiency pigment in rat livers also emits a "golden-yellow autofluorescence" in situ but both blue emitting and yellow emitting fluorophores were noted by spectroscopy after Folch extractions of isolated pigment (Porta, 1988b). A re-evaluation of the relative importance of the putative malonaldehyde derived fluorophore in pigmentogenesis is thus required.

The common denominator in the forms of pigment discussed above, is the implied peroxidation of lipid consequent to either degeneration of fat, antioxidant deficiency and/or diets high in polyunsaturated lipid. We propose that the term ceroid should be limited to lipopigments with this general aetiology so that when the term is used, the implication of these aetiological associations can be assumed. To use the term outside these limitations for all lipopigments associated with pathological phenomena is dangerous because of the implication of lipid peroxidation. To do so may distort reasoned thought and research into the pathogenesis of the pigment in question. Such perturbation of research effort is exemplified by the history of research into the so-called ceroid-lipofuscinoses, inherited storage diseases of children and a variety of animals (see below).

AGE PIGMENT

Age pigment may be found in many cells of the body but is most commonly noted in long lived post mitotic neurons and in cardiac muscle. Yellow pigment in nerve cells was first described by Hannover in 1842 and correlated with age by Koneff in 1886. It was only in 1922 that Borst introduced the name lipofuscin for this type of pigment (Porta and Hartcroft, 1969). It has also been referred to as "fat containing wear and tear" or "yellow pigment". The term lipofuscin is currently used both as a synonym for age pigment and as a generic name for lipopigments of the type discussed in this paper but not necessarily associated with age. The term age pigment has likewise been used in both contexts. It would be preferable if the latter was restricted to lipofuscin pigment which is thought to be primarily associated with the ageing process of cells and which is uncomplicated by local or environmental factors.

The yellow/brown granules of age pigment are not outstanding in routine haematoxylin and eosin stained paraffin sections of brain. However, by the use of PAS, Sudan black and other stains, its progressive accumulation can be followed from a relatively young age. During this long period, cytosomes undergo physical and/or chemical change associated with different staining and tinctorial properties (Pearse, 1985). These have generally been ascribed to progressive oxidation of lipid and/or protein with the formation of complex cross-linked malonaldehyde-protein polymers. There is no direct evidence demonstrating a link between free radical initiated lipoperoxidation and lipofuscin (age pigment) formation and accumulation but it is strongly suggested by analogy from the in vitro and in vivo studies on ceroidogenesis (Porta, 1987). However the proposed Schiff base polymers are blue emitting whereas those of age pigments emit more in the yellow range. There is thus a need to reassess the theory of lipofuscin (age pigment) formation and the chemistry of age pigment autofluorescence (Eldred, 1987; Eldred and Katz, 1988).

A new line of investigation was opened when it was shown that leupeptin and chloroquine inhibition of proteolysis lead to accumulation of lipofuscin-like inclusions in cells of brain (Ivy et al., 1984). This suggested that proteins could be a primary precursor of lipofuscin pigments. Oxidative damage to proteins and subsequent cross-linking has also been suggested as an initiating cause of pigmentogenesis (Davies, 1988). Whereas oxidatively damaged proteins are recognised and selectively degraded by proteolytic systems, their susceptibility to degradation declines with cross-linking. Such cross-linking need not necessarily be induced by oxidative damage as other methods, e.g. non enzymic glycosylation may also occur and the product may fluoresce when excited by UV light (Pangor et al., 1984)..

Support for primary damaged proteins being involved in pigmentogenesis comes from observations in our laboratory associated with age pigment in the equine thyroid gland. This tissue was chosen for study because it accumulates pigment with age and is readily available. In addition, follicular cells synthesize large amounts of a single protein thyroglobulin, secrete it and after storage as colloid, endocytose and catabolise it in their lysosomal system to form the thyroid hormones T3 and T4.

The insolubility of lipofuscins is often stated but this is relative and based mainly on lipid solvents and pigments in tissue sections. In pigment isolated from the equine thyroid gland, up to 80% of measurable protein is soluble in 3% lithium dodecyl sulphate/10% mercaptoethanol at $100^{\circ}C$ X 5min. On polyacrylamide gel electrophoresis (PAGE), the soluble fraction separates as distinct bands, some of which co-electrophorese with bands in homogenates of whole tissue. This indicates that they are not polymers or fragments of polymers resulting from covalent cross linking. The lipids of this form of age pigment approximate 10-25% and consist predominantly of cholesterol, dolichol and phospholipids with negligible triglyceride. Metals account for 1-2% of the pigment mass. The percentage of metals is variable with no obvious correlation with total amount of pigment. Their presence however is indicative that pigment cytosomes are still active lysosomes at least in the sequestration of metals. It appears that the pigment cytosomes are the repository of accumulated metals but whether there is a causal association of metals with age pigment remains to be determined. At least they are a complicating factor that deserves further study.

It was noted that in two young horses, 5 and 6 years old, pigmentogenesis was exacerbated in cells of a small percentage of follicles. These were follicles whose colloid contained clumps of granular slightly basophilic material. This colloid fluoresced intensely when excited by UV light whereas normal colloid did not [Figure 3]. The pigment in these follicles also fluoresced but with a more orange emission. It stained minimally with Sudan black and colloid not at all. This, and the known and expected composition of pigment and colloid, makes it unlikely that lipid was primarily involved in pigmentogenesis. Whether the intracellular pigment resulted from endocytosis of altered colloid or developed from intracellular components is not known. Previous electronmicroscopic observations have shown that pigment can arise from, and within, intracellular colloid droplets (Dalefield et al., 1988). Although pigmentogenesis in these follicles appears to reflect disordered cellular function and hence was a pathological process, it may reflect an exacerbation of the process by which thyroid age pigment usually forms.

Of the pigments discussed in this paper, age pigment is the least understood. Its ubiquitous presence in normal aged individuals implies

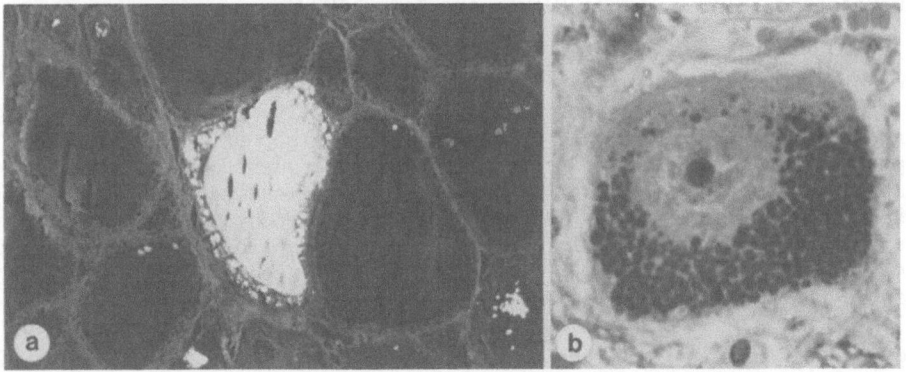

Figure 3. Fluorescence microscopy of the thyroid gland from a 5 year old horse showing a follicle containing fluorescent colloid and fluorescent pigment in follicular cells.

Figure 4. Luxol fast blue cytosomes in a neuron from a sheep with ceroid-lipofuscinosis

that it is not a result of a pathological process. That it is a normal age related phenomenon is supported by the fact that it is found in aged individuals of species with a great diversity of expected life spans. Life span is genetically programmed so it is to be expected that age pigment biogenesis also has a genetic control. Against this argument of normality are the many observations and experiments involving in vivo or in vitro systems, whereby pigment accumulation is augmented or exacerbated by a range of environmental factors. Are the myriad of molecular insults or abnormalities that lead to pigment formation genetically preordained and inevitable? If not then age pigment accumulation is an age related disease.

It is still not known whether accumulation of age pigment has a deleterious effect on cells and particularly neurons. Whereas this is of no significance in veterinary medicine, the relative time course association between advancing senility of humans and pigment accumulation warrants a working hypothesis that the two are connected. Given current research emphasis on ageing and age related diseases, age pigment research presently has a low profile or research priority. Is it because the real issues are not being addressed?

The accumulation of age pigment is a storage phenomenon that is not dissimilar to that in storage diseases due to inborn errors of lysosomal catabolism. In the latter there is an adage that the dominantly stored chemical species will reflect the underlying biochemical anomaly. There are few analytical studies reported for age pigment yet the technology and chemistry available should be able to provide useful information. There are specialised tissues such as the thyroid gland (see above) and the retinal pigment epithelium discussed elsewhere in this volume, that lend themselves to study. In these, the function and metabolism of specialised cells are well understood simplifying the experimental and deductive process. By a better understanding of pigmentogenesis in these specialised tissues, a more general unifying concept might be developed.

162

THE CEROID-LIPOFUSCINOSES

The term ceroid-lipofuscinosis was introduced as a descriptive name by Zeman and Dyken (1969) for a group of inherited storage diseases of children previously classified with the gangliosidoses as the amaurotic idiocies. They were also known by a plethora of eponymic names of which Batten's disease remains as a generic term. Three forms, infantile, late infantile and juvenile, are mainly recognised but variant forms and an adult form also exist (Lake, 1984; Dyken, 1988). Although frequently referred to as the neuronal ceroid-lipofuscinoses (NCL) because secondary degenerative disease is limited to nervous tissue, they are in fact generalised diseases. The term ceroid-lipofuscinosis derives from the histochemical and fluorescent properties of the lipopigment stored within neurons and a wide variety of other cells, which resemble those of ceroid and lipofuscin. However, they are usually colourless or only mildly yellow in routine paraffin sections but grossly brown on isolation. This pigment is frequently referred to in the literature as ceroid. By analogy it was postulated that the pathogenesis of this group of storage diseases involved abnormal peroxidation of lipid (Zeman, 1974). The reporting of a partial deficiency of a "myeloperoxidase" (Armstrong et al., 1974) initially supported this hypothesis but, it has not stood the test of time or experimentation, despite considerable investigation.

The ceroid-lipofuscinoses also occur in diverse domestic animal species of which the South Hampshire sheep has been extensively studied as a model of its human counterparts (Jolly, et al., 1980; 1988; 1989; Graydon et al., 1984; Mayhew et al., 1985). The results of systematic analyses of isolated lipopigment are reported by Palmer et al., 1986a,b; 1988, 1989 and elsewhere in this volume. They show that the lipids and their fatty acids are probably normal for a lysosome derived cytosome. Of the proteins, accounting for approximately 70% of the cytosome mass, 74% is the lipid binding subunit of mitochondrial ATP synthase known also as DCDD binding protein, subunit c or subunit 9. It is concluded that the pathogenesis of ovine ceroid-lipofuscinosis is not associated with lipid peroxidation but is in fact a proteolipid proteinosis (Palmer et al., 1989). Proteolipids are unusual proteins with a high proportion of hydrophobic amino acids that causes them to have some lipid like properties. The best known proteolipid is the intrinsic protein of myelin. As with myelin, the stored lipopigment is thought to exist as a complex three dimensional structure in association with lipids (Jolly et al., 1988, 1989). It is probably these proteolipids, perhaps with their intimate association with lipids that gives the characteristic luxol fast blue staining of both pigment (Figure 4) and myelin. We suggest that the strong Sudan black staining of this pigment may have a similar basis. Recognition that the dominantly stored species was proteolipid explained many of the recorded physical properties of this so-called lipopigment. However the intense fluorescence of pigment cytosomes in situ is not yet fully explained. On fractionation of the component parts of the pigment cytosomes, the fluorescence of the whole appears to be greatly diminished. A number of minor fluorophores only can be demonstrated in the lipid soluble fraction and it is suggested that the fluorescence of cytosomes arises from environmentally dependent fluorescence of proteins (Palmer et al., 1986a).

Recent work also summarised by Palmer et al., in this volume, has shown that the late infantile and juvenile forms of the analogous human disease also store cytosomes rich in the lipid binding subunit c of mitochondrial ATP synthase but the infantile form does not.

In retrospect, the descriptive name "ceroid-lipofuscinosis", the reference in the literature to the pigment as ceroid, and the consequential implication that ceroid was derived from peroxidation of lipid, caused undue emphasis on one aspect of research into this group of diseases. This is a telling reason for not using the term ceroid for pathological lipopigments in general.

CONCLUSION

Despite nearly a century and a half of observation, description and experimentation, the lipofuscin pigments remain an enigma. Their pathogenesis is uncertain as is their relation, if any, to age related functional changes of the nervous system. Their terminology is confused and misleading. There has been a preoccupation with the nature and significance of the fluorophore(s) and their presumed derivation from peroxidised lipid to the detriment of other aspects of research. There has been a failure to separate early events leading to residual body formation and the consequential changes associated with maturation of the pigment, both ascribed to peroxidation. The time scale of the latter does not appear to have been questioned in regard to the putative mechanism.

In recent years there have been three international meetings on the so-called ceroid-lipofuscinoses and this is the third meeting on lipofuscin pigments. However, the problems with terminology are still compounded by assumptions on pathogenesis that are not fully justified by the information available. The array of theoretical considerations and peripheral experiments that can be interpreted to support these assumptions must be supported by more direct experimentation. The insolubility of these pigments is overstated and this may account for the paucity of analytical data on isolated pigment cytosomes. Much of the assumption on relative insolubility is based on lipid solvent extractions of tissue sections. In ceroid-lipofuscinosis, pigment cytosomes are mainly sobulibilised in sodium dodecyl sulphate/mercaptoethanol (Palmer et al., 1986b; Palmer et al., this volume) and the same general methodology solubilises up to 80% or more of the proteins in equine thyroid age pigment (see above). A range of techniques are available to analyse the constituents of these various pigments. If an understanding of the nature of age pigment is to be worthwhile then it is also necessary to explore its accumulation in relation to other senile changes in neurons and their processes.

It is submitted that a uniform terminology should be adopted. The terms lipopigment and lipofuscin retain a general usefulness as generic terms to describe yellow/brown pigments that stain for lipid and which fluoresce. The addition of qualifying adjectives or phrases could be informative. The term age pigment should be used only in the context of age related pigment not obviously associated with complicating factors. It is unequivocal when used thus.

Of all the pigments, those clearly derived from degenerating lipid are the best understood in terms of aetiology. For this pigment, preceroid and ceroid are the preferred terms particularly where there is epidemiological or other inference that peroxidation of lipid is involved. To use it for any pathological lipopigment is superfluous and may be misleading; lipopigment or lipofuscin will do.

The term ceroid-lipofuscinosis as a descriptive name for a group of inherited storage diseases of children replaced a plethora of eponymic names and as such perhaps had merit. However, as discussed above, it unduly influenced the course of research. The full pathogenesis of this group of diseases has still to be determined but they are now well and truly distanced from the perceived pathogenesis of ceroid. The dominant chemical species is the DCCD binding subunit c of mitochondrial ATP synthase. These diseases have been described as proteo-lipid proteinoses (Palmer et al., 1989, Jolly et al., 1989), but this still lacks precision and perhaps it is premature to promote yet another name change. The term proteolipid is used here for highly hydrophobic proteins that extract with lipids in chloroform/methanol. The stepwise analytical approach exemplified by this study of ovine ceroid-lipo-fuscinosis has shown that not all so-called lipopigments are primarily derived from lipid. It has also demonstrated that a particular fluoro-phore is not necessarily central to understanding their pathogenesis and that they may be more soluble than anticipated.

Perusal of the literature in the past decade shows variable usage of the terms discussed in this paper to the extent that they sometimes confuse rather than clarify. A rationalisation of terminology is required in the interests of clear understanding and communication and, as far as possible, this should reflect underlying mechanisms or associations rather than a variety of conflicting historical usages. However, it should not be forgotten that these are residual body pigments, a term which exemplifies their heterogenous, and at times likely mixed nature.

ACKNOWLEDGMENTS

We would wish to acknowledge the help of colleagues who contributed to the ideas and results discussed above. The original work discussed was supported by Grant NS.11238 from the National Institute of Neurological and Communicative Disorders and Stroke, and the New Zealand Medical Research Council Grant 89/91.

REFERENCES

Armstrong, D., Dimmitt, S., and van Wormer, D.E., 1974, Peroxidase deficiency in granulocytes. Arch. Neurol., 30:144.

Barker, I.K., and Van Dreumel, A.A., 1985, The Alimentary System, in: "Pathology of Domestic Animals," Vol 2, K.V.E. Jubb, P.C. Kennedy and N. Palmer, eds., Academic Press Inc, New York.

Brooks, H.V., Rammell, C.G., Hoogenboom, J.J.L., and Taylor, D.E.S., 1985, Observations on an outbreak of nutritional steatitis (yellow fat disease) in fitch (Mustella putorius furo). N.Z. Vet.J. 33:141.

Carpenter, K.L.H., Ball, R.Y., Ardeshna, K.M., Bindman, J.P., Enright, J.H., Hartley, S.L., Nicholson, S., and Mitchinson, M.J., 1988, Production of ceroid and oxidised lipids by macrophages in vitro, in: "Lipofuscin - 1987 : State of the Art," I.Zs.-Nagy, ed., Excerpta Medica, Amsterdam, New York and Oxford.

Cordes, D.A., and Mosher, A.H., 1966, Brown pigmentation (lipofuscinosis) of intestinal muscularis. J. Pathol. Bacteriol., 92:197.

Dalefield, R.R., Jolly, R.D., Craig, A.A., Martinus, R.D., and Palmer, D.N., 1988. Age pigment in the thyroid of aged horses, in: "Lipofuscin - 1987 : State of the Art," I.Zs.-Nagy, ed., Excerpta Medica, Amsterdam, New York and Oxford.

Davies, K.J.A., 1988, Protein oxidation, protein cross linking and proteolysis in the formation of lipofuscin: rationale and methods for the measurement of protein degradation, in: "Lipofuscin - 1987: State of the Art," I.Zs.-Nagy, ed., Excerpta Medica, Amsterdam, New York and Oxford.

Dyken, P.R., 1988, Reconsideration of the classification of the neuronal ceroid-lipofuscinoses, Amer. J. Med. Genet. Supp. 5:69.

Eldred, G.E., 1987, Questioning the nature of the fluorophores in age pigments, in: "Advances in Age Pigment Research," E. Aloj Totara, P. Glees, and F.A. Pisanti, eds., Pergamon Press, Oxford and New York.

Eldred, G.E., and Katz, M.L., 1988, Possible mechanism for lipofuscinogenesis in the retinal pigment epithelium and other tissues, in: "Lipofuscin - 1987 : State of the Art," I.Zs.-Nagy, ed., Excerpta Medica, Amsterdam, New York and Oxford.

Gedik, P., and Totivić, V., 1983, Lysosomes and lipopigments, in: "Cellular Pathobiology of Human Disease," B.F. Trump, A. Laufer and R.T. Jones, eds., Gustav Fischer, New York.

Graydon, R.J., and Jolly, R.D., 1984, Ceroid-lipofuscinosis (Batten's disease): Sequential electrophysiologic and pathologic changes in the retina of the ovine model, Invest. Ophthalmol. Visual Sci., 25:294.

Ivy, G.O., Schottler, F., Wenzel, J., Baudry, M., and Lynch, G., 1984, Inhibitors of lysosomal enzymes : accumulation of lipofuscin-like dense bodies in the brain, Science, 226:985.

Jolly, R.D., Janmaat, A., West, D.M., and Morrison, I., 1980, Ovine ceroid-lipofuscinosis - a model of Batten's disease, Neuropathol. Appl. Neurobiol. 6:195.

Jolly, R.D., Barns, G., Bube, A., and Palmer, D.N., 1987, Ovine ceroid-lipofuscinosis: Chemical constituents of the lipopigment, their pathogenic significance and similarities to age pigment, in: Advances in Age Pigment Research," E. Aloj Totara, P. Glees, and F.A. Pisanti, eds., Pergamon Press, Oxford and New York.

Jolly, R.D., Shimada, A., Craig, A.S., Kirkland, K.B., and Palmer, D.N., 1988, Ovine ceroid-lipofuscinosis II: Pathology interpreted in light of biochemical observations, Amer. J. Med. Genet. Supp. 5:159.

Jolly, R.D., Shimada, A., Dopfmer, I., Slack, P.M., Birtles, M.J., and Palmer, D.N., 1989, Ceroid-lipofuscinosis (Batten's disease): Pathogenesis and sequential neuropathological changes in the ovine model, Neuropathol. Appl. Neurobiol. In Press.

Lake, B.D., 1984, Lysosomal enzyme deficiencies, in: "Greenfields Neuropathology," J. Hume Adams, J. Corsellis and L.W. Duchen, eds., 4th edit., Edward Arnold, London.

Mayhew, I.G., Jolly, R.D., Pickett, B.T., and Slack, P.M., 1985, Ceroid-lipofuscinosis (Batten's disease): Pathogenesis of blindness in the ovine model, Neuropathol and Appl. Neurobiol. 11:283.

Palmer, D.N., Husbands, D.R., Winter, P.J., Blunt, J.W., and Jolly, R.D., 1986a, Ceroid-lipofuscinosis in sheep I. Bis(monoacylglycero)-phosphate, dolichol, ubiquinone, phospholipids, fatty acids, and fluorescence in liver lipopigments, J. Biol. Chem. 261:1766.

Palmer, D.N., Barns, G., Husbands, D.R., and Jolly, R.D., 1986b, Ceroid-lipofuscinosis in sheep II. The major component of the lipopigment in liver, kidney, pancreas, and brain is low molecular weight protein, J. Biol. Chem. 261:1773.

Palmer, D.N., Martinus, R.D., Barns, G., and Jolly, R.D., 1988, Ovine ceroid-lipofuscinosis: Lipopigment composition is indicative of a lysosomal proteinosis, Amer. J. Med. Genet. Supp. 5:141.

Palmer, D.N., Martinus, R.D., Cooper, S.M., Midwinter, G.G., Reid, J.P., and Jolly, R.D., 1989, Ovine ceroid-lipofuscinosis. The major lipopigment protein and lipid binding subunit of mitochondrial ATP synthase have the same NH_2-terminal sequence, J. Biol. Chem. 264:5736.

Pangor, S., Ulrich, P.C., Benesath, F.A., and Cerami, A., 1984, Aging in proteins : isolation and identification of a fluorescent chromophore from the reaction of polypeptides with glucose, Proc. Nat. Acad. Sci. U.S.A., 81:2684.

Pearse, A.G.E., 1985, Histochemistry Theoretical and Applied, Vol. 2, Churchill Livingstone, Edinburgh, London.

Porta, E.A., 1987, Tissue lipoperoxidation and lipofuscin accumulation as influenced by age, type of dietary fat and levels of Vitamin E in rats, in: "Advances in Age Pigment Research," E.Aloj Totaro, P. Glees, and F.A. Pisanti, eds., Pergamon Press, Oxford and New York.

Porta, E.A., 1988a, Role of oxidative damage in the ageing process, in: Cellular Antioxidant Defence Mechanisms," C.K. Chow, ed., Vol. 3, C.R.C. Press, Inc., Boca Raton, Florida.

Porta, E.A., 1988b, Differential features between lipofuscin (age pigment) and various experimentally produced "ceroid pigments", in: "Lipofuscin - 1987: State of the Art", I.Zs.-Nagy, ed., Excerpta Medica, Amsterdam, New York and Oxford.

Porta, E.A., and Hartcroft, W.S., 1969, Lipid pigments in relation to aging and dietary factors (Lipofuscins), in: "Pigments in Pathology," M. Wolman, ed., Academic Press Inc., New York and London.

Pritchard, D.H., Jolly, R.D., Howell, L.J., and Fairley, R.A., 1983, Ceroid-lipidosis : An acquired storage type disease of liver and associated lymph node. Vet. Pathol., 20:242.

Schroder, R., 1980, The lipopigments in human brain tissue necroses : I. Ceroid, Acta Neuropathol. (Berl), 52:141.

Schröder, R., and Reinartz, B., 1980, The lipopigments in human brain tissue necroses : II. Hemoceroid, Acta Neuropathol. (Berl), 52:147.

Smith, H.A., and Jones, T.C., 1966, Veterinary Pathology, 3rd Edit., Bailliere, Tindall and Cassell, London.

Thompson, S.W., 1969, Lipogenic pigments related to treatment with exogenous lipid, in: "Pigments in pathology," M. Wolman, ed., Academic Press Inc., New York and London.

Yager, J.A., and Scott, D.W., 1985, The skin and appendages, in: "Pathology of Domestic Animals," Vol. 1, J.V.F. Jubb, P.C. Kennedy and N. Palmer, eds., Academic Press Inc., New York.

Zeman, W., 1974, Presidential address: Studies on the neuronal ceroid-lipofuscinoses, J. Neuropathol. Exp. Neurol., 33, 1.

Zeman, W., and Dyken, P., 1969, Neuronal ceroid-lipofuscinosis (Batten's disease): Relationship to amaurotic familial idiocy?, Pediatrics, 44:570.

DISCUSSION

IVY: In my previous studies on leupeptin done in collaboration with Dr. Wolfe (Soc. Neurosci. Abstr., 10:885, 1984) we found increased dolichols in the brain correlating with the increased lipopigment-like substances. This is similar to the increases of dolichols found in aging and other neurological diseases. This is not surprising since dolichol is associated with lysosomal membranes, and the pigment is accumulating in lysosomes. What I do find surprising is that in other conditions, as for example in Tay-Sachs disease, dolichols do not increase in affected tissues. I noticed in the graph you presented that in some instances dolichols went down. Why didn't the dolichols increase with ceroid?

JOLLY: I don't exactly know why, but Dr. Porta has done some studies on this, and he can tell us.

PORTA: Well, this may depend on the model used for ceroid formation and on the tissue where ceroid pigments accumulate. In the epididymal fat pad crushed model of rats, we found in fact that total dolichols substantially increased and somewhat correlated with a substantial increase in ceroid pigments in this fat tissue (In: Lipofuscin-1987: State of the Art, Zs.Nagy [ed.], Excerpta Medica, Amsterdam, 1988). In the liver of vitamin E deficient rats, however, relatively less amounts of ceroid accumulated, and the levels of total dolichols were not significantly higher than those in the supplemented animals. I will expand on this in my presentation.

DAWSON: You seem to be implying that the fluorescence seen in Batten's disease is somehow peripheral to the disease, and this appears to be a confusing issue. What then causes the autofluorescence in the Batten's disease in humans, sheep, and dogs? Do you really think this is unrelated to the main pathogenesis of the disease?

JOLLY: I think that the main pathogenesis is the specific storage of the subunit C of mitochondria synthase, but we did not find that by showing that this substance is fluorescent. The implications of our work is that fluorescence comes from this particular subunit in a particular lipid environment. Eldred and Katz have questioned this interpretation, and I am not going to argue with the fluorescent chemists. Wherever the fluorescence comes, it seems to be very minor. Maybe if we knew what is making the fluorescence in precise detail, we may add knowledge to the story. But, in my opinion, there has been an excessive preoccupation in chasing the fluorophores in the pigment material. As you know, there has been an assumption, later proved wrong, that malonaldehyde binding to proteins was forming the fluorescent polymers. All this preoccupation did not add to much about the pathogenic mechanisms in the pigment formation.

WITKOP: It seems to me that in the classification and terminology of these autofluorescent pigments I have been walking in circles. Initially, I called the pigment in Hermansky-Pudlak syndrome, lipofuscin. Then, I was corrected and supposed to call it ceroid, because it was present in an abnormal or disease state. Now, I come back here and found out that I cannot call it ceroid anymore. Anyway, I like your proposed terminologic scheme, except for one thing. I don't like to call it lipofuscinosis, because I don't think that the evidence for lipid is very good. If you feed our patients with saturated or unsaturated fat diets, there is no difference in the amount of ceroid accumulating in tissues or being excreted in the urine. The material that accumulates in Hermansky-Pudlak syndrome and which can be isolated, does not stain for fat, but still fluoresces, still is PAS positive, but we cannot find any fat component in it.

JOLLY: I could not agree more with you.

GOEBEL: You said that the electron lucent parts of lipofuscin and also in the pigment of the thyroid gland are not lipids. What do you think they are?

JOLLY: I think they are endocytic vacuoles which contain partly degraded colloid. The reason I don't think they are lipid is that the density of these pigment granules is such that these vacuoles remain after chloroform extraction, and in all the chemical analysis, we only see cholesterol and dolichols.

EFFECTS OF LOVASTATIN AND LEUPEPTIN ON CEROIDOGENESIS OF VITAMIN E-DEFICIENT AND-SUPPLEMENTED YOUNG RATS

Eduardo A. Porta[1], Alberto J. Monserrat[2], Alejandro Berra[2] and Modesto C. Rubio[3]

[1]Dept. Pathology, Sch. Med., Univ. Hawaii, Honolulu, HI, USA.
[2]Center of Experimental Pathology, Faculty of Medicine, Univ Buenos Aires, Argentina. [3]Inst. Pharmacol. Investigations CONICET, Buenos Aires, Argentina

SUMMARY

Previous studies in young normal rats have shown that intracerebral administration of the proteinase inhibitor, leupeptin, caused a rapid accumulation of lipofuscin-like pigment in lysosomes of brain cells (Ivy et al., 1984a). On the other hand, we have recently found that the administration of lovastatin, an inhibitor of HMG-CoA reductase, reduced the ceroid-like pigment and dolichol contents in the crushed epididymal fat pad of rats (Porta et al., 1988). In order to study now the possible modulating effects of these enzyme inhibitors on ceroidogenesis associated with vitamin E deficiency, two main groups of weanling Wistar female rats were respectively fed ad libitum a vitamin E-deficient basal diet, or the same diet supplemented with 16 mg% of dl - α- tocopherol acetate. The vitamin E-deficient and -supplemented rats were further subdivided and received for 8 weeks their diets alone or with 2, 1, or 0.5 g of lovastatin/kg of diet. Other subgroups were treated with constant peritoneal infusion of 0.5 mg/day of leupeptin by means of osmotic minipumps (Alzet 2002) consecutively implanted at days 15, 30, and 45.

Lovastatin treatment to vitamin E-deficient rats was associated with dose-dependent toxicity, resulting in 100%, 75%, and 50% mortality at concentrations of 2, 1, and 0.5 g/kg diet, respectively. This mortality was mainly due to extensive hepatic necrosis. Food intake and growth rates were reduced, while the relative weights of liver, kidneys, spleen, heart and brain, as well as the serum levels of GPT and GOT were significantly increased over the values of the untreated vitamin E-deficient control rats. The volumetric densities of ceroid pigment and the dolichol contents in liver and kidneys were not significantly modified. Lovastatin toxicity was partially prevented by vitamin E supplementation. However, in these supplemented rats, lovastatin treatment did not modify the volumetric densities of hepatic and renal ceroid, although the contents of hepatic and renal dolichol were significantly increased. No correlations could be found between levels of hepatic or renal ceroid and total dolichol content in vitamin E-deficient and supplemented rats.

Leupeptin treatment to vitamin E-deficient rats only slightly reduced food intake and growth rates, and did not significantly modify the relative organ weights or the serum levels of cholesterol, GOT and GPT. Although in both vitamin E-deficient and -supplemented rats the leupeptin treatment consistently showed a tendency to increase the volumetric densities of hepatic and renal ceroid pigment, the differences with the control untreated rats were not statistically significant. The levels of hepatic and renal dolichol contents were not significantly modified by leupeptin treatment in vitamin E-deficient or -supplemented rats.

KEYWORDS: Ceroid, lovastatin, leupeptin, vitamin E deficiency, dolichol, liver, kidneys, serum enzymes.

INTRODUCTION

Primary or secondary vitamin E deficiency typically results in the accumulation of ceroid pigment in diverse animal tissues (Porta and Hartroft, 1969). In the liver of mice and rats fed vitamin E-deficient diets, ceroid predominantly accumulates, at least initially, in kupffer cells (Mason and Emmel, 1945; Porta and Hartroft, 1963; Maeda, 1967; Dabholkar and Ogawa, 1978), while in the kidney the pigment is mainly found in the epithelial cells of proximal convoluted tubules (Mason and Emmel, 1945; Mason et al., 1946). Ultrastructurally, the ceroid granules are localized in lysosomes and residual bodies (Porta and Hartroft, 1964; Lampert et al., 1964; Howes et al., 1964).

Early studies indicated that ceroid granules of vitamin E-deficient animals displayed a yellow brown natural color, and when excited with UV light emitted a yellow-bronze fluorescence (Martin and Moore, 1939; Moore and Wang, 1943, 1947). In the retinal pigment epithelium (RPE) of vitamin E-deficient rats, ceroid showed a fluorescent peak emission from 590 to 650 nm, which corresponded to the yellow-orange region of the visible spectrum (Katz et al., 1984). Although in tissue sections the fluorescence of mature ceroid is not appreciably reduced by polar and nonpolar solvents (Moore and Wang, 1947; Hartroft and Porta, 1965), at least some of the fluorophores from the intact RPE of vitamin E-deficient rats have been extracted with chloroform-methanol, and chromatographically separated into several components (Katz and Eldred, 1989). Furthermore, some water-methanol and chloroform soluble fluorophores have been recently extracted from ceroid granules isolated form livers of vitamin E-deficient mice (Porta et al., 1988). It should be noted, however, that in both tissues the major fractions of ceroid fluorophores are not solubilized by chloroform-methanol mixtures. Although autofluorescence is one of the most specific and consistent physical properties of ceroid, the fluorophores constitute only a very small fraction of the whole granules.

Our present information on the chemical composition of vitamin E deficiency-related ceroid is quite limited and has been derived from the histochemical reactions of this pigment in situ (Mason and Emmel, 1945; Mason et al., 1946; Elftman et al., 1949; Einarson, 1953; Gedigk and Fischer, 1959). While on this basis is generally accepted that the pigment granules essentially consist of a variable mixture of oxidized and polymerized lipids and proteins, the detailed composition of these major constituents has not been yet studied.

In addition to ceroid, vitamin E deficiency produces structural and functional alterations in the cell membranes of many animal tissues and organs, and the available data on these alterations, as well as on the function of vitamin E, permits to conceive the pathogenic factors involved in ceroidogenesis. Since vitamin E is the major chain breaking antioxidant in the body (Burton and Ingold, 1986), and is an integral part of all cell membranes (Diplock and Lucy, 1973), the deficiency of this vitamin renders the membrane constituents vulnerable to oxidative damage. Before undergoing substantial secondary modifications, the peroxidized lipids can be catabolized by membrane-associated phospholipases and transformed by cytosolic glutathione peroxidase (Sevanian et al., 1983), while the peroxidized proteins are breakdown in part by cytosolic proteinases (i.e. calpain), but mostly by lysosomal proteinases (Barret, 1980). Although oxidative damage increases enzymatic proteolysis (Wolff et al, 1983; Davis, 1988), and perhaps lipolysis, there is also a progressive increase in protein cross linking and formation of protein oligomers which further react with lipid oxidation products to form increasingly undigestible lipoprotein aggregates (ceroid) in lysosomes.

In several tissues of vitamin E-deficient animals, and predominantly in dystrophic muscles, there is a significant increase in the activity of many lysosomal enzymes such as cysteine-proteinases, β-glucoronidase, ribonuclease, aryl sulphatase, acid phosphatase (Zalkin et al., 1962; Bunyan et al., 1967a), as well as in cytosolic calpain (Dayton et al, 1979). It would appear, therefore, that the formation and aggregation of ceroid may not be due to any obvious enzyme inhibition. However, it has been recently reported that in vitamin E-deficient rats the serum level of a cysteine proteinase inhibitor was significantly increased (Minakata et al., 1984). Although the possible implications of this inhibitor on ceroidogenesis is presently unknown, it has been shown in recent years that the administration of leupeptin, a well known inhibitor of cysteine proteinases, produced in normal rats the accumulation of substantial amounts of lipofuscin-like pigment in lysosomes of brain cells (Ivy et al.,

1984a; Ivy and Gurd, 1988). We decided, therefore, to determine whether the leupeptin treatment could modulate ceroidogenesis in liver and kidneys of vitamin E-deficient rats.

Another aspect explored in the present studies was the possible influence of lovastatin. This widely used hypocholesterolemic compound in man is a competitive inhibitor of 3-hydroxy-3-methylglutaryl coenzyme A (HMG-CoA) reductase, the rate limiting enzyme that catalyzes the conversion of HMG-CoA to mevalonate, which in turn, is an early step in the formation of cholesterol and other isoprenoid compounds, such as ubiquinone and dolichol (Alberts et al., 1980). It has been reported, for example, that lovastatin reduced the cholesterol and dolichol synthesis in rat hepatocytes in vitro (Keller, 1986), and the levels of ubiquinone in cardiac homogenates of guinea pigs (Guillory, personal communication), and in brain homogenates of dogs in vivo (Berry et al., 1988). Our interest in lovastatin stems from the observations that dolichol is present in nearly all cell membranes, particularly in lysosomes (Wong et al, 1982), and has been found in high proportions in the ceroid granules isolated from cerebral cortex of patients with Batten's disease (Ng Ying Kin et al., 1983), and from liver, kidney, pancreas and brain of sheep similarly affected (Palmer et al., 1986a, b). It has been also reported that the levels of dolichol increased with age in the brain of normal human subjects and rats (Wolfe et al., 1982; Pullarkat and Reha, 1982; Sakakihara and Volpe, 1985; Ng Ying Kin et al., 1983), and in various tissues of mice (Pullarkat et al., 1984). Additionally, increased brain levels of dolichol have been found in the brain of patients with Alzheimer's disease (Wolfe et al, 1982), and rats treated with leupeptin (Ivy et al., 1984b). It has been considered, therefore, that the increased levels of tissue dolichol could be due to corresponding increases in the amounts of lipofuscin and ceroid pigments. Finally, we have recently found that dolichol increased in the crushed epididymal fat pad of rats in which ceroid is rapidly formed, and that the treatment of these animals with lovastatin substantially reduced the level of dolichol and the amount of ceroid (Porta et al., 1988).

MATERIAL AND METHODS

Female Wistar weanling rats (34.51 ± 0.66 g initial body weight, Mean ± SEM) were randomly assigned to two main dietary groups, and were offered ad libitum either a semisynthetic basal diet deficient in vitamin E, or the same diet supplemented with vitamin E. These dietary groups were further subdivided and received for 8 weeks their respective diets alone, or supplemented with either 2, 1, or 0.5 g of lovastatin per kg of diet, or were treated with constant peritoneal infusion of 0.5 mg/day of leupeptin by means of osmotic minipumps (Alzet 2002, Alza Corp., Palo Alto, CA) consecutively implanted at days 15, 30, and 45.

The basal vitamin E-deficient diet (Table 1) originally formulated by Drapper et al. (1964), was used in powdered form (U.S. Biochemical Corp., Cleveland, OH), and the supplementation was done by adding 160 mg of dl-α-tocopherol acetate per kg of diet. The lovastatin ((Merk Sharp & Dohme, Rahway, NJ) was incorporated into the diets of the two main dietary groups at the above mentioned levels. Leupeptin (U.S. biochemical Corp.) was diluted with saline to a concentration of 40 mg/ml, and each minipump was filled with 8 mg of leupeptin. These pumps deliver approximately 0.5 mg of leupeptin per day.

All the animals had free access to drinking tap water, and were individually housed in suspended wire-bottomed cages in an air conditioned room (~25⁰ C, 50% relative humidity). Body weights and food consumption were recorded daily.

At sacrifice, blood was drawn from the aorta, and analyzed for total cholesterol (Allain et al., 1974), and for GOT and GPT (Reitman and Frankel, 1957). A complete autopsy was performed and weights of liver, kidneys, brain, spleen and heart were recorded. Total dolichol was determined by reverse-phase HPLC (Yamada et al, 1985) in livers and kidneys. For the visualization of ceroid, sections of liver and kidneys were fixed in buffered formalin, and stained with HE, Oil red O, Ziehl Neelson and PAS. Unstained sections were examined under fluorescent microscopy. Since in the liver ceroid was almost exclusively found in kupffer cells, and in kidneys predominantly in the epithelial cells of segments 1 and 2 of the proximal convoluted tubules, the volumetric densities of ceroid granules were only determined in these cells using the method of Reichel et al. (1968) in sections stained with PAS. Although similar determinations of ceroid were done in unstained sections under fluorescent microscopy, and the values were comparable to those obtained in PAS-stained sections, only data from the latter will be presented.

Statistical analysis of numerical data was done by the Duncan's test (1955).

Table 1. Composition of the Basal Diet[1]   %

Vitamin Free Casein	20.0
Tocopherol Stripped Corn Oil	10.0
Glucose Monohydrate	65.0
Salt Mixture[2]	4.0
Vitamin Mixture[3]	1.0

[1]U.S. Biochemical Corp., Cleveland, OH. [2]Contains: (g/kg) NaCl, 108.09; $K_3C_6H_5O_7 \cdot H_2O$ 236.53; K_2HPO_4, $CaHPO_4 \cdot 2H_2O$, 355.56; $CaCO_3$, 163.56; $MgCO_3$, 40.89; $FeC_6H_5O_7 \cdot 3H_2O$, 16.00; $CuSO_4 \cdot 5H_2O$, 0.18; $MnSO_4 \cdot H_2O$, 1.38; KI, 0.04; $ZnCO_3$, 0.44. [3]Contains: (g/kg) Ascorbic acid, 45.00; Choline Chloride, 75.00, Calcium Pantothenate, 3.0; Inositol, 5.0; Menadione, 2.25; Niacin, 4.5; PABA, 5.0; Pyridoxine HCl, 1.0; Riboflavin, 1.0; Thiamine HCl, 1.0, Vit. A Acetate, 900.000 U.; Calciferol, 100.000 U., and (in ug/kg) Biotin 20; Folic Acid, 90, and Vit. B-12, 1.35.

RESULTS

Lovastatin Studies

Mortality

In vitamin E-deficient rats, the lovastatin treatment at the levels of 2, 1, and 0.5 g/kg of diet was associated with 100%, 75%, and 50% mortality, respectively. In each instance, death was mainly due to extensive and massive hepatic necrosis, and no significant pathologic changes were seen in all the other organs examined grossly or macroscopically. There was no mortality in the untreated vitamin E-deficient and supplemented rats, or in the supplemented ones treated with lovastatin. Because there were no survivors in the vitamin E-deficient rats treated with 2 g of lovastatin per kg of diet, no morphologic or biochemical consideration could be done with other groups, and this group, as well as the group supplemented with vitamin E and treated with lovastatin at the level of 2 g/kg of diet, were eliminated from the study.

Food Consumption (Table 2)

The averages of absolute daily food intakes of the vitamin E-deficient and supplemented untreated rats were not significantly different. Lovastatin treatment reduced the absolute food intakes in vitamin E-deficient rats, but not in the supplemented ones. For example, the average absolute daily intake of the vitamin E-deficient rats treated with 1 g of lovastatin per kg of diet was significantly lower than in untreated vitamin E-deficient rats ($p < 0.005$), and lower than in vitamin E-supplemented untreated rats, or treated with 1 and 0.5 g of lovastatin per kg of diet ($p < 0.025$). Figure 1 shows the absolute food intakes of the treated and untreated vitamin E-deficient groups throughout the experiment. The averages of relative daily food intakes were also included in Table 2, because these data served to calculate the actual average daily intakes of lovastatin per kg of body weight. Thus, in vitamin E-deficient rats fed the diets containing 1 or 0.5 g of lovastatin per kg, the averages of daily lovastatin intakes were 90 mg and 52 mg per kg of body weight, respectively. In vitamin E-supplemented rats fed the diets containing 1 g or 0.5 g of lovastatin, the averages of daily lovastatin intakes were 108 mg and 58 mg per kg of body weight, respectively.

Body Weights (Figures 2 and 3)

All groups gained weight, but at different rates. In vitamin E-deficient rats treated with lovastatin at the level of 1 g/kg diet, the average daily growth rate, and the final body weight (0.75 g/day, and 88.00 ± 10.00 g) were significantly lower ($p < 0.001$)

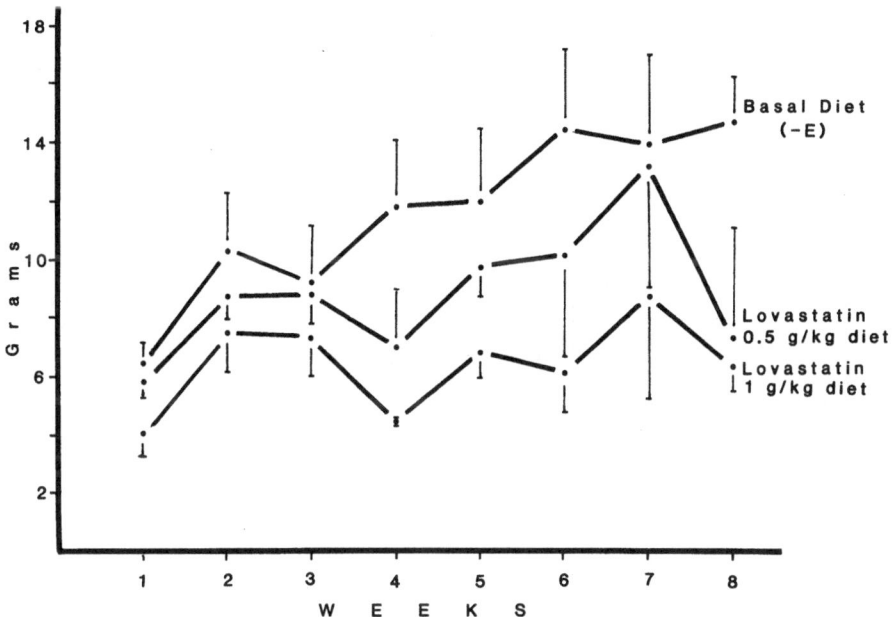

Fig. 1. Changes in abolute food intakes in lovastatin-treated and untreated vitamin E-deficient rats.

Table 2. Effects of Lovastatin Treatment on Absolute and Relative Food Intakes

Vitamin E-Deficient Rats		
Treatment	Absolute Intake[1]	Relative Intake[2]
Untreated (6)	11.11 ± 1.11	9.21 ± 0.51
Lovastatin 1 g/kg (2)	6.07 ± 0.56[3]	8.99 ± 0.71
Lovastatin 0.5 g/kg (4)	8.30 ± 1.10	10.36 ± 0.97
Vitamin E-Supplemented Rats		
Untreated (5)	9.33 ± 0.84	9.56 ± 1.40
Lovastatin 1 g/kg (4)	10.10 ± 1.10	10.77 ± 0.75
Lovastatin 0.5 g/kg (4)	10.79 ± 1.15	11.67 ± 1.32

[1]g/day. [2]g/100 body weight/day. [3]Significantly lower than in untreated Vitamin E-Deficient rats (p < 0.025) and than in Vitamin E-Supplemented untreated rats or treated with lovastatin at the level of 0.5 g/kg of diet (p < 0.025) Number of rats in parenthesis.

than those of the corresponding untreated control group (2.54 g/day, and 113.75 ± 15.77 g) were also significantly lower (p < 0.005) than those of untreated rats.

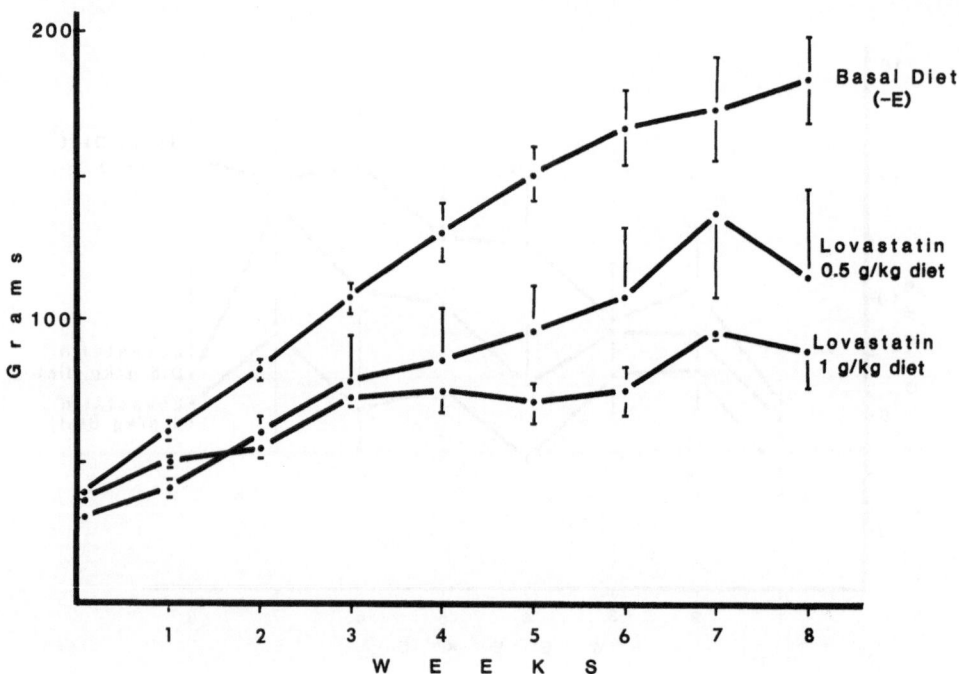

Fig. 2. Changes in body weights in lovastatin-treated and untreated vitamin E-deficient rats.

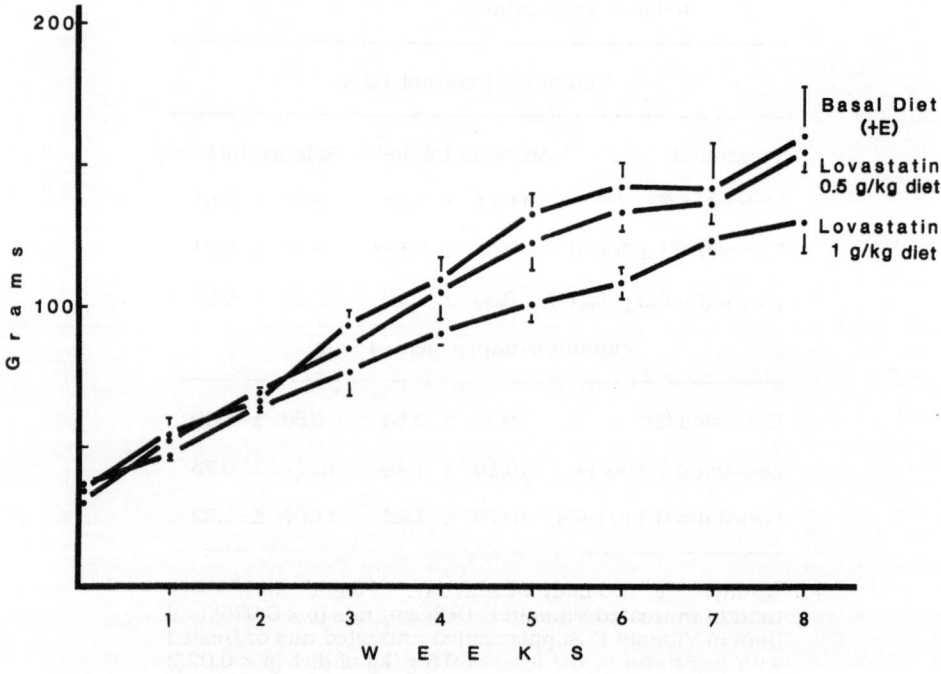

Fig. 3 Changes in body weights in lovastatin-treated and untreated vitamin E-supplemented rats.

In vitamin E-supplemented rats treated with lovastatin at the level of 1 g/kg diet, the average daily growth rate and final body weight (1.70 g/day, and 129.00 ± 5.14 g) were significantly lower than those in the corresponding untreated control group (2.20 g/day, and 158.00 ± 10.57 g) However, the growth rate and final body weight of the group treated with lovastatin at the level of 0.5 g/kg diet (2.23 g/day, and 154.00 ± 3.58 g) were not significantly different from those in the untreated control group.

Organ Weights (Table 3)

In vitamin E-deficient rats treated with lovastatin at the level of 1 g or 0.5 g/kg diet, the relative weights of all the organs were heavier than in the untreated rats. In the vitamin E-supplemented rats treated with lovastatin at the level of 1 g/kg diet, the relative weights of the organs were not significantly different from those of the untreated control rats, but in those treated with lovastatin at the level of 0.5 g/kg diet, the relative weights of liver, kidneys, and hearts were significantly heavier than in the controls.

Volumetric Densities of Ceroid (Figures 4 and 5)

Neither in the vitamin E-deficient rats nor in the supplemented ones, the lovastatin treatment at the two levels significantly influenced the volumetric densities of ceroid granules in hepatic kupffer cells or in the renal epithelial cells of proximal convoluted tubules. Somewhat unexpectedly, it was also found that the volumetric densities of ceroid in kupffer and renal cells of the untreated vitamin E-deficient and supplemented groups did not differ significantly.

Serum GOT, GPT and Cholesterol (Table 4)

In vitamin E-deficient rats, lovastatin treatment at 1 g and 0.5 g/kg diet significantly increased the serum levels of activity of GOT and GPT over the value of the corresponding untreated rats. However, in vitamin E-supplemented rats, differences between treated and untreated groups were not statistically significant. The serum level of activity of GOT in the untreated vitamin E-deficient rats was significantly higher than that in the untreated rats supplemented with vitamin E, while the levels of GPT did not differ significantly between these two groups.

In vitamin E-deficient rats, the serum level of cholesterol was significantly reduced by lovastatin treatment at the level of 1 g/kg diet, but was significantly increased by lovastatin at the level 0.5 g/kg diet. In vitamin E-supplemented rats, the level of serum cholesterol in rats treated with lovastatin at the level of 1 g/kg diet was not significantly different from that in the untreated control group. No significant differences in serum levels of cholesterol were found between untreated vitamin E-deficient and supplemented rats.

Hepatic and Renal Total Dolichol (Table 5)

Although in vitamin E-deficient rats treated with lovastatin at 1 g or 0.5 g/kg diet the levels of hepatic and renal dolichol were higher than in the untreated control rats, the differences were not statistically significant. Conversely, in vitamin E-supplemented rats treated with lovastatin at 1 g or 0.5 g/kg diet, the hepatic and renal levels of dolichol were significantly higher than in the corresponding control group. In the untreated rats, the level of hepatic and renal total dolichol in vitamin E-deficient rats did not differ statistically from that in the supplemented ones.

Leupeptin Studies

Mortality

One rat from the vitamin-deficient group and another from the supplemented group died during surgical implantation of the minipumps into the abdominal cavity.

Food Consumption (Table 6)

The leupeptin treatment did not affect the absolute or relative food intakes of vitamin E-deficient and supplemented groups.

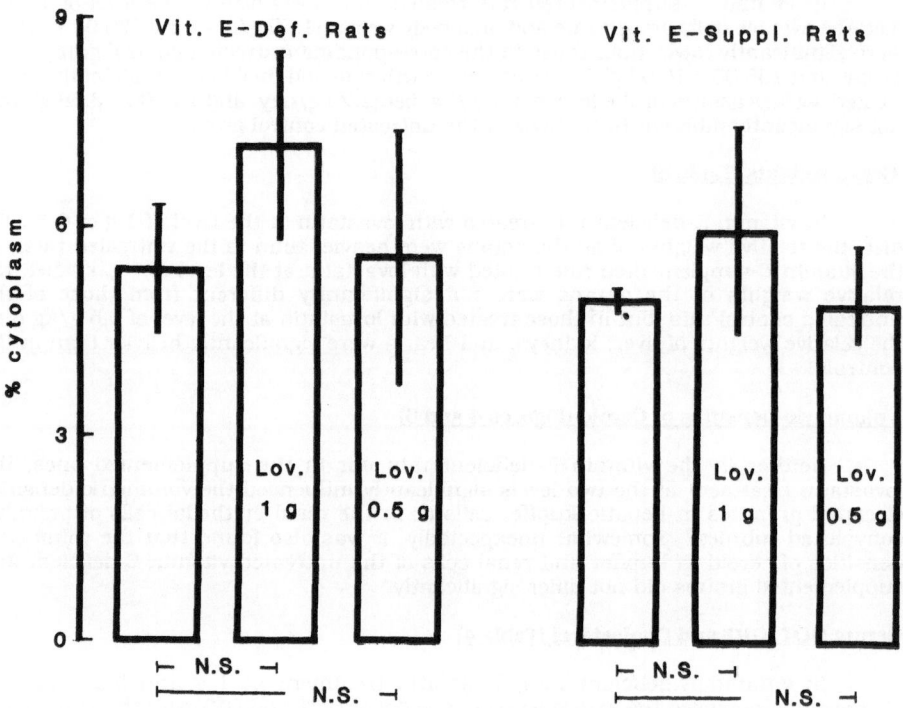

Fig. 4. Volumetric densities of ceroid granules in the cytoplasm of kupffer cells in lovastatin-treated and untreated vitamin E-deficient and -supplemented rats.

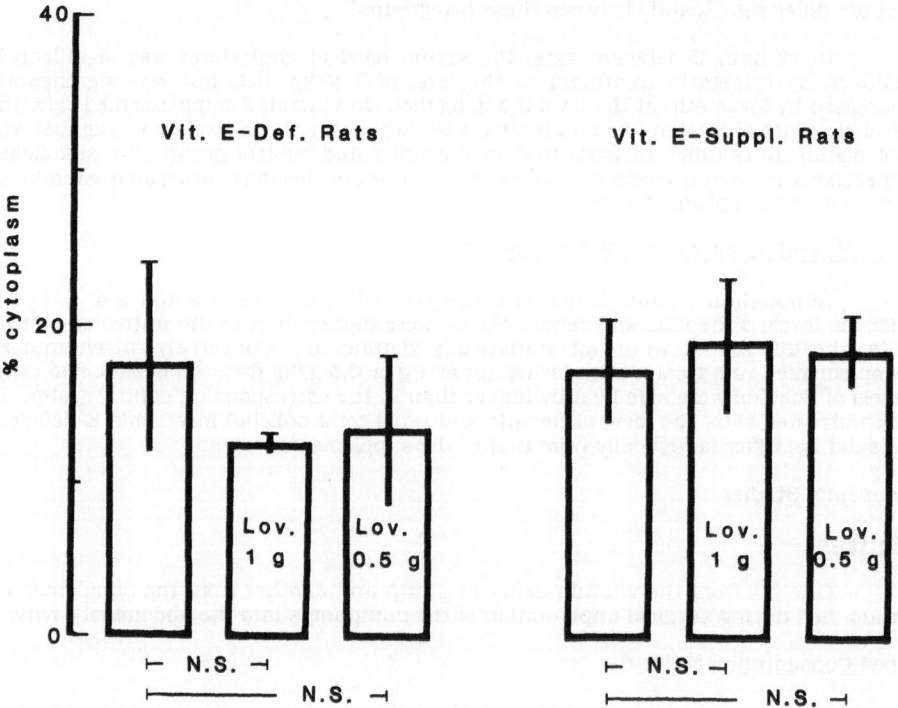

Fig. 5 Volumetric densities of ceroid granules in the cytoplasm of epithelial cells of renal proximal tubules in lovastatin-treated and untreated vitamin E-deficient and -supplemented rats.

Table 3. Effects of Lovastatin Treatment on Relative Organ Weights[1]

Vitamin E-Deficient Rats

Treatment	Liver	Kidneys	Heart	Spleen	Brain
Untreated (6)	3.73 ± 0.20	0.79 ± 0.05	0.33 ± 0.01	0.21 ± 0.02	0.68 ± 0.03
Lovastatin 1 g/kg (2)	6.47 ± 0.03[2]	1.42 ± 0.14[2]	0.51 ± 0.03[2]	0.42 ± 0.07[2]	1.24 ± 0.12[2]
Lovastatin 0.5 g/kg (4)	6.54 ± 0.67[2]	1.42 ± 0.10[2]	0.43 ± 0.02[2]	0.31 ± 0.03[2]	1.04 ± 0.12[2]

Vitamin E-Supplemented Rats

Treatment	Liver	Kidneys	Heart	Spleen	Brain
Untreated (5)	4.29 ± 0.17	0.79 ± 0.02	0.36 ± 0.01	0.19 ± 0.01	0.77 ± 0.04
Lovastatin 1 g/kg (4)	4.20 ± 0.23	0.79 ± 0.02	0.40 ± 0.02	0.23 ± 0.04	0.74 ± 0.03
Lovastatin 0.5 g/kg (4)	5.36 ± 0.10[2]	1.00 ± 0.02[2]	0.42 ± 0.01[2]	0.20 ± 0.01	0.72 ± 0.04

[1]g/100 g body weight. [2]Significantly higher ($p < 0.05$) than in the corresponding untreated control rats. Number of rats in parenthesis.

Table 4. Effects of Lovastatin Treatment on Serum GOT (AST)[1] GPT (ALT)[1] and Cholesterol[2]

Vitamin E-Deficient Rats

Treatment	GOT	GPT	Cholesterol
Untreated (6)	150.00 ± 19.47	12.42 ± 2.49	116.50 ± 3.80
Lovastatin 1 g/kg (2)	322.50 ± 16.50[3]	42.75 ± 3.25[3]	100.00 ± 1.00[3]
Lovastatin 0.5 g/kg (4)	386.00 ± 19.28[3]	51.50 ± 5.10[3]	136.00 ± 7.40[3]

Vitamin E-Supplemented Rats

Treatment	GOT	GPT	Cholesterol
Untreated (5)	94.80 ± 19.67	6.00 ± 0.42	105.00 ± 5.60
Lovastatin 1 g/kg (4)	88.50 ± 6.18	14.63 ± 1.23	98.00 ± 23.00
Lovastatin 0.5 g/kg (4)	71.50 ± 6.40	13.50 ± 2.18	N.D.

[1]IU/liter. [2]mg/dl. [3]Significantly different ($p < 0.05$) from the corresponding untreated controls. N.D. = not determined. Number of rats in parenthesis.

Table 5. Effects of Lovastatin Treatment on Hepatic and Renal Total Dolichol[1]

Vitamin E-Deficient Rats

Treatment	Liver	Kidneys
Untreated (6)	64.17 ± 4.73	25.50 ± 0.98
Lovastatin 1 g/kg (2)	155.00 ± 17.56	39.50 ± 1.06
Lovastatin 0.5 g/kg (4)	177.00 ± 37.00	47.50 ± 1.75

Vitamin E-Supplemented Rats

Treatment	Liver	Kidneys
Untreated (5)	47.20 ± 11.00	13.60 ± 0.67
Lovastatin 1 g/kg (4)	274.25 ± 35.00[2]	46.00 ± 2.91[2]
Lovastatin 0.5 g/kg (4)	463.25 ± 77.30[2]	69.25 ± 2.77[2]

[1]$\mu g/g$ fresh tissue.   [2]Significantly higher (p < 0.001) than in the corresponding untreated control rats. Number of rats in parenthesis.

Table 6. Effects of Leupeptin Treatment on Absolute and Relative Daily Food Intakes

Vitamin E-Deficient Rats

Treatment	Absolute Intake[1]	Relative Intake[2]
Untreated (6)	11.11 ± 1.11	9.21 ± 0.51
Leupeptin (3)	9.56 ± 0.78	9.84 ± 1.02

Vitamin E-Supplemented Rats

Treatment	Absolute Intake[1]	Relative Intake[2]
Untreated (5)	9.33 ± 0.84	9.56 ± 1.40
Leupeptin (3)	9.29 ± 1.15	9.33 ± 0.99

[1]g/day.   [2]g/100 body weight/day.   Number of rats in parenthesis.   Differences between groups were not statistically different.

Body Weights (Figures 6 and 7)

All the animals gained weight during the experiment, although the vitamin E-deficient rats treated with leupeptin lost some weight during the last 2 weeks. The average daily growth rates of the untreated vitamin E-deficient and supplemented rats were 2.54 g/day and 2.20 g/day, respectively, while in those treated with leupeptin the growth rates were 2.05 g/day and 2.49 g/day, respectively. These differences were not statistically significant.

In vitamin E-deficient rats, the final body weight of the untreated rats was slightly, but significantly higher than that in rats treated with leupeptin, while in the vitamin E-supplemented rats, the final body weights of treated and untreated rats were statistically similar.

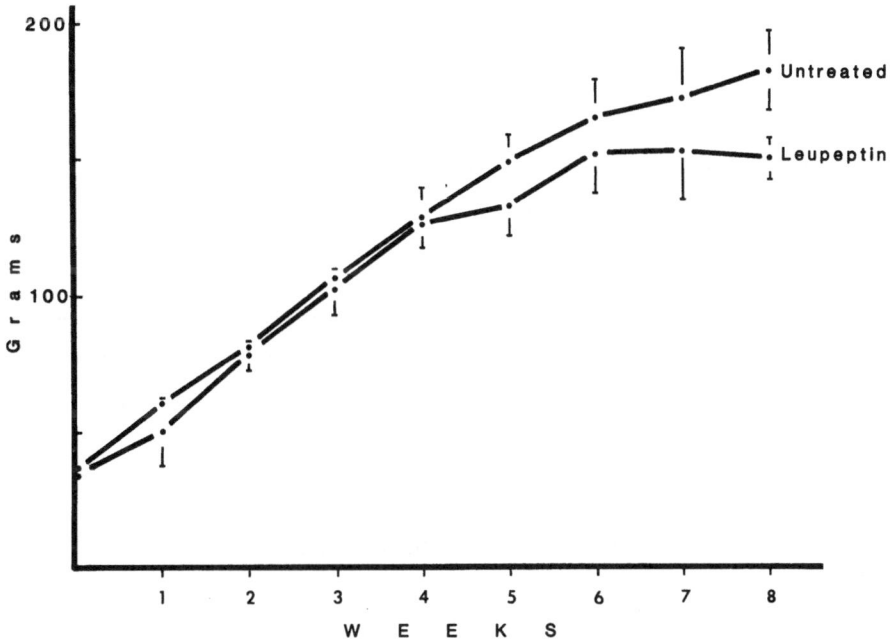

Fig. 6. Changes in body weights in leupeptin-treated and untreated vitamin E-deficient rats.

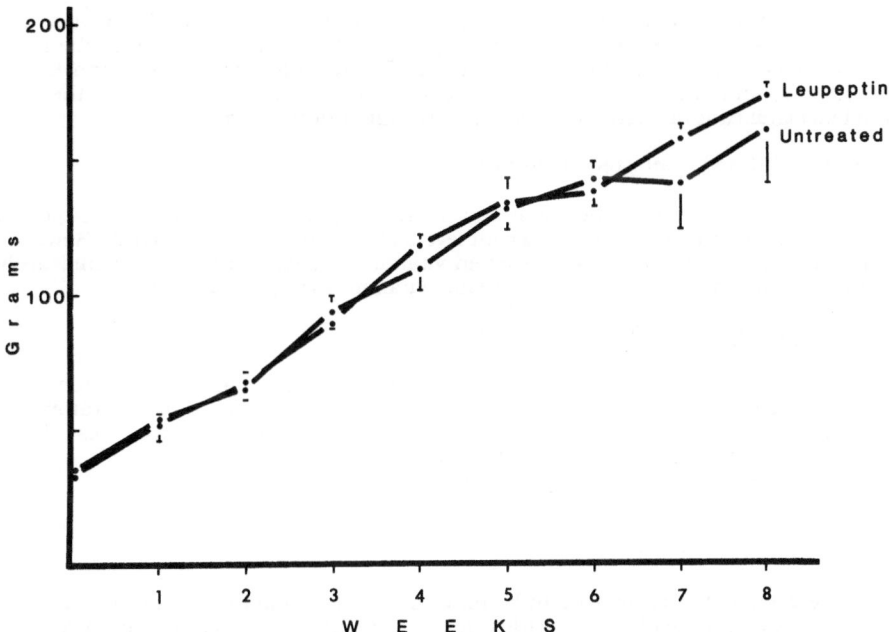

Fig. 7 Changes in body weights in leupeptin-treated and untreated vitamin E-supplemented rats.

Organ Weights (Table 7)

Although in vitamin E-deficient rats the leupeptin treatment had a tendency to increase the relative weights of liver, kidneys, heart, spleen and brain, the differences with the untreated control rats were not statistically significant. Similarly, in vitamin E-supplemented rats no significant difference in relative organ weights were found between leupeptin-treated and untreated rats.

Table 7. Effects of Leupeptin Treatment on Relative Organ Weights[1]

			Vitamin E-Deficient Rats		
Treatment	Liver	Kidneys	Heart	Spleen	Brain
Untreated (6)	3.73 ± 0.20	0.79 ± 0.05	0.33 ± 0.01	0.21 ± 0.02	0.68 ± 0.03
Leupeptin (3)	4.28 ± 0.31	0.91 ± 0.06	0.36 ± 0.01	0.25 ± 0.01	0.76 ± 0.01
			Vitamin E-Supplemented Rats		
Untreated (5)	4.29 ± 0.17	0.79 ± 0.02	0.36 ± 0.01	0.19 ± 0.01	0.77 ± 0.04
Leupeptin (3)	3.87 ± 0.31	0.79 ± 0.01	0.34 ± 0.01	0.21 ± 0.02	0.68 ± 0.05

[1]g/100 g body weight.   Number of rats in parenthesis.   Differences in organ weights between groups were not statistically significant.

Volumetric Densities of Ceroid (Figures 8 and 9)

In vitamin E-deficient and supplemented rats treated with leupeptin, the volumetric densities of ceroid granules in hepatic kupffer cells, as well as in the renal epithelial cells of the proximal convoluted tubules were always higher than in the corresponding untreated control groups. However, the statistical analysis of the data showed no significant differences between treated and untreated rats.

Serum GOT, GPT and Cholesterol (Table 8)

Neither in vitamin E-deficient nor in the supplemented rats the leupeptin treatment significantly modified the serum levels of activity of GOT and GPT. However, the values of GOT in treated and untreated vitamin E-deficient rats were significantly higher than those in the treated and untreated vitamin E-supplemented rats.

Hepatic and Renal Total Dolichol (Table 9)

Although in the vitamin E-deficient rats the leupeptin treatment had a tendency to decrease the levels of hepatic and renal dolichol, while in the vitamin E-supplemented rats the opposite effect was observed, the differences between treated and untreated rats were not statistically significant.

DISCUSSION

Effects of Vitamin E Deficiency

The preferential early accumulation of ceroid in kupffer cells and in epithelial cells of proximal convoluted tubules found in this study confirmed previous observations cited in the introduction. The reasons for these particular cell localizations are not totally clear. Similar preferential localizations have been previously observed in nutritionally normal rats treated with chloroquine (Porta et al.,

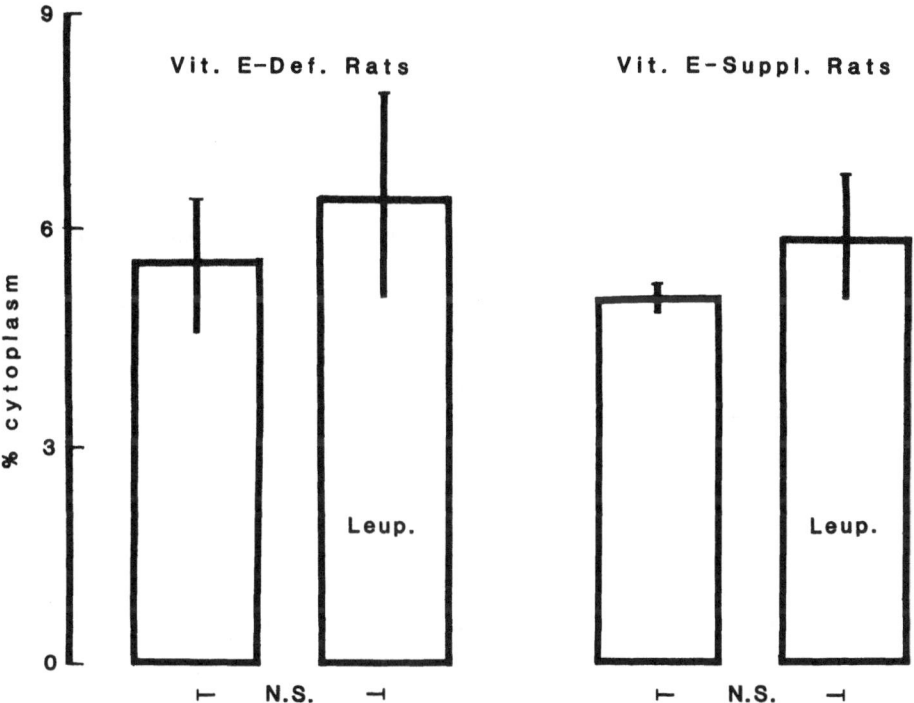

Fig. 8. Volumetric densities of ceroid granules in the cytoplasm of kupffer cells in leupeptin-treated and untreated vitamin E-deficient and -supplemented rats.

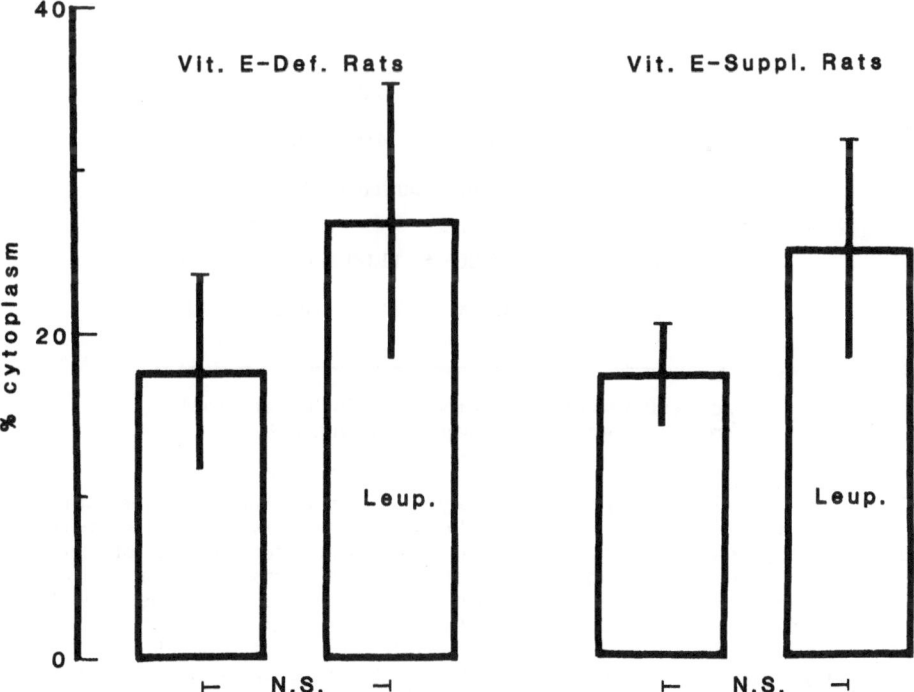

Fig. 9 Volumetric densities of ceroid granules in the cytoplasm of epithelial cells of renal proximal tubules in leupeptin-treated and untreated vitamin E-deficient and -supplemented rats.

Table 8. Effects of Leupeptin Treatment on Serum GOT (AST)[1]
GPT (ALT)[1] and Cholesterol[2]

Vitamin E-Deficient Rats

Treatment	GOT	GPT	Cholesterol
Untreated (6)	150.00 ± 19.47	12.42 ± 2.49	116.50 ± 3.80
Leupeptin (3)	190.00 ± 25.17	24.15 ± 9.90	113.70 ± 9.90

Vitamin E-Supplemented Rats

Treatment	GOT	GPT	Cholesterol
Untreated (5)	94.80 ± 19.67	6.00 ± 0.42	105.00 ± 5.60
Leupeptin (3)	104.00 ± 12.00	15.33 ± 2.01	117.00 ± 8.30

[1]IU/liter. [2]mg/dl. Number of rats in parenthesis. Differences betweeen groups were not statistically significant.

Table 9. Effects of Leupeptin Treatment on Hepatic and Renal Total Dolichol[1]

Vitamin E-Deficient Rats

Treatment	Liver	Kidneys
Untreated (6)	64.17 ± 4.73	25.50 ± 0.98
Leupeptin (3)	24.67 ± 4.91	9.83 ± 0.75

Vitamin E-Supplemented Rats

Treatment	Liver	Kidneys
Untreated (5)	47.20 ± 11.00	13.60 ± 0.67
Leupeptin (3)	103.00 ± 24.58	38.00 ± 1.00

[1]µg/g fresh tissue. Number of rats in parenthesis. Differences between groups were not significant.

1988). Kupffer cells are very active phagocytes rich in lysosomal enzymes, and it is possible, therefore, that in vitamin E deficiency these cells phagocytize increased amounts of peroxidized products from degenerating membranes of neighbor hepatocytes or from other organs. Alternatively, or perhaps concomitantly, the previously mentioned increase in the serum levels of a cysteine proteinase inhibitor in vitamin E-deficient rats (Minakata et al., 1984) may also play a role. We don't know, however, if this inhibitor is increased from early stages of the deficiency, or if the lysosomal proteinases of kupffer cells are particularly sensitive to this inhibitor. The epithelial cells of proximal convoluted tubules of the kidneys are also rich in lysosomes due to the well known resorptive capacity of these tubules for the substances of the glomerular filtrate. Therefore, in vitamin E-deficient rats, these cells may probably reabsorb increasing amounts of soluble material from degenerating tissues.

In this study, the dietary deficiency of vitamin E per se (untreated rats) did not affect the food intake, growth rates, relative weights of the organs, or the serum levels of cholesterol, but significantly increased the serum level of activity of GOT. Vitamin E deficiency also showed a tendency to increase the serum level of activity of GPT, as well as the hepatic and renal concentrations of total dolichol. Increased levels of GOT have been previously observed in early and advanced stages of vitamin E deficiency in rats (Porta et al., 1968; Boguth and Sernetz, 1968; Machlin et al., 1978; Minakata et al., 1984). Although the level of activity of serum GOT was significantly reduced by vitamin E supplementation, it was still higher than that found in normal rats fed conventional commercial rations (Mitruka and Rawnsley, 1970). Hepatic and renal dolichol contents have not been previously studied in vitamin E-deficient animals, and our results suggested a modest increasing effect of this deficiency in both organs. Since dolichol is particularly concentrated in lysosomes (Wong et al., 1982), the apparent increase in the liver and kidneys of our vitamin E-deficient rats may be related to an increase in ceroid containing lysosomes and residual bodies.

Although vitamin E deficiency was associated with the presence of ceroid in kupffer cells and epithelial cells of renal proximal tubules, the supplementation of vitamin E at the level of 16 mg/100 g of diet did not significantly reduce this accumulation. This level of supplementation is about six times higher than the recommended level by the National Research Council (National Academy of Sciences, 1978). It is evident now, as it was also in previous studies in mice (Porta et al., 1988), that in order to prevent ceroid formation, the diets commonly used in studies of vitamin E deficiency should contain much higher amounts of dl-α-tocopherol acetate.

Effects of Lovastatin

In vitamin E-deficient rats the effects of lovastatin were unexpected and devastating, and resulted in very high mortality rates due to massive hepatic necrosis, even at the lower dosis of 0.5 g/kg diet. At this dosis, for example, the mortality was 50%, and the survivors showed poor growth rate (1.51 g/day), as compared with that in the control rats (2.54 g/day). The level of activity of GOT significantly increased more than two-fold, and those of GPT more than four-fold, and the serum concentration of cholesterol was also significantly elevated. The hepatic and renal concentrations of dolichol were also increased, although the differences with those of the control rats were not statistically significant.

Although in vitamin E-supplemented rats, the lovastatin treatment, even at the dosis of 2 g/kg diet, did not result in any mortality, it significantly reduced, however, the growth rate at the dosis of 1 g/kg diet, but not at the dosis of 0.5 g/kg diet. At the dosis of 1 and 0.5 g/kg diet, lovastatin did not produce any significant increase in the serum levels of activities of GOT and GPT, and there was no histological evidence of organ damage. Somewhat surprisingly, at the dosis of 1 and 0.5 g/kg diet, the concentrations of hepatic dolichol were significantly elevated (6- and 10-fold, respectively), and those of renal dolichol were also elevated (3- and 5-fold, respectively). The concentration of serum cholesterol in vitamin E-supplemented rats treated with lovastatin at the dosis of 1 g/kg of diet did not significantly differ from that in the corresponding control rats. Although previous studies have shown that lovastatin or related compounds have a significant hypocholesterolemic effect in humans, monkeys, rabbits, and dogs (Tobert et al., 1982; The Lovastatin Study Group II, 1986; Koruda et al, 1979; Kroon et al, 1982; Alberts et al, 1980) no such lowering effects have been found in rats and mice (Endo et al, 1979; Edwards et al, 1983). This lack of effect has been attributed to a possible induction and stabilization of these compounds on HMG-CoA reductase (Edwards et al, 1983; Singer et al., 1984). Our results on serum cholesterol appear, therefore, in line with previous observations. The significant elevations of hepatic and renal dolichol in vitamin E-supplemented rats are difficult to interpret. Our previous study in nutritionally normal adult rats showed a lowering effect of lovastatin on dolichol contents of the intact and crushed epididymal fat pads (Porta et al., 1988). It has been also recently shown that prolonged lovastatin administration to normally fed dogs resulted at least in a modest decrease in brain dolichol content (Berry et al, 1988). Further studies are needed to clarify these apparent discrepancies.

The most important problem is to find now the reasons for the lethal massive hepatic necrosis associated with lovastatin treatment in our vitamin E-deficient rats. Although in weanling and mature rats the deficiency of vitamin E is associated with elevated levels of activity of GOT and GPT which suggest hepatic damage, in our previous studies, the hepatocytes of these animals did not show any significant ultrastructural alterations, other than a modest accumulation of ceroid pigment in lysosomes, and hepatic necrosis was only observed when the dietary deficiency of vitamin E was associated with selenium deficiency (Porta et al., 1968). Furthermore, in weanling rats fed a vitamin E-deficient diet the hepatic levels of Se-dependent glutathione peroxidase were not significantly reduced (Porta et al, 1977), and while in mature vitamin E-deficient rats these hepatic levels have been found reduced (Chen et al., 1980), no hepatic necrosis was reported. The vitamin E-deficient diet used in the present study is not supplemented with extra amounts of selenium because it contains 20% casein, and at this level the selenium present in casein is sufficient to prevent hepatic necrosis, even when fed to rats for as long as 22 weeks (Machlin et al., 1977).

Of particular interest are the recent findings of Berry et al. (1988) indicating that lovastatin treatment (180 mg/kg/day) to nutritionally normal dogs produced a significant (60%) reduction in the serum level of α-tocopherol, and that 30% of these animals developed hemorrhagic encephalopathy. Because the supplementation of these dogs with extra amounts of vitamin E did not prevent this damage, these authors concluded that the brain lesions were not due to the low serum levels of α-tocopherol. At any rate, an important aspect emerging from studies on lovastatin treatment in man and dogs is the fact that this drug particularly reduces the LDL fraction of serum cholesterol (Alberts et al, 1980; The Lovastatin Study Group II, 1986), and since most of the circulating vitamin E is found in this fraction (Traber and Kayden, 1984), the delivery of this vitamin to tissues may be seriously impaired. It is possible, therefore, that lovastatin treatment in our vitamin E-deficient rats may have accentuated this deficiency. However, even if this was the case, it would be still difficult to explain the development of hepatic necrosis on the sole basis of severe vitamin E deficiency.

In summary, lovastatin treatment did not modify the accumulation of ceroid in vitamin E-deficient rats, and was associated for still unknown reasons with a dose-dependent lethal hepatic necrosis.

<u>Effects of Leupeptin</u>

Although the peritoneal infusion of leupeptin for 45 days at the dosis of 0.5 mg/day to vitamin E-deficient rats showed a weak but consistent tendency to increase ceroid in kupffer and renal tubular cells, the analysis of the data by the Duncan's test indicated that the differences with control rats were not statistically significant. Since Ivy et al. (1984a) have previously shown that similar amounts of leupeptin infused into the cerebral ventricles of young rats for 14 days induced the formation of vast amounts of lipofuscin-like pigments in brain cells, several possibilities may be entertained to explain the differences with our present results. First, the dosis used in our study might have been insufficient to produce a significant effect on ceroidogenesis for various reasons. The peritoneal cavity is much larger than the ventricular system of the brain, and consequently the amounts of leupeptin reaching the target cells in our rats must have been relatively small. Furthermore, in young rats the lysosomes and their contents of hydrolytic enzymes appear more abundant in kupffer and renal proximal tubular cells than in brain cells. Additionally, there is evidence that the deficiency of vitamin E increases the activity of cysteine proteinases in the liver of rabbits (Zalkin et al., 1962), and in the liver and kidneys of chicken (Bunyan et al., 1976b). It seems, therefore, that much higher amounts of leupeptin should be needed to produce a significant inhibitory effect on thiol proteinases of liver and kidneys of rats whether deficient or supplemented with vitamin E. Secondly, our prolonged infusion of leupeptin might have produced a transient inhibitory effect on cysteine proteinases followed by an adaptive increase. Although unlikely, this possibility cannot be ruled out without sequential enzymatic determinations. Finally, at least the ceroidogenesis in vitamin E-deficient rats may have minimal, if any, relation to a possible inhibition of cysteine proteinases. In fact, and as mentioned before, the activity of these enzymes generally increases in several tissues of vitamin E-deficient animals, and ceroid continues building-up despite the increased serum levels of the cysteine proteinase inhibitor reported by Minakata et al. (1984). Even if larger dosis of leupeptin to vitamin E-deficient rats may eventually result in the increase of ceroid, this may not necessarily prove that the inhibition of cysteine proteinases have a significant role in this type of ceroidogenesis. All kinds of ceroid pigments, whether they are produced by vitamin E deficiency, administration of diverse enzyme inhibitors, or excessive accumulation of fat in tissues due to various factors (i.e. CCl_4, choline deficiency, alcohol abuse, etc.) end-up in lysosomes. The combination of two or more factors may obviously increase the amounts of lysosomal pigments, but the mechanisms of formation for each of the blended lysosomal pigments may be totally different.

In vitamin E-deficient rats, the leupeptin treatment showed a consistent tendency to decrease the dolichol contents of liver and kidneys, but the differences with the levels in the control rats were not statistically significant. Conversely, in vitamin E-supplemented rats treated with leupeptin, the contents of renal and hepatic dolichol were consistently higher than in the corresponding controls. Statistically, however, these differences were again not significant. In the study of Ivy et al. (1984b), the infusion of leupeptin to normally fed young rats was associated with a significant increase of brain dolichol, and this was attributed to the increase in lipofuscin-like pigments. Although in our vitamin E-supplemented rats, the leupeptin associated increases in ceroid pigment and dolichol were not statistically significant, our data are not in real conflict with the results of Ivy et al. (1984b). The lack of significant effect of leupeptin on dolichol contents in our vitamin E-supplemented rats may be perhaps a matter of dosage, site of infusion, tissue distribution, strain and sex of rats, etc. In any case, one of the intriguing facts in this study was that, despite the lack of statistical significance, at least the tendencies of leupeptin to influence the contents of hepatic and renal dolichol in vitamin E-deficient and supplemented rats were in total contrast. These opposite tendencies could not be attributed to differences in the amounts of ceroid between groups or to any other reasons obvious to us. We have to conclude, therefore, that under the conditions of this experiment leupeptin treatment neither influences the amounts of ceroid nor the contents of dolichol in liver and kidneys of vitamin E-deficient or supplemented rats.

ACKNOWLEDGMENTS

The authors are indebted to Alfred W. Alberts of Merk Sharp & Dohme for his generous gift of Lovastatin,and to Miss Valerie Song for her assistance in the editing and typing of the manuscript. These studies have been supported in part by the CONICET of Argentina.

REFERENCES

Alberts, A.W., Chen, J., Kuron, G. et al., 1980, Mevilonin: A highly potent competitive inhibitor of hydroxymethylglutaryl-coenzyme A reductase and a cholesterol-lowering agent, Proc. Natl. Acad. Sci., USA, 77:3957-3961.

Allain, C.C., Poon, L.S., Chan, C.S.G. et al., 1974, Enzymatic determination of total serum cholesterol, Clin. Chem., 28:470-475.

Barret, A.J., 1980, The many forms and functions of cellular proteinases, Federation Proc., 39:9-14.

Berry, P.H., MacDonald, J.S., Alberts, A.W. et al., 1988, Brain and optic system pathology in hypocholesterolemic dogs treated with a competitive inhibitor of 3-hydroxy-3-methylglutaryl coenzyme A reductase, Am. J. Path., 132:427-443.

Boguth, W. and Sernetz, M., 1968, Aktivitaten von aldolase und glutamatoxalacetatetransaminase in plasma von vitamin E mangelratten nach verabreidung von α-tocopherol isoprenanalogen, Int. Z. Vitamin Forsch, 38:320-327.

Bunyan, J., Green, J., and Diplock, A.T., 1967a, Lysosomal enzymes and vitamin E deficiency. 3. Liver necrosis and testicular degeneration in the rat, Brit. J. Nutr., 21:147-154.

Bunyan, J., Green, J. and Diplock, A.T., 1967b, Lysosomal enzymes in vitamin E deficiency. 1. Muscular dystrophy, encephalomalacia and exudative diathesis in the chick, Brit. J. Nutr., 21:127-136.

Burton, G.W. and Ingold, K.U., 1986, Vitamin E: Application of the principles of physical organic chemistry to the exploration of its structure and function, Acc. Chem. Res., 19:194-201.

Chen, L.H., Thacker, R.R., and Chow, C.K., 1980, Tissue antioxidant status and related enzymes in rats with long-term vitamin E deficiency, Nutr. Rep. Int., 22:873-881.

Dabholkar, A.S. and Ogawa, K., 1978, Ultracytochemical changes in the distribution of acid phosphatase in the liver cells of vitamin E-deficient and vitamin E-supplemented rats, Acta Histochem. Cytochem., 11:195-204.

Davies, K.J.A., 1988, Protein oxidation, protein cross-linking, and proteolysis in the formation of lipofuscin: Rationale and methods for the measurement of protein degradation, in: "Lipofuscin-1987: State of the Art", I. Zs.-Nagy, ed., Excerpta Medica, New York, pp. 109-132.

Dayton, W.R., Scholmeyer, J.V., Chan, A.C. et al., 1979, Elevated levels of a calcium-activated muscle proteinase in rapidly atrophying muscle from vitamin E-deficient rabbits, Biochem. Biophys. Acta, 584:216-230.

Diplock, A.T. and Lucy, J.A., 1973, The biochemical modes of action of vitamin E and selenium: A hypothesis. FEBS Letters, 29:205-210.

Draper, H.H., Bergam, J.G., Chin, M. et al., 1964, A further study of the specificity of the vitamin E requirement for reproduction, J. Nutr., 84:395-400.

Duncan, D.B., 1955, Multiple range and multiple F test, Biometrics, 11:1-42.

Edwards, P.A., Lau, S.F., and Fogelman, A.M., 1983, Alterations in the rates of synthesis and degradation of rat liver 3-hydroxy-3-methylglutaryl coenzyme A reductase produced by cholestyramine and mevilonin, J. Biol. Chem., 258:10219-10222.

Einarson, L., 1956, Deposits of fluorescent acid-fast products in the nervous system and skeletal muscles of adult rats with chronic vitamin E deficiency, J. Neurol. Neurosurg. Psychiat. 16:98-109.

Elftman, H., Kaunitz, H., and Slanetz, C.A., 1949, Histochemistry of uterine pigment in vitamin E-deficient rats, Ann. New York Acad. Sci., 52:72-79.

Endo, A., Tsujita, Y., Koruda, M. et al., 1979, Effect of ML-236B on cholesterol metabolism in mice and rats: Lack of hypocholesterolemic activity in normal animals, Biochem. Biophys. Acta, 575:266-276.

Gedigk, P. and Fischer, R., 1959, Uber die Entstehung von Lipopigment in Muskelfasern. Untersuchungen bein experimentellen Vitamin E-mangel der Ratte und an Organen des Menschen, Virchows Arch. Path. Anat., 332:431-468.

Hartroft, W.S. and Porta, E.A., 1965, Ceroid, Amer. J. Med. Sci., 250:324-345.

Howes, E.L., Price, H.M., and Blumberg, J.M., 1964, The effects of a diet producing lipochrome pigment (ceroid) on the ultrastructure of skeletal muscle in the rat, Am. J. Path., 45:599-631.

Ivy, G.O., Schottler, F., Wenzel, J. et al., 1984a, Inhibitors of lysosomal enzymes: Accumulation of lipofuscin-like dense bodies in the brain, Science, 226:985-987.

Ivy, G.O., Wolfe, L.S., Houston, K. et al., 1984b, Lysosomal enzyme inhibitors cause the accumulation of ceroid lipofuscin and dolichol in rat brain, Soc. Neurosc., Abstr., 10:885.

Ivy, G.O. and Gurd, J.W., 1988, A proteinase inhibitor model of lipofuscin formation, in: "Lipofuscin-1987: State of the Art", I. Zs.-Nagy, ed., Excerpta Medica, New York, pp. 83-106.

Katz, M. L. and Eldred, G.E., 1989, Retinal light damage reduces autofluorescent pigment deposition in the retinal pigment epithelium, Invest. Ophthalmol. Vis. Sci. 30:37-43.

Katz, M.L., Robison, W.G. Jr., Hermann, R.K. et al., 1964, Lipofuscin accumulation resulting from senescence and vitamin E deficiency: Spectral properties and tissue distribution, Mech. Age. Dev., 25:149-159.

Keller, R.K. 1986, The mechanism and regulation of dolichol phosphate synthesis in rat liver, J. Biol. Chem. 261:12053-120509.

Koruda, M. Tsujita, Y., Tanzawa, K. et al., 1979, Hypolipidemic effects in monkeys of ML-236B, a competitive inhibitor of 3-hydroxy-3-methylglutaryl coenzyme A reductase, Lipids, 14:585-589.

Kroon, P.A., Hand, K.M., Huff, J.W. et al., 1982, The effects of mevilonin on serum cholesterol levels of rabbits with endogenous hypercholesterolemia, Atherosclerosis, 44:41-48.

Lampert, P., Blumberg, J.M., and Peutschew, A., 1964, An electron microscopic study on dystrophic axons in the gracile and cuneate nuclei of vitamin E-deficient rats, J. Neuropath. Expt. Neurol., 23:60-77.

Machlin, L.J., Gabriel, E., Spiegel, H.E. et al., 1978, Plasma activity of pyruvate kinase and glutamic oxalacetic transaminase as indices of myopathy in the vitamin E-deficient rat, J. Nutr., 108:1963-1968.

Machlin, L.J., Filipski, R., Nelson, J. et al., 1977, Effects of a prolonged vitamin E deficiency in the rat, J. Nutr., 107:1200-1208.

Maeda, R., 1967, The origin and characteristics of ceroid, Acta Path. Jap., 17:439-456.

Martin, A.J.P. and Moore, T., 1939, Some effects of prolonged vitamin E deficiency in the rat, J. Hyg. (London), 39:643-650.

Mason, K.E. and Emmel, A.F., 1945, Vitamin E and muscle pigment in the rat, Anat. Rec., 92:33-60.

Mason, K.E., Dam, H., and Granados, H., 1946, Histological changes in adipose tissue of rats fed a vitamin E-deficient diet high in cod liver oil, Anat. Rec., 94:265-288.

Minakata, K., Asano, M., Takahashi, Y. et al, 1984, Increase in α-cysteine proteinase inhibitor level in serum of vitamin E-deficient muscular dystrophic rats, Nutr. Rep. Int., 29:445-459.

Mitruka, B.M. and Rawnsley, H.M., 1970, Clinical, Biochemical and Hematological Reference Values in Normal and Experimental Animals, Masson Publishing, U.S.A., Inc., New York, pp. 121-126.

Moore, T. and Wang, Y.L., 1943, The fluorescence of tissues in avitaminosis E, Biochem. J., Abst., 37:i.

Moore, T. and Wang, Y.L., 1947, Formation of fluorescent pigment in vitamin E deficiency, Brit. J. Nutr., 1:53-64.

National Research Council, 1978, Nutritional Requirements of Laboratory Animals, 3rd edition, National Academy of Sciences, Washington, D.C., pp.21-22.

Ng Ying Kin, N.M.K., Palo, J., Haltia, M. et al., 1983, High levels of brain dolichols in neuronal ceroid-lipofuscinosis and senescence, J. Neurochem., 40:1465-1473.

Palmer, D.N., Husbands, D.R., Winter, P.J. et al., 1986a, Ceroid-lipofuscinosis in sheep. I. Bis (monoacylglycerol) phosphate, dolichol, ubiquinone, phospholipids, fatty acids and fluorescence in liver lipopigment lipids, J. Biol. Chem., 261:1766-1772.

Palmer, D.N., Barns, G., Husbands, D.R. et al, 1986b, Ceroid-lipofuscinosis in sheep. II. The major component of the lipopigment in liver, kidney, pancreas, and brain is low molecular weight protein, J. Biol. Chem., 261:1773-1777.

Porta, E.A., Chin, B.K.F., and Joun, N.S., 1977, Glutathione peroxidase system and microsomal lipoperoxidation in prenecrotic stages of dietary hepatic necrosis in rats, J. Nutr., 107:1852-1858.

Porta, E.A., de la Iglesia, F.A., and Hartroft, W.S., 1968, Studies on dietary hepatic necrosis, Lab. Invest., 18:283-297.

Porta, E.A. and Hartroft, W.S., 1963, Early deposition of ceroid in kupffer cells in mice fed hepatic necrogenic diets, Can. Med. Assoc. J., 38:1167-1169.

Porta, E.A. and Hartroft, W.S., 1964, The distribution of ceroid and acid phosphatase in prenecrotic stages of dietary hepatic necrosis in rats and mice, Tijd. v. Gastro-Enterol., 7:200-209.

Porta, E.A., Mower, H.F., Moroye, M. et al., 1988, Differential features between lipofuscin (age pigment) and various experimentally produced "ceroid pigments", in: "Lipofuscin-1987: State of the Art", I. Zs.-Nagy, ed., Excerpta Medica, New York, pp. 341-372.

Pullarkat, R.K. and Reha, H., 1982, Accumulation of dolichols in brains of elderly, J. Biol. Chem., 257:5991-5993.

Pullarkat, R.K., Reha, H., and Pullarkat, P.S., 1984, Age-associated increase of free dolichol levels in mice, Biochem. Biophys. Acta, 793:494-496.

Reichel, W.J., Hollander, J., Clark, J.H. et al., 1968, Lipofuscin pigment accumulation as a function of age and distribution in rodent brain, J. Gerontol., 23:71-78.

Reitman, S. and Frankel, S., 1957, A colorimetric method for the determination of serum glutamic oxalacetic and glutamic pyruvic transaminases, Am. J. Clin. Path., 28:56-63.

Sakakihara, Y. and Volpe, J.J., 1985, Dolichol in human brain: Regional and developmental aspects, J. Neurochem., 44:1535-1540.

Sevanian, A., Muakkassah-Kelly, S.F., and Montestruque, S., 1983, Influence of phospholipase A_2 and glutathione peroxidase on the elimination of membrane lipid peroxides, Arch. Biochem. Biophys., 223:441-452.

Singer, I.I., Kawka, D.W., Kazazis, D.M. et al., 1984, Hydroxymethylglutaryl-coenzyme A reductase-containing hepatocytes are distributed periportally in normal and mevilonin-treated rat livers, Proc. Natl. Acad. Sci. USA, 81:5556-5560.

The Lovastatin Study Group II. 1986, Therapeutic response to lovastatin (mevilonin) in nonfamiliar hypercholesterolemia, JAMA, 256:2829-2834.

Tober, J.A., Bell, G.D., Birtwell, J. et al., 1982, Cholesterol-lowering effect of mevilonin, an inhibitor of 3-hydroxy-3-methylglutaryl coenzyme A reductase, in healthy volunteers, J. Clin. Invest. 69:913-919.

Traber, M.G. and Kayden, H.J., 1984, Vitamin E is delivered to cells via the high affinity receptor for low-density lipoprotein, Am. J. Clin. Nutr., 40:747-751.

Wolfe, L.S., Ng Ying Kin, N.M.K., Palo, J. et al, 1982, Raised levels of cerebral cortex dolichols in Alzheimer's disease, Lancet, II:99.

Wolff, S.P., Garner, A., and Dean, R.T., 1986, Free radicals, lipids and protein degradation, Trends Biochem. Sci., 11:27-31.

Wong, T.K., Decker, G.L., and Lennarz, W.J., 1982, Localization of dolichol in the lysosomal fraction of rat liver, J. Biol. Chem., 257:6614-6618.

Yamada, K., Yokohama, H., Abe, S. et al., 1985, High-performance liquid chromatographic method for the determination of dolichols in tissues and plasma, Anal. Biochem., 150:26-31.

Zalkin, H., Tappel, A.L., Cadwell, K.A. et al., 1962, Increased lysosomal enzymes in muscular dystrophy of vitamin E-deficient rabbits, J. Biol. Chem., 2678-2684.

DISCUSSION

ICHIKAWA: Was there any evidence of kidney damage in your vitamin E deficient rats treated with lovastatin? I am asking you this question, because in our vitamin E deficient rats, the depletion of glutathione caused only kidney, but not liver damage.

PORTA: The microscopic examination of the kidneys did not reveal any abnormality.

KITANI: Do you have any evidence that lovastatin suppresses hepatocytic cholesterogenesis in your rats?

PORTA: Although we don't have direct evidence, at least our data on serum cholesterol in normal (vitamin E supplemented) rats strongly suggested that lovastatin treatment at the level of 1 g/kg of diet did not affect hepatocytic cholesterogenesis. As mentioned in my presentation, the hypocholesterolemic effect of lovastatin and other related inhibitors of HMG-CoA reductase, such as compatin, varies in different animal species. While the chronic administration of these compounds effectively reduced serum cholesterol in humans, monkeys, dogs, and rabbits, they are ineffective in rats and mice (Berry et al., Am. J. Path., 132:427, 1988).

KITANI: Do you have any explanation why prolonged leupeptin treatment decreased hepatic dolichols? Wouldn't you rather expect an increase?

PORTA: Yes, since leupeptin showed a tendency to increase ceroid pigment in the liver, I was also expecting an increase in hepatic dolichols. However, neither in vitamin E deficient nor in supplemented rats, the leupeptin treatment significantly modified the levels of hepatic dolichols in relation to the untreated controls. Because the prolonged leupeptin treatment moderately increased the serum levels of activity of GOT and GPT in some of the vitamin E deficient and supplemented rats, I initially thought that a possible hepatotoxic effect of leupeptin might have probably prevented any increase in dolichols. However, the microscopic examination of livers of leupeptin-treated rats did not show any histological alteration. Therefore, I don't have any explanation for the lack of effect of leupeptin on hepatic total dolichols, as it may be the case for leupeptin-associated increase in brain dolichols reported by Ivy and collaborators.

KITANI: I asked you this question because we also could not find any increase in hepatic dolichols in our leupeptin-treated rats. As a hepatologist myself, I was also concerned with a possible hepatotoxic effect of leupeptin, and although I did not mention it yesterday in my presentation, we did also find some slight increase in serum GOT in our leupeptin-treated rats. At first, I thought that it might have been some hepatotoxicity of leupeptin at high doses, particularly since with Dr. Zs-Nagy we have previously found that leupeptin may affect the permeability of the hepatocytic plasma membrane. Anyway, the possible hepatotoxic effect of leupeptin is not clear and is probably a matter of dosage.

PORTA: Yes, it could be a matter of dosage, duration of treatment, or other factors, but since you mentioned some alterations in the hepatocytic plasma membrane, I noticed in your presentation that even in your rats treated with high doses of leupeptin, the diverse liver functions studied, and which also depend on the permeability and morpho-functional integrity of this membrane, were essentially unaffected. So, a possible primary or specific toxic effect of leupeptin to this membrane would appear unlikely.

HALL: I enjoyed your presentation. If I am correct, you measured total dolichols, not just free dolichols, didn't you?

PORTA: Yes, that's correct. Although we originally planned to measure also dolichyl monophosphate and free dolichol, it was not possible.

HALL: My view of the situation of dolichol increase in the brain in relation to lipopigment formation, but not in the liver, is that in the brain we are dealing mainly with postmitotic cells that gradually accumulate free dolichols with very little dolichol esters, at least in humans. So, when we get this age-related increase of dolichols in the brain it is because there is a gradual deposition of lipofuscin, which is probably the major source of free dolichols. In the human liver, however, as much as 90 percent of dolichol is present as dolichol esters, and it is probably a completely different process from the formation of lipofuscin. I don't know whether free dolichol does accumulate within the liver with age, because hepatocytes are not postmitotic cells. So, in the liver, it may be more interesting to measure free dolichol if we expect to see any increase. But, even then, I wouldn't be surprised if there is not an increase.

PORTA: I you refer to the situation of hepatic dolichol in relation to age-pigment or lipofuscin accumulation, I also wouldn't be surprised to find no dolichol increase in livers of old individuals. Contrary to what is generally believed, lipofuscin (age-pigment) does not significantly increase with age in the livers of normal humans and rats (Kranz, Acta Hepatogastroenterol., 14:345, 1968; Porta et al., Mech. Age. Dev., 15:297, 1981). However, since the liver is so sensitive to nutritional disorders and other diseases that result in the formation and accumulation of pathological ceroid pigments, and the incidence of these diseases obviously increases with age, the aging liver usually, but not always, may contain more ceroid than the liver of young adults.

HALL: The other point I want to make is that your results suggested that lovastatin did not probably inhibit HMG-CoA reductase. There are experiments which seem to indicate that dolichol synthesis, at least in human fibroblasts, was virtually insensitive or much less sensitive than cholesterol synthesis to the effect of lovastatin. So, it is not surprising to find no inhibitory effects of lovastatin.

PORTA: Well, as I mentioned before, the inhibitory effects of lovastatin on HMG-CoA reductase and on the synthesis of metabolites of the mevalonate pathway may depend on the experimental conditions, and there is evidence that vary in different animal species, as well as in different cells of the same animal. So, it is possible that human fibroblasts in vitro may react differently from human or rat hepatocytes in vitro or in vivo. Keller has shown, for example, that lovastatin inhibits both cholesterol and dolichol synthesis in rat hepatocytes in vitro (J. Biol. Chem., 261:12053, 1986), but we did not find at least indirect evidence for such an effect on the livers of rats in vivo. On the other hand, we have previously found that lovastatin reduced total dolichol concentrations in the epididymal fat pad of rats (Porta et al., In: Lipofuscin-1987: State of the Art, I. Zs-Nagy (ed.), Excerpta Medica, Amsterdam, 1988).

IVY: I think that there might be an alternative interpretation for the failure to increase dolichols in the liver of leupeptin-treated rats, as indicated in your study and in that of Dr. Kitani and contrary to the increase, which we have previously observed in the brain of leupeptin-treated rats. In the brain, we had a very dense accumulation of lipofuscin-like pigment, and although I have not quantitated this pigment, it seems that there is much more accumulation in brain cells than in liver cells. So, probably in liver you may be below the level of detection in your dolichol assay.

PORTA: Yes, this is quite possible. Although in our rats the prolonged leupeptin treatment showed a tendency to increase the volumetric density of ceroid granules in kupffer cells, the amount of ceroid in these cells was apparently lower than in the neurons of your leupeptin-treated rats. It might be perhaps necessary to have more ceroid in order to detect any possible increase in dolichol.

SO - CALLED MEMBRANOCYSTIC LESION (MCL) - A NEW VARIANT

OF CEROID TYPE LIPOPIGMENT

Milan Elleder

1st Department of Pathology
Charles University, School of Medicine
Prague, Czechoslovakia

SUMMARY

Structures very close morphologically to the so-called membranocystic formations of the Nasu-Hakola's disease and identical in histochemical properties with them were found in several
other metabolically unrelated conditions such as cerebrotendinous
xanthomatosis (perivascularly in the brain) and in human atheromatous plaques. This with some other literary data points to unspecific nature of the membranocystic lesion (MCL) which also
has been resisting satisfactory classification in terms of path-
obiochemistry. Evidence is presented suggesting the MCL is lipopigment in nature. This is based on its lipid histochemical properties dominated by prominent autofluorescence and marked sudanophilia resistent to lipid extraction procedures. Ultrastructural pattern of the MCL was membranous, being dominated by mostly
individual trilaminar membranes about 15 nm thick which could be
also occasionally identified in various intralysosomal ceroid
type lipopigments. It is supposed that the MCL lipopigment is
formed mainly extracellularly from the lipid rich debris.

INTRODUCTION

Membranocystic lesion (MCL) first described by Nasu et
al., (1973) is the histological hallmark of the Nasu-Hakola disease, called accordingly, membranous lipodystrophy due to lack
of agequate biochemical criteria. The present day delineation
of the disease is thus purely morphological and is based on a
combination of sudanophilic leukodystrophy and neroaxonal dystrophy with extensive regression of the subcutaneous and visceral adipose tissue. The latter is gradually replaced by numerous
undulating eosinophilic membranes lining cystic spaces. The nature of these cystic membranous formations was entirely obscure.
According to lipid histochemistry they were classified as a
complex glycolipid (Tanaka 1980). Yagishita et al. (1976) spoke
about a metabolic lipid disorder of mesenchymal cells resulting

KEY WORDS : Membranocystic lesion. Nasu-Hakola's disease,
lipopigment

in the formation of peculiar membranous structures derived from the endoplasmic reticulum. Lectin studies showed a high concentration of galactose (Kitajima et al., 1988).

Recent advance in our knowledge of MCL includes the finding that it is nosologically unspecific as its structures have been described in the regressing white adipose tissue due to chronic arterial obstruction (Machinami 1983,1984). In this communication I should like to add another argument pointing to the nosological nonspecifity of MCL together with histochemical and ultrastructural body of evidence of its lipopigment nature.

MATERIAL AND METHODS

The following material was available for the study. Brain from one case of cerebrotendinous xanthomatosis (formaldehyde fixed samples cut as frozen,paraffin or resin embedded sections), bone marrow and lung from one case of Nasu - Hakola disease paraffin and resin embedded samples (kindly provided by Dr.Yagishita, Japan) and a series of human aortic atheromatous plaques (from ten autopsies). The latter were available in unfixed frozen state (cut in the cryostat) and after paraffin or resin embedding.

The following histochemical techniques were used: autofluorescence in UV light, Oil Red O and Sudan Black B (both in 70% ethanol), PAS,Cresyl violet,carbolfuchsin (acidoresistence), permanganate (acidified)-aldehydefuchsin sequence,coupled tetrazonium reaction (CTR) for aromatic amino acid residues,and Luxol Fast Blue.Delipidization was carried out using chloroform methanol in the conventional form (2:1 v/v) or with water added (chloroform-methanol-water 60:30:4) or acidified (chloroform-methanol-conc.HCl 20:10:0.1).Duration of extraction : 30 minutes at room temperature. For details of the techniques used see Pearse (1968) and Elleder (1977). Trypsin (0.1% in 50 mMol Tris buffer) and Pronase (1mg/ml,0.1 Mol phosphate buffer 7.2) were used for proteolytic digestion (1 hour at 37°C) after 3 minute fixation of cryostat sections with neutralized 10% formaldehyde.

Activities of acid phosphatase and nonspecific esterase were studied with azocopulation techniques in cryostat sections. Glial fibrillar acidic protein (GFAP), vimentin, lysozyme (Ly), alpha-1-antitrypsin (AAT) and alpha-1-antichymotrypsin (AACT) were detected by immunostaining techniques.Details of the enzyme and immunohistological techniques are given elsewhere (Elleder et al., 1989).

For lectin histochemistry the following panel of biotinylated lectins (Sigma) detected with ABC complex technique (Vector Laboratories) was used: Arachis hypogea (2.5 ug/ml), Bandeirea simplicifolia (BS-I;2 ug/ml), Concanavalia ensiformis (1-2.5 ug/ml), Dolichos biflorus (1 ug/ml), Glycine max, Phaseolus vulgaris, Triticum vulgaris, Pisum sativum, Helix pomatia (each 1 ug(ml), Ulex europaeus and Ricinus communis (each 0.5 ug/ml).

Paraffin embedded sections were used throughout the study. Generally the procedure reported by Alroy et al.(1984) was adhered to with some modification (incubation lenght: overnight at 4 C). Some futher details are mentioned in the forthcoming publication (Elleder, 1989).

Details of the processing for electron microscopy are given elsewhere (Elleder et al.,1989).Selected samples of paraffin embedded tissues were examined ultrastructurally after removal of paraffin (xylene) and after additional extraction with the conventional chloroform methanol mixture (overnight, room temperature). Further steps and details are given in our previous publication (Elleder and Smid, 1977).

Ultrastructure of ceroid type lipopigments in a series of lysosomal storage diseases (three cases of acid lipase deficiency, two cases of Niemann-Pick type B, splenic lipidosis in two cases of thrombocytolytic thrombocytopenia) was revised for comparison.

RESULTS

MLC in the Brain of Cerebrotendinous Xanthomatosis (CTX)

Details of the clinical,chemical and neuropathological findings of the case will be described elsewhere (Elleder et al., 1989). The neuropathological picture comprised two unilateral lesions,a cavity about 1 cm in diameter in the left basal ganglia and a larger whitish tough lesion in the depth of the right cerebellar hemisphere.One of the histological features were numerous perivascular granulomas composed of lipid phagocytes containing a mixture of apolar anisotropic lipid giving strong staining for cholesterol (Schultze) and a number of ceroid lipopigment granules. Frequent necroses transformed many granulomas into lipid rich debris and a conspicuous tendency to fibrosis coverted them to a fibroatheromatous lesions. Within these there was a number of medium sized cavities surrounded by a colla-

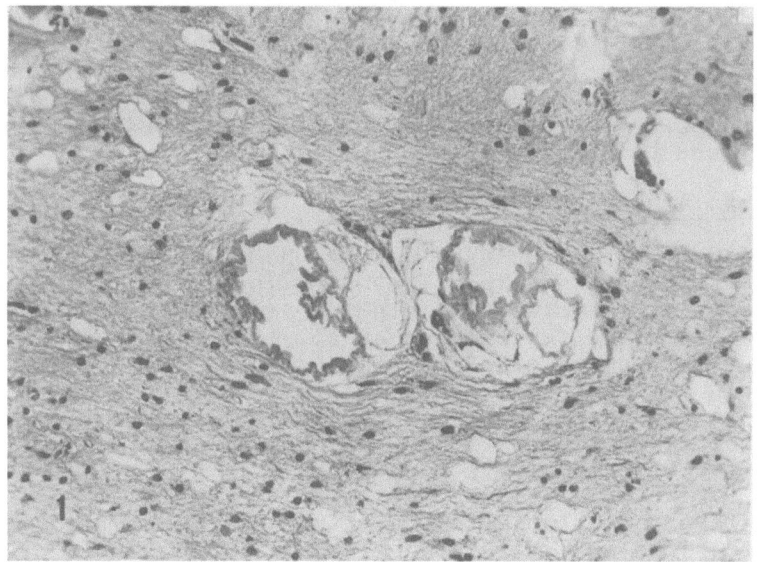

FIGURE 1. Histological appearance of MCL localized in the perivascular fibroatheromatous region of the brain in CTX. H   E , 150.

genous or glial fibrillar condensed layer. Many of the cavities were lined with eosinophilic refractile undulating membranes indistinquishable from the so-called membranocystic lesion of Nasu-Hakola's disease (Fig 1.). They were stained red with Trichrome stain, negative for elastic (Orcein) and glial fibrils (Kanzler-Arendt). Immunostaining for GFAP,AAT.AACT,Ly and vimentin was negative. The lipid histochemical staining pattern is demonstrated in Table 1. and in Fig.2 - 4.

Ultrastructurally(Figs.5,7),a typical and highly developed MCL consisted of an intricate, relatively loose, network of conspicuous trilaminar membranes ranging in thickness from 11.5 to 15.8 nm. The membranes were separated by a lucent space or by a homogeneous substance of low electrondensity. Quite rarely the space between the parallel membranes was reduced to 6 nm and filled with the above mentioned amorphous (lipidic?) substance. Membrane fusion was exceptional and when occured the resulting period was about 10 nm. The membranes were arranged mostly into parallel, cystic or vesicular formations. The MCL interior was inconstantly lined by a single discontinuous membrane, the outer border of MCL being poorly identifiable and merging with the collagenous or glial fibrillar cavity surrounding layer, or directly abutting on the cholesterol crystals.Frequently,the membranes were stippled with small dense granules (calcification?).

Very often there were structures highly suggestive of incipient MCL formation. They were seen as a patchy loose network of short membranes identical in dimmensions to those in the mature MCL, localized within or on the periphery of the amorphous low density extracellular deposits. The conspicuous thick membranes were frequently in direct connection with a similarly loose network of progressively thinner membranes. The latter merged ultitely with the amorphous deposits (Fig.9).

The interior of the MCL cavity was mostly electronlucent or contained patchy amorphous material mostly of low electrondensity,probably lipidic in nature.There was also a loose membranous network partly derived from necrotic cells,and larger cell remnants including ceroid granules released from the lipid phagocytes.The granules were either intact or in various stages of structural breakdown with the densely packed constituting membranes unfolding (see below, and Fig.10).

MCL in Human Atheromas

Membranocystic structures were found in each of the ten atheromatous plaques examined.They were localized in the central necrotic part of the plaque within the lipid rich tissue debris, as already described (Mitchinson et al.,1985).They lined a cystic space either empty or containing (seldom) a small amount of apolar lipid. Often they lined a small cluster of cholesterol crystals. In some macrophages loaded with apolar lipid, there were ceroid rims on the periphery of some of the intracellular droplets, staining similarly (see below).

The staining characteristics of MCL corresponding to lipopigment are shown in Table 1.The extractibility tested in unfixed cryostat sections was remarkably low, the autofluorescence (Fig.3) and sudanophilia being visibly unchanged by extraction with chloroform-methanol (2:1 v/v) or chloroform-methanol-water mixture (60:30:4) for 30 minutes at room temperature. Only the

Figure 2.
Prominent autofluorescence of MCL
in the brain of CTX. Note the
heterogeneity in MCL thickness.
Paraffin section mounted into
glycerin, x 270.

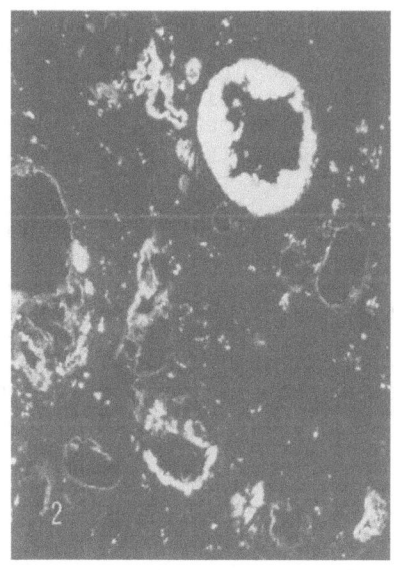

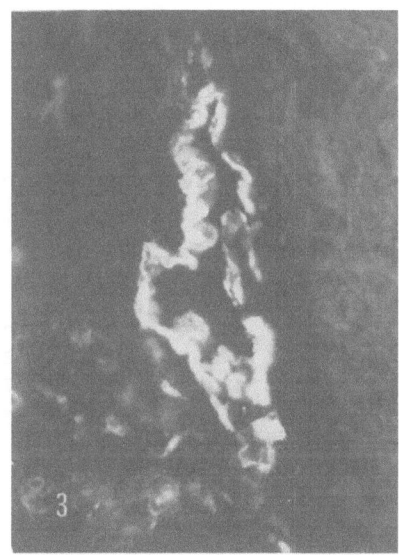

Figure 3.
Human atheroma. Cryostat section
mounted into glycerin. Prominent
autofluorescence of MCL, x 680.

Figure 4.
Prominent autofluorescence of MCL
in the Nasu-Hakola's disease. Lung.
Paraffin section. Glycerin, x 270.

acidified mixture (chloroform-methanol-conc.HCl 20:10:0.1) decreased the staining and autofluorescence significantly but the MCL structures still remained easily discernible. They were resistent to digestion with both trypsin and pronase (checked by autofluorescence).There were no acid phosphatase or nonspecific estrase activities detectable in the MCL structures.

The basic features of the ultrastructure (Figs.6,8), i.e. the appearance of the membranes including their thickeness and arrangement,and the low tendency to association into periodic crystaloid formations were identical with those of MCL found in the case of CTX (see above).The only difference which could be recognized was a generally higher amount of intermembranous substance which on average was of higher electrondensity.The fine structural pattern of the MCL in atheromas was not changed by the delipidization procedure.

MCL in Nasu-Hakola's Disease

The reference samples of the bone marrow and lung tissue with numerous MCL structures were kindly provided by Dr.Yagishita.Lipid histochemical staining was identical to that in our material (see Table 1).What seems to be most important is the fact that even the reference MCL structures emited strong yellow autofluorescence in UV light (Fig.4).

The fine structure was largely different from that in our series,as there were many relatively thick fuzzy membranous formations lacking inner substructure,often fusing with the intermembranous densities.Often there were tubular formations about 16 nm in diameter which due to frequent focal microcystic dilatations resembled rather two relatively loosely attached individual membranes than the highly organized trilaminar structures dominating in our cases.The latter were also present,but were in minority.The ultrastructure was,similarly as in the case of the MCL in human atheromas unchanged even by vigorous lipid extraction of the paraffin embedded sample.

Lectin histochemistry of MCL. Neither of the lectins used (see Methods) displayed detectable binding to any of the MCL lesions examined. The only exception was slight and irregular staining with Ricinus communis (RCA-I) lectin (affinity for B-D galactose).Similarly the classical ceroid type lipopigments in the liver histiocytes of acid lipase deficiency,and sphingomyelinase deficiency type B (spleen) displayed negative staining or there was positivity which was indinstinquishable from that in histiocytes without ceroid.Details will be given in the prepared publication.

Comparison with Ultrastructure of other Ceroid Variants

As it became clear that, in terms of lipid histochemistry, the MCL can be classified as lipopigment, an attempt was made to compare also its fine structure with other variants of the group.

CTX. The lipid phagocytes displayed a number of mostly lucent cytoplasmic vacuoles partly membrane-limited and of many ceroid granules.The latter were composed of densely packed la-

Table 1. Staining properties of MCL

	CTX	atheroma	Nasu-Hakola disease
Fettrot	+/++	++/+++	+/++
Sudan Black B	++/+++	++/+++	++/+++
autofluorescence	+++[a]	+++[a]	+++[a]
carbolfuchsin[b]	+/++	+/++	+/++
PAS	++/+++	++/+++	++/+++
PAF[c]	+/++	+/++	++
CTR[d]	++/+++	++/+++	++/+++
Luxol Fast Blue	+/++	+/++	+/++

[a] yellow, (see also Fig.2 - 4, respectively)
[b] stronger staining was always seen in the prolonged variant
[c] permanganate-aldehydefuchsin method
[d] coupled terazonium reaction

mellae with a period of about 10 nm. When released extracellularly into the cell debris (see above), they frequently became
loosened and partly dissociated into individual trilaminar membranes about 13.5 nm thick (Fig.10.

Ceroid in the protracted variant of acid lipase deficiency.
Liver histiocytes (Kupffer cells and portal histiocytes) are
always an excellent source of ceroid which is stored in these
cells in great quantities together with cholesterol esters in
lysosomes. The ceroid was characteristically present mainly in
the form of ring-like accretions in all four cases examined.
Ultrastructurally it was medium dense and amorphous. A careful
scrutiny of the ceroid ultrastructure revealed the presence of
membranes indistinquishable from those in MCL (thickness 11.5-
15.8 nm) mostly isolated, hapzardly or parallely arranged,and
always buried in the prevailing amorphous substance (Fig.11).

In the so-called sea-blue histiocytes in sphingomyelinase
deficiency type B (two cases) and in the splenic lipidosis in
thrombocytopenia and hyperlipoproteinaemia (two cases of each)
the ceroid was represented by a core of concentrically arranged
membranes with the period range 5.5 - 6.6 nm and of loosely
arranged concentric membranes on the periphery. The thickness
of the individual membranes was 10.5 - 11.1 nm. There was a
varying amount of an amorphous substance of medium electrondensity both in the core and on the periphery of the ceroid granules. The lipid storing histiocytes which represented the bulk
of the storage histiocyte population displayed lucent lysoso-

Figure 5. Electronogram of MCL in CTX. Bar = 1 μ

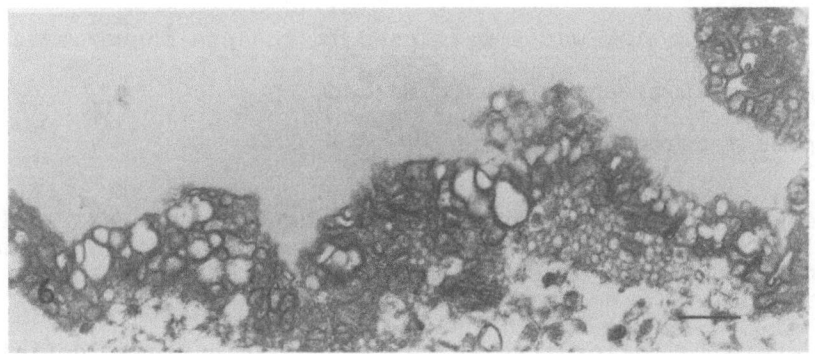

Figure 6. Electronogram of MCL in human atheroma. Bar = 1 μ

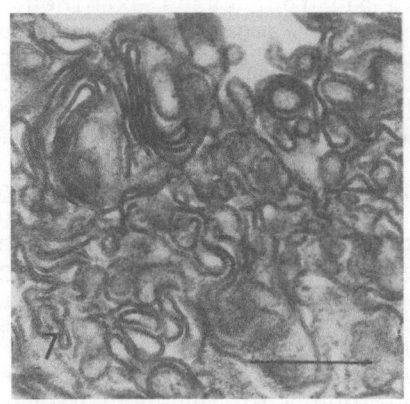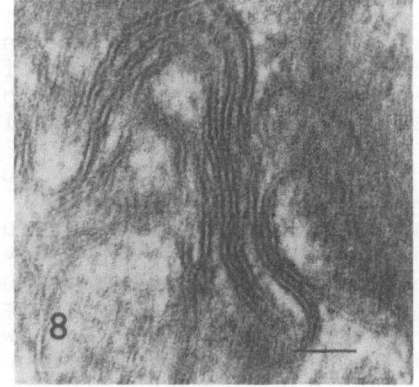

Figure 7. Membranous network in MCL (CTX). Bar = 1 μ
Figure 8. Detail of the MCL membrane (atheroma). Bar = 100nm

mes with loosely arranged narrow stacks of concentric membranes with a period of 5.5 nm.

DISCUSSION

It is the author's suggestion that the attempts so far to qualify MCL as an overproduction of preexisting cellular membranes should be substituted by classifying the membranous lesion as lipopigment. The conclusion is based on its histochemical properties as it meets both fundamental criteria: autofluorescence in UV and apolarity, both resisting the conventional lipid extraction procedures as well as the other important criterion - acidoresistence.Its striking apolarity allows to classify it as a ceroid variant (Elleder 1977,1981; Fullmer and Lillie 1976). conclusion is corroborated by the fact that the extracellular lipopigment in human atheroma, which belongs to the MCL group

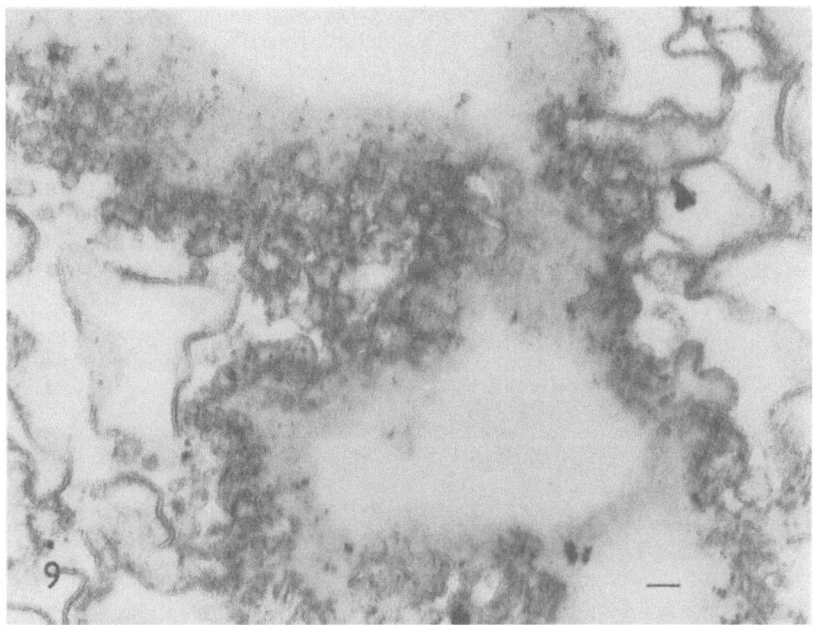

Figure 9. Brain.CTX.Extracellular lipid with a membranous rim on the periphery interpreted as the initial phase of MCL formation. Note the heterogeneity of the membrane thickeness. Bar = 100nm

(see above), is generally called ceroid (Burt 1952,Mitchinson et al.,1985,Schornagel,1965,Györkey et al.,1967,Sinapius and Gunkel 1964). Even if the existing nomenclature (ceroid,lipofuscin) stems from attempts to classify lipopigments primarily with regard to the intracellular metabolic events (see Sohal and Wolfe, 1986) this extracellular lipopigment of MCL is very close to histiocytic ceroid (see below and Gedigk and Totovič,1983) and to experimental in vitro induced lipopigments usually called ceroid (Casselman 1951, Porta 1963).

The unusual membranous substructure is compatible with the

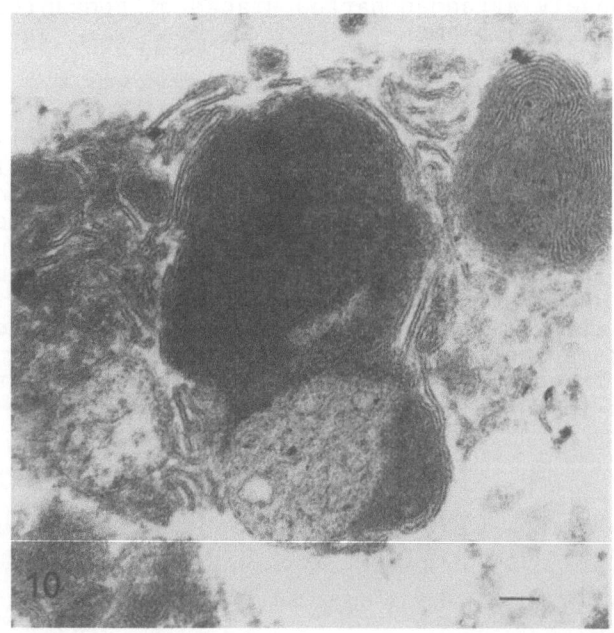

Figure 10. Brain.CTX.Loosely arranged ceroid lysosomal
membranes (and amorphous structures) released
from the remnants of a necrotic lipid phago-
cyte. Perivascular tissue debris. Bar = 100nm

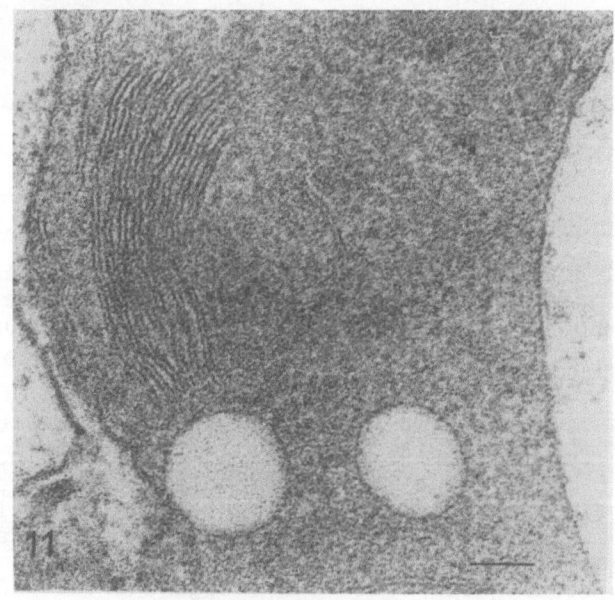

Figure 11. Liver histiocyte in protracted variant of acid
acid lipase deficiency. Lyososmal ceroid with
membranous formations buried into an amorphous
substance. Bar = 100nm

lipopigment nature of MCL.Identical membranous deposits were found in histiocytic ceroid in the case of CTX and in acid lipase deficiency, albeit few compared with the amorphous ceroid accretions in the latter.Laminated ceroid was also described not only in lipid phagocytes in atherosclerotic plaques (Ball et al., 1987) but also extracellularly (Gyöerkey et al.,1968).Ceroid deposits in other diseases examined resembled the MCL but the constituting membranes were thinner in average (8.2.-11.2 nm).

In should be emphasized that the ultrastructure in our own MCL series (CTX,atheromas) differed substantially from that in the Nasu-Hakola's disease,the latter being dominated by much simpler membranes,albeit with great tendency to apposition and formation of tubular structures,the part of which resembled superficially the highly organized trilaminar membranes dominating in the former.

Review of the data published on experimental ceroid production from pure lipids in vitro failed to find any evidence of a similar ultrastructure of induced ceroid,however (Porta, 1963,Kajihara et al.,1975). Nevertheless, it should be born in mind,the experimental conditions may be quite different from those leading to the MCL formation from the heterogeneous lipid rich tissue debris in vivo.Rare membranous structures were described in macrophages exposed to artificial lipoproteins (Ball et al.,1988).The lipopigment fine structure in neuronal ceroid-lipofuscinoses differes in many respects.It displays much higher tendency to membrane aggregation into fingerprint crystaloids or into fragmentary stacks of curvilinear or rectilinear patterns (Carpenter et al.,1977,Towfighi et al.,1973).Isolated longer membranous profiles are exceptionally seen in neuronal ceroid lipofuscinosis.When present in our cases their width r range was 8.9-11.2 nm.Linear elements were recognized even in classical neurolipofuscin and classified as tubules (Boellaard and Schlote,1986) of 8 nm width (Braak 1978).

Generally,the lysosomal membranous ceroid deposits displayed much greater tendency to constituent membrane aggregation when compared with MCL where this phenomenon was exceptional. The degree of packing may be directly proportional to the tightness of the compartment where the lipopigment is produced.The highest degree of packing was observed with intralysosomally segregated polar lipids in our series of classical lipidoses where simple isolated membranes were absent (gangliosidoses, sulphatidosis,Fabry's disease;unpublished personal observation). This is in keeping with the published data where isolated polar lipid membranes were observed only under artificial conditions (Samuels et al.,1963).

Despite all the differences it could be concluded that the membranous structure is very frequent in the family of lipopigments and therefore that in MCL may be seen as one of the numerous variants. Different dimensions and aggregation patterns of the membranes may reflect chemical differences in the precursors or in their lysosomal processing or in the presence of different concomitantly segregated compounds.This might also explain differences between the membranous patterns in our material and in Nasu-Hakola's disease.

As for the of the mechanism of MCL ceroid production the most plausible explanation seems to be that the ceroid membranes are synthesized directly de novo from lipid rich debris, i.e. extracellularly in the cell free millieu. The process is most probably catalyzed by factor/s released from necrotic cells. The arguments in favour of this hypothesis are as follows.

(a) All the situations leading to MCL formation are connected with <u>necrosis</u> of lipid rich tissues irrespective of the organ or tissue localization. Clear-cut examples are our case of CTX (with perivascular fibroatheromatous lesions, within which MCL arose), and the necrotic center of human atheromas which are a reliable source of MCL for all the prospective studies. The situation in the Nasu-Hakola's disease is not so clear-cut at first glance.The status of the affected adipose tissue is usually described,using general nonspecifying terms,as regression,reduction or replacement by MCL (Nasu et al.,1973,Yagishita et al.,1976). The term necrosis was used exceptionally (Järvi,1970).The presence of necrosis was proved by Wood (1978) but was considered to be a phenomenon secondary to MCL formation.It is highly probable that the necrosis of adipose tissue, which undoubtedly exists in the Nasu-Hakola disease,is most probably the primary event caused by an as yet unknown mechanism secondarily leading to MCL development.The finding of MCL in the adipose tissue in chronic arterial obstruction pieces well together with this assumption.The cell necrosis thus seems to be a factor of pathogenetic importance in the formation of MCL.

What remains difficult to explain is the presence of the MCL in the lung and in the liver sinusoids (Amano et al., 1987, Nasu et al.,1973, Yagishita et al.,1985) but hematogenous dissemination of the process could be hypothesized, even if fat embolism in the lung is not considered as an unambiguous explanation (Yagishita et al.,1985)

(b) There is no convincing evidence of a proliferation of any preexisting physiological cell membrane system of any kind which could be could be seen as a <u>precursor structure</u> of MCL.Neither in CTX (see above) nor in human atheromas (Ball et al.,1987;Mitchinson et al.,1985;Györkey et al.,1968) were such structures in lipid phagocytes observed. As for the Nasu-Hakola's disease,no rare focal proliferation of smooth endoplasmic reticulum in adipocytes (Yagishita et al.,1976) was found by Nasu et al.(1973).The questionable connection with the plasma membrane of adipocytes as reported by Tanaka (1980)is suggested to be significant by Wood (1973) even if, in our view,the illustrations are not convincing.All these findings are either questionable or too rudimentary to explain the mass of extracellular MCL membranes in necrotic tissue in Nasu-Hakola's disease or in other conditions, a finding incompatible with increased cell membrane production by viable cells.Also the thickness of MCL membranes in our series was almost twice that of normal cell membranes (Bloom and Fawcett,1986).

The only preformed cellular source of MCL may be the lysosomal ceroid membranes of lipid phagocytes released from damaged cells and subjected to the remodelling process.Partial participation of this mechanism cannot be excluded due to granular inhomogeneity of some MCL found in CTX. This conclusion,however may be generally acceptable only provi-

ded intensive intracellular accumulation of ceroid is a
regular concomitant phenomenon such as in atherosclerotic
plaques (see also Ball et al.,1987,Mitchinson et al.,1985).
This,however,is not the case in Nasu-Hakola's disease
(Nasu et al.,1973).

(c) Regarding the nature of the precursor compound it is tempting to think of a lipid as the most probable candidate due
to the occurrence of the process in the lipid rich tissues,
and to the high degree of apolarity of the MCL.However,the
remarkable resistence to most potent lipid extraction procedures is surprising and stands in contrast to other ceroid variants.These include deposits in the majority of
neuronal ceroid lipofuscinoses,which display much higher
degree of extractibility (Elleder 1981).It is also tempting to speculate that the main precursor lipid may differ
in various states and that this difference might be responsible for the ultrastructural heterogeneity (triacyl-
glycerols in adipose tissue affections,cholesterolesters
in CTX and in human atheroma). The presence of galactose
(Kitajima et al.,1988) and of aromatic aminoacid residues
(Table 1) point to participation of nonlipids,which might
copolymerize,stemming from the cell remnats or from some
local intercellular structures.However, minimal results
obtained with the battery of lectins speak for the paucity
of the hexoses detectable by this approach. It should be
pointed out that essentially the same results were obtained in classical histiocytic ceroids (see also Elleder,
1989).These findings stand in contrast to an intense PAS
positivity of MCL and histiocytic ceroids generally and
rise a question whether autooxidized lipids do not participate substantially in the staining as could be infered
from experimental induction of ceroid in vitro (Casselman
1951,Porta 1963). It is worth mentioning in this connection
that even blood lipoproteins should be taken into consideration as a possible MCL precursor owing to the fact that
they are presumed to be a precursor of intracellular histiocytic ceroid (Ball et al.,1987,Mitchinson et al.,1988).

CONCLUSIONS AND PERSPECTIVES

All the results obtained point to the lipopigment nature
of the MCL and to its nosological unspecifity. It is closely
associated with lipid rich tissue debris within which it arises
in the form of membranes (simple or trilaminar) constructed from
lipids and nonlipids.The process seems to be analogous to that
leading to ceroid deposition in the lysosomal compartment of hi-
stiocytes.It is probably controlled by the same enzymes,capable
of lipid peroxidation (Mason et al.,Stossel et al.,1974),rele-
ased extracellularly.Lipid oxidation capacity has been described
even in other cell types,e.g.endothelium (Steinbrecher et al.,
1984).The process of MCL production may,in some instances,e.g.,
in atherosclerotic plaques,proceed partly or predominantly intralysosomally,the intracellular product being remodelled extracellularly as suggested by Ball et al. (1987).

However,important details of the process are still missing.
This may be partly overcome by more extensive search for MCL incidence in the pathology of adipose tissue and in lipid metabolic disorders generally,as according to the reports published

so far the occurrence of MCL could easily be taken as relatively rare and,accordingly,the mechanism as specific and different
from that in other ceroid variants.However,the high incidence
of LMC in human atheroma suggests it might be a more common
phenomenon than expected. Ultrastructural analysis of further
cases may show the full range of the MCL fine structure and help
to recognize whether there are precursor dependent patterns.
Future investigation may also show wether the MCL should be generally taken as a synonym for extracellular ceroid. Our first
results seem to be promising in this sense (unpublished).

In the light of these findings the Nasu-Hakola's disease
should be classified as an extensive primarily extracellular
ceroidosis consenquent to the massive adipose tissue breakdown.
According to the nonspecific nature of the membranous ceroid
structures,which might be taken as a "marker" of adipocytes'
necrosis,it is the biochemical defect leading both to a major
regression of white adipose tissue and to leukodystrophic
changes in the brain which should be searched for.Nevertheless,
the diagnostic validity of the MCL in Nasu-Hakola's disease
remains high,but its presence must always be assessed in con-
juction with other manifestations of the disease, each of them
being individually nonspecific.

REFERENCES

Alroy, J., Ucci, H.A., Pereira, E.A, 1984, Lectins: histochemical
probes for specific carbohydrate residues, In: "Advances in Immunohistochemistry", R.A. DeLellis,ed.,Masson Inc.,New York.
Amano, N., Iwabuchi, K., Sakai, H., Yagishita, S., Iton, Y.,
Isyeki, E., Yokoi, S., Arai, N., Kinoshita, J., 1987, Nasu-Ha-
kola's disease membranous lipodystrophy, Acta Neuropathol.(Berl).
74:294.
Ball,R.Y.,Carpenter,K.L.H.,Mitchinson,M.J.,1987,What is the si-
gnificance of ceroid in human atherosclerosis? Arch.Pathol.Lab.
Med. 111:1134.
Ball,R.Y.,Carpenter,K.L.M.,Mitchinson,M.J.,1988, Ceroid accu-
mulation by murine peritoneal macrophages exposed to artificial
lipoproteins: ultrastructural observations.Br.J.exp.Path.69:43.
Bloom,W.,Fawcett,D.W.,1986, A textbook of histology, Saunders,
Philadelphia.
Boellaard,J.W.,Schlote,W.,1986, Ultrastructural heterogeneity
of neuronal lipofuscin in the normal human cerebral cortex.
Acta Neuropathol.(Berl.) 71:285.
Braak,E.,1978, On the structure of the human striate area.
Lamina IVcB.Cell Tiss.Res. 188:217.
Burt,R.C.,1952, The incidence of acid-fast pigment (ceroid) in
aortic atherosclerosis. Am.J.Clin.Pathol. 22:135.
Carpenter,S.,Karpati,G.,Andermann,F.,Jacob,J.C.,Andermann,F.,
1977, The ultrastructural characteristics of the abnormal
cytosomes in Batten-Kufs' disease. Brain 100:137.
Elleder,M.,1977, Lipidhistochemistry - a critical survey, Acta
histochemica (Jena) Suppl.-Band, XIX:239.
Elleder,M.,1981, Chemical characterization of age pigments. In:
"Age Pigments" R.S.Sohal,ed.,Elsevier,Amsterdam.
Elleder,M.,1989, Lectin histochemistry of lipopigments with spe-
cial regard to neuronal ceroid - lipofuscinosis. Results with
Concanavalin A. Histochemistry, in press.

Elleder,M.,Michalec,C.,Jirasek,A.,Khun,R.,Ranny,M.,1989,Membra-
nocystic lesion (MCL) in the brain of cerebrotendinous xanthomatosis. Histochemical and ultrastructural study with evidence
of its ceroid nature,Virchows Arch (Pathol.Anat.),in press.
Elleder,M.,Šmíd,F.,1977,Lysosomal non-lipid component of Gau-
cher's cells. Virchows Arch.B Cell Path. 26:133.
Fullmer,H.M.,Lillie,R.D.,1976, Histopathologic technic and
practical histochemistry. Chapter 11. Endogenous pigments,
Mc-Graw-Hill, New York.
Gedigk,P.,Totovič, V.,1983, Lysosomes and lipopigments, In:Cel-
lular pathobiology of human disease,B.Trump,A.Laufer,R.T.Jones,
Gustav Fischer, New York, Stuttgart.
Györkey,T.,Shimamura,T.,O'Neal,R.M.,1968, The fine structure
of ceroid in human atheroma. J.Histochem.Cytochem. 15:732.
Järvi,O.,1970, A new entity of phacomatosis:a bone lesions
(hereditary polycystic osteodysplasia). Acta pathol.microbiol.
Scand.Suppl.215:27 (abstract).
Kajihara,H.,Totovič,V.,Gedigk,P.,1975, Ultrastructure and morp-
hogenesis of ceroid pigment.II.Late changes of lysosomes in
Kupffer cells of rat liver after phagocytosis of unsaturated
lipids. Virchows Arch. B Cell Path. 19:239.
Kitajima,I.,Suganuma,T.,Murata,F.,Nagamatsu,K.,1988, Ultrastructural demonstration of Maclura pomifera agglutinin binding
sites in the membranocystic lesions of membranous lipodystrophy
(Nasu-Hakola disease). Virchows Arch.(Pathol.Anat.) 413:475.
Machinami,R.,1983, Membranous lipodystrophy-like changes in
ischemic necrosis of the legs. Virchows Arch. (Pathol.Anat.)
399:191.
Machinami,R.,1984, Incidence of membranous lipodystrophy-like
change among patients with limb necrosis caused by chronic
arterial obstruction. Arch.Pathol.Lab.Med., 108:823.
Mason,R.J.,Stossel,T.P.,Vaughan,M.,1972, Lipids of alveolar
macrophages,polymorphonuclear neutrophil leukocytes and their
phagocytic vesicles. J.Clin.Invest. 51:2399.
Mitchinson,M.J.,Hothersall,D.C.,Brooks,P.N.,deBurbure,C,Y.,
1985, The distribution of ceroid in human atherosclerosis.
J.Pathol. 145:177.
Nasu,T.,Tsukahara,Y.,Terayama,K.,1973, A lipid metabolic disease - "membranous lipodystrophy" - an autopsy case demonstrating numerous peculiar membrane-structures composed of compound lipid in bone and bone marrow and various adipose tissues. Acta pathol.Jap. 23:539.
Pearse,A.G.E.,1968, Histochemistry.Theoreticall and Applied,
Volume 1, J.A.Churchill, London.
Porta,E.A.,1963, Experimental electron microscopic study of
the sequential stages of in vitro formation of ceroid. Exp.
mol.Pathol. 2:219.
Samuels,S.,Korey,S.,Gonatas,J.,Terry,R.D.,Weiss,M.,1963.Studies
in Tay-Sachs disease.IV.Membranous cytoplasmic bodies.J.Neuro-
path.exp. Neurol. 22:81.
Schornagel,H.E.,1956, The occurrence of iron and ceroid in
coronary arteries. J.Path.Bact. 72:267.
Sinapius,D.,Gunkel,R.D.,1964, Ceroid bei Lipoidose und Athero-
sklerose der Aorta. Morphologie und Histochemie. Frankfurt. Z.
Path. 78:485.
Sohal,R.S.,Wolfe,L.S.,1986, Lipofuscin: characteristics and
significance. In: "Progress in Brain Research", D.F.Swaab,
E.Fliers,M.Mirmiran,W.-A.van Gool and F.van Haaren,eds.,Else-
vier, Amsterdam.
Steinbrecher,U.P.,Parthasarathy,S.,Leako,D.S.,1984,Modification
of low density lipoprotein by endothelial cells involves lipid

peroxidation and degradation of low density lipoprotein phospholipids. Proc.Natl.Acad.,Sci USA 81:3883.

Sternberg,L.A.,Hardy,P.H.,Cuculis,J.J.,Meyer,H.G.,1970, The unlabelled antibody enzyme method of immunohistochemistry. Preparation and properties of soluble antigen-antibody complex (horse-radish peroxidase) and its use in identification of spirochetes. J.Histochem.Cytochem. 18:315.

Stossel,T.P.,Mason,R.J.,Smith,A.L.,1974, Lipid peroxidation by human blood phagocytes, J.Clin.Invest. 54:638.

Tanaka,J.,1980, Leukoencephalopathic alteration in membranous lipodystrophy. Acta Neuropathol. (Berl.) 50:193.

Towfighi,J.,Baird,H.W.,Gambetti,P.,Gonatas,N.K.,1973, The significance of cytoplasmic inclusions in late infantile and junevile amaurotic idiocy. An ultrastructural study. Acta Neuropathol (Berl.) 23:32.

Yagishita,S.,Ito,Y.,Ikezaki,R.,1976, Lipomembranous polycystic osteodysplasia. Virchows Arch. (Pathol.Anat.) 372-245.

Wood,C.,1978, Membranous lipodystrophy of bone. Arch.Pathol. Lab.Med. 102:22.

DISCUSSION

TAKAHASHI: On the basis of the terminology proposed by Dr. Jolly, would you call this pigment preceroid or ceroid?

ELLEDER: It is really ceroid, but undoubtedly in the initial stages of this granulomatous formation, it was preceroid. I agree with Dr. Jolly in that pigments derived from lipids should be called ceroid, and those derived from other precursors should be termed differently. The precursors of this particular pigment are unquestionably lipids, so I prefer to call it ceroid.

TAKAHASHI: Does this pigment accumulate in lysosomes?

ELLEDER: No, this is totally an extracellular form of ceroid that most probably is formed and controlled by enzymatically mediated processes. All the lysosomal enzymes have been released in these necrotic foci, and are, therefore, available at the site of pigment formation and so capable of converting the lipids to the pigment material.

GOEBEL: I have two questions; the first is that you said that the material starts out from triglycerides and fat tissue, which by electron microscopy is somewhat electron lucent. Then, you get the membranes which are highly organized structures. Is there other material entering into the homogeneous translucent lipids to organize them into laminar structures? The second question is whether there is a transient stage or this is an end product which can be phagocytized by macrophages and degrade it?

ELLEDER: I am afraid I cannot answer your question because I also would like to know why they are so highly organized. Now whether macrophages could engulf it, I don't know because the cells in these foci are all necrotic and difficult to identify. It would be perhaps possible to isolate this material and offer it to macrophages in vitro to see what happens.

JOLLY: I am anxious to go back to my lab and look at these cells in nutritional steatitis. I would think that is logical to expect that these membranous structures are more than just lipids. They must have proteins to organize and modify the lipids. This might point once again to the fact that proteins are also involved in the formation of ceroid.

ELLEDER: I agree totally with that.

Zs.-NAGY: It would be interesting to explore whether failure of autophagocytosis is involved in the formation of these membranocystic lesions, and then to understand which are the main points of the autophagocytosis failure.

ELLEDER: I don't think that failure of autophagocytosis has anything to do with this pigment. This model essentially originated in a necrotic process and is different, for instance, from lipogranulomas where the process of phagocytosis may be studied. Furthermore, although the pigment is similar to intralysosomal ceroid, it is totally extracellular ceroid.

GOEBEL: From a generalized point of view of histopathology, with the exception of atheroma, you cited examples of very rare conditions in which this process occurs. Is this a particular type of necrosis? I am asking this because in any fat necrosis, whether in the brain or elsewhere, this process or structure is not seen.

ELLEDER: I think that the best answer to your question and comments is that people don't care to detect it, because in routine pathology everything occurring in adipose tissue is unimportant. I am sure that if you look more in detail to the necrosis of fat tissue, you would be surprised how frequently this lesion is present. This is, in fact, a very common lesion and has been repeatedly found in ordinary ischemic brain lesions removed by surgeons.

HALL: If you get these structures building up in ischemic organs, there is a very simple experimental system to induce them. It should be possible to look at the process of pigment formation in a very simple manner, if the observation you made is general.

207

CEROID IN GENETIC DISORDERS

LYSOSOMAL STORAGE OF THE DCCD REACTIVE PROTEOLIPID SUBUNIT OF MITOCHONDRIAL ATP SYNTHASE IN HUMAN AND OVINE CEROID LIPOFUSCINOSES

D. N. Palmer[1,2], I. M. Fearnley[2], S. M. Medd[2], J. E. Walker[2], R. D. Martinus[1], S. L. Bayliss[1], N. A. Hall[3], B. D. Lake[3], L. S. Wolfe[4] and R. D. Jolly[1]

[1] Department of Veterinary Pathology and Public Health, Massey University, Palmerston North, New Zealand

[2] Medical Research Council Laboratory of Molecular Biology, Hills Road, Cambridge, CB2 2QH, England

[3] Institute of Child Health, 30 Guildford Street, London, WC1 1EH, England

[4] Montreal Neurological Institute and Hospital, McGill University, 3801 University Street, Montreal, Quebec, Canada, H3A-2B4

ABSTRACT

The ceroid lipofuscinoses (Batten's disease) are a group of neuro-degenerative lysosomal storage diseases of children and animals that are recessively inherited. In the diseased individuals fluorescent storage bodies accumulate in a wide variety of cells, including neurons. The material stored in the cells of sheep affected with ceroid lipfuscinosis is two-thirds protein. The stored material does not arise from lipid peroxidation or a defect in lipid metabolism, and the lipid content is consistent with a lysosomal origin for the storage bodies. The major protein stains poorly with Coomassie blue dye and is soluble in organic solvents. It has an apparent molecular weight of 3,500 and its amino acids sequence is identical to that of the dicyclohexylcarbodiimide (DCCD) reactive proteolipid, subunit c, of mammalian mitochondrial ATP synthases. Apart from removal of mitochondrial import sequences, it has not been modified post-translationally. At least 50% of the mass of the storage bodies is composed of this protein. A minor protein sequence related to the 17-kDa subunit of vacuolar H^+-ATPase is also found in storage bodies isolated from pancreas.

As in humans and cattle, the ovine protein is the product of two expressed genes named P1 and P2. In normal and diseased animals there are no differences in sequence between P1 cDNAs or P2 cDNAs, nor do levels of mRNAs in liver for P1 or P2 differ substantially between normal and diseased animals. Both normal and diseased sheep also express a spliced pseudogene encoding amino acids 1 to 31 of the mitochondrial import presequence. The peptides they encode differ by one amino acid; arginine-23 is changed to glutamine in the diseased sheep.

Storage bodies isolated from brains and pancreas of children affected with the juvenile and late infantile forms of ceroid lipofuscinosis also contain large amounts of material that is identical to subunit c of ATP synthase. However, the protein is not present in storage bodies isolated from brains of patients affected with the infantile form of the disease, and these storage bodies contain other unidentified proteins. It is possible that the cause of ovine, juvenile and late infantile ceroid lipofuscinoses is related to a defect in degradation of the subunit c of mitochondrial ATP synthase.

INTRODUCTION
The neuronal ceroid-lipofuscinoses (Batten's disease) are a group of recessively inherited lysosomal storage diseases of children that lead to blindness, seizures, dementia and premature death. Three main forms have been described on basis of the age of onset of clinical features, namely the infantile, late infantile and juvenile forms. A number of eponyms have also been used for each of these diseases. Other variant types and an adult form (Kufs' disease) have also been reported (Rider and Rider, 1988; Wisniewski et al., 1988; Boustany et al., 1988; Dyken, 1988; Lake, 1984; Eto et al., 1988). Collectively they are the most common lysosomal storage diseases, and their incidence has been estimated to be in the range of 1 in 12,500 to 1 in 25,000 live births (Rider and Rider, 1988; Zeman, 1976). Ceroid lipofuscinoses also occur in a number of animals (Palmer et al., 1988). In particular a flock of affected sheep have been bred, maintained and studied as a model of the human diseases (Jolly et al., 1980; 1982; Graydon and Jolly, 1984; Mayhew et al., 1985).

The characteristic features of these diseases are retinal and brain atrophy and the accumulation of a fluorescent storage bodies in neurons and a wide variety of other cells. The visual similarity of these storage bodies in the light microscope to the lipopigments ceroid and lipofuscin gave rise to the name ceroid lipofuscinosis (Zeman and Dyken, 1969), but the biochemical anomalies underlying this group of diseases remain undetermined, although a variety of hypotheses have been advanced. It has been proposed that they arise because of abnormal peroxidation of lipids (Zeman, 1974; Siakotos et al., 1988) perhaps due to a deficiency of a peroxidase (Armstrong et al., 1974), or because of disturbances of fatty acid metabolism (Svennerholm et al., 1975; Pullarkat et al., 1982), or because of defects in retinoic acid and dolichol metabolism (Wolfe et al., 1977; Ng Ying Kin et al., 1983), or in iron metabolism (Gutteridge et al., 1982). More recent suggestions include defective metabolism of dolichol-linked oligosaccharides (Hall and Patrick, 1985), that might cause a lack of cathepsin D activity (Pullarkat et al., 1988), a defective thiol endoprotease causing faulty recycling of lysosomal membranes (Wolfe et al., 1987), the inhibition of cathepsin B activity by accumulating abnormal peroxides (Dawson and Glaser, 1987) and defective processing of amyloid precursor protein (Wisniewski and Maslinska, 1989). It has also been suggested that the infantile form be considered as a separate type of disease and the name "Polyunsaturated Fatty Acid Lipidosis" has been proposed for it (Svennerholm, 1976).

Most of these hypotheses are based upon the presumed nature of the material in storage bodies or of compounds found at "elevated levels". Until the present work, there has been little systematic study of the composition of storage bodies in the human diseases, but such studies have been made in the ovine disease. These experiments on the ovine and human storage bodies are reviewed below, and particular attention is drawn to the relevance of these findings in understanding both the ovine and human diseases.

Composition of storage bodies

Storage bodies have been isolated from fresh tissues of affected sheep using a combination of osmotic shock, sonication and low speed centrifugation. They contain very little contaminating material (Palmer et al., 1986b; 1988). They also were isolated from brains and pancreas of children affected with the juvenile, late infantile and infantile diseases. As in the case of the sheep bodies, the human bodies retained an ultrastructural appearance that was indistinguisable from that observed in electron micrographs of tissues (Fig. 1).

Storage bodies from sheep have been analysed in order to determine the nature of the stored component, in the expectation that it might provide clues to the nature of the defect (Fig. 2). They contain normal lysosomal phospholipids, including bis(monoacylglycero)phosphate, and the neutral lipids, cholesterol, dolichol, ubiquinone and dolichyl esters. These two lipid classes are present in similar amounts and make up less than one third of the mass of the storage bodies (Palmer et al., 1986 a, b). They contain 1-2 % by weight of metal ions and the metal composition is also indicative of a lysosomal origin of the storage bodies (Palmer et al., 1988). Dolichyl pyrophosphate linked oligosaccharides are a minor component (Hall et al., 1989). Most of the mass of the bodies is protein. Compositions of the storage bodies in the human diseases have not been established as thoroughly. However, they appear to be similar in both lipid and protein contents (Wolfe, et al., 1987; Palmer et al., 1988) and in density to the ovine bodies. Dolichyl pyrophosphate linked oligosaccharides are a minor constituent (Hall et al., 1989).

Analysis of Storage Body Proteins

Ovine storage body proteins are difficult to solubilise, but they dissolve in solutions of lithium or sodium dodecyl sulphate in the presence of high concentrations of mercaptoethanol (Palmer et al., 1986b, 1988). Analysis of the proteins by polyacrylamide gel electrophoresis revealed that a major component has an apparent Mr of 3,500 (Palmer et al., 1986b, 1988; Fearnley et al., 1989). Similarly, storage bodies from victims of the juvenile and late infantile diseases also contain a component of the same size (Fig. 3), but it was not detectable in storage bodies isolated from a brain affected with the infantile form of the disease.

Storage bodies from all sources except the infantile diseased brain contained few components of high molecular weight, and these were present in very minor amounts except for a band from the ovine pancreatic bodies that migrated with an apparent Mr of 14,800. The amount of this band was always much less than that of the 3,500 Mr component and varied from preparation to preparation and from gel to gel. A band at 24,000 Mr was also seen when larger amounts were loaded onto the gel (Fearnley et al., 1989). Both of these higher molecular weight forms have been shown by sequencing to be oligomers of the protein with an Mr of 3,500.

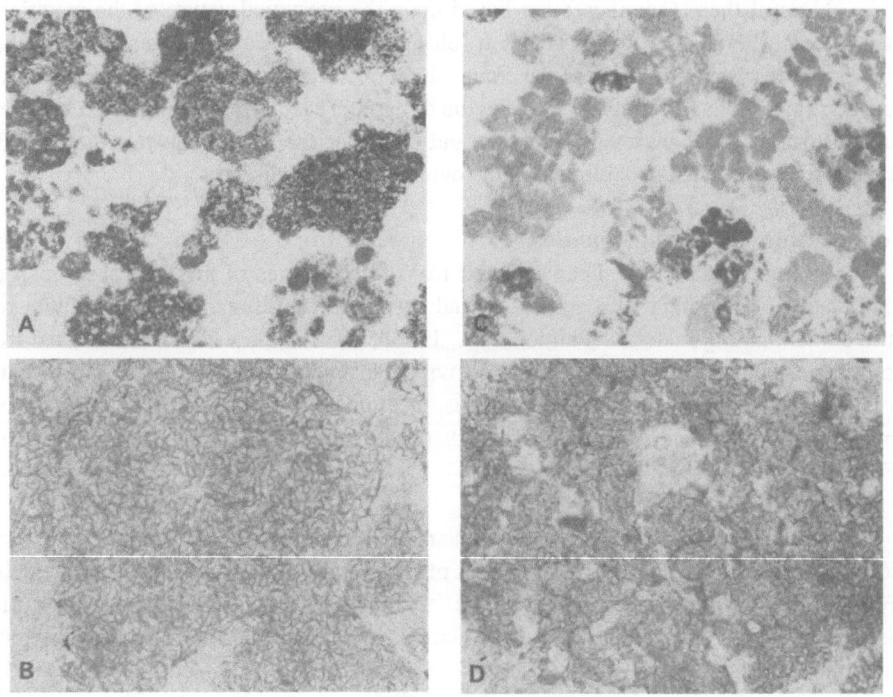

Figure 1. Electron micrographs of storage bodies associated with Batten's disease isolated from (A), a juvenile affected brain; (B), a late infantile affected brain; (C), an infantile affected brain; (D), a late infantile affected pancreas.

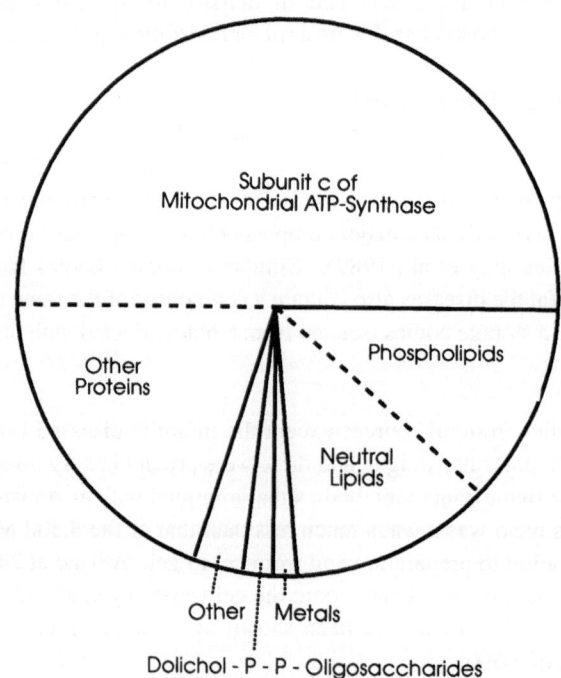

Figure 2. Composition of the storage bodies in ovine ceroid lipofuscinosis

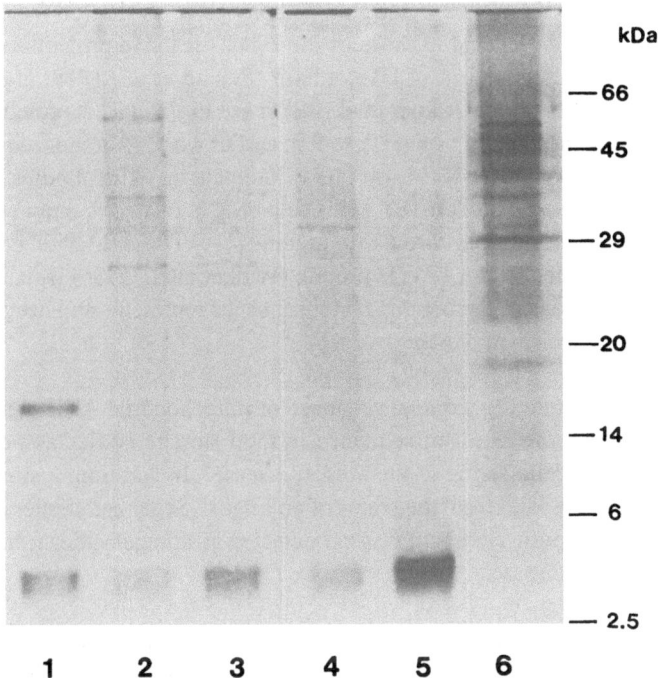

kDa

— 66

— 45

— 29

— 20

— 14

— 6

— 2.5

1   2   3   4   5   6

Figure 3. Polyacrylamide gel electrophoresis in the presence of lithium dodecyl sulphate of storage body proteins. The gel was silver stained. The storage bodies were isolated from the following sources; track 1, ovine affected pancreas; track 2, juvenile affected pancreas; track3 late infantile affected pancreas; track 4, juvenile affected brain; track 5, late infantile affected brain; track 6, infantile affected brain.

| 1 | | | | | | | | | 10 | | | | | | | | | | 20 |
Asp-Ile-Asp-Thr-Ala-Ala-Lys-Phe-Ile-Gly-Ala-Gly-Ala-Ala-Thr-Val-Gly-Val-Ala-Gly-

| 21 | | | | | | | | | 30 | | | | | | | | | | 40 |
Ser-Gly-Ala-Gly-Ile-Gly-Thr-Val-Phe-Gly-Ser-Leu-Ile-Ile-Gly-Tyr-Ala-Arg-Asn-Pro-

| 41 | | | | | | | | | 50 | | | | | | | | | | 60 |
Ser-Leu-Lys-Gln-Gln-Leu-Phe-Ser-Tyr-Ala-Ile-Leu-Gly-Phe-Ala-Leu-Ser-Glu-Ala-Met-

| 61 | | | | | | | | | 70 | | | | | | | | | | 80 |
Gly-Leu-Phe-Cys-Leu-Met-Val-Ala-Phe-Leu-Ile-Leu-Phe-Ala-Met

Figure 4. The amino acid sequence of the major protein constituent of storage bodies associated with lipofuscinosis in humans and sheep determined by direct protein sequence analysis. The sequence is identical to that of subunit c, the DCCD reactive proteolipid, of mitochondrial ATP synthase of cattle, humans and sheep.

Sequence analysis of total ovine storage body proteins showed that the N-terminal amino acids 1-40 were identical to those of the DCCD reactive proteolipid, subunit c, of human and bovine mitochondrial ATP synthase (Sebald et al., 1979; Gay and Walker, 1985; Dyer and Walker, 1989; Palmer et al., 1989; see Fig. 4). This protein has also been called 'the lipid binding subunit', or ATPase-9 in earlier work. A minor sequence Ala-Pro-Glu-Tyr-X-Ser-X-X-Ala-Met-Val- was also detected in ovine bodies isolated from pancreas, but not from any other tissue. This sequence is related to amino acids 7-17 of a proteolipid isolated from bovine chromaffin granule H+-ATPase (Mandel et al., 1988), and to the N-terminal sequence of a 17 kDa protein (Walker et al., 1986) isolated from mouse gap junction preparations. Whether this minor sequence represents an intrinsic components of storage bodies is unclear at present.

Similarly, the same N-terminal sequence of mitochondrial ATP synthase subunit c was shown to be the predominant sequence in total storage bodies associated with the juvenile and late infantile forms of the human disease. In addition, a similar result was obtained with storage bodies from the brains of an English Setter and tissues of Devon cattle affected with ceroid lipofuscinosis. It was not detected in storage bodies from the brain of a victim of the infantile disease.

The c-subunit of mitochondrial ATP synthase is very hydrophobic, and is classified as a proteolipid because of its lipid-like solubility in chloroform methanol mixtures (Folch and Lees, 1951). The term proteolipid does not imply that lipid is attached covalently to the protein. In common with the protein present in the storage bodies it migrates on polyacrylamide gels with an apparent Mr of about 3,500, stains poorly with Coomassie blue dye and forms aggregates (Walker et al., 1989). About 70 % of the ovine storage body proteins can be extracted into chloroform:methanol, and the amino acid composition of this fraction was very similar to that of the complete subunit c (Fearnley et al., 1989), and this has been established to be so by direct protein sequencing (see Fig. 4). These experiments showed that the protein is identical to the subunit c that is assembled into the mitochondrial ATP synthase complex, and that the protein has not been modified in any detectable way. Mitochondrial import sequences that are present at the N-terminal end of precursors of the subunit c have been removed by proteolysis at precisely the same position in the polypeptide chain. Similar experiments conducted on human storage bodies have shown that the complete subunit c is stored in the juvenile and late infantile diseases. The proteolipids in storage bodies associated with the infantile form of the disease have yet to be identified.

From sequence data, it can be calculated that subunit c constitutes at least 73 % of the protein in the ovine storage bodies (Palmer et al., 1989; Fearnley et al., 1989). Since the bodies are at least two thirds protein this means that subunit c accounts for greater than 50% of the total storage body mass. Similar calculations made for the bodies from the juvenile and late infantile ceroid lipofuscinoses gave values ranging from 20 to 85 % of storage body protein.

Studies of genes for subunit c of mitochondrial ATP synthase
In humans and cattle, subunit c has two nuclear genes, called P1 and P2, that encode precursor proteins containing the same mature protein, but with different N-terminal presequences to direct their import into mitochondria (Gay and Walker, 1985; Dyer et al., 1989; Dyer and Walker, 1989). These presequences are removed during entry into the organelle. In addition, numerous spliced and partly spliced pseudogenes related to P1 or P2

have been discovered in man and cattle (Dyer & Walker, 1989) and many are also present in sheep. In cattle, studies of the distribution of mRNAs for P1 and P2 in various tissues have demonstrated that both the P1 and P2 genes are expressed in all tissues, but the ratio of their expression differs according to the embryonic origin of the tissue. The ratio of P1:P2 is 1:1 in tissues of mesodermal origin and 1:3 in those of endodermal origin (Gay and Walker, 1985).

By constructing cDNA libraries and using the polymerase chain reaction (PCR) on total cDNA, we have cloned cDNAs for the P1 and P2 genes from normal sheep and sheep with ceroid lipofuscinosis. By sequence analysis of these clones it has been shown that the coding region sequences of the P1 and P2 cDNAs cloned from the sheep with ceroid lipofuscinosis are identical with those of the normal sheep. Therefore, both precursor proteins are unchanged in normal and diseased animals, and the disease is not caused by mutation in the mitochondrial import sequences. We have also compared the expression of P1 and P2 genes in normal and diseased sheep liver by studying the amounts of mRNA for P1 and P2 in the tissue by Northern hybridisation. These experiments have shown that there is no gross difference in the levels of the mRNAs for P1 and P2 in normal sheep and sheep with ceroid lipofuscinosis. Therefore, it appears to be unlikely that the c-subunit is being synthesised in the diseased animals in greater amounts than are normally required for assembly in mitochondria of the multi-subunit enzyme, ATP synthase.

One unexpected observation was made in the course of the cDNA cloning experiments. A mutated cDNA related to P2 was isolated from the cDNA library made from sheep with ceroid lipofuscinosis. By the use of oligonucleotide primers specific for the mutated P2 transcript the PCR technique was used to show that this cDNA arose by expression of a spliced P2 pseudogene, and that normal sheep also express a closely related pseudogene. The pseudogene from the normal sheep encodes a 31 amino acid peptide identical with the first 31 amino acids of the P2 precursor, but the peptide encoded by the pseudogene from the sheep with ceroid lipofuscinosis is one amino acid different, Arg-23 being changed to Gln. Spliced pseudogenes are believed to be retroposons that are copied from mRNA during evolution by reverse transcriptase, and the resulting cDNA is then introduced into the genomic DNA. Expression of a pseudogene, as in the present case, requires that potential promoter elements are present to the 5' side of the site of insertion.

The peptides that are encoded in the transcribed pseudogenes conceivably could interfere with mitochondrial import by blocking import receptors or by inhibiting proteolytic cleavage of the import presequences. This hypothesis can be investigated by transforming these sequences in expression vectors into appropriate cell lines and observing what effects, if any, ensue. In both normal and diseased animals the pseudogenes are apparently transcribed in very low amounts compared with the transcription of the normal genes, and hence it is perhaps unlikely that their products perform a significant function in the cell. It follows that the observed difference in amino acid sequence between the products of the transcribed pseudogene in normal and diseased sheep is also unlikely to be the cause of ovine ceroid lipofuscinosis.

A further argument against the product of the transcribed pseudogene playing a significant causative role in ceroid lipofuscinosis is that it appears to be likely that the ovine and the late infantile and juvenile forms of Batten's disease are related to a common cellular defect. Given that the spread of pseudogenes probably took place after speciation, this would require that the transcribed pseudogene arose twice in two independent events.

Summary and conclusions

1.   We have shown that the storage body proteins associated with ovine ceroid lipofuscinosis and with the late infantile and juvenile forms of Batten's disease are almost entirely of a single hydrophobic polypeptide. This polypeptide is identical to the mature c-subunit of the mammalian ATP synthase, a multi-subunit enzyme found in mitochondria.

2.   Its mitochondrial import sequences have been removed, which implies that the protein has passed through the mitochondrial import machinery prior to its accumulation in lysosomes.

3.   This protein is not present in storage bodies found in the infantile form of Batten's disease, and other unidentified proteins accumulate.

4.   As in humans and cattle there are two different unspliced genes P1 and P2 in sheep that encode different precursors of the same mature c-subunit. The coding sequences of these genes in sheep with ceroid lipofuscinosis are identical to those found in normal animals. They are also transcribed at approximately the same levels in normal and diseased sheep. Therefore, ovine ceroid lipofuscinosis is not caused by the misdirection of precursors of the c-subunit to lysosomes rather than to mitochondria due to a mutation in the presequences of either P1 or P2 ; nor is it a consequence of a increase in the level of synthesis of either precursor in the diseased animal relative to the normal sheep.

5.   A spliced pseudogene related to the P2 unspliced gene is transcribed at low levels in both normal sheep and sheep with ceroid lipofuscinosis. These transcripts encode the first 31 amino acids of the import presequence and the peptide produced in the diseased sheep differs in a single amino acid from that in normal animals. It is possible that these peptides could interfere with mitochondrial import by blocking import receptors or by inhibiting proteolytic cleavage of the import presequences. However, the very low levels of transcription make this a rather unlikely eventuality, and so the amino acid sequence difference between the normal and diseased sheep pseudogene products is also unlikely to be the cause of ceroid lipofuscinosis. Nor is it likely that humans have the same expressed pseudogene, yet the cellular defect in the juvenile and late infantile diseases is closely related to that in the ovine disease.

6.   The possibility remains that a defect in the mitochondrial import pathway causes one or both subunit c precursors to bypass mitochondria and to be processed to the mature protein by a lysosomal protease, but we find this hard to rationalise with the apparent normality of mitochondria in the diseased cells. Assembly of subunit c into the ATP synthase complex is another possible process affected in animals with ceroid lipofuscinosis, but this process is essential to the formation of a functional ATP synthase complex and if disrupted would probably be lethal.

7.   A plausible hypothesis is that ceroid lipofuscinosis and the juvenile and infantile forms of Batten's disease are caused by a lesion in the degradative pathway of subunit c of mitochondrial ATP synthase. Mitochondrial protein degradation and turnover are poorly understood. Lysosomes are thought to digest whole mitochondria by a process of autophagocytosis and to posses many proteases, or cathepsins, with a wide range of specificities. The results of turnover studies on a variety of mitochondrial proteins have shown that turnover rates vary enormously. Even the proteins of the mitochondrial inner membrane, of which subunit c is an example, can differ widely in their turnover rates. An ATP dependent protease has been localised in the mitochondrial matrix and is capable of

hydrolysing proteins to amino acids, although it is not known whether this protease or any other mitochondrial proteases are involved in the degradation of subunit c.

ACKNOWLEDGEMENTS

This work was supported in part by the United States National Institute of Communicative Disorders and Stroke grant number NS11238 (D.N.P.,R.D.M.,R.D.J.), a New Zealand MRC -Wellcome Trust Travel Fellowship and the Children's Brain Diseases Foundation, San Francisco (D.N.P.), the Medical Research Council of Great Britain (N.A.H.) and the Medical Research Council of Canada, grant number MT-1345 (L.S.W.).

REFERENCES

Armstrong, D., Dimmett, S. and Van Wormer, D. E., 1974, Studies in Batten's disease. I. Peroxidase deficiency in granulocytes, Arch. Neurol. 30: 144-152.

Boustany, R. N., Alroy, J. and Kolodny, E. H., 1988, Clinical classification of neuronal ceroid lipofuscinosis subtypes, Am. J. Medical Genet. Suppl. 5: 47-58.

Dawson, G. and Glaser, P., 1987, Apparent cathepsin B deficiency in neuronal ceroid lipofuscinosis can be explained by peroxide inhibition, Biochem. Biophys. Res. Commun. 147:267-274.

Dyer, M. R., Gay, N. J. and Walker, J. E., 1989, DNA sequences of a bovine gene and of two related pseudogenes for the proteolipid subunit of mitochondrial ATP synthase, Biochem. J. 260:249-258.

Dyer, M. R. and Walker, J. E., 1989, Sequence analysis of members of the human gene family for the proteolipid subunit of mitochondrial ATP synthase, J. Mol. Biol. in press.

Dyken, P. R., 1988, Reconsideration of the classification of the neuronal ceroid-lipofuscinoses, Am. J. Medical Genet. Suppl. 5:69-84.

Eto, Y., Tsuda, T., Ohhashi, T., Yamaguchi, S. and Okuno, A., 1988, Clinical and biochemical studies of Japanese neuronal ceroid-lipofuscinoses, Am. J. Medical Genet. Suppl. 5:59-68.

Fearnley, I. M., Walker, J. E., Jolly, R. D., Martinus, R. D., Kirkland, K. B., Shaw, G. J. and Palmer, D. N., 1989, The major protein stored in ovine ceroid lipofuscinosis is identical to the DCCD-reactive proteolipid of mitochondrial ATP synthase, Biochem J. submitted for publication.

Folch, J. and Lees, M., 1951, Proteolipides, A new type of tissue lipoproteins. Their isolation from brain, J. Biol. Chem. 191:807-817.

Gay, N. J. and Walker, J. E., 1985, Two genes encoding the bovine mitochondrial ATP synthase proteolipid specify precursors with different import sequences and are expressed in a tissue -specific manner, EMBO J. 4: 3519-3524.

Graydon, R. J. and Jolly, R. D., 1984, Ceroid-lipofuscinosis (Batten's disease): Sequential electrophysiologic and pathologic changes in the retina of the ovine model, Invest. Ophthalmol. Visual Science 25:294-301.

Gutteridge, J. M. C., Rowley, D. A., Halliwell, B. and Westermarck, T., 1982, Increased non-protein bound iron and decreased protection against superoxide-radical damage in cerebrospinal fluids from patients with neuronal ceroid lipofuscinoses, Lancet II:459-460.

Hall, N. A., Lake, B. D., Palmer, D. N., Jolly, R. D. and Patrick, A. D., 1989, Glycoconjugates in storage cytosomes from ceroid-lipofuscinosis (Batten's disease) and in lipofuscin from old age brain, this volume.

Hall, N. A. and Patrick, A. D., 1985, Dolichol and phosphorylated dolichol content of tissues in ceroid lipofuscinosis, J. Inherit. Metab. Dis. 8: 178-183.

Jolly, R. D., Janmaat, A., Graydon, R. J. and Clemett. R.S., 1982, Ceroid-lipofuscinosis: The ovine model., in: "Ceroid-lipofuscinosis (Batten's disease)," D. Armstrong, N. Koppang and J. A. Rider, eds., Elsevier Biomedical Press, Amsterdam.

Jolly, R. D., Janmaat, A., West, D. M. and Morrison, I., 1980, Ovine ceroid-lipofuscinosis - a model of Batten's disease, Neuropathol. Appl. Neurobiol. 6:219-228.

Jolly, R. D., Shimada, A., Dopfmer, I., Slack, P. M., Birtles, M. J. and Palmer, D. N., 1989, Ceroid Lipofuscinosis (Batten's disease): Pathogenesis and sequential neuropathological changes in the ovine model, Neuropathol. Appl. Neurobiol. 15:in press.

Lake, B. D., Lysosomal deficiencies, in: "Greenfield's Neuropathology," J. Hume-Adams, J. A. N. Corsellis and L. W. Duchen, eds., Edward Arnold, London.

Mandel, M., Moriyama, Y., Hulmes, J. D., Pan, Y. E., Nelson, H. and Nelson, N., 1988, Cloning of a cDNA encoding the proteolipid of chromaffin granules implies gene duplication in the evolution of H^+-ATPases, Proc. Natl. Acad. Sci. U.S.A. 85:5521-5524.

Mayhew, I. G., Jolly, R. D., Pickett, B. T. and Slack, P. M., 1985, Ceroid-lipofuscinosis (Batten's disease): Pathogenesis of blindness in the ovine model, Neuropath. Appl. Neurobiol. 11:273-390.

Ng Ying Kin, N. M. K., Palo, J., Haltia, M. and Wolfe, L. S., 1983, High levels of brain dolichols in neuronal ceroid lipofuscinosis and senescence, J. Neurochem 40:1465-1473.

Palmer, D. N., Barns, G., Husbands, D. R. and Jolly, R. D., 1986b, Ceroid-lipofuscinosis in sheep. II. The major component of the lipopigment in liver, kidney, pancreas and brain is low molecular weight protein, J. Biol. Chem. 261:1773-1777.

Palmer, D. N., Husbands, D. R., Winter, P. J., Blunt, J. W. and Jolly, R. D., 1986a, Ceroid-lipofuscinosis in sheep. I. Bis (monoacylglycero)phosphate, dolichol, ubiquinone, phospholipids, fatty acids and fluorescence in liver lipopigment lipids., J. Biol. Chem. 261:1766-1772.

Palmer, D. N., Martinus, R. D., Barns, G. and Jolly, R. D., 1988, Ovine ceroid-lipofuscinosis I: Lipopigment composition is indicative of a lysosomal proteinosis, Am. J. Medical Genet. Suppl. 5:141-158.

Palmer, D. N., Martinus, R. D., Cooper, S. M., Midwinter, G. G., Reid, J. C. and Jolly, R. D., 1989, Ovine ceroid lipofuscinosis: The major lipopigment protein and the lipid binding subunit of mitochondrial ATP synthase have the same NH_2-terminal sequence, J. Biol. Chem. 264:5736-5740.

Pullarkat, R. K., Kim, K. S., Sklower, S. L. and Patel, V. K., 1988, Oligosaccharyl diphosphodolichols in the ceroid-lipofuscinoses, Am. J. Medical Genet. Suppl. 5:243-252.

Pullarkat, R. K., Reha, H., Patel, V. K. and Goebel, H. H., 1982, Docosahexaenoic acid levels in brains of various forms of ceroid lipofuscinosis, in: "Ceroid Lipofuscinosis (Battens disease)," D. Armstrong, N. Koppang and J. A. Rider, eds., Elsevier Biomedical Press, Amsterdam, New York, Oxford.

Rider, J. A. and Rider, D. L., 1988, Batten Disease: Past, present and future, Am. J. Medical Genet. Suppl. 5:21-26.

Sebald, W., Wachter, E. and Tzagoloff, A., 1979, Identification of amino acid substitutions in the dicyclohexylcarbodiimide-binding subunit of the mitochondrial ATPase complex from oligomycin resistant mutants of *Saccharomyces cerevisiae*, Eur. J. Biochem. 100:599-607.

Siakotos, A. N., Bray, R., Dratz, E., van Kuijk, F., Sevanian, A. and Koppang, N., 1988, 4-Hydroxynonenal: a specific indicator for canine neuronal-retinal ceroidosis, Am. J. Medical Genet. Suppl. 5:171-182.

Svennerholm, L., 1976, Polyunsaturated fatty acid lipidosis: A new nosological entity, in: "Current Trends in Sphingolipidoses and Allied Disorders," B. W. Volk and L. Schnek, eds., Plenum Press, New York.

Svennerholm, L., Hagberg, B., Haltia, M., Sourander, P. and Vanier, M. T., 1975, Polyunsaturated fatty acid lipidosis. II. Lipid biochemical studies, Acta Pediatr. Scand. 64:489-496.

Walker, J. E., Fearnley, I. M. and Blows, R. A., 1986, A rapid solid phase microsequencer, Biochem. J. 237:73-84.

Walker, J. E., Lutter, R. and Runswick, M. J., 1989, Identification of the subunits of F_1F_0-ATPase from bovine heart mitochondria, Biochemistry submitted for publication.

Wisniewski, K. E. and Maslinska, D., 1989, Immunoreactivity of ceroid lipofuscin storage pigment in Batten disease with monoclonal antibodies to the amyloid beta protein, New Eng. J. Med. 20:256-257.

Wisniewski, K. E., Rapin, I. and Heaney-Kieras, J., 1988, Clinico-pathological variability in the childhood neuronal ceroid-lipofuscinoses and new observations on glycoprotein abnormalities, Am. J. Medical Genet. Suppl. 5:27-46.

Wolfe, L. S., Ivy, G. O. and Witkop, C. J., 1987, Dolichols, Lysosomal membrane turnover and relationships to the accumulation of ceroid and lipofuscin in inherited diseases, Alzheimer's disease and aging, Chemica Scripta 27: 79-84.

Wolfe, L. S., Ng Ying Kin, N. M. K., Baker, R. R., Carpenter, S. and Andermann, F., 1977, Identification of retinoyl complexes as the autofluorescent component of the neuronal storage material in Battens disease, Science 195:1360-1362.

Zeman, W., 1976, The neuronal ceroid-lipofuscinoses, in: "Progress in Neuropathology," H. M. Zimmerman, ed., Grune and Stratton, New York.

Zeman, W., 1974, Studies in the neuronal ceroid-lipofuscinoses, J. Neuropathol. Exp. Neurol. 33:1-12.

Zeman, W. and Dyken, P., 1969, Neuronal ceroid-lipofuscinosis (Batten's disease): Relationship to amaurotic family idiocy?, Pediatrics 44: 570-583.

DISCUSSION

ELLEDER: Was there any sign that the extraction with chloroform-methanol was improved by the addition of water? I am asking this because if you extracted your isolated body, it was probably suspended in a water containing medium.

PALMER: You touched a very important point. Once the subunit C aggregates in a fashion, it can become quite nonextractable, even with these procedures, and one has to be very careful; otherwise, you could easily end up with dissolving everything. If you deal with fresh tissue, it is not so hard, and you follow our published procedures, but once the body has a chance to change in a conformational manner, it becomes quite difficult and resistant. For instance, these entire bodies dissolve quite well in formic acid, but when you try to put the sample back on a gel, it just sits on top of the sticking gel, and it will not go anywhere, whatever you do. So, this experience is part of the nature of the protein structure.

IVY: Your results on the products of juvenile and late infantile forms of the disease are the same. Would you comment on this, because the storage products are the same, and yet there are differences in the clinical symptoms? How can this be explained?

PALMER: Yes, they are the same, but might be different mutations late in the same condition. Whether this is a mutation in the same protein, or a different kind of mutation, I don't know. It is also possible that for some reason, as for instance the protein is aggregating in such a form that becomes resistant to proteolysis, as I mentioned this morning.

IVY: In the sheep, you showed the morphology of the bodies in the brain. Is the same morphology found in all other tissues?

PALMER: No, for instance, in the pancreas you can have even in the same bodies curvilinear areas and granular patterns. I am sure these are just different manifestations of the same protein.

ELLEDER: Do you think that because the storage of this mitochondrial enzyme, the enzyme turnover and function might be altered in NCL?

PALMER: No, there is really no evidence for that, and it would be incompatible with life.

DAWSON: Have you try to incubate this protein with extracts from normal and affected tissue to see if there is impaired degradation of it?

PALMER: Yes, we tried. This is not easy, and we had problems with this experiment. We have tried with labelled tyrosine, but the problem is that at the moment you finish this reaction, it is in a state with which we cannot work. It is highly insoluble.

DAWSON: So, there is evidence that its catabolism is somewhat specific.

PALMER: Yes.

PORTA: Your findings on the composition of the pigment and the identification of the mitochondrial subunit as the major protein component are, in my opinion, the most important advances on molecular aspects of the storage material in NCL in the last decade. Because of the great importance of your findings, did you take any possible measures to rule out any possible mitochondrial contamination in your preparative procedures?

PALMER: Well, it was not an easy problem, but there were no suggestions of possible contamination. Another thing that is important to realize is that the proportion of subunit C in ATP synthase is considered to be about 6% of the enzyme; so, you cannot have a 6% contamination leading to a 50% of the component.

GLYCOCONJUGATES IN STORAGE CYTOSOMES FROM CEROID-LIPOFUSCINOSIS (BATTEN'S DISEASE) AND IN LIPOFUSCIN FROM OLD-AGE BRAIN

NA Hall, BD Lake, DN Palmer[1], RD Jolly[1] and AD Patrick

Departments of Clinical Biochemistry and Histopathology, Institute of Child Health, 30 Guilford Street, London, WC1N 1EH, England [1]Department of Veterinary Pathology and Public Health, Massey University, Palmerston North, New Zealand

SUMMARY

The ceroid-lipofuscinoses (CL) are a group of inherited diseases characterised by the accumulation, in brain, of autofluorescent storage cytosomes which have similar histochemical staining properties to lipofuscin, the neuronal wear and tear pigment of old-age brain. The storage cytosomes stain strongly with periodic acid-Schiff reagent (PAS), indicating the presence of carbohydrate. In brain from each childhood form of CL, concentrations of phosphorylated dolichol (Dol-P) are 10- to 20- fold higher than in age-matched controls. Brain Dol-P concentrations are also increased between 2 and 5- fold in several different lipidoses and in elderly subjects. Much of the Dol-P which accumulates is located within the storage cytosomes. Dol-P constitutes 2-3 % of the dry weight of storage cytosomes from juvenile and late-infantile CL, and 0.3-0.7% of storage cytosomes from infantile CL, ovine CL and of lipofuscin isolated from old age brain. The bulk of the Dol-P in CL brain and in isolated storage cytosomes is present as dolichyl pyrophosphoryl oligosaccharides (Dol-PP-OS). The constitutions of the oligosaccharide moieties differ in the various forms of the disease. Histochemical analysis of frozen sections of unfixed brain after extraction by various lipid solvents indicates that the major part of the PAS positive intraneuronal material in CL brain and in old-age brain has the extraction properties of Dol-PP-OS. Carbohydrate represents 4-7 % of the dry weight of CL storage cytosomes and of lipofuscin. The major monosaccharide components are mannose, N-acetyl glucosamine, glucose and galactose. Depending on the form of the disease studied, up to 40% of this material can be accounted for by Dol-PP-OS. Polyacrylamide gel electrophoresis of storage cytosomes followed by lectin blotting demonstrates several low molecular weight components which bind concanavalin A. These do not coelute with the major protein components and may well

Lipofuscin and Ceroid Pigments
Edited by E. A. Porta
Plenum Press, New York, 1990

225

be Dol-PP-OS. We conclude that Dol-PP-OS are concentrated in storage cytosomes in CL and are one of their major glycoconjugate components.

KEYWORDS: Ceroid-lipofuscinosis, storage cytosome, lipofuscin, dolichyl pyrophosphoryl oligosaccharides, dolichol, carbohydrate, concanavalin A

INTRODUCTION

The ceroid-lipofuscinoses (CL) (Batten's disease) are a group of neurodegenerative disorders of unknown etiology in humans and in animals, transmitted by autosomal recessive modes of inheritance. They are characterised by the accumulation of tertiary lysosomes (storage cytosomes) containing storage material with the staining properties of ceroid or lipofuscin in a variety of tissues, particularly within neurons in brain. Three different childhood forms of the disease, juvenile (JCL), late-infantile (LICL) and infantile (ICL) can be distinguished on the basis of clinical course, neurophysiological findings and ultrastructural appearance of the storage material (Lake, 1984). Several animal forms of the disease are known, notably in sheep (OCL) (Jolly et al, 1980) and English setters (Koppang, 1970). The storage material is autofluorescent and stains with lipid stains and with periodic acid-Schiff reagent. The intraneuronal storage cytosomes found in CL brain have similar histochemical staining properties to lipofuscin, the neuronal wear and tear pigment which builds up progressively with increasing age in normal brain.

CL shows similarities to the group of lysosomal storage diseases in which the lack of a specific lysosomal hydrolase results in high concentrations of the resulting catabolic intermediates in diseased tissues. These compounds are generally highly concentrated in the tertiary lysosomes which consequently accumulate. Until recently, there was little to indicate which areas of metabolism might be affected in CL, and relatively little was known about the chemistry of the accumulating storage cytosomes.

With the successful isolation of storage cytosomes from CL brain, free dolichol was shown to be a component, though the major part of these bodies could not be characterised, (Ng Ying Kin et al, 1983). Dolichol was reported to be increased in CL brain (compared to age-matched controls) but also to rise as a normal consequence of the ageing process. Indeed, concentrations of dolichol are 10 to 20-fold higher in elderly subjects than in young children (Ng Ying Kin et al, 1983; Andersson et al, 1987). Subsequent reports suggested that the rise in dolichol in CL brain was at most 3-fold (Hall and Patrick, 1985; Wolfe et al, 1988; Pullarkat, 1987).

The metabolically active form of dolichol is phosphorylated dolichol (Dol-P), which is an essential cofactor in the synthesis of N-linked glycoproteins (Kornfeld and Kornfeld, 1985). An increase of at least 10-fold in brain concentrations of total Dol-P (measured after hydrolysis of all Dol-P-containing compounds) has been shown to be a consistent feature in canine CL (Keller et al, 1984), in all three human childhood forms of the disease and in the adult form of the disease (Hall and Patrick, 1985; Wolfe et al, 1988; Pullarkat, 1987). There have been no reports of the analysis of Dol-P in subcellular fractions of CL brain to determine whether the accumulating Dol-P is located in storage cytosomes.

It is possible that brain Dol-P concentrations increase in conditions which cause the accumulation of post-lysosomal residual bodies. Dol-P concentrations in normal elderly subjects are between 2-fold and 5-fold higher than in young subjects (Andersson et al, 1987; Wolfe et al, 1988). High concentrations of Dol-P, comparable to those found in CL brain, have been reported in two cases of G_{M1} gangliosidosis and in one case of sialidosis (Wolfe et al, 1988).

Most of the Dol-P which accumulates in brain from JCL and LICL has the solubility properties of Dol-PP-OS (Hall and Patrick, 1985). Though Dol-PP-OS containing short oligosaccharide chains are soluble in chloroform/methanol (2:1), the major components in CL brain, which have longer oligosaccharide chains, require a solvent consisting of chloroform/methanol/water (1:1:0.3 v/v) (CMW1103) for their extraction. The structures of the oligosaccharide chains are consistent with their being closely related to the dolichol-linked oligosaccharides involved in the glycosylation of proteins within the endoplasmic reticulum (Hall and Patrick, 1988a, 1988b). The predominant components in JCL are $Man_{4-7}GlcNAc_2$ and $Glc_3Man_{7-9}GlcNAc_2$ (N. A. Hall, J. E. Thomas-Oates, A. Dell and A. D. Patrick unpublished results).

Storage cytosomes of high purity have been isolated from brain from sheep with OCL. Dolichol is a minor component comprising less than 1 % of their dry weight (Palmer et al, 1986); Dol-P concentrations have not previously been reported. The major component is protein (Palmer et al, 1986), in particular, subunit c of mitochondrial ATP synthase, previously referred to as the lipid-binding subunit (Palmer et al, 1989). The same protein has been identified as a major component of isolated storage cytosomes from JCL and LICL (Palmer et al, 1990). In this study we have investigated the composition of storage cytosomes isolated from brain from several different forms of CL, and of lipofuscin from old-age brain.

MATERIALS AND METHODS

Materials

Acetic anhydride, jack bean α-mannosidase, gelatin (225 Bloom), concanavalin A coupled to horseradish peroxidase, and protein standards for determination of molecular weight were obtained from Sigma (Poole, Dorset). Sodium [^{3}H]borohydride (0.6-12.1 Ci/m mol) was from New England Nuclear (Stevenage, Herts.). HPLC columns, either a 15cm X 4.6mm 3μ Spherisorb ODS 2 column for the analysis of monosaccharides and oligosaccharides, or a 25cm X 4.6mm 5μ Apex ODS column for the analysis of dolichol and Dol-P were obtained from Jones Chromatography (Llanbradach, Mid Glamorgan). A guard column

(5cm X 2.3mm) was packed with CO:PELL ODS from Whatman (Maidstone, Kent). HPLC-grade acetonitrile, methanol and propanol, scintillation-grade 2,5-diphenyloxazole and standard grade Triton X-100 were obtained from Koch-Light (Haverhill, Suffolk). Mixed xylenes, and pyridine (gold label) were from Aldrich (Gillingham, Dorset). All other solvents and chemicals were from BDH (Poole, Dorset). Chloroform and methanol were redistilled before use. Water for HPLC was purified using a Milli Q system (Millipore, Harrow, Middlesex). HPLC solvents were filtered through a 0.5 μm filter before use. Prestained protein standards and 4-chloro-1-naphthol were from Biorad (Watford, Hertfordshire).

Tissue samples, Isolation of Storage Cytosomes, and Analysis of Dol-P

Tissues were obtained at autopsy and stored at -70°C prior to use. Storage cytosomes were isolated from fresh brain of a sheep affected with CL, aged 15 months, as described previously [Palmer et al, 1986]. Storage cytosomes were isolated from frozen human brain using a modification of these methods. For each brain it was necessary to monitor purifications by light, electron and fluorescence microscopy and adapt the precise procedure appropriately to generate pure fractions. All tissues and isolates were stored at -70°C. Dolichol and Dol-P concentrations were determined as described previously (Hall and Patrick, 1987). Dry weight was determined gravimetrically.

Electron Microscopy

Purified storage cytosomes or lipofuscin were resuspended in 10 % bovine serum albumin in saline and centrifuged at 2,000 X g for 10 minutes. The supernatant was discarded and a solution of 2.5 % glutaraldehyde in 0.1 M cacodylate buffer containing 2.5 mM $CaCl_2$, pH 7.4, carefully added to the pellet. After fixation at room temperature for at least 2 hours, the fixed pellet was briefly washed in 0.1 M cacodylate buffer containing 2.5 mM $CaCl_2$, pH 7.4, and post-fixed in 1 % osmium tetroxide in 0.1 M cacodylate buffer containing 2.5 mM $CaCl_2$, pH 7.4, at 4°C for 1 hour. The pellet was transferred through buffer to 70 % v/v aqueous ethanol and dehydrated in acidified dimethoxypropane for 2 minutes before impregnation and embedding in Araldite. Ultrathin sections were contrasted with uranyl acetate and lead citrate.

Histochemistry

Cryostat sections of brain cortex from each frozen sample were cut at 10 μm onto glass slides and air dried. To study the extraction properties of the staining material, dried sections were treated for 30 minutes with either chloroform/methanol (2:1 v/v) (50 ml) or with CMW1103 (50 ml) and then air dried. Unfixed sections were then examined for autofluorescence under darkground conditions with a mercury lamp with a UG5 excitation filter (300-370 nm) and a 410 nm barrier filter. Alternatively, sections were stained with the periodic acid-Schiff reaction, after fixing with formol-calcium for 10 minutes (Filipe and Lake, 1983) and were then examined by light microscopy.

Extraction of Dol-PP-OS and Analysis of Oligosaccharides

Dol-PP-OS were extracted from tissue homogenates or suspensions of storage cytosomes as described (Hall and Patrick, 1988a). Briefly, samples were delipidated by extraction with butanol/diisopropyl ether (2:3), the lower phase plus interfacial pellet was adjusted to produce a two phase solvent mixture of chloroform/methanol/water (3:2:1, v/v) and the lower phase/interfacial pellet was washed with methanol/water (1:1). The lower phase/interfacial pellet was then adjusted to give a solvent of composition CMW1103 and reextracted twice with CMW1103. Oligosaccharides were released from the pooled CMW1103 extracts by mild acid hydrolysis, labelled with sodium ([3H]borohydride, desalted and analysed by TLC as described (Hall and Patrick, 1988a) except that the solvent for TLC was butanol/acetic acid/water (2:1:1, v/v).

To determine the sensitivity of oligosaccharides to digestion with α-mannosidase, radiolabelled oligosaccharides were purified by TLC as described (Hall and Patrick, 1989) and incubated in 100 mM sodium citrate, pH 5.5, with 9 units/ml α-mannosidase. After incubation for 24 hours, the samples were desalted, peracetylated, and analysed by HPLC as described (Hall and Patrick, 1989).

Analysis of Monosaccharide Composition

Monosaccharide composition was analysed by hydrolysing suspensions of storage cytosomes in 2M HCl for 4 hours at 100°C in tubes flushed with nitrogen prior to securing lids. After drying under nitrogen at 60°C and redrying three times after addition of aliquots of water, the residue was dissolved in water. Internal standard (ribose, 10 nmoles) was added to aliquots followed by NaOH to give a final concentration of 15 mM, the pH was checked, and if necessary adjusted to pH 10. Sodium [3H]borohydride (600 mCi/m mol) (20 mM final concentration) was added and samples were incubated for 3 h at 30°C and recovered as described (Hall and Patrick, 1989). Aliquots were analysed by TLC on Merck HPTLC plates. After two successive developments in ethyl acetate:pyridine:acetone: water:acetic acid (20:10:7.5:7.5:1.5 by vol), plates were dried, sprayed with En3Hance and exposed against preflashed Kodak X-omat XAR-5 film at -70°C. Aliquots were peracetylated as described (Hall and Patrick, 1989) and the peracetylated monosaccharides were dissolved in mobile phase, and analysed by HPLC, eluting with acetonitrile:water (25:75 v/v) at a flow rate of 1 ml/min. Radioactivity was detected in the eluant with a flow radioactivity detector, scintillant flow rate 2.6 ml/min.

TLC could resolve galactosamine from glucosamine, and fucose from glucose or mannose, although resolution of glucose, galactose and mannose was incomplete. HPLC provided accurate quantification of components and was able to provide good resolution of mannose, glucose, galactose and N-acetyl glucosamine, although it could not completely resolve fucose from glucose, nor N-acetyl galactosamine from N-acetyl glucosamine. Response factors for HPLC were determined by analysing mixtures of standards after hydrolysis.

Table 1. Concentrations of Dolichol and Phosphorylated Dolichol (μg/g Wet Weight) in Brain from Various Pathological Conditions.

Disease	Age	(n)	phosphorylated dolichol	dolichol
Infantile CL	7-13	(4)	100.6 +/- 24.2	63.1 +/- 21.6
Late-infantile CL	4-12	(6)	120.0 +/- 84.9	109.5 +/- 37.2
Juvenile CL	21-25	(3)	164.5 +/- 69.8	301.0 +/- 80.8
Control (non-lipidosis)	4-12	(4)	7.7 +/- 2.7	58.9 +/- 14.5
G_{M1} Gangliosidosis	1-4	(4)	23.6 +/- 8.5	25.1 +/- 16.4
G_{M2} Gangliosidosis	3	(1)	27.5	15.3
Niemann-Pick Type C	9-10	(2)	22.9, 41.4	139.6, 149.1
Control (old age)	63-79	(4)	40.8 +/- 15.0	362.0 +/- 96.4

Electrophoresis and Detection of Concanavalin A-Binding Components

Polyacrylamide gels (17.5 %) with a 4 % stacking gel (140x170x1.5mm) were run in a Pharmacia GE-2/4 LS apparatus by the method of Laemmli (1970), except that lithium dodecyl sulphate (LDS) was used in place of sodium dodecyl sulphate (SDS). Samples were diluted to approximately 1 mg dry weight /ml, and an equal volume of a solution containing 2% LDS, 2% dithiothreitol, 20 % glycerol, 0.02 % bromophenol blue in 62.5 mM Tris-HCl, pH 6.8 was added; after boiling for 5 minutes and cooling, 10 μl was loaded. Gels were run at room temperature at 180 V for 30 minutes and then 260 V for approximately two hours until bromophenol blue had reached the bottom of the gel. Prestained protein standards were separated in one channel to monitor electrophoretic transfer.

For the detection of components able to bind to concanavalin A, the separated proteins were transferred to Hybond-C (Amersham, Aylesbury, Buckinghamshire) in 192 mM glycine, 25 mM Tris, 20 % methanol in a Biorad TransBlot apparatus at 110 Volts for 150 minutes. After transfer the gel was silver stained to check that all protein components had been transferred. Concanavalin A binding components were detected using a modification of the method of Faye and Chrispeels (1985). The nitrocellulose filter was washed 3 times for 10 minutes with 20 mM Tris-HCl, pH7.4, containing 500 mM NaCl (TBS) and then incubated overnight at 4°C in blocking solution (3% gelatin in TBS). The blot was then incubated overnight at 4°C in concanavalin A-peroxidase conjugate (40 μg/ml protein in TBS in the presence of 1% gelatin, 1 mM $MnCl_2$, 1 mM $CaCl_2$. It was then warmed to room temperature, rinsed twice briefly and washed 4 times for 15 minutes each time with TBS containing 0.1 % Tween 20, 1 mM $MnCl_2$, 1 mM $CaCl_2$. After rinsing the blot briefly with TBS, colour development was with 4-chloro-1-naphthol (Biorad), following the manufacturers instructions.

For silver staining, gels were fixed in 12 % trichloroacetic acid for one hour and then developed with the Bio-Rad silver stain kit, following the manufacturers instructions. Gels were then decolourised by incubation for 10 minutes in a solution containing 10g/l potassium ferricyanide, 16 g/l sodium thiosulphate, and given 5 successive 20 minute washes in water, until the yellow colour had disappeared. They were then stained following the Bio-Rad kit instructions from incubation in silver reagent onwards.

RESULTS

The concentrations of Dol-P and of dolichol were estimated in a number of different brains (Table 1). Concentrations of Dol-P were higher in each childhood form of CL than in any of the other subjects studied. The elderly subjects, and each of the three types of lipidosis did, however, reveal an increase in Dol-P concentrations compared to young non-lipidosis controls. Brain concentrations of non-phosphorylated dolichol generally increased with the ages of the subjects. Estimates of the dolichol contents of non-lipidosis controls, old-age controls and G_{M1}- and G_{M2}-gangliosidoses were consistent with reported normal ranges for the ages of the subjects (Ng Ying Kin et al, 1983; Andersson et al, 1987), whereas estimates for LICL, JCL and Niemann-Pick disease type C were higher than expected.

Storage cytosomes were isolated from each form of CL

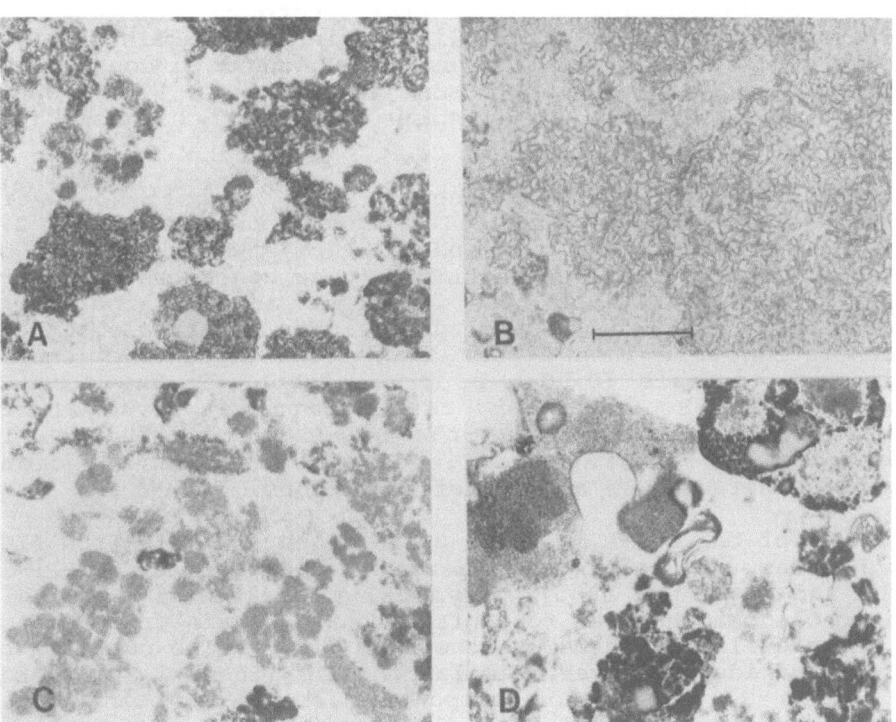

Fig. 1. Electron micrographs of storage cytosomes isolated from brain from JCL (A), LICL (B), ICL (C), and of lipofuscin isolated from old-age brain (D). The scale mark indicates 1 μm. Low power prints of areas chosen at random were assessed for purity by morphometry: all preparations were judged to be at least 75 % pure.

Table 2. Concentrations of Phosphorylated Dolichol and of Dolichol (μg/mg Dry Weight) in Storage Cytosomes Isolated from CL Brain and in Lipofuscin Isolated From Old Age Brain.

Disease	phosphorylated dolichol	dolichol
Infantile CL	3.0, 3.7	1.9, 3.3
Late-infantile CL	32.7	7.4
Juvenile CL	21.2, 18.5	17.5, 13.4
Ovine CL	4.7	2.5
Lipofuscin	7.5, 6.2	50.4, 34.5

Where two preparations were analysed, both were from the same patient; the first value is the lighter fraction and the second the denser fraction prepared by differential pelleting. Concentrations of phosphorylated dolichol in unfractionated brain from four young non-lipidosis controls were determined to be 0.039 (+/- 0.011 μg/mg dry weight) and for a case of juvenile CL 1.05 μg/mg dry weight.

using a combination of differential pelleting, disruption of membranes by sonication and caesium chloride flotation or density centrifugation. A similar protocol was used to generate purified samples of lipofuscin from brain from an elderly subject. Electron microscopy was used to confirm the purity of isolated storage cytosome preparations. Fig. 1 shows representative parts of preparations from each childhood form of CL and from old-age brain.

The concentrations of Dol-P and of dolichol were estimated in samples of isolated storage cytosomes (Table 2). Dol-P comprised between 1.8 and 3.3% of the dry weight of storage cytosomes from JCL and LICL. Lower concentrations were found in OCL and ICL storage cytosomes and in lipofuscin, ranging between 0.3 and 0.8% of dry weight. Nevertheless these concentration were much higher than found in unfractionated brain from controls (0.004% of dry weight). In the four CL preparations the dolichol concentrations were either comparable to or lower than the Dol-P concentrations. In lipofuscin, by contrast, dolichol was a major component comprising between 3.5 and 5 % of dry weight.

Most of the Dol-P in total unfractionated brain from LICL and JCL has the solvent extraction properties of Dol-PP-OS (Hall and Patrick, 1985). Typical profiles of dolichol-derived oligosaccharides for unfractionated brain from each of the three childhood forms are shown in Fig. 2. The patterns of oligosaccharides were very similar in JCL (Lanes 1 and 9) and LICL (Lane 2): the major components of these being $Man_{4-7}GlcNAc_2$ and $Glc_3Man_{7-9}GlcNAc_2$. Similar patterns were found in three other LICL and two other JCL brains analysed.

Most of the Dol-P in ICL brain also had the extraction properties of Dol-PP-OS. However, the pattern of oligosaccharides in ICL brain (Lane 3), was very different from that seen in LICL and JCL. This pattern was consistently found in three other ICL brains. The components were analysed by TLC and by HPLC of peracetylated derivatives. Their chromatographic properties in relation to standards, indicated that the major oligosaccharides were Hexose$_{6-9}$GlcNAc$_2$. All the dolichol-derived oligosaccharides from ICL brain could be broken down to Man$_1$GlcNAc$_2$ by digestion with α-mannosidase. We conclude, therefore, that the major components were Man$_{6-9}$GlcNAc$_2$.

The two control brains analysed in this study from subjects aged 12 (Lane 4) and 63 years (Lane 5) had much lower levels of dolichol-derived oligosaccharides, making their detection more difficult. Nevertheless they appear to have a different pattern again, having predominantly short oligosaccharide chains, probably corresponding to Man$_{2-5}$GlcNAc$_2$ with small amounts of the longer chain length oligosaccharides.

The extraction of Dol-P from OCL storage cytosomes by various lipid solvents was analysed. As with unfractionated JCL brain, only a small proportion had the extraction properties of free dolichol monophosphate. CMW1103 was required to extract the major part of Dol-P, indicating that it is present in the form of Dol-PP-OS.

The oligosaccharide portions of Dol-PP-OS in isolated storage cytosomes from OCL and JCL and from lipofuscin were analysed (Fig. 2, lanes 6-8). The pattern of oligosaccharides in JCL storage cytosomes (lane 7) appears to be similar to that seen in unfractionated brain from the same case (lane 9). The pattern in OCL storage cytosomes (lane 6) is similar to that found in JCL storage cytosomes, except that the largest oligosaccharides (Hexose$_{10-12}$GlcNAc$_2$ are absent. The oligosaccharides derived from Dol-PP-OS from OCL storage

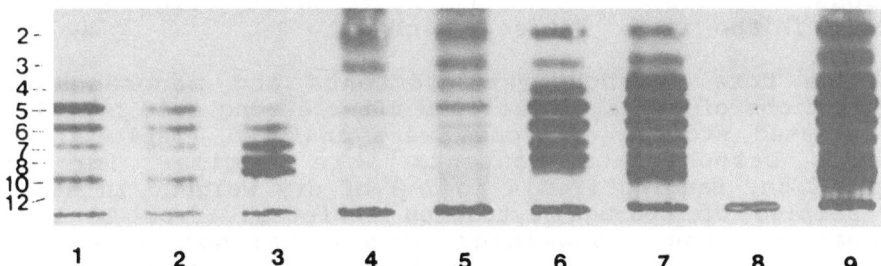

Fig. 2. Thin layer chromatographic separation of dolichol-derived oligosaccharides from total brain (lanes 1-5, 9) and isolated storage cytosomes (lanes 6-8). Brain samples are juvenile CL (lanes 1 and 9), late-infantile CL (lane 2), infantile CL (lane 3), control age 12 (lane 4) and control age 63 (lane 5). Storage cytosomes isolated from ovine CL (lane 6), juvenile CL (lane 7) and lipofuscin from the 63 year old subject (lane 8). Loadings were adjusted to enable optimum detection of the different component oligosaccharides. Thus different equivalent weights of tissue or storage cytosomes were loaded in each channel. The numbers used to label bands in lane 1 indicate the number of hexose residues linked to two N-acetyl glucosamine residues in each band.

233

Table 3. Total Carbohydrate Content (% of Dry Weight) and Monosaccharide Composition of Storage Cytosomes and Lipofuscin.

Disease	Total Carbohydrate	Monosaccharides (% of total sugars)			
		Man	Gal	Glc	HexNAc
Infantile CL	6.3	37.3	21.7	16.6	24.4
Late-infantile CL	7.3	54.5	<5	14.7	30.8
Juvenile CL	6.9	61.7	6.5	15.1	16.7
Ovine CL	4.3	52.4	17.8	12.6	16.9
Lipofuscin (1)	4.5	33.2	38.7	13.6	14.3
(2)	6.2	19.4	49.7	25.4	5.6

For lipofuscin, two preparations from the same patient were analysed:- (1) is lighter fraction, and (2) is denser fraction prepared by differential pelleting. Man is mannose, Gal is galactose, Glc is glucose, HexNAc is the sum of N-acetyl glucosamine and N-acetyl galactosamine (TLC analysis indicated that for each sample N-acetyl glucosamine was the predominant amino sugar).

cytosomes, like those from unfractionated ICL brain, can be broken down to $Man_1GlcNAc_2$ by digestion with α-mannosidase. This observation, in combination with their chromatographic behaviour, indicates that these oligosaccharides are predominantly $Man_{4-8}GlcNAc_2$. It was not possible to determine if this pattern was the same as that found in total unfractionated OCL brain as this tissue was not available.

Only a small amount of lipofuscin was available for analysis of its Dol-PP-OS components: as a result only very faint bands were seen in the separation of these oligosaccharides (lane 8). Nevertheless the major part of the oligosaccharides appeared to have short oligosaccharide chains, in the range $Hexose_{2-5}GlcNAc_2$.

The total carbohydrate contents and monosaccharide compositions of storage cytosomes after strong acid hydrolysis of isolated storage cytosomes were analysed (Table 3). The total carbohydrate contents were similar for each preparation, ranging from 4 to 7 % of dry weight. In each of the samples of storage cytosomes isolated from CL brain, mannose was the predominant monosaccharide; appreciable amounts of N-acetyl glucosamine and glucose were also present. Galactose concentrations were low in storage cytosomes from JCL and LICL, but contributed roughly 20 % of carbohydrate detected in ICL and OCL. In lipofuscin, however, galactose was the predominant monosaccharide. Fucose and N-acetyl galactosamine contributed a small proportion of the monosaccharides in each of the samples analysed.

In each preparation there was sufficient mannose, N-acetyl glucosamine and glucose to provide the carbohydrate portions of the Dol-PP-OS in that sample. In JCL and LICL, assuming all Dol-P was present as Dol-PP-OS, this would account for between 20 and 40 % of the total carbohydrate in those storage cytosomes. In the other samples Dol-PP-OS would

account for less than 10 % of their total carbohydrate
content. These results indicate that there are other
glycoconjugates present within the storage cytosomes.

In order to identify other glycoconjugate components, and
to investigate whether any of the major protein components of
storage cytosomes were glycoconjugates, storage cytosomes were
separated by electrophoresis, transferred to nitrocellulose,
and analysed for their ability to bind to concanavalin A.
Fig.3 shows such a separation, as well as a parallel
separation silver stained to reveal protein components. The
lectin blot reveals that the predominant staining was of
components with an apparent molecular weight of 6-16 kDa. The
pattern in JCL (lane 2) was similar to that in LICL (lane 3):
both had a narrow band at 16 Kda, as well as a broad band at

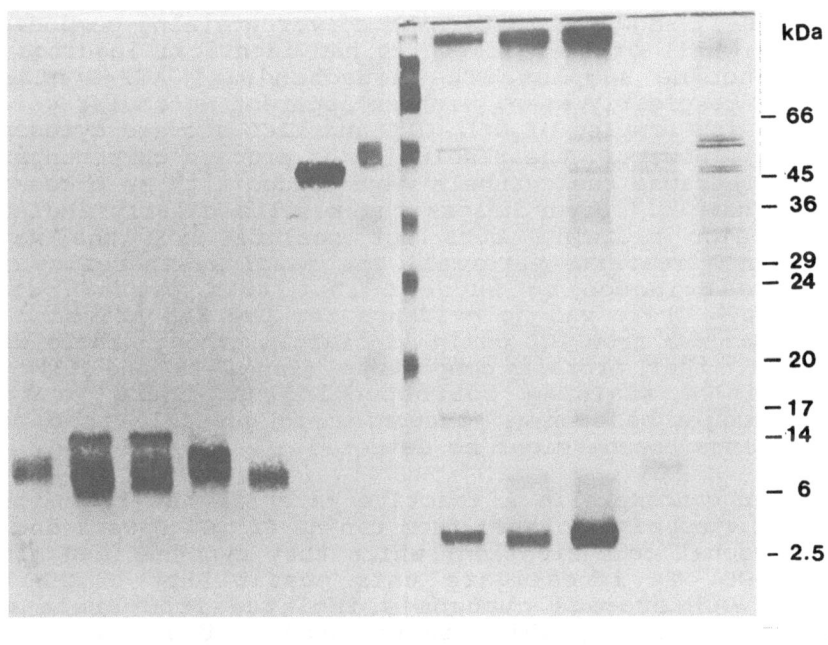

Fig. 3. Electrophoretic separation of storage cytosomes
isolated from brain. Components were identified either after
transfer to nitrocellulose by their ability to bind
concanavalin A (lanes 1-7) or by silver staining of fixed gels
(Lanes 8-13). Samples were storage cytosomes isolated from
ovine CL (lanes 1 and 9), juvenile CL (lanes 2 and 10), late-
infantile CL (lanes 3 and 11), infantile CL (lanes 4 and 12),
and lipofuscin isolated from a 63 year old subject (lanes 5
and 13). Between 4 and 7 μg dry weight of sample was separated
in each lane. Lane 6 contains 2.5 μg of ovalbumin. Lanes 7 and
8 contain a mixture of six prestained molecular weight markers
with apparent molecular weights of 110, 84, 47, 33, 24, 16
kDa. The mobilities of unstained molecular weight markers,
identified by silver staining, are indicated to the right of
the electrophoretogram.

6-14 kDa. The other samples had single broad bands of staining, those for OCL (lane 1) and ICL (lane 4) had similar mobilities, whilst that in the lipofuscin sample (lane 5) had a slightly lower apparent molecular weight.

Ovalbumin (lane 6) was included as a positive control to reveal staining of a glycoprotein. In another separation of a crude tissue homogenate, blotted and processed in the same manner, a number of concanavalin A binding components were revealed (not shown). Incubation of a blot with concanavalin A in the presence of 200 mM methyl mannoside virtually abolished staining with ovalbumin and completely eliminated staining with JCL storage cytosomes. A channel containing bovine serum albumin did not stain under these conditions. Thus staining with concanavalin A was shown to be specific for the presence of ligand. Similar experiments carried out by fixing gels and incubating them with I^{125} concanavalin A revealed a similar pattern, though resolution and sensitivity were poorer than revealed here.

Lanes 8-13 show the pattern of silver staining components in the other half of the gel, which had identical loadings of storage cytosome samples. The mitochondrial ATP-synthase subunit can be clearly seen, with an apparent molecular weight of 3.5 kDa, in samples of OCL, JCL and LICL storage cytosomes (lanes 9-11). However, the resolution of protein components is not optimal because the channels were loaded with an excessive amount of sample. Nevertheless the results clearly indicate that the major protein does not coelute with the major concanavalin A reactive material. The results, therefore, do not provide evidence to suggest that this protein is a glycoprotein. There was no evidence for the 3.5 kDa band in storage cytosomes from ICL or in lipofuscin, though there were a number of other protein components (lanes 12 and 13). No concanavalin A staining corresponding to these protein components could be seen, though it is possible that with higher loadings, bands might be detected.

Similar concanavalin A reactive material has been found in storage cytosomes isolated from canine CL (G. Dawson and G. O. Ivy, personal communication) which they hypothesised might be Dol-PP-OS. To investigate this possibility a CMW1103 extract of OCL storage cytosomes isolated from brain was analysed. This sample, which represented just 1.1 % of the total dry weight of the storage cytosomes, was at least 30 % Dol-PP-OS. After electrophoresis, transfer, and incubation with concanavalin A this sample revealed a strongly staining band with an apparent molecular weight of 10-14 kDa, as well as a small amount of aggregation at the start of the separating gel. Thus it is likely that the material located by concanavalin A in lanes 1-5 is Dol-PP-OS since it comigrates with an authentic standard. The different patterns seen in the various samples could result from separation on the basis of size. The largest Dol-PP-OS ($Hex_{10-12}HexNAc_2$-PP-Dol) found in JCL and LICL brain might contribute the band of 16 kDa apparent molecular weight, while the other band in these samples might represent the remaining $Man_{4-8}GlcNAc_2$-PP-Dol components. The size ranges in the other samples would be consistent with such an explanation.

Table 4. The Extraction Properties of Neuronal Storage Substances.

Extraction Staining Disease	None PAS	UV	C/M 2:1 PAS	UV	CMW1103 PAS	UV
Infantile CL	++	++	++	++	+/-	++
Late-Infantile CL	++	++	++	+	-	+/-
Juvenile CL	++	++	++	+	-	+/-
Ovine CL	++	+	+	+	-	-
Canine CL	++	++	+	++	-	+/-
Lipofuscin	++	++	++	++	-	++
G_{M1} Gangliosidosis	+++	-	-	-	-	-

PAS denotes periodic acid-Schiff, UV denotes autofluorescence, intensity is in arbitrary units scale - to +++. Sudan black staining gave similar results to those for autofluorescence.

Histochemical examination of brain sections after extraction by lipid solvents (Table 4, Fig. 4) shows that the PAS positive material, present in neurons and glial cells of each form of CL or in old-age brain is not extracted by chloroform/methanol (2:1 v/v) in contrast with the gangliosides of G_{M1}-gangliosidosis which are readily extracted under these conditions. However, extraction of PAS positive material by CMW1103 is demonstrated by the grossly diminished PAS staining and the appearance of holes where the neuronal and glial storage material had been removed by this treatment. Extraction with CMW1103 also abolishes autofluorescence in OCL, canine CL, JCL and LICL but has no effect on ICL or old-age brain.

DISCUSSION

Dol-P is a component of storage cytosomes from all three human forms of CL and from OCL, comprising between 0.3 and 3.3

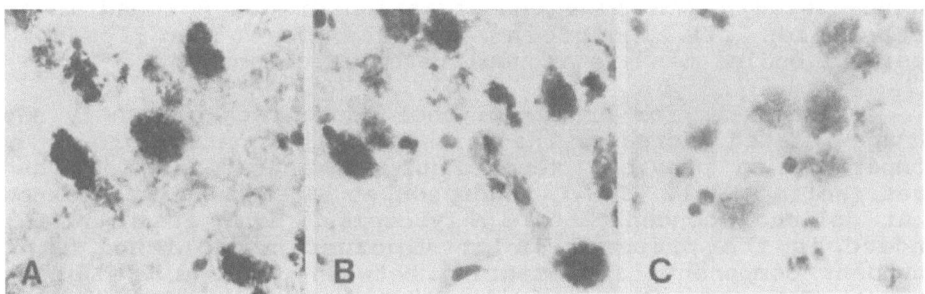

Fig. 4. Cryostat sections of unfixed brain from juvenile CL (a) unextracted, (b) extracted with chloroform/methanol (2:1), or (c) with chloroform/methanol/water (1:1:0.3), stained by periodic-acid Schiff method.

% of the dry weight of the bodies. These concentrations are between 75 and 800-fold higher than we found in unfractionated brain from normal individuals and much higher than has been previously reported in isolated subcellular organelles. Even in endoplasmic reticulum, the functional location of Dol-PP-OS, Dol-P concentrations are much lower than in CL storage cytosomes: they have been estimated to be 0.6 μg/mg protein in rat brain microsomes (Rust et al, 1988). There have been no reports of Dol-P concentrations in lysosomes derived from brain, but estimates of 0.18-0.20 μg/mg protein in isolated rat liver lysosomes (Edlund et al, 1986) are also much lower than those found in storage cytosomes.

It is likely that the bulk of the Dol-P in storage cytosomes is present in the form of Dol-PP-OS. In OCL and JCL storage cytosomes, the average molecular weights of the lipid and oligosaccharide portions of the molecules indicate that on a weight basis the total concentrations of Dol-PP-OS are roughly double those of Dol-P. Thus in the storage cytosomes from JCL brain, Dol-PP-OS would comprise roughly 4 % of dry weight, and in the case of LICL, which had the highest concentrations of Dol-P, Dol-PP-OS could represent as much as 7% of their dry weight (assuming that all Dol-P was present as Dol-PP-OS).

In CL brain, Dol-P compounds clearly concentrate within storage cytosomes. It is not clear, however, whether this accumulation is specific to the disease. Brain concentrations of Dol-P are consistently higher in conditions which involve hypertrophy of lysosomes (G_{M1}- and G_{M2}-gangliosidosis, Niemann-Pick disease type C and old-age brain) than in controls. Though brain Dol-P concentrations were much lower in the three G_{M1}-gangliosidosis brains studied here than in those studied by Wolfe et al (1988), they are nevertheless elevated compared to normal controls. The Dol-P concentrations of brain cortex grey matter reported here in subjects aged 63-79 years (41 μg/mg dry weight) are 5-fold higher than in young controls and are in good agreement with estimates reported by Wolfe et al (1988) for subjects aged 55-82 years (37 μg/mg dry weight, 3-fold higher than in the young controls used in that study). The results also agree with those reported by Andersson et al (1987) for subjects aged 62-91 years (26-44 μg/mg dry weight depending on the region; 2-fold higher than the young controls in that study). The high concentration of Dol-P found in the preparation of lipofuscin suggests that post-lysosomal residual bodies tend to accumulate Dol-P compounds.

The results for brain dolichol content consistently show an age-related increase. The values for the control groups are comparable to previous reports of concentrations for these ages (Wolfe et al, 1988; Andersson et al, 1987). It is known that dolichol concentrates in lysosomes (Wong et al, 1982). Indeed in the preparation of lipofuscin, dolichol is an abundant component, representing between 3.5 and 5 % of dry weight. This estimate is similar to that reported by Pullarkat (1987) for isolated lipofuscin (4.3 % of dry weight) and would be consistent with a substantial proportion of the dolichol which accumulates in old-age brain being located within lipofuscin. The dolichol content of storage cytosomes was lower in the samples from younger subjects (ICL and LICL, both aged 8 years) than in the JCL sample (aged 25 years). This might indicate that the accumulation of dolichol in post-lysosomal residual bodies is a slow process, occurring over a number of years. Thus, the observation that estimates for G_{M1}- and G_{M2}-gangliosidosis appear normal (compared to age-matched controls), but that estimates in LICL, JCL and Niemann-Pick C are two- to three-fold higher than expected may indicate that in the first group of diseases, there was insufficient time for dolichol accumulation before death ensued.

The oligosaccharide portions of the Dol-PP-OS components differ in the various forms of CL. In the infantile form of the disease, the pattern of oligosaccharides would be consistent with a defect in the release of α-linked mannose residues from Dol-PP-OS. The predominant metabolite is $Man_9GlcNAc_2$, with decreasing amounts of $Man_{8-5}GlcNAc_2$. Since the full-length Dol-PP-OS glycosylation intermediate is $Glc_3Man_9GlcNAc_2$, this pattern could be generated if there were a defect in the release of α-mannose residues from these compounds. The report by Krusius et al (1986) showing increased high-mannose glycopeptides in ICL brain and in isolated storage cytosomes also provides evidence for such a defect. On the other hand, Dol-PP-OS concentrations are not as high in storage cytosomes from ICL brain as in the two other human forms of CL. Furthermore, the analysis of the monosaccharide composition of isolated storage cytosomes from ICL in this study did not indicate that there was an accumulation of mannose-containing glycoconjugates in these preparations compared to other forms of CL or lipofuscin. Nor was there evidence for any major concanavalin A reactive components in these storage cytosomes, after lectin blotting, apart from the component tentatively identified as Dol-PP-OS. It is possible that the alteration in oligosaccharide pattern reflects some of the secondary pathological changes that occur in ICL, in particular, differences between the metabolism of glial cells (which are the dominant cell type in ICL brain), and neurons (which were virtually absent from ICL brain).

The mechanism which generates particular oligosaccharide patterns in the different forms of CL is at present unknown. Essentially a similar pattern of oligosaccharides has been seen in each of the seven JCL and LICL brains studied to date. This pattern is very different from that seen in ICL brain. The observation that the oligosaccharide pattern appears very similar in unfractionated JCL brain and in isolated storage cytosomes rules out the possibility that the largest (glucose-containing) oligosaccharides, $Hex_{10-12}HexNAc_2$, are present in endoplasmic reticulum, but that the shorter oligosaccharides $Man_{4-8}GlcNAc_2$ are present in storage cytosomes.

The pattern of Dol-PP-OS in OCL brain storage cytosomes appears to be intermediate between the two human CL patterns. On the one hand the relative abundance of oligosaccharides up to $Man_8GlcNAc_2$ is similar to that found in JCL and LICL (though the higher molecular weight oligosaccharides seen in those two human forms are absent). On the other hand, in common with ICL brain, all Dol-PP-OS within OCL storage cytosomes are entirely composed of α-linked mannose after the core trisaccharide common to all these structures. At present it is not possible to interpret the significance of these observations, though the difference between OCL and human diseases may partly reflect species differences in catabolic enzymes acting on Dol-PP-OS. The absence of high molecular weight oligosaccharides in lipofuscin suggests that these lysosomes are effective at degrading the carbohydrate portion of Dol-PP-OS.

The observation that the solvent extraction properties of PAS material from storage cytosomes of neurons and glial cells, in situ, is similar to that of Dol-PP-OS from CL storage cytosomes suggests that the major part of this material is Dol-PP-OS. In the storage cytosome preparations isolated from LICL and JCL, Dol-PP-OS constituted 4-7 % of the dry weight of these bodies and accounted for 20-40 % of their carbohydrate content. In ICL and OCL storage cytosomes they constitute a smaller proportion of carbohydrate content. It is possible that storage cytosomes contain other compounds which react with PAS and have similar extraction properties to Dol-PP-OS. It is also possible, however, that the major part of the carbohydrates in ICL and OCL storage cytosomes and in lipofuscin are not present in a form which enables them to react with PAS. Thus, although Dol-PP-OS may represent only about 10 % of their carbohydrate content they may be the major PAS positive component.

Lectin blotting of electrophoretically separated storage cytosomes failed to reveal any components that bind to concanavalin A, other than those tentatively identified as Dol-PP-OS. However, it is possible that other glycoconjugates present in storage cytosomes have structures that cannot bind to this lectin. In particular, in lipofuscin the high galactose content would probably be associated with structures unlikely to bind concanavalin A.

In summary, these results have demonstrated that the high levels of Dol-P found in CL are associated with their accumulation in storage cytosomes. This would be their expected location if the disease results from the lack of a lysosomal enzyme involved in their catabolism. However, as they constitute at most 7 % of the dry weight of the storage cytosomes they are clearly not the dominant stored metabolite, particularly in OCL, LICL and JCL, where subunit c of mitochondrial ATP-synthase is the predominant component. Their accumulation may be a secondary event and may be a general feature of pathological conditions in which lysosomes accumulate. Based on quantitative monosaccharide analysis, there are clearly a number of other glycoconjugates in storage cytosomes which remain to be identified.

ACKNOWLEDGEMENTS

We are grateful for the generous financial support of the Medical Research Council of Great Britain (N.A.H., A.D.P.) and grant no. NS 11238 of the United States National Institute of Neurological and Communicative Disorders and Stroke (R.D.J., D.N.P). We thank Dr G. Besley, Dr. G. Dawson, Dr. M. Haltia, Dr. M. Lynch, Dr A. Siakotos and Prof. G. Slavin for making tissues available to us. We thank Dr J. E. Thomas-Oates and Dr A. Dell for fast atom bombardment-mass spectrometry analysis of oligosaccharide samples and Mrs V. V. Smith for skilled microscopical technique.

REFERENCES

Andersson, M., Appelkvist, E.L., Kristensson, K., and Dallner, G. (1987) J. Neurochem, **49,** 685-691

Edlund, C., Ganning, A.E. and Dallner, G. (1986) Chem. Biol. Interact. **57,** 255-270

Faye, L., and Chrispeels, M.J. (1985) Anal. Biochem. **149,** 218-224

Filipe, M. I. and Lake, B.D. (1983) Histochemistry in Pathology. Churchill Livingstone, Edinburgh.

Hall, N.A., and Patrick, A.D. (1985) J. Inher. Metab. Dis. **8,** 178-183

Hall, N.A., and Patrick, A.D. (1987) Clinica Chimica Acta **170,** 323-330

Hall, N.A., and Patrick, A.D. (1988a) Amer. J. Med. Genet. Supplement **5,** 221-232

Hall, N.A., and Patrick, A.D. (1988b) Biochem. Soc. Trans. **16,** 1031-1032

Hall, N.A., and Patrick, A.D. (1989) Anal. Biochem. **178,** 378-384

Jolly, R.D., Janmaat, A., West, D.M., and Morrison, I., (1980) Neuropathol Appl Neurobiol **6,** 195-209

Keller, R.K., Armstrong, D., Cram, F.C., and Koppang, N. (1984) J. Neurochem. **42,** 1040-1047

Koppang, N. (1970) J. Small Anim. Pract. **10,** 639-644

Kornfeld, R., and Kornfeld, S. (1985) Ann. Rev Biochem. **54,** 631-64

Krusius, T., Viitala, J., Palo, J., and Maury, C.P.J. (1986) J. Neurol. Sci. **72,** 1-10

Laemmli, U.K. (1970) Nature **224,** 680-685

Lake, B.D. (1984) in Greenfield's Neuropathology (Hume-Adams, J., Corsellis, J.A.N., and Duchen, L.W., eds.), pp. 491-572, Edward Arnold, London.

Ng Ying Kin, N.M.K., Palo, J., Haltia M. and Wolfe L.S. (1983) J. Neurochem. **40,** 1465-1473

Palmer, D.N., Barns, G., Husbands, D.R., and Jolly, R.D. (1986) J. Biol. Chem. **261,** 1773-1777

Palmer, D.N., Martinus, R.D., Cooper, S.M., Midwinter, G.G., Reid, J.C. and Jolly, R.D. (1989) J. Biol. Chem. **264,** 5736-5740

Palmer, D.N., Martinus, R.D., Bayliss, S., Jolly, R.D., Hall, N.A., Lake, B.D., Fearnley, I.M., Medd, S.M. and Walker, J.E. (1990) Chapter in this volume.

Pullarkat, R.K. (1987) Chemica Scripta **27,** 85-88

Rust, R.S., Sakakihara, Y. and Volpe, J.J. (1988) Dev. Neurosci. **10,** 25-33

Wolfe, L.S., Gauthier, S., and Durham, I. (1988) in Lipofuscin-1987: State of the Art (Zs.-Nagy, I., ed.), pp. 389-411, Elsevier, Amsterdam

Wong, T.K., Decker, G.L., and Lennarz, W.J. (1982) J. Biol. Chem. **250,** 6614-6618

DISCUSSION

HERVONEN: You showed that in normal brains of 63 to 69 year old individuals, there was a high amount of dolichols. Have you ever looked at the brain of even older subjects 90 years or over?

HALL: No, but other people have looked, and this is documented quite extensively in the literature.

HERVONEN: Where do you put Alzheimer's disease in this comparison?

HALL: Alzheimer's disease has significantly increased amounts of dolichols compared to ceroid lipofuscinosis, and also has significantly increased content of phosphorylated dolichols. So, if it is really the case, as some people believe that lipofuscin granules also greatly increase in Alzheimer's disease, then you can say that you have more residual bodies containing a normal concentration of dolichol and phosphorylated dolichol.

IVY: Is there any normal system breaking down dolichols in vivo?

HALL: Dolichol oligosaccharides are synthesized in the endoplasmic reticulum. There is a catabolic enzyme within the end of the endoplasmic reticulum and possibly in the Golgi apparatus which releases all of the saccharide phosphates; so, it will break down dolichol polyphosphates into dolichol monophosphates.

IVY: Let me put the question in a different way. Is it possible that the following hypothesis is consistent with your data? The dolichols do not turn over, that they irreversibly build up with age, and that they build up at different rates depending on the particular disorder we are looking up?

HALL: Yes, I can go along with that.

ELLEDER: If I understood your results, the Dol-PP-OS that accumulates in lipofuscin or age-pigment is different or contains less amounts of hexoses than that accumulating in NCL. Is that correct?

HALL: Yes, but I would like to do further analysis before giving a definite answer to this question.

LECTIN HISTOCHEMICAL STUDY OF LIPOPIGMENTS:

RESULTS WITH CONCANAVALIN A

Milan Elleder[1], Hans.H. Goebel[2], Nils Koppang[3]

(1) Hlava's Institute of Pathology, School of
Medicine, Prague, Czechoslovakia
(2) Division of Neuropathology, University of
Mainz, Federal Republic of Germany
(3) National Veterinary Institute, Oslo, Norway

SUMMARY

Concanavalin A (Con A) binding to lipopigments (LPs) of
the lipofuscin type was proved to be due to the high content of
mannose. Two mannose bearing compounds could be recognized due
to their different organic solvent solubility. One was best soluble in modified chloroform-methanol-water mixture (10:10:3)
and corresponded most probably to the oligosaccharyl disphosp-
hodolichol (oligo-PP-Dol) described to be significantly increased in LPs of inherited type. The second one, organic solvent
insoluble corresponded to a glycoprotein (GP). The ratio of the
two components varied. The deposition of the typical lipofuscin (age pigment) was dominated by the GP component. Its amount
was greatest in neurolipofuscin (especially in the olivary nucleus) but very little in hepatocytic lipofuscin. In human neuronal ceroid lipofuscinoses (of early juvenile, and juvenile
types) both components were found in large quantities in the
storage granules of the affected neurons. The "protein type variant" of the storage material (Elleder,1978) displayed the highest degree of lipid-bound mannose accumulation, the GP component
being absent. In the late infantile, infantile and Kufs variants
studied in paraffin sections only, the GP component was detectable, too as in the case of the secondary neuronal LP in mucopolysaccharidoses and gangliosidoses. In the canine model of
NCL lipid bound mannose clearly predominated, the GP component
being in low amount on average. Neither of the Con A reactive
glycoconjugates could be indentified as the component responsible for autofluorescence. However, both are most probably responsible for PAS positivity of lipofuscins. There were no detectable Con A reactive glycoconjugates in the histiocytic ceroid.

Key words: lipopigments, lipofuscins, ceroids, mannose, glyco-
           protein, dolichol, Concanavalin A

Lipofuscin and Ceroid Pigments
Edited by E. A. Porta
Plenum Press, New York, 1990

INTRODUCTION

Despite the wide application of lectin histochemistry in storage diseases (Alroy et al.,1988b) lysosomal deposits of the LP type have not received due attention yet.However, the recent reports (Hall and Patrick,1988;Pullarkat et al.,1988) succeeded in proving high accumulation of mannose in the form of oligosaccharyl diphosphodolichol (oligo-PP-Dol), which probably represents the bulk of the esterified dolichol fraction found to accumulate in cells increasingly generating LPs of the lipofuscin type (Wolfe et al.,1988). As these findings are suggested by the above quoted authors to reflect metabolic error of pathogenetic importance, this sugar moiety has immediately attracted high attention and occupies at present high position on the list of priorities of the current analytical programmes. In this communication we would like to present the results of the first part of our research including a study of Concanavalin A (Con A) sugar receptors represented mostly by mannose (Alroy et al.,1988b).

MATERIAL AND METHODS

Paraffin embedded bioptic and postmortem tissues (human and animal, specified in Results) fixed by immersion in formaldehyde (buffered or unbuffered) were used throughout the study. The sections were cut about 5 μm thick and mounted on glass slides covered either with L-polylysine (Sigma) or silicagel, deparrafinized and subjected to the staining procedure.For the staining the ABC technique (Alroy et al.,1984) was employed using biotinylated Concanavalin A (Sigma,Lot.117F-8090) diluted 1:800 (2.5 ug/ml) with TBS (Tris-buffered saline) and supplemented with 1 Mmol $CaCl_2$. Details of the slightly modified procedure and of all the preparative steps are given in the forthcomming publication (Elleder, 1989).

Blocking and digestion procedures. Vicinal glycols were blocked by acetylation (Lillie and Fullmer,1976) with pyridine extraction as a control, and with HIO_4 preoxidation (1% aqueous, 10-15 minutes, at room temperature). Part of the HIO_4 treated sections were subsequently exposed to the sodium borohydride reduction procedure (Katsuyama and Spicer,1978). Both blocking procedures (acetylation and HIO_4 oxidation) were employed after the trypsin digestion step. The following sugars were used for inhibition of Con A binding: alpha-D-methylmannopyranoside (Sigma), alpha-methylglucopyranoside, both in 0.1 and 0.2 M final concentrations and according to the prescription given by Alroy et al.(1984). Endogenous biotin was blocked with the Avidin/Biotin blocking kit (Vector). A series of sections was exposed to digestion with alpha-D-mannosidase (Sigma, Batch No.N-2133) according to Hall and Patrick (1988) and with Neuraminidase (type X,Sigma,Batch No.N-2123) according to Alroy et al.,(1984).

The extraction procedures were carried out at room temperature for 1 hour, unless otherwise stated. The following mixtures were used:chloroform - methanol 2:1 (v/v), chloroform-methanol-water 10:10:3 ("1103",according to Hall and Patrick,1988). Some sections were extracted with anhydrous acetone (Elleder and Lojda,1971) or with absolute isopropanol. Further details of the extraction procedure are detailed elsewhere (Elleder,1989). The results of Con A were correlated with PAS staining and with autofluorescence.

RESULTS

Notes on the method. As the staining intensity was generally higher with the prolonged lectin step (overnight, 4° C) without any increase in background staining, this procedure was used throughout the whole study. As for the temperature during the lectin step it caused moderate background staining only when increased to 37°C. Both the room and decreased temperatures(4°C) gave the same results. However, only the latter one was selected and used throughout the study.

The staining of paraffin sections posed no essential problems except for occasional staining of collagen, which was much stronger in cryostat than in paraffin sections. Since it was completely inhibited by periodic acid preoxidation and by alpha-methyl mannopyranoside it may have been caused by specific Con A binding as described (Söderström,1987).

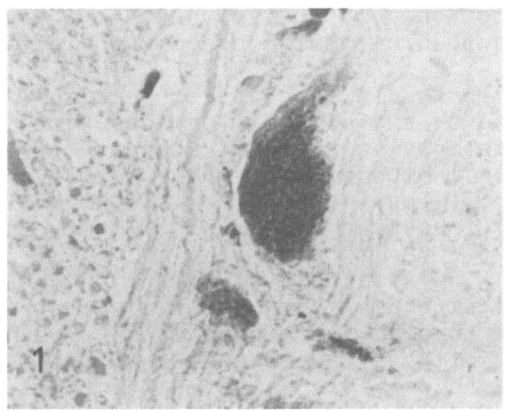

Fig.1. Human neurolipofuscin stained
with Con A. Medulla oblongata, x 400

In cryostat sections the only procedure which gave reproducible results was the modification excluding any drying of the freshly cut section, i.e. before staining. Unless this condition was met the results were inconsistent with a great tendency toward decrease of staining. In practice this meant that the sections could not be stored but had to be used immediately after cutting and fixation.

The binding of Con A was localized to the LP granules,which displayed some variation in staining intensity and even heterogeneity of binding inside the granules. In some affected cells, especially in neurons, part of the intervening cytoplasm was stained, too, thus obscuring the granular appearance of affected cells.

The staining in a series encompassing neurolipofuscin, NCL of various types, myocardium, liver and peripheral neurons (controls and HNCL) was completely blocked by acetylation, and pre-oxidation with 1% HIO_4. Sodium borohydride reduction applied after HIO_4 failed to restore the staining. Inclusion of alpha-D-methylmannopyranoside into the incubation medium extinguished the the staining. Alpha-D-methylglucopyranoside decreased the staining slightly. Pretreatment with neuraminidase (37°C, 1 hour) caused no significant change in staining intensity. Incubation with alpha mannosidase was also without any substantial effect in paraffin sections. There was slight decrease in staining in cryostat sections of HNCL and CNCL neuronal deposits. Prolongation of the digestion led to substantial descrease in staining even in controls incubated in buffer only.

Human LPs of Acquired Type (Lipofuscin and Ceroid)

The staining of <u>lipofuscin</u> type LPs in paraffin sections was positive to various degree. The brain neurolipofuscin (twelve bioptic and postmortem samples) displayed, on average, moderate staining degree with maximum in the olivary neurons (Fig.1). The staining of myocardial lipofuscin was also relatively strong (two postmortem samples), similarly as in the lipofuscin granules of the epidydimal epithelium (two bioptic samples) and in skin eccrine glands. In the latter the staining was mostly concentrated on the granule periphery. The weakest binding of Con A, which was just detectable or entirely negative was a feature of hepatocytic lipofuscin (twelve bioptic and postmortem samples.

Staining of lipofuscin in cryostat sections was not convincingly decreased by any of the lipid preextraction procedures used. There was only very slight decrease of staining of the neurolipofuscin in the olivary nucleus, but sometimes, especially in the cortical neurons the staining after extractions was slightly increased.

Staining with Con A correlated well with the PAS positivity which was the strongest in the olivary neurolipofuscin and the weakest in the hepatocytic pigment. There was no correlation with the pigment autofluorescence intensity. Both PAS and autofluorescence were entirely resistant to the lipid extraction procedures.

<u>Ceroid</u> type LP in histiocytes (so called "sea blue histiocytes in the cholesterol ester storage disease, sphingomyelinase deficiency type B, acquired type of storage in the spleen, and in the so called melanosis coli) either resisted staining with Con A or, when stained, the intensity never exceeded that in the surrounding pigment free histiocytes. There was no correlation either with PAS positivity, which was always relatively strong, or with autofluorescence which was invariably of very high intensity. Both proved resistant to the extraction mixtures used.

Neuronal LPs Secondary to Inherited Lysosomal Storage

Samples of the brain (paraffin embedded) from a series of patients with <u>gangliosidoses</u> and <u>mucopolysaccharidoses</u> (MPS) were examined for any correlation between the staining with Con A and

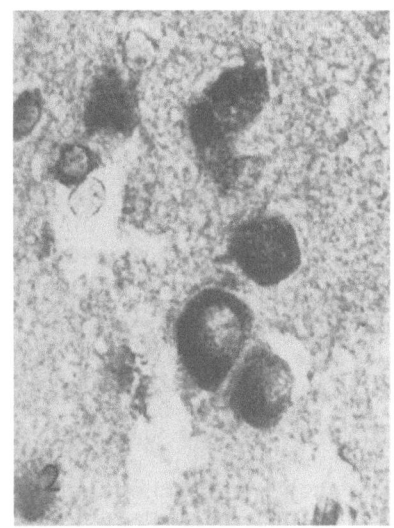 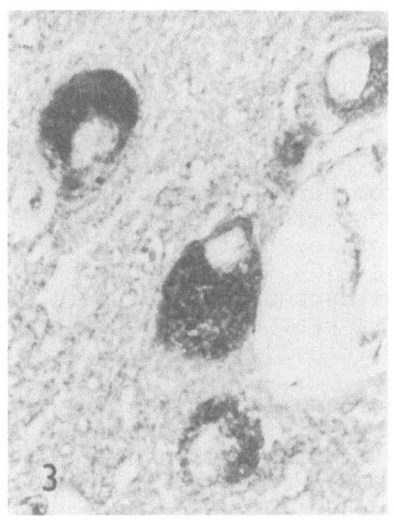

Fig.2. Cortical neuron in MPS type I moderately distended
with Con A stained lipopigment (male,1yr). Paraffin
section. x 600

Fig.3. Cortical neuron in MPS type I strongly distended
with Con A stained lipopigment (male,8yr). Paraffin
section. x 600

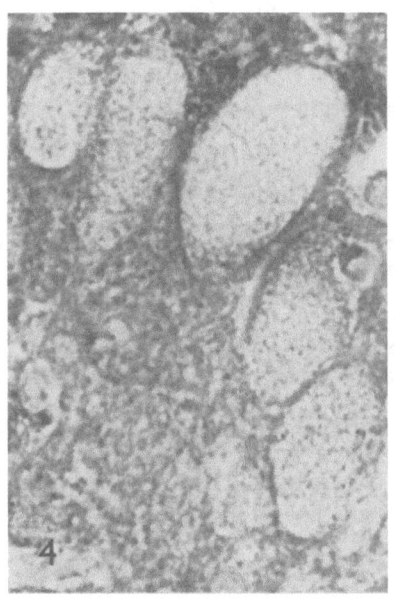 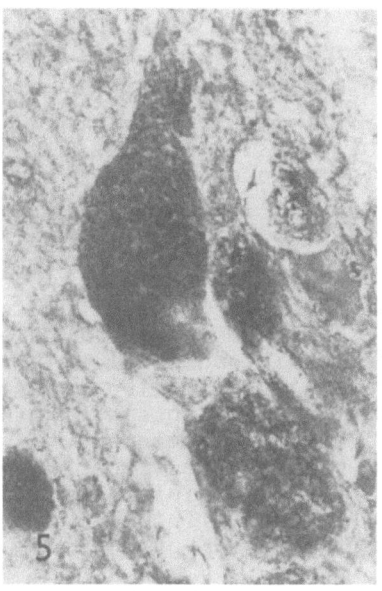

Fig.4. GM gangliosidosis. Ballooned lipid storing neuron
free of lipopigment, unstained with Con A. Paraffin
section. x 600

Fig.5. The same case as in Fig.4. Brain stem neuron loaded
with Con A stained lipopigment granules. Paraffin
section. x 600

the presence of secondarily induced LP. The results showed positive correlation, i.e. the only positive staining was present in
neuronal perikaryan harboring LP granules. This was best demonstrable in the group of mucopolysaccharidoses (three cases of
MPS type I, and three cases of type IIIA) in which the very
strong granular staining of neuronal bodies was uniform throughout the neuronal population (Fig.3) and correlated well with
the prominent PAS positivity, sudanophilia and autofluorescence
of the pigment granules. The intensity of LP deposition and
therefore even staining with Con A was age-independent (age
range 6 - 36 years). Moderately distended neuronal bodies in
mucopolysaccharidosis type I patient succumbing at an early
age of 1 year displayed also strong staining of the accumulated LP (Fig.2).

In gangliosidoses (two cases of GM type, aged 1 and 4 years;
three cases of GM type, aged 22 and 29 months,and 4 years) staining of the ballooned neuronal bodies was minimal or entirely
absent (Fig.4). Only in some regions especially in the case with
protracted course the neurons displayed slight to moderate staining given by the simultaneous LP deposition (Fig.5). However,
the stronest, mainly granular staining was observed in hypertrophic astrocytes on the periphery of their dense cytoplasm.
The staining was lipid- and lipopigment independent. The micro-
glial lipid phagocytes, containing cholesterol esters, were
stained relatively strongly. Those with LP loaded cytoplasm
displayed decreased staining or were entirely unstained.

It should be pointed out that the staining was carried out
in paraffin sections additionally extracted with the conventional chloroform-methanol mixture to exclude any parallel staining
staining glycolipids.

Human Neuronal Ceroid Lipofuscinosis (HNCL)

The results, demonstrated in Fig.6-9 and 14-15, clearly
point to the presence of two different mannose bearing components: the extraction resistent (GP type) was found in all the
HNCL variants examined (Fig.6-9), including the Kufs form. The
staining was always relatively strong with the exceptions mentioned below. The extraction sensitive component, being generally much better extractible with the 1103 extraction mixture
than with the conventional chloroform-methanol 2:1, was easily
detectable histochemically in the two variants available in frozen state: juvenile (two cases) and early juvenile. It appeared
as a strong blackish staining with Con A (see also Fig.14), reduced by the extraction down to the level comparable with that
seen in paraffin sections. Owing to the unavailability of frozen
material from the infantile, late infantile and adult variants
the incidence of lipid bound mannose could not be established
for the whole HNCL group.

Distribution of the granular Con A - stained structures
corresponded with the distribution of the stored LP, i.e. pre-
dominantlly the neuronal perikarya were stained, the staining
of glia being minimal. The only exception was the infantile HNCL
variant where the hypertrophic astrocytes, enormously increased
in amount displayed very high staining intensity (see Fig.6).
The pattern of staining corresponded partly to that seen in the
hypertrophic astrocytes in gangliosidoses (see above) suggesting

thus that substantial part of the staining may have been caused by phenomena other than storage.

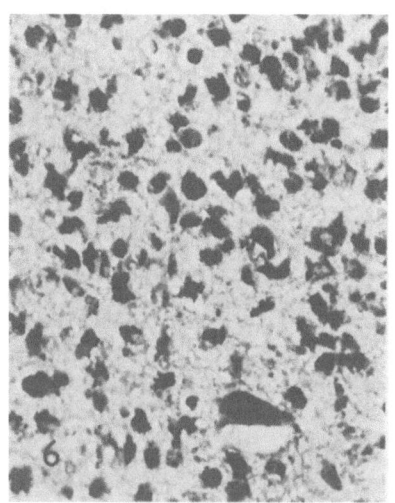 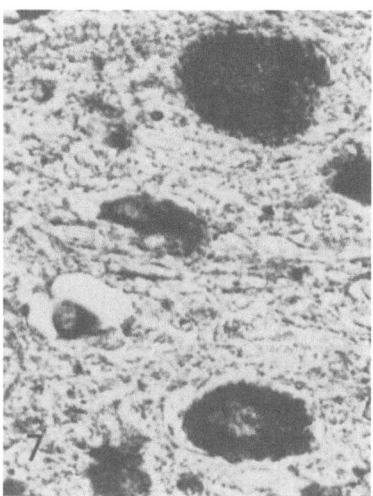

Fig.6. Infantile HNCL. Cortical neurons and hypertrophic astrocytes strongly stained with Con A. Paraffin section. x 270

Fig.7. Late infantile HNCL. Paraffin section. Strong staining with Con A. x 600

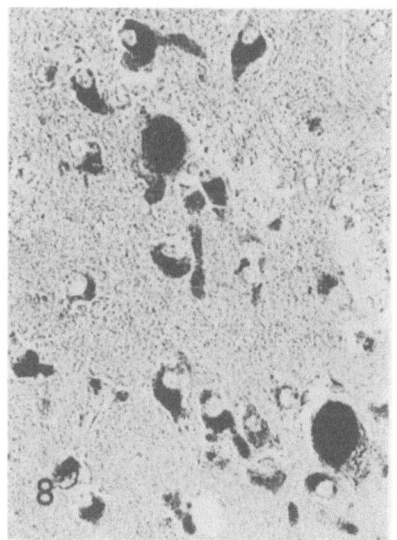 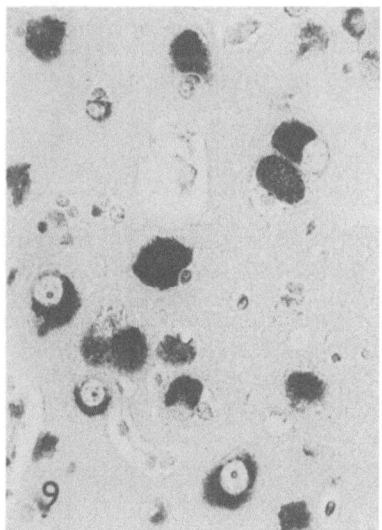

Fig.8. Early juvenile HNCL. Paraffin section. Strong staining of the stored LP with Con A. x 270

Fig.9. Juvenile HNCL. Paraffin section. Strong staining of the stored LP with Con A. x 270

Matching the Con A with the PAS staining showed the same behaviour in the extraction tests,i.e.part of the PAS staining could be eliminated by the "1103" solvent mixture, part of the staining persisted corresponding to the well known PAS positivity of the stored material in paraffin sections. Prominent autofluorescence was not visibly changed by any extraction procedure, however.

The material of the so-called "protein" - type myoclonous bodies (Seitelberger et al.,1967) called also type II stored LP material (Elleder 1978) was studied in nigral and dentatal nuclei, and in basal ganglia in three cases of the late infantile form of HNCL. They were consistently unstained with Con A in paraffin sections (Fig.10,11). This was in accord with the absence of PAS staining in this LP variant, present often in the form of spheroidal neuronal inclusions (for further details of staining, see Elleder,1978). Contrasting with the paucity or absence of staining in lipid depleted sections (paraffin or 1103 extracted cryostat sections), there was very strong staining of this stored LP variant in unextracted cryostat sections (Fig.14), pointing to a high degree accumulation of the lipid bound Con A reactive glycoconjugate.

Canine Neuronal Ceroid Lipofuscinosis (CNCL)

The results are shown in Figs.12,16 and 17. They differed from the human analogue by the absence or, on average, very low amount of the extraction resistant Con A stained GP component,i.e. by the paucity of storage granule staining in paraffin sections.The staining, concentrated in the intervening cytoplasm, was sometimes stronger around the deposits with but rare instances of interior staining (Fig.12). The bulk of the staining was extraction sensitive (see Figs.16,17). Correlation with PAS staining was roughly positive,i.e.many neuronal LP deposits were unstained or stained only partly in paraffin sections. However, autofluorescence was uniform and strong even in paraffin sections, being thus completely unrelated to the PAS positive component.

Ovine Neuronal Ceroid Lipofuscinosis (ONCL)

In the brains of 16 months old animals (No 9180 and 9185, kindly provided by Dr.R.D.Jolly), examined in paraffin sections only, the results were practically the same as in CNCL, i.e. the storage granules were variably stained with great tendency to decrease or even absence of staining in larger deposits (Fig.13). The correlation with PAS staining and with autofluorescence was the same as in CNCL.

Results in Control Neurons

Peripheral neurons (appendical) in the control series of six cases showed intensive almost uniform granular staining with Con A, which was age independent and unrelated to lipofuscin deposition which was absent or very low (occasional individual autofluorescent granules). By pattern and intensity the staining strongly resembled that seen in neurons in HNCL.

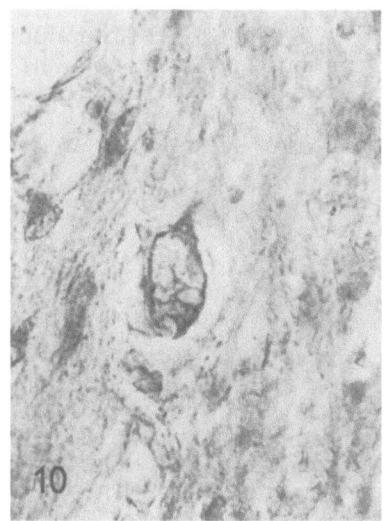 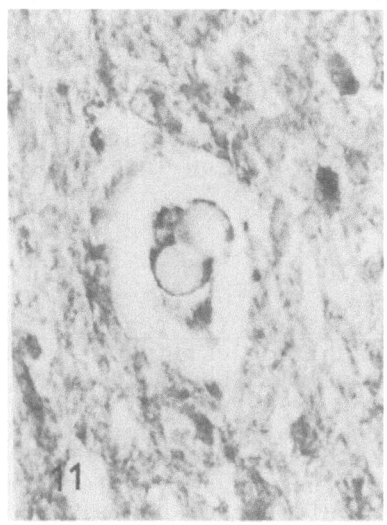

Fig.10. Late infantile HNCL. Thalamic neuron loaded with
lumps of the "protein-type" stored LP unstained
with Con A. x 600

Fig.11. The same case. Dentatal neuron with a couple of
spheroids composed of the "protein-type" material,
again unstained with Con A. x 600

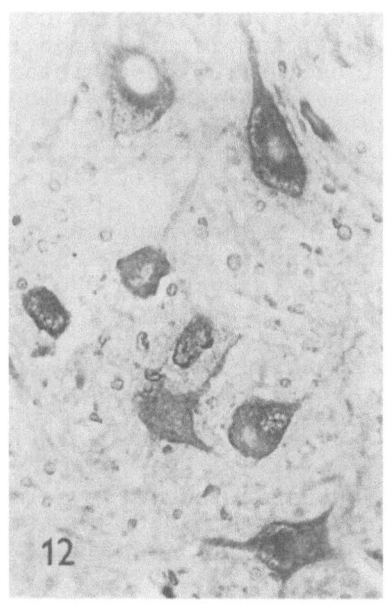 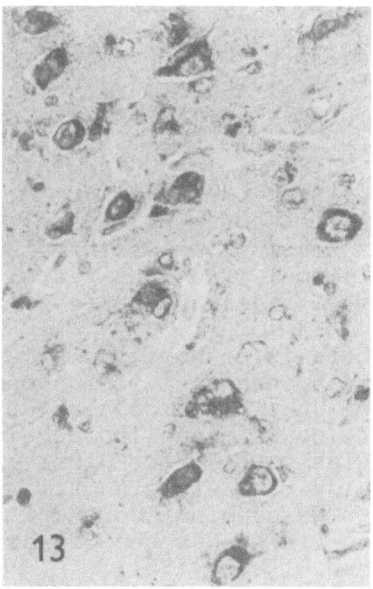

Fig.12. CNCL. Paraffin section. Globular neuronal deposits
are unstained (cf. with Fig.17 and with Fig.10).
x 600

Fig.13. ONCL. Paraffin section. Con A. Note the frequent
absence of staining of the neuronal deposits. x 270

Cerebral neurons examined in brain biopsies from three patients (aged 3 months, 3 and 5 years) with unidentified neurological diseases with minimal neuronal pathology displayed highly heterogeneous staining. About half of the neuronal population was entirely unstained. The rest exhibited granular staining of various intensity again unrelated to the LP deposition, which was very low in intensity.

DISCUSSION

Nature of the reacting sugar. Results of all the blocking tests applied justify to conclude that the vic.glycols in mannose are mainly responsible for staining of LPs of the lipofuscin with Con A. Results of the blocking tests identify the staining mechanism as type I reaction (Katayama and Spicer, 1978). It fits well with the well known property of Con A to bind mannosyl, glucosyl and N-acetyl-glucosaminyl residues present in N-linked GPs (Kornfeld and Kornfeld, 1985).

Nature of the reacting component. In contrast to immunohistology which permits (more or less) to recognize the specifity' of the target macromolecule,lectin histochemistry which demonstrates unspecific monosaccharides provides details merely about basic building unites, leaving thus the biological specifity of the whole reacting glycoconjugate to speculations only. However, using of additional control steps enabled to conclude that the nature of the reacting component is twofold.Part of the staining which could be prevented by lipid extraction must have been due to lipid bound mannose and the only compound which comes into consideration is the oligo-PP-Dol, recently described to be considerably increased in human NCL (Hall and Patrick 1985,1988; Pullarkat et al.,1988). This dolichol derivative is an intermediate in N-glycoprotein synthesis and is localized physiologically in the rough endoplasmic reticulum (for further details see Kornfeld and Kornfeld 1985). Its oligosaccharide chain transferred en block to the protein backbone is especially rich in mannose which is the principal hexose, followed by β-glucose and -N-acetylglucosamine. The dolichol in this form represents probably a substantial part of the dolichol-phosphate and esterified dolichol fractions recently recorded in CNCL and HNCL (Keller et al.,1984; Wolfe et al.,1988).

There is probably no other lipid species which would explain the presence of lipid-bound Con A-reactive glycoconjugate with the possible exception of a glucose-containing ceramid hexoside or GM_2 ganglioside. The latter, due to the terminal N-acetylgalactosamine (O'Brien,1983), might (?) display Con A reactivity (N-acetylglucosamine is considered to be the main reactive hexosamine, however: see Alroy et al.,1988b). Neither of these lipids has been described significantly increased in HNCL (see Elleder,1981).

The extraction resistent part of the staining is given most probably by GP bound mannose. With some exceptions (liver lipofuscin variant LP in HNCL) it seems to be a constant component of lipofuscin type pigments. Its metabolic significance is unclear before all its relationship to the lipid bound mannose, which if existed would point to a complex disorder of GP metabolism as suggested (Pullarkat et al.,1988;Wisniewski and Szumanska,1986; Wisniewski et al.,1988c). Worth proving is whether the GP component might represent accumulation of lysosomal membrane GPs suggested to be the crucial pathogenetic process in LP formation (Wolfe et al.,1987).

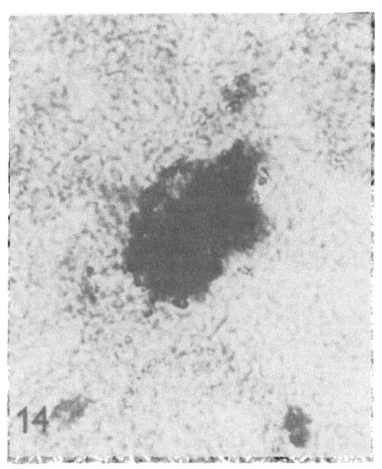 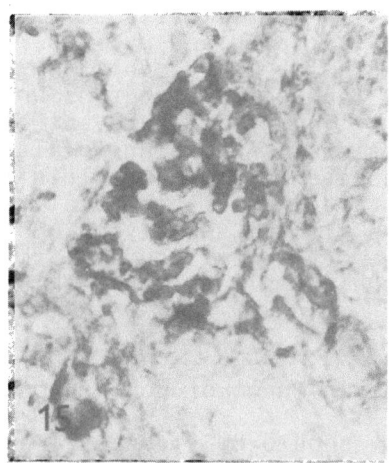

Fig.14. Early juvenile HNCL. Nigral neuron. Cryostat section. Massive staining of the deposits with Con A. x 600

Fig.15. The same specimen and region. Cryostat section pre-extracted with 1103 mixture. Note the remarkable decrease in staining with Con A. x 600

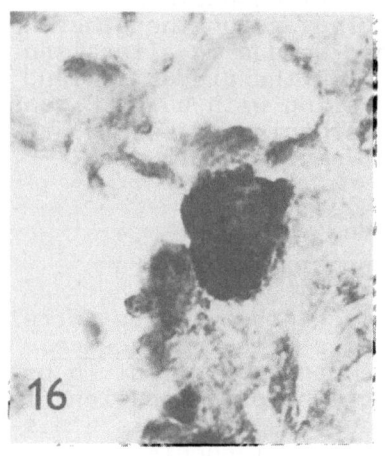 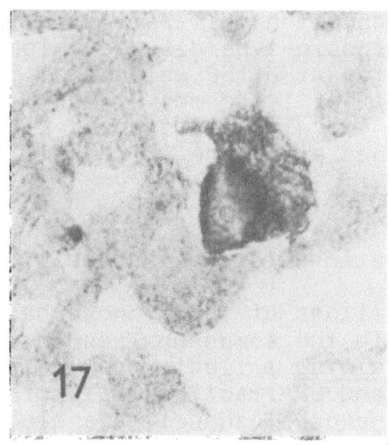

Fig.16. CNCL. Cryostat section. Cortical neuron stained strongly with Con A. x 600

Fig.17. The same specimen after preextraction with 1103 mixture. The staining with Con A is markedly decreased (cf.with Fig.12). x 600

The literary data suggesting an increase in GP bound mannose in lipofuscin are extremely scarce. According to Hoogwinkel et al. (1986) most of the mannose of the isolated LP granules was lipidbound. On the contrary Krusius et al.(1986) described increase of GP bound mannose in isolated storage granules of the infantile variant of HNCL.

Another problem is whether there is any interrelationship between the disease-related Con A-reactive GP and that normally present in the extralysosomal compartment (Tartakoff and Vassalli,1983;Virtanen and Laurila,1980). Existence of a large amount of the latter is evident from our results, namely in the peripheral (appendical) neurons which seem to be exceptionally rich in this type of GP. Relevant literary data are scarce and controversional. According to Alroy et al.(1988a) human cerebral neurons are negative or stain weakly with Con A. Schwechheimer et al.(1984) described relatively strong staining of the brain neurons without specifying whether lipofuscin participated. As for the peripheral neurons, there are no published data dealing with their stainability with Con A.

Different Amount of Con A Receptors in Lipofuscin and in Ceroid

It should be stressed that we understand ceroid to denote the pigment variant as defined by Gedigk and Totovič (1983), i.e.,as that formed in histiocytes during endocytosis, thus differring from lipofuscins supposed to originate in other cell types by autophagocytosis.Our findings showing surprising difference between the two Lp types in the Con A-reactive monosaccharide content suggest also existence of a biochemical difference between them. The absence of Con A staining of ceroid and an overall decrease in staining of the histiocytes bearing the pigment as compared to normal (Kreipe et al.,1986; Ree 1983,1986) or lipid storing counterparts suggests absence of both GP and lipid - bound mannose.This might mean that the ceroid mass could be generated within the preformed lysosomal compartment without any concomitant alteration in the GP turnover. On the other hand, a lipofuscin-type disorder might be seen as due to alteration of that part of GP metabolism coupled with the production and turnover of lysosomes.Other theories dealing with both LP generation and with differences between lipofuscin and ceroid are mentioned elsewhere (Elleder,1981;Sohal and Wolfe,1986).

Implications for Lectin and Lipid Histochemistry of LPs

Neither of the mannose bearing components could be correlated with the component responsible for the autofluorescence. PAS staining reflects most probably the mannose content in both lipid and GP fraction of lipofuscins but the participation of other hexoses can not be excluded. In ceroids, there is a discrepancy between absence of Con A reactive components and the well known PAS positivity of these pigments (Lillie and Fullmer, 1976). This, however, may be caused by the presence of other hexoses or by reactivity of quite different chemical groups, probably lipid-derived, as could be inferred from the results of experimental induction of ceroid from pure lipids in vitro (Casselmann,1951; Porta,1963).

For better recognition of the nature of the Con A reacting glycoconjugate we reccommend using lipid extraction techniques. It proved to be essential in recognizing the double nature of the mannose compound (see above) and helped to recognize impressive variations of both components in the lipopigment generation process. It may explain discrepancy between our results in CNCL and those of Wisniewski et al.(1988a) as regards the nature of the main Con A reactive glycoconjugate in the storage granules. Its easy extractibility speaks in favour of the lipid component being responsible (see Results).

As for the lectin histochemistry in general, it should be pointed out that in interpreting any positive lysosomal staining with Con A and probably even with other lectins, presence of lipofuscin should taken into consideration and checked with appropriate techniques.

Conclusions and Perspectives

It could be concluded that in the formation of LP of the lipofuscin type in various acquired disorders the dominating mannose component is a GP probably chiefly located in the lysosomal compartment. The only exception is the hepatocytic lipofuscin with the borderline values of stainable Con A-reactive GP component. The explanation for this is missing but the finding contrasts with very high values of unesterified dolichol in the liver (Rupar and Carroll,1978). The oligo-PP-Dol is most probably not accumulated in excess in this group , as it could be inferred from the questionable results of the histochemistry.

On the other hand, the genetically conditioned neurolipofuscinoses examined here (HNCL,CNCL) displayed increase of both mannose bearing components, especially of the lipid-bound one. Increase in the latter one was often the major finding, typically in neurons in HNCL storing the second type LP (Elleder 1978; Seitelberger et al.1967) and in CNCL. It could also be concluded that the type of LP stored in CNCL and even in the ovine model resemble by tinction in many points the so-called protein variant of the LP in HNCL (Elleder,1978).

All this means that the only biochemical difference between the acquired and inherited forms of lipopigment generation processes is the accumulation of oligo-PP-Dol in the latter, provable by both biochemical analysis (Hall and Patrick,1988;Pullarkat et al.,1988) and by lectin histochemistry (see above). However, it remains to be established whether the other variants of HNCL, and especially whether the accelerated variants of LPs secondarily generated in various neurolipidoses and mucopolysaccharidoses, also display this phenomenon. This may tell us whether oligo-PP-Dol reflects a special metabolic error or only a further secondary biochemical disorder caused by accelerated lipofuscin deposition.

At any rate, the high level accumulation in the storagebobodies, classified as tertiary lysosomes, of the oligo-PP-Dol normally present in the endoplasmic reticulum (Lennarz 1980) is surprising. No doubt, its explanation may provide important information about the biochemical background of the lipofuscin generation process. One approach to the problem may be to examine systematically all the members of the dolichol group,i.e. free dolichol and its phosphorylated and especially glycosylated variants, in all the above mentioned conditions of acquired and inherited (primary and secondary) states connected with lipofuscin deposition.

Worth following is the striking difference between the histiocytic ceroid and the group of lipofuscins in the amount of the Con A-reactive glycoconjugates as it may help to disclose an important biochemical difference between the two basic lipopigment variants.

REFERENCES

Alroy, J., Ucci, A.A., Periera, E.A., 1984, Lectins: histochemical probes for specific carbohydrate residues. In: "Advances of Immunohistochemistry", DeLellis R.A.,ed., Masson inc., New York.
Alroy, J., Adelman, L.S., Warren, C.D., 1988a, Lectin histochemistry of gangliosidosis. II. Neurovisceral tissues from patients with Sandhoff's disease. Acta Neuropathol.(Berl.) 76:359.
Alroy, J., Ucci, A.A., Periera, E.A., 1988b, Lectin histochemistry: an update, In: "Advances in Immunohistochemistry", DeLellis R.A., ed., Raven Press, New York.
Casselmann, W.G.B., 1951, The in vitro preparation and histochemical properties of substances resembling ceroid. J.Exp.Med. 94:549.
Elleder, M., 1978, A histochemical and ultrastructural study of stored material in neuronal ceroid-lipofuscinosis. Virchows Arch B (Cell Path.) 28:167.
Elleder, M., 1981, Chemical characterization of age pigments. Chapter 7, In: "Age pigments", Sohal R.S., ed., Elsevier, Amsterdam.
Elleder, M., Lojda, Z., 1971, Studies in lipid histochemistry. VI. Problems of extraction with acetone in lipid histochemistry. Histochemie 28:68.
Gedigk, P., Totovič, V., 1983, Lysosomes and lipopigments. Chapter 8, In: "Cellular pathobiology of human disease", B.J.Trump, A.Laufer, R.T.Jones, eds., Gustav Fischer, New York.
Hall, N.A., Patrick, A.D., 1988, Accumulation of dolichol-linked oligosaccharides in ceroid-lipofuscinosis (Batten disesa). Amer.J.Med.Genet., Supplement 5:221.
Hoogwinkel, G.J.M., Blaauboer, A.J., Novak, L., Trippelvitz, L.A.W., 1986, On the composition of autofluorescent accumulation products: ceroid and lipofuscin. In: "Enzymes of lipid metabolism". II. Freys Z., Dreyfus H., Massarelli R., Gatt S., eds., Life Science Series 126, Plenum Press, New York.
Katsuyama, T., Spicer, S.S., 1978, Histochemical differentiation of complex carbohydrates with variants of the Concanavalin A horseradishe peroxidase method. J.Histochem.Cytochem. 26:233.
Keller, R.K., Armstrong, D., Crum, F.C., Koppang, N., 1984, Dolichol and dolichol phosphate levels in brain tissue from English setters with ceroid lipofuscinosis. J.Neurochem. 42:1040.
Kornfeld, R., Kornfeld, S., 1985, Assembly of asparagine-linked oligosaccharides. Ann.Rev.Biochem. 54:631.
Kreipe, H., Radzun, H.J., Schumacher, U., Parwaresch, M.R., 1986, Lectin binding surface glycoprotein pattern of human macrophage population. Histochemistry 86:201.

Krusius, T., Viital, J., Palo, J., Maury, C.P.J., 1986, Enrichment of high mannose type glycans in nervous tissue glycoproteins in neuronal ceroid - lipofuscinosis. J.Neurol.Sci. 72:1.

Lennarz, W.J., 1980, The biochemistry of glycoproteins and proteoglycans, Plenum Press, New York.

Lillie, R.D., Fullmer, H.M., 1976, Histopathologic technic and practical histochemistry, McGraw-Hill, New York.

O'Brien, J.S., 1983, The gangliosidoses, In: "The metabolic basis of inherited disease, Stanbury J.B., Wyngaarden J.B., Fredrickson D.S., Goldstein J.L., Brown M.S., eds., McGraw-Hill, New York.

Porta, E.A., 1963, Experimental electron microscopic study of the sequential stages of in vitro formation of ceroid. Exp.mol. Pathol. 2:219.

Pullarkat, R.K., Kim, K.S., Sklower, S.L., Patel, V.K., 1988, Oligosaccharyldisphosphodolichols in the ceroid-lipofuscinoses. Amer.J.Med.Genet., Supplement 5:243.

Ree, H.J., 1983, Lectin histochemistry of malignant tumors. II. Concanavalin A: a new histochemical marker for macrophage-histiocytes in follicular lymphoma. Cancer 51:1639.

Ree, H.J., 1986, Concanavalin A - binding histiocytes in Hodgin's disease. A predictor of early relapse. Cancer 58:87.

Rupar, C.A., Carroll, K.K., 1978, Occurence of dolichol in human tissues. Lipids 13:291.

Schwechheimer, K., Weiss, G., Schnabel, P., Möller, P., 1984, Lectin target cells in human central neurons and in the pituitary gland. Histochemistry 80:165.

Seitelberger, F., Jacob, H., Schnabel, R., 1967, The myoclonic variant of cerebral lipidosis. In: "Inborn errors of sphingolipid disorders, S.M.Aronson, B.W.Volk, eds., Pergamon Press, Oxford.

Sohal, R.S., Wolfe, L.S., 1986, Lipofuscin: characteristics and significance, In: "Progress in Brain Research", Swaab D.F., Fliers E., Mirmiran M., Van Gool W.A., Van Haaren F., eds., Elsevier, Amsterdam.

Söderström, K.O., 1987, Lectin binding to collagen strands in histologic tissue sections. Histochemistry 87:557.

Tartakoff, A.M., Vassalli, P., 1983, Lectin-binding sites as markers of Golgi subcompartments: proximal-to-distal maturation of oligosaccharides. J.Cell Biol. 97:1243.

Virtanen, I., Ekblom, P., Laurila, P., 1980, Subcellular compartmentization of saccharide moieties in cultured normal and malignant cells. J.Cell Biol. 85:429.

Wisniewski, K.E., Maslinska, D., Kitaguchi, T., 1988a, Studies of lectin binding sites in brains of English setters in different stages of ceroid lipofuscinosis. Ann.Neurol.24,309 (abstr.)

Wisniewski, K.E., Maslinska, D., Kitaguchi, T., 1988b, High level of mannose type glycopeptides in the brains of human neuronal ceroid lipofuscinosis (HNCL) and canine experimental modell (CNCL). J.Neuropathol.exp.Neurol. 47:327 (abstr.)

Wisniewski, K.E., Szumanska, G., 1986, The ultrastructural observation and histochemical localization of some glycoconjugates in neuronal ceroid-lipofuscinosis. Xth Intl.Congr.Neuropathol. Stockholm, No 799:2 (Abstract).

Wisniewski, K.E., Rapin, I., Heaney-Kieras, J., 1988c, Clinico pathological variability in the childhood neuronal ceroid-lipofuscinoses and new observations on glycoprotein abnormalities. Amer.J.Med.Genet., Supplement 5:27.

Wolfe, L.S., Ivy, G.O., Witkop, C.J., 1987, Dolichols, lysosomal membrane turnover and relationship to the accumulation of ceroid and lipofuscin in inherited diseases. Alzheimer disease and aging. Chemical Scripta 27:79.

Wolfe, L.S., Gauthier, S., Durham, H.D., 1988, Dolichols and phosphorylated dolichols in the neuronal ceroid lipofuscinoses, other neuronal storage diseases and Alzheimer disease, induction of autolysosomes in fibroblasts. In: "Lipofuscin - 1987 : State of the Art, Zs.-Nagy, ed., Akadémiai Kiadó, Budapest, and Elsevier, Amsterdam.

DISCUSSION

ALHO: I have been working with immunohistochemistry for several years, so I have a technical question. What is the specificity of your antiserum? It cross-reacts with similar kinds of glycoproteins?

ELLEDER: This is a standard and widely accepted technic with concanavalin A. We succeeded in proving specificity for mannose. If you work with mannose, you can block the reaction with mannoside. There are no other sugars, except glucose, that can react in this way. We also blocked this reaction with several other agents. I think, therefore, that I have done everything possible to block the reaction and to locate the specificity of mannose.

HALL: I enjoyed very much your presentation. It is very nice to see how well different analysis correlate with each other, and how well chemical and histochemical analysis can reveal the same structures. I just would like to clarify that con A is very well defined in its specificity. There are about 7,000 publications on this. It will bound glycopeptides, glycoproteins, glycolipids, and it is specific for particular types of structures, but provided the structure of glycan is appropriate, it will bind any class of molecules with mannose or glucose in proper conformation.

PHOSPHOLIPASES AND THE MOLECULAR BASIS FOR THE FORMATION OF CEROID IN BATTEN DISEASE

Glyn Dawson, Sylvia A Dawson, and
Aristotle N. Siakotos

Departments of Pediatrics and Biochemistry
University of Chicago, IL   60637 and
Department of Pathology, Indiana University
School of Medicine, Indianapolis, IN   46223

SUMMARY

Lysosomal ceroid/lipofuscinosis storage in human, canine, and ovine forms of neuronal ceroidlipofuscinosis is predominantly in neurons and retinal pigment epithelial cells. Despite problems in indentifying individual storage materials, it is believed that non-enzymic oxidation of unsaturated fatty acids in phospholipids and inhibition of lysosomal proteolysis, leading to massive deposition of autofluorescent pigment, is the cause of the disease. We have, therefore, studied cellular phospholipases and find a marked deficiency of lysosomal phospholipase A_1 (PLA_1) in canine NCL brain. Other lysosomal hydrolases, and cytosolic/mitochondrial forms of phospholipase A_2 are completely normal. We believe that the PLA_1 deficiency leads to transient lysosomal storage of phospholipids containing peroxy fatty acids which are then chemically converted to hydroxynonenal, a potent inhibitor of a thiol-dependent enzymes. Inhibition of proteases is believed to be intrinsic to the formation of lipofuscin. An inherited deficiency of a thiol protease (the lysosomal cathepsin H) in two siblings with NCL can also lead to build up of peptides which are then cross-linked and converted into ceroid-containing curvilinear bodies. Thus there is evidence for molecular and genetic heterogeneity in Batten disease.

INTRODUCTION

Neuronal ceroid lipofuscinosis (NCL) (Batten disease) is the term describing a group of childhood diseases where the major pathognomic feature is the accumulation of autofluorescent pigment (called ceroid or lipofuscin) in cells, especially neurons. The pigment, which appears as membranous fingerprint or curvilinear bodies by electron microscopy is believed to be

---

KEY WORDS

Neuronal ceroidlipofuscinosis, retinal pigment epithelium, lysosomal phospholipase A_1, 4-hydroxynonenal, thiol proteases, cathepsin H.

Lipofuscin and Ceroid Pigments
Edited by E. A. Porta
Plenum Press, New York, 1990

enclosed within lysosomal membranes and the disorders are generally considered to be one of lysosomal storage (Zeman and Donahue, 1963; Lenn and Dawson. 1973). Considerable clinical heterogeneity exists within the disease but four broad types are generally accepted (Zeman & Dyken, 1969):

Infantile: This disorder is especially common in Finland, with onset (seizures, dementia and blindness) around 1 year of age and rapid progression to a decerebrate stage, and a severe loss of neurones.

Late infantile: The late infantile type starts around the age of 3 years with seizures and rapidly progresses to blindness and dementia. Myoclonic seizures may, or may not, be present and the disease has been subclassified on this basis.

Juvenile: This form of Batten disease is especially common in Northern Europe and the United States, and is characterized by developmental regression and seizures at age 7. Blindness and dementia soon follow as described by Stengel (1826) in his case report of four siblings with the disease in the Norwegian mining town of Roros.

Adult: This is typically called Kuf's pre-senile dementia and inheritance may be of the dominant type rather than the recessive mode of inheritance typical of the first three forms.

All these forms of NCL are currently given the generic name of Batten disease after the neuropathologist who originally described the pigment storage disease. At present there are several theories to explain the pathology of Batten's disease, but none have yet gained general acceptance and there is currently no reliable biochemical diagnostic test for the disease.

STORAGE MATERIAL

There is general agreement that the NCL storage bodies contain a range of neutral lipids, glycolipids, phospholipids and proteins (Zeman and Siakotos, 1973; Palmer et al., 1986) with the major abnormal lipids being dolichol phosphate oligosaccharides, containing from 5 to 9 mannose residues (Hall and Patrick, 1988), in human, sheep and dog forms of NCL. These dolichol phosphate sugars are necessary intermediates in glycoprotein synthesis and their mode of recycling or catabolism,after their functional role in the endoplasmic reticulum is complete, is largely unknown. Initially, the appearance of increased amounts of dolichol in urine sediment from NCL patients was thought to be of diagnostic value but this test has been discontinued because other degenerative/storage diseases can also result in increased urinary dolichol levels (Wolfe et al., 1986). The storage of dolichol sugars is not thought to be a clue to the primary cause of NCL but may be an important clue to the pathogenesis. We have determined the activity of enzymes involved in dolichol synthesis and found them to be normal in NCL. Thus a primary defect in glycoprotein synthesis appears unlikely because it would affect all tissues and most likely be lethal in utero.

The most dramatic report of abnormal storage material in NCL is from the ovine form of NCL and concerns a 4KDa peptide and its proposed origin from the lipid binding region of mitochondrial ATPase (Palmer et al., 1989). The peptide or peptides have not yet been purified and sequenced and storage bodies are thought to contain mitochondrial fragments. Characterization and detection of the "storage" peptide in canine NCL is further complicated by the fact that both dolichol phosphate sugars, and the major abnormal storage material isolated from NCL dog brain migrate in the 4KDa region on SDS-polyacrylamide gels. The storage material is lipid-soluble and reacts strongly with concanavalin A, confirming it as Dol-P-P-mannose (Ivy & Siakotos, unpublished data).

In addition to these individual compounds, there is a lot of evidence that lipofuscin or ceroid is a cross-linked protein-lipid polymer whose production is somehow related to the uncontrolled generation of aldehydes, peroxides, free radicals and other cross-linking reagents (Siakotos et al., 1972). This process, which also occurs with increasing frequency as normal individuals age, is the subject of this Symposium and is the thread which connects Batten disease with the aging process.

HYPOTHESIS FOR BATTEN DISEASE

The storage material in Batten disease is believed to consist of a chloroform-methanol soluble fluorophore (probably polyisoprenoid-related) and an insoluble (protein) fraction. The proteinaceous material could result from an inherited lysosomal protease deficiency (as suggested by Ivy et al. 1984) and Palmer et al. (1986, 1989)) or from cross-linking of proteins by aldehydes, thereby preventing their normal catabolism. One compound which is capable of causing such cross-linking is 4-hydroxynonenal (HNE), and this aldehyde has been shoen by Esterbauer et al.(1986) to form a chromolipid fluorophore with excitation maximum at 360nM and emission maximum at 430nM. HNE forms both Schiff (C-N) bases, which can be cleaved by hydroxylamine, and C-S bonds which are resistant to cleavage (Cadenas et al., 1983). The targets for C-S bond formation are thiol (SH) groups and therefore thiol proteases (the lysosomal cathepsins) would be logical targets for HNE-mediated inhibition. Siakotos et al. (1988) have reported elevated HNE levels in tissue from dogs with NCL (which seems equivalent to juvenile human NCL) but studies on human NCL samples have not consistently revealed such high levels of HNE and most of the elevated HNE is believed to be trapped in non-hydrolysable C-S bonds. Esterbauer et al. (1985) have shown in vitro that the preferred substrate for generation of 4-hydroxy-nonenal (HNE) is a peroxidized fatty acid on an intact phospholipid molecule. Van Kuijk et al. (1987) have independently proposed that HNE can arise chemically from peroxidized unsaturated fatty acids and that a major function of cellular phospholipase A_2 is to excise these peroxidized fatty acids from damaged membrane phosphoglycerides.

Our hypothesis is that peroxidation of membrane phospholipids is a normal event in lysosomes and that a lysosomal phospholipase exists which excises these peroxidized fatty acids and makes them available for catabolism. A deficiency of LYSOSOMAL phospholipase activity should result in the accumulation of peroxidized PL which breaks down to glycero-aldehydes and HNE. This HNE will then rapidly complex to thiol groups (for example in the active site of cathepsins B, D and H) and amino groups to produce the insoluble complex which is the hallmark of lipofuscin and ceroid. The autofluorescent pigment may represent the further chemical/enzymatic conversion products of this unsaturated aldehyde.

The Phospholipase A$_2$ group of enzymes are typically extracellular, calcium-dependent, active at neutral pH (6.8-8.0), and release polyunsaturated fatty acids (PUFA) from phospholipids. Cellular phosholipases are much less well-studied, although Fu et. al (1980) reported both phospholipase A$_2$ (hydrolysis of 2-[1-^{14}C] linoleoyl-1-acyl-sn-glycero-3-

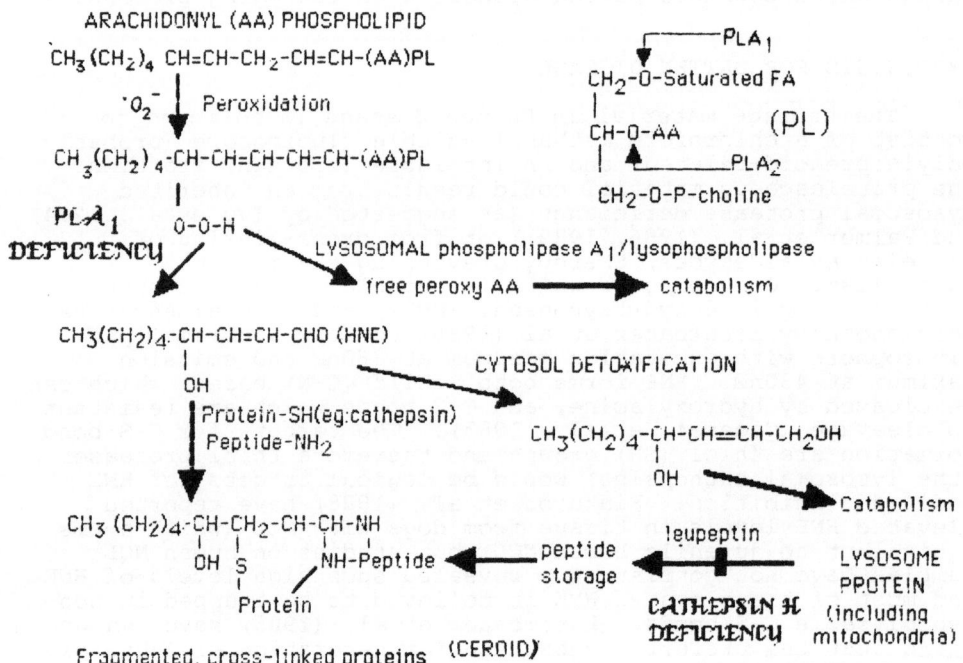

Fig. 1. Proposed scheme for formation of ceroid in NCL as a result of impaired phospholipid and/or acid protease catabolism.

phosphoethan-olamine) and phospholipase A$_1$ (hydrolysis of 1-[^{14}C] palmitoyl-2-acyl-sn-glycero-3-phosphoethanolamine) activity at lysosomal pH (4.4) in acetone powders from dog cerebral hemispheres. Phospholipase A$_2$ (optimum pH 8.5) has been purified from rat liver mitochondria (DeWinter et al , 1982) and phospholipase A$_1$ (pH 3.7) has been purified from rat liver and kidney lysosomes by Hostetler et. al., (1982). This latter enzyme is of interest because it appeared to be the major lysosomal phospholipase and was associated with both

endogenous protein PLA$_1$ inhibitors and a lyso-phospholipase
which removed the PUFA from lyso-PC (Hostetler et al., 1986).
**Thus, phospholipase A$_1$ could be the "candidate" enzyme
for the lysosomal defect in NCL.** Beckman et al. (1987)
have also studied rat liver microsomal (pH 7.5) phospholipases
and conclude that A$_1$ is also the major activity in these cell
preparations. They proposed that it plays a role in protecting
membranes against peroxidative injury by generating substrates
(fatty acyl peroxides) which can be reduced by glutathione
peroxidase. Thus, although the function of lysosomal
phospholipases is unknown, other than to degrade phospholipid
into fatty acids and glycerol, they are clearly distinct from
the well-studied calcium-dependent cytosolic phospholipase A$_2$
enzymes (which liberate the PUFA necessary for prostanoid
synthesis) and important for preventing lipofuscin formation.

REDUCED PHOSPHOLIPASE A ACTIVITY IN BATTEN DOG BRAIN

Homogenates prepared from autopsy cerebral cortex showed
two pH optima for hydrolysis of [^{14}C] dioleoylphosphatidyl-
choline (Fig. 2). The pH 4.4 optimum form is presumed to be
lysosomal phospholipase A$_1$ whereas the pH 7.4 optimum form is
presumed to be cytosolic/mitochondrial phospholipase A$_2$ and was
dependent upon calcium for activity. When canine neuronal
ceroid lipofuscinosis (CNCL) cerebral cortex homogenates were
assayed for activity, there was a profound deficiency in PLA$_1$
(pH 4.4) activity but the PLA$_2$ (pH 7.4) and other lysosomal
hydrolase activities (N-Acetyl-β-D-hexosaminidase, β-
mannosidase, acid phosphatase and, α-fucosidase, for example)
were normal or elevated.

Optimum conditions were established for the lysosomal
phospholipase A$_1$ assay. The buffer was 50mM sodium acetate pH
4.4 containing 300mM NaCl, 1mM HgCl$_2$, 4mM EDTA Triton X-100
(2mg/ml). The substrate, liposomes, were formed from
phosphatidylcholine (plus 1,2,-[^{14}C] dioleoylphosphatidyl-
choline) (0.4mM). 100μl of this was mixed with 100 μl of brain
extract and incubated for up to 4hr. The reaction was stopped
with 5 vols. of CHCl$_3$:CH$_3$OH (2:1 v/v) and [^{14}C] fatty acid and
[^{14}C] lyso PC were resolved from the [^{14}C]PC substrate by thin-
layer chromatography in CHCL$_3$:CH$_3$OH:H$_2$O:CH$_3$COOH

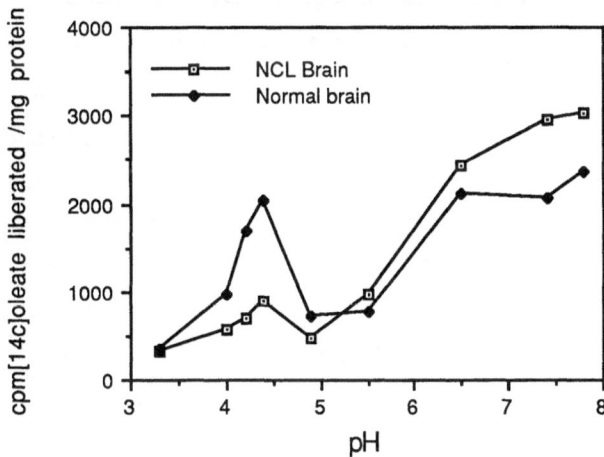

Fig. 2.   Brain phospholipase A activity: pH profile showing
biphasic peak of activity in normal dogs($\blacklozenge$); note the deficient
acidic phospholipase A$_1$, and normal phospholipase A$_2$ activity in
NCL dogs ($\square$).

(65:50:5:1 v/v). Time-course studies indicated that liberation of [^{14}C] oleate was linear for up to 2hr and clearly showed deficient phospholipase activity in CNCL cerebral cortical homogenates (Fig. 3).

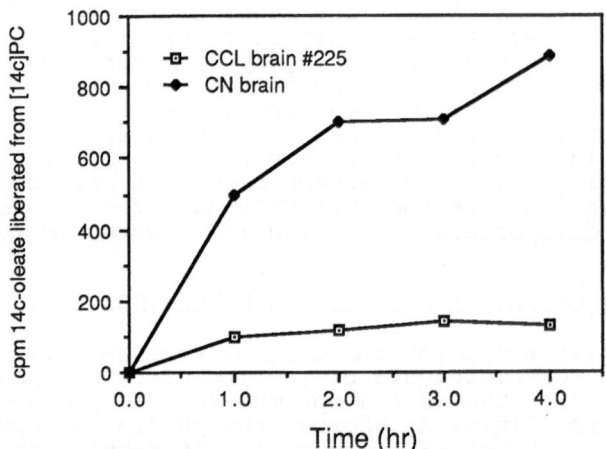

Fig. 3. Time course of liberation of [^{14}c]oleate from [^{14}C]dioleoylphosphatidylcholine by dog brain homogenates showing marked deficiency of phospholipase A$_1$ in NCL dog (□) compared to normals (◆).

The deficiency of phospholipase A$_1$ was confirmed in three affected dogs but as yet we have not had access to carrier CNCL brain. Confirmation of the primary nature of the defect was shown by the intermediate level of activity in heterozygote carriers. We attempted carrier detection by isolating platelets from three non-carrier dogs, three confirmed heterozygotes and three homozygotes, and the results are shown in Table 1.

TABLE 1

Deficiency of phospholipase activity in
platelets from English setters with NCL

Dog	[^{14}C] Oleic acid liberated cpm/mg protein)	
Normal	(n = 3)	410±79
Carrier	(n = 3)	218±80
NCL affected	(n = 3)	133±41

As can be seen from Table 1, the three types can be separated, but there was some overlap between carriers and affected dogs. The residual activity (30% of control PLA$_1$ activity) in CNCL platelets may reflect phospholipase A$_2$ contamination or other activity. Complete resolution will require either a more specific substrate such as 1-[^{14}C] stearoyl linoleyl phosphatidylcholine/-phosphatidylethanolamine or some initial subcellular fractionation to enrich the lysosomal fraction.

In summary, we can measure lysosomal PLA$_1$ activity in dog brain and platelets. The enzyme is quite distinct from other phospholipases (which are normal in CNCL tissue), and a promising candidate for the primary defect.

IS CANINE NCL RELATED TO HUMAN NCL?

Assay of phospholipase A_1 activity in human lymphocytes, monocytes, macrophages and platelets indicated that the latter cells were the most enriched but that activity was low and results were variable. However, in one family (R_0) with late infantile NCL we could clearly distinguish the affected child, carrier parent and unaffected (presumably non-carrier) siblings (Fig. 4).

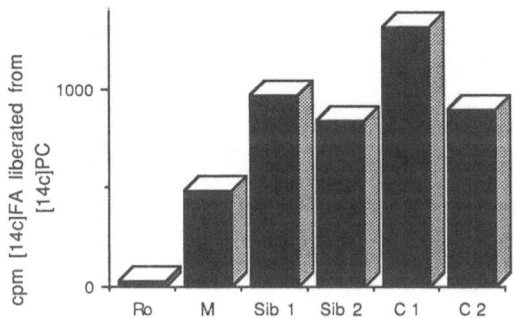

Fig. 4.   Release of [^{14}C]oleate from [^{14}C]dioleoylphos-
          phatidylcholine by human platelet homogenates showing
          impaired acidic (pH 4.4) phospholipase A_1 activity in
          proband (R_0) compared to mother (M), two older,
          unaffected siblings (Sib1 and Sib2), and two unrelated
          controls (C_1 and C_2).

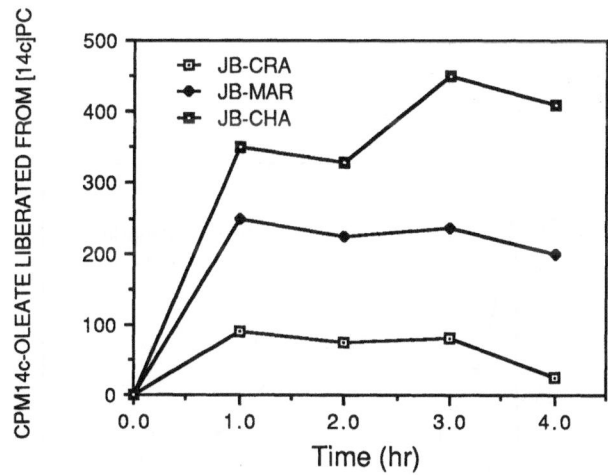

Fig. 5.   Time course of release of [^{14}C]oleate from
          [^{14}C]dioleoylphosphatidylcholine by human lymphocytes
          showing impaired acidic (pH 4.4) phospholipase A_1
          activity in proband (CRA), intermediate activity in
          presumed heterozygote father (MAR), and higher activity
          in the presumed non-carrier unaffected elder sibling
          (CHA).

In 5 other families, the results were too variable to permit a reliable diagnosis. Lymphocytes from one juvenile human NCL patient (Ge family, in which diagnosis was based on clinical findings, plus curvilinear bodies in biopsy tissue) studied thus far showed deficient PLA₁, while an unaffected sibling showed normal PLA₁ activity and the patients' father showed an intermediate level of activity (Fig. 5). It therefore appears that such cells will be important for studying enzyme defects in NCL.

OTHER ENZYME DEFECTS IN NCL: GENETIC HETEROGENEITY?

Although quantitation of HNE is difficult, we have clearly shown an elevation of HNE in Batten disease tissue (Siakotos et al., 1988), and an example of CNCL data is shown in Table 2. Increased HNE could result from PLA₁ deficiency or from a defect in the glutathione peroxidase or aldehyde dehydrogenase pathways.

TABLE 2

Elevation of 4-Hydroxynonenal in Neutrophils and
Plasma from English Setter Dogs with NCL

Dog	Free HNE (Neutrophils)	Catalytically (Rh)-cleaved (C-S-Bound) HNE (Plasma)
	HNE (nM)	
NCL Affected Dogs	106.31±20	130±30
NCL Carrier Dogs	71.75±10	23±10
Non-Carrier Normal Dogs	2.18± 1	22± 9

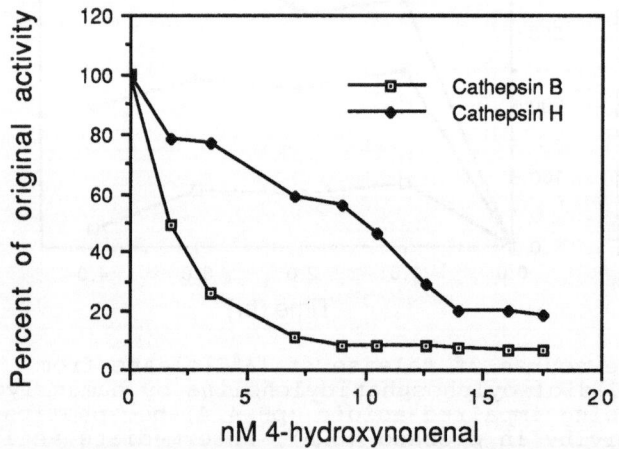

Fig. 6.   Inhibition of cathepsin B and cathepsin H activity following incubation of human skin fibroblasts with increasing concentrations of 4-hydroxynonenal.

The lysosomal thiol proteases (cathepsins) are a logical target for HNE inhibition, and addition of HNE to cells clearly shows that cathepsins (e.g., B and H) can be specifically inhibited by nM concentrations of HNE.

A wide range of other lysosomal hydrolases was unaffected by this treatment, suggesting that HNE is a specific inhibitor of lysosomal thiol proteases. Measurement of HNE in tissues is technically difficult because of its instability and propensity to form complexes with proteins. We believe that this process initiates lipofuscin formation in neurons, which have a high rate of membrane turnover. Esterbauer et al. (1986) have described the formation of fluorescent chromolipids from HNE.

Preliminary reports of cathepsin H deficiency in fibroblasts and brain tissue from patients with Batten disease or NCL has not been substantiated in the 10 or so patients we have studied, but the possibility of a primary cathepsin defect producing the same clinical symptoms as a primary phospholipase deficiency remain strong - for the reasons stated above. Cathepsin assays in general are complicated by the lack of absolutely-specific substrates and overlapping specificities of the unknown number of cathepsins which might exist. Cathepsins B, D, H, L and S have been cloned and sequenced and are believed to be responsible for much of the bulk protein turnover of cells. However, many other cathepsins are believed to exist (e.g., C and G), and little is known of the subcellular localization of the different cathepsins. Thus, cathepsin D is thought to be lysosomal, but by use of subcellular fractionation and a specific antibody to cathepsin D, we have shown it to exist in compartments other than lysosomes (Hancock et al., unpublished data). Cathepsin B is assayed with Z-Arg-Arg Methylcoumarin (MCA), cathepsin H with Arg-MCA and cathepsin L with Z-Phe-Arg-MCA, (Barrett, 1981), and these fluorometric assays are sensitive and reliable. EDTA is often included to inhibit any Ca^{2+}-dependent protease activity and both DTT and cysteine have been used to stabilize some cathepsins while inhibiting others such as dipeptidyl peptidase I. This peptidase is believed to complete protein degradation by digesting oligopeptides, but it is not known if it is found in lysosomes. The existence of many proteases in disrupted (sonicated) cell extracts, which are not physiologically localized to the lysosome, may explain the variability of cathepsin data for NCL tissue. In addition, endogenous inhibitors (cystatins) exist for most cathepsins.(Barrett, 1987), and further complicate in vitro assays.

To resolve the problem of possibly non-specific hydrolysis of Arg-MCA, we focussed on two siblings (EC and SC) with onset of NCL symptoms at age 3, who showed characteristic curvilinear bodies in brain biopsies (Lenn and Dawson, 1973). Fluorometric assay of cathepsin H activity showed only 25% of normal. cDNA probes for cathepsins B and H (Chan et al., 1985) were used to demonstrate normal levels of mRNA for cathepsin B, but dramatically low levels of mRNA for cathepsin H (Fig. 7), suggesting a primary defect in the cathepsin H gene (which has been mapped to chromosome 15). It can be seen that another late infantile NCL patient (D.W.) expressed normal levels of cathepsin H mRNA. Of course, many patients with storage diseases express mRNA for a particular enzyme, but the enzyme is either prematurely degraded or sequestered in the wrong compartment. Thus, patient D.W. could still have a functional cathepsin H deficiency, although we have no in vitro data to support this.

CONCLUSIONS

We have shown a dramatic lysosomal phospholipase A_1 deficiency in NCL dogs, and believe this to be the primary defect. We have found a similar phospholipase A_1 deficiency in some human patients with NCL, and believe it to be a primary defect, since we can detect intermediate levels in known heterozygotes (mothers and fathers). Further studies will require better substrates, for example 1-[^{14}C]stearoyl phosphatidylcholine, to eliminate the small amount of background activity. We have provided an explanation for how a PLA_1 deficiency can lead to the formation of ceroid/lipofuscin, and believe that neurons and retinal pigmentary epithelium are primarily affected because of the high level of membrane turnover in these cell types. Future studies will involve purification and cloning of the gene for lysosomal phospholipase A_1 to study the molecular basis for NCL, and whether the gene is located on the long arm of chromosome 16, as indicated by linkage studies on 26 families with juvenile NCL (Eiberg, Gardiner, Mohr, 1989).

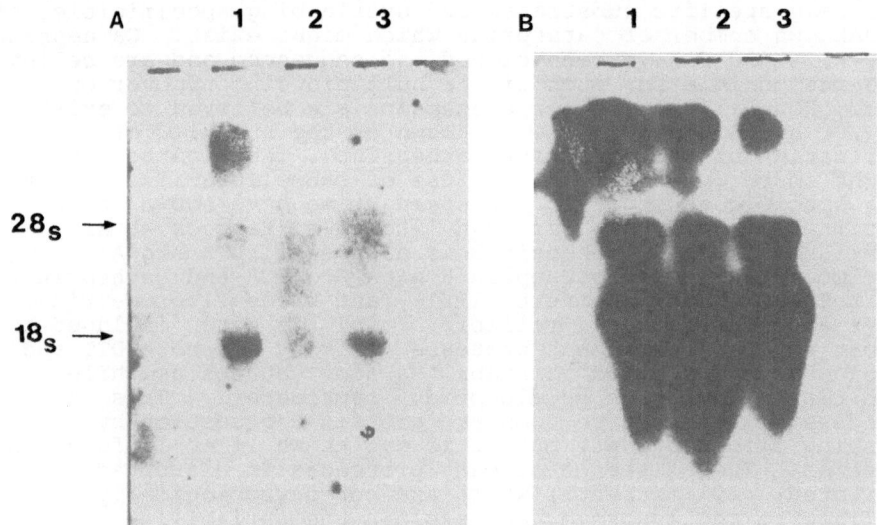

Fig. 7.   Autoradiograph of Northern blot hybridization of human skin fibroblast RNA for cathepsin H mRNA (A) and cathepsin B mRNA (B). Each lane contains 20µg of total ribosomal mRNA. Lane 1, control; Lane 2, patient EC; lane 3, late infantile NCL patient D.W. The same mRNA preparations, following electrophoresis and transfer blotting onto nitrocellulose, were sequentially exposed to ^{32}P-labeled cDNA probes for cathepsins H and B.

We believe that the acid protease deficiency, although extremely important for the formation of ceroid/lipofuscin, is a secondary effect in these patients, caused by the excess generation of the potent cross-linking reagent 4-hydroxynonenal. Similarly, we believe that the storage of dolichol phosphate sugars is a secondary phenomenon related to the organellar origins of the membraneous curvilinear bodies. However, we cannot rule out the possibility that certain cases of neuronal ceroidlipofuscinosis result from a primary acid protease (cathepsin) deficiency, and present data for two such examples. Once again, the direct enzyme assay, using fluorometric MCA substrates, is not totally specific, and we cannot demonstrate a total cathepsin H deficiency in any NCL patient. In one NCL family (EC and SC) we have been fortunate in finding a defect which results in the failure to synthesize stable mRNA for cathepsin H, suggesting that it is the primary defect in these two siblings. At present, our understanding of the role of individual lysosomal cathepsins (B,D,H,L,S, etc.) is extremely limited and it is not clear why the deficiency of a single cathepsin should result in lipofuscin formation, since apparent substrate specificities overlap. Understanding the cell biology of Batten disease will tell us a lot about the role of phospholipids and cathepsins in the cell, and how impairment in these processes leads to organellar disruption, and the formation of ceroid and lipofuscin.

ACKNOWLEDGMENTS

We would like to acknowledge the support and encouragement of the Children's Brain Research Foundation and the National Institutes of Child Health and Human Development (HD-06426) in these studies. We have also been aided by the help and advice from Dr. Shu Chan (Howard Hughes Institute, University of Chciago) and Dr. Jay Tischfield (Indiana University Medical Center, Department of Medical Genetics).

REFERENCES

Barrett, A.J., 1981, Cathepsins, Methods in Enzymol. 80:771-778.
Barrett, A.J., 1987, The cystatins: a new class of peptidase inhibitors, TIBS, -:193-196.
Beckman, J.K., Borowitz, S.M., and Burr, I.M., 1987, The role of phospholipase A activity in rat liver microsomal lipid peroxidation, J. Biol. Chem. 262:1479-1481.
Cadenas, E., Muller, A., Brigelius, R., Esterbauer, J., and Seis, H., 1983, Mechanism of action of 4-hydroxynonenal:depletion of glutathiones, Biochem. J. 214:479-487.
Chan, S.J., San Segundo, B., McCormick, M.B., and Steiner, D.F., 1985, Proc. Natl. Acad. Sci. USA 83:7721-7725.
DeWinter, J.M., Vianen, G.M., and van den Bosch, H., 1982, Purification of rat liver mitochondrial phospholipase A_2, 712:332-341.
Eiberg, H., Gardner, R.M., and Mohr, J., 1989, Batten disease (Spielmeyer-Sjogren disease) and haptoglobins (HP): Indication of linkage and assignment to chr. 16, Clin. Genet. 36 (in press).
Esterbauer, H., Zollner, H., and Lang, J., 1985, Metabolism of the lipid peroxidation product 4-hydroxynonenal by isolated hepatocytes and by liver cytosolic fractions, Biochem. J. 228:363-373.
Esterbauer, H., Koller, E., Slee, R.G., and Koster, J.F., 1986, Possible involvement of the lipid-peroxidation product of 4-hydroxynonenal in the formation of fluorescent chromolipids, Biochem. J. 239:405-409.
Fu, S.C., Mozzi, R., Krakowka, S., Higgins, R.J., and Horrocks, L.A. (1980) Plasmalogenase and phospholipase A_1, A_2, and L_1 activities in white matter in canine distemper virus-associated demyelinating encephalitis Acta Neuropath. (Berl.) 49:13-18.
Hall, N.A., and Patrick, A.D., 1988, Accumulation of dolichol-linked oligosaccharides in ceroid lipofuscinosis (Batten Disease), Amer. J. Med. Genet., Suppl.:5:221-232.

Hostetler, K., Yazaki, P.J., and van den Bosch, H., 1982, Purification of lysosomal phospholipase A, <u>J. Biol. Chem.</u> 257:13367-13373.

Hostetler, K.Y., Gardner, M.F., and Giordano, J.R., 1986, Purification of lysosomal phospholipase A, and demonstration of proteins that inhibit phospholipase A in a lysosomal fraction from rat kidney cortex, <u>Biochem.</u> 25:6456-6461.

Ivy, G.O., Schottler, F., Wenzel, J., Baudry, M., Wolfe,L.S., and Lynch, G., 1984, Inhibition of lysosomal enzymes produces a rapid and massive accumulation of lipofuscin-like dense bodies in the CNS, <u>Science</u> 226:985-987.

Lenn, N.J., and Dawson, G., 1973, On the significance of curvilinear bodies in late infantile lipidosis, <u>Amer. J. Ment. Def.</u> , 77:597-606.

Ng Ying Kin, N.M.K., Palo, J., Haltia, M., and Wolfe, L.S., 1983, High levels of brain dolichols in neuronal ceroid-lipofuscinosis and senescence, <u>J. Neurochem.</u> 40:1465-1473.

Palmer, D.N., Husbands, D.R., Winter, P.J., Blunt, J.W., and Jolly, R.D., 1986, Ceroid lipofuscinosis in sheep, <u>J. Biol. Chem.</u> 261:1772-1777.

Palmer, D.N., Martin, R.D., Cooper, S.M., Midwinter, G.G., Reid, J.C., and Jolly, R.D., 1989, Ovine ceroid lipofuscinosis is a proteolipid proteinosis, <u>J. Biol. Chem.</u>, 264:5736-5740.

Siakotos, A.N., Bray, R., Dratz, E., van Kuijk, F., Sevanian, A, and Koppang, N., 1988, 4-Hydroxynonenal: A specific indicator for canine neuronal-retinal ceroidosis, <u>Amer. J. Med. Genet.</u> Suppl. 5:171-181.

Siakotos, A.N., Goebel, H.H., Patel, V., Watanabe, I., and Zeman, W., 1972, The morphogenesis and biochemical characteristic of ceroid isolated from cases of neuronal ceroid-lipofuscinosis, <u>in</u>: "Sphingolipids, Sphingolipidoses and Allied Disorders", Plenum Publshing Co., New York, 53-61p.

Stengel, O.C., 1826, Beretning om et maèrkeligt Sygdomstilfaelde nos' fire Sodskende i naerheden of Roros, <u>Eyr</u> 1:347-352.

van Kuijk, F.J.G.M., Sevanian, A., Handelman, G.J., and Dratz, E.A., 1987, A new role for phospholipase A$_2$: Protection of membranes from lipid peroxidation damage, <u>TIBS</u> 12:31-34.

Wolfe, L.S., Palo, J., Santavuori, P., Andermann, F., Andermamm, E., Jacob, J.C., and Kolodny, E., 1986, Urinary sediment dolichols in the diagnosis of neuronal ceroid lipofuscinosis, <u>Ann. Neurol.</u>, 270-274.

Zeman, W. and Siakotos, A.N., 1973, The neuronal ceroid-lipofuscinosis, <u>in</u>: Lysosomes and Storage Diseases, Academic Press, New York, pp. 519-551.

Zeman, W., and Donahue, S., 1963, Fine structure of the lipid bodies in juvenile amaurotic idiocy, <u>Acta Neuropath.</u>, 3:144-149.

Zeman, W. and Dyken, P., 1969, Neuronal ceroid-lipofuscinosis is (Batten's disease): relationship to amaurotic family idiocy?, <u>Pediatrics</u> 44:570-583.

DISCUSSION

PALMER: Do you have any comments about the inducibility of cathepsins?

DAWSON: I think that the only evidence of the inducible nature is in tumors. For instance, cathepsin L is massively induced in transformed cells and so is the major protein of many tumors.

KATZ: According to the scheme you presented with the hydroxynonenal mechanism secondarily affecting cathepsin proteins, you will predict that tissues with high amounts of polyunsaturated fatty acids and which are more susceptible to oxidation and more easily form aldehydes, would be then more susceptible to protease inhibition and therefore to pigment accumulation. Is this the case in people with this disease that tissues with high contents of PUFA will then accumulate more pigment than in other tissues with less contents?

DAWSON: I thought that this disease is primarily neuronal, but on the basis of present evidence, the storage of pigment occurs in many tissues. I think that the accumulation is more prominent and is earliest seen in neurons.

JOLLY: I would like to comment on that. We are interested in the pancreas in these diseases, and since this organ produces some 40 different enzymes, it seems that pigment formation relates rather with protein.

Zs.-NAGY: Has anybody looked for the presence of flavins in the storage granules? I am asking this question because we found that the hepatocytic plasma membrane has a particular autofluorescence after being exposed for 20 - 50 minutes to oxygen at atmospheric pressure, and the analysis done in Tokyo by colleagues of Dr. Kitani showed that it is due to riboflavin covalently bound to the proteins of this membrane.

DAWSON: I think that Dr. Palmer may more properly comment on this.

PALMER: First, I looked for flavins, and I did not find any. Second, we thoroughly analyzed, some two years ago, the lipids of the storage granules from various tissues in the ovine NCL, and there was no evidence of loss or depletion of PUFA.

HALL: I want to ask a couple of questions in relation to the enzyme studies you did. Firstly, the cathepsin H study was done in a single patient. Is that correct?

DAWSON: Yes, we looked at the fibroblasts, and the activity was 20 to 30% of the normal.

HALL: Since the tissue has been stored for about 10 - 15 years, I wonder how the tissue was stored and whether you have the control tissues similarly stored to compare their activities.

DAWSON: It was stored in the cold along with other tissues of comparable ages.

HALL: Another question is in relation to your studies with dogs. You studied 9 different dogs, and from what I saw, the levels you found in heterozygous were not significantly different from the values you found in the affected dogs. In which way will you try to ensure that the normal heterozygous and affected dogs came from similar genetic populations, and not from different ones?

DAWSON: The heterozygous and affected dogs were much more inbred than the normals.

HALL: And if so, then the population might have had a low activity in the expression of phospholipase A_1, and could account for roughly half that activity in those subjects as compared with the 3 control dogs.

DAWSON: Yes, that is a possibility.

COMPARISON OF THE CLINICAL COURSES IN PATIENTS WITH JUVENILE NEURONAL CEROID LIPOFUSCINOSIS RECEIVING ANTIOXIDANT TREATMENT AND THOSE WITHOUT ANTIOXIDANT TREATMENT

P. Santavuori, H. Heiskala, T. Autti, E. Johansson and T. Westermarck

Children's Hospital, University of Helsinki 00290 Helsinki, and Helsinki Central Institution for the Mentally Retarded, 02400 Kirkkonummi, Finland Department of Physical Biology, Gustaf Werner Institute (E.J.), Uppsala, Sweden

SUMMARY

Juvenile neuronal ceroid-lipofuscinosis (JNCL) is a progressive encephalopathy characterized by a neural and extraneural accumulation of ceroid and lipofuscin like storage cytosomes and by an autosomal recessive inheritance. It begins with a gradual loss of vision at the age of 4-7 years and is accompanied by epilepsy, a loss of motor function, and a progressive dementia (Santavuori 1988).

We have studied 26 Finnish JNCL patients treated with vitamins E, B_2, B_6 and sodium selenite (antioxidant treatment) by using a JNCL disease specific scoring system introduced by Kohlschütter et al. (1988). Scores were given for the problems of vision, intellect, language, motor function, as well as epilepsy, and compared with the data of

Lipofuscin and Ceroid Pigments
Edited by E. A. Porta
Plenum Press, New York, 1990

17 German JNCL patients not treated with antioxidants (Kohlschütter et al. 1988).

Loss of vision began at the same time among the Finnish and the German JNCL patients. However, loss of intellectual, language, and motor functions and total blindness occured later among the group of Finnish JNCL patients treated with antioxidants. Courses of the epileptic seizures were rather heterogenous and slightly favouring the Finnish patients.

This study supports the theory that antioxidant treatment retards JNCL disease. The study design, however, contains many possible biases, so that the results must be interpreted cautiously.

KEY WORDS: Neuronal ceroid lipofuscinosis - juvenile (JNCL), antioxidant treatment

INTRODUCTION

Juvenile neuronal ceroid lipofuscinosis (JNCL) is a progressive encephalopathy charcterized by neural and extraneural accumulation of ceroid- and lipofuscin like storage cytosomes. JNCL is inherited recessively in autosomes. The disease begins with a gradual loss of vision at the age of 4-7 years, followed by slurring of speech as a first sign of progressive motor disability (mostly extrapyramidal but also cerebellar and pyramidal) and by a gradual mental impairment and epileptic seizures (generalized or complex partial type) (Santavuori 1988, Santavuori et al. 1988).

The pathogenesis of the NCLs is unknown, but several hypotheses have been proposed. Among these are 1) abnormalities in the glycoprotein turnower, possibly in enzymes involved in the degradation or intracellular recycling of lysosomal membrane and possibly linked to dolichol metabolism (Krusius et al. 1986, Hall and Patrick 1988, Pullarkat 1988, Wisniewski et al. 1988, Wolfe 1988), 2) error in lysosomal proteolysis (Ivy et al. 1984, Jolly et al. 1988, Palmer et al. 1988), 3) deficiency in phospholipase A2 activity (Dawson and Glaser 1988), 4) secondary deficiency of prebeta-lipoprotein (Bennet et al. 1986, 1988), 5) secondary lipid peroxidation (Halliwell and Gutteridge 1984 (2), Dawson and Glaser 1987, Heiskala et al. 1988).

In the 70's a disturbance in the peroxidation of polyunsaturated fatty acids was proposed by Zeman et al. (1970). Prompted by this an experimental antioxidant treatment with "Zeman formula" was initiated in the Finnish JNCL patients. Later the treatment was developed by Westermarck et al. (1988) into its present form, which contains vitamins E, B_2, B_6 and sodium selenite. Experiences have shown that this form of antioxidant medication is able to retard the progress of JNCL to some extent (Santavuori et al. 1985, 1988).

In order to study further the antioxidant treatment among these patients we scored the clinical parameters of the Finnish JNCL patients treated with antioxdants using a JNCL disease specific scoring system previously used and developed by Kohlschütter et al. (1988) enabling comparisons to German patients.

PATIENTS AND MATERIAL

The patients comprised all the Finnish JNCL patients fulfilling the following criteria: a) diagnoses were confirmed by clinical and histopathological (rectal biopsy) characteristics (Santavuori 1988), b) the treatment was initiated before any marked clinical symptoms, c) follow-up period was at least 8 years.

Kohlschütter's et al. (1988) scoring method was used. Scores of 0 e.g. for total loss of function, 1 e.g. for severe problem, 2 e.g for slight, but readily recocnized problem or 3 e.g. for normal function were assigned to the following variables: patient's vision, intellect, language and motor function, yearly during retrospective observation period. Epileptic symptoms were also scored according to their severity.

The antioxidant treatment included: Sodium selenite (0.05-0.01 mg/Se/kg b.w.), vitamin E as alpha tocopherol acetate (0.014-0.05 g/kg b.w.), vitamin B_2 (0.025-0.05 mg/kg b.w.) and vitamin B_6 (0.063-0.8 mg/kg b.w.).

RESULTS

The clinical courses of our patients (FIN) and German ones (FRG) are shown in Figure 1.

Loss of vision began to occur at the same time among the Finnish and the German patients. However, total blindness began at the mean age of 9 and 13 years, respectively. Similarly, loss of intellectual, language and motor functions occured later among the Finnish JNCL patients treated with antioxidants. Courses of the epileptic seizures were rather heterogenous and slightly favouring the Finnish patients.

DISCUSSION

Figure 1. shows that except for the debut of visual failure, the Finnish patients treated with antioxidants seem to have clinical problems much later on than the non-treated German patients. The variability of the scores, however, is great among our patients. The results of Kohlschütter et al (1988) are similar in this respect.

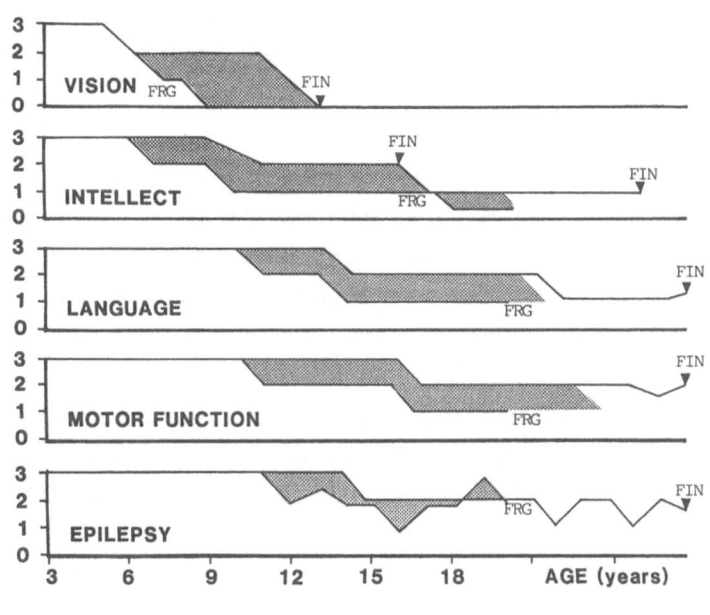

Figure 1.

LEGEND: Clinical courses of JNCL patients, depicted by medians (solid lines) FIN = Finland (26 patients) and FRG = Germany, federal rebublic (17 patients) and the differences (shaded area) of single scores. A score of 3 signifies normal function, a score of 0 maximal dysfunction. The steps in **Vision** are: normal - poor, but orientation good - poor, orientation diffucult - blind; in **Intellect**; normal - abstract reasoning (mathematics) has become difficult - dementia, clearly evident - apparent total loss; in **Language**: normal - minor difficulties recognized - hardly understood - no verbal contact; in **Motor function**: normal - slight handicap recognized - mostly wheel chair, some motility preserved -immobile, bedridden; in **Epilepsy** (only Grand Mal): no seizures - 1-2/year - <1/month,<12/year - >12/year (Kohlschütter et al. 1988).

While considering a disease with an unknown pathogenesis both geno- and phenotypic variation can lead to differing clinical courses. The clinical variability of NCLs in general is known to be great with numerous variant forms (Boustany et al. 1988, Dyken 1988, Kohlschütter et al. 1988, Philippart 1988, Santavuori 1988). It can be argued that Finnish and German diseases are not identical in this respect. However, the fact that loss of vision begins at the same age among the two groups, supports the concept of similar disease.

In order to circumvent the difference in ancestry, we also compared the present results to the data of 53 Finnish JNCL patients not treated with antioxidants (Santavuori et al. 1985). These patients had visual failure at the same time, mental retardation between 8-18 years and the following milestones (average): mild-moderate speech failure 12.9 years, severe speech failure 15.9 years, epilepsy 11.1 years, Parkinsonian walk 13.7 years, walking lost 17.3 years. The comparison speaks in favour of antioxidant treatment but is somewhat biased since the data of the 53 patients dates mostly from pre 70's (historical comparison) and thus precluding the use of the method of Kohlschütter et al. Differences in the rehabilitation practices and treatments, other than antioxidants, may have influenced the outcome.

It is known that a histological disease begins already in utero (Rapola et al. 1988). However, the so called preclinical (a disease histologically confirmed through an index sibling) antioxidant treatment seemed to retard the disease process only in one out six patients (Santavuori et al 1988). It can be hypothesized that a certain phase of disease may be susceptible to a secondary peroxidation. Westermarck et al. (1989) have shown that mononuclear cells (containing storage material) of JNCL patients have iron and zinc concentrations that favour peroxidation phenomenon, table 1. Heiskala et al. (1988) have shown that CSF complexable iron was elevated among NCL patients at lowered pH values, whereas, at an assay pH value of 7.4 the values did not correlate with the presence of NCL or its disease severity. In these studies the concentrations of Fe and Zn were, however, unaltered by antioxidant treatment.

While interpreted cautiously, these results speak in favor of the antioxidant treatment as a part of the treatment protocol. Antioxidant treatment is the only nonsymptomatic

Table 1

ELEMENTAL CONCENTRATIONS IN MONONUCLEAR CELLS COMPARING NCL-GROUP TO CONTROLS NCL-GROUP DIVIDED TO INCL AND JNCL GROUPS AND IF SUPPLEMENTED WITH ANTIOXIDANTS						

	CONTROLS	P	NCL-GROUP			
FE (μg/g) SD/test	<0.5 (all)	***** rank-t.	5.8 0.7			
ZN (μg/g) SD/test	15 2.0	***** t-t.	3.4 0.5			

	INCL	P	JNCL	AO+	P	AO-
FE (μg/g) SD/test	6.1 0.6	NS t-t.	5.7 0.8	5.9 0.9	NS t-t.	5.8 0.6
ZN (μg/g) SD/test	3.7 0.5	NS t-t.	3.3 0.4	3.5 0.5	NS t-t.	3.4 0.4

LEGEND: Statistical tests: t-t. = Student's T-test (two-tailed), rank-t. = Wilcoxon rank sum test (two-tailed), SD = standard deviation, ***** = $P < 0.00001$, **** = $P < 0.001$, NS = nonsignificant

form of treatment suggested so far - if a treatment aimed at hypothetical secondary peroxidation can be called nonsymptomatic. We are planning to expand this study into a multinational one in order to increase the reliablity of these observations.

REFERENCES

Bennet MJ, Hosking GP, Gayton R, Thompson G, Galloway JH, Cartwright IJ (1988): Therapeutic modification of membrane lipid abnormalities in juvenile neuronal ceroid-lipofuscinosis (Batten disease). Am J Med Genet, Suppl 5:275-284.

Bennet MJ, Gillis WS, Hosking GP, Galloway JH, Cartwright IJ (1986): Lipid abnormalities in serum in Batten's disease. Dev Med Child Neurol 28:815-817.

Boustany RMN, Alroy J, Kolodny EH (1988): Clinical classification of neuronal ceroid-lipofuscinosis subtypes. Am J Med Genet, Suppl 5:47-58.

Dawson G, Glaser P (1987): Apparent cathepsin B deficiency in neuronal ceroid lipofuscinosis can be explained by peroxide inhibition. Biochem Biophys Res Commun 147(1):267-274.

Dawson G, Glaser PT (1988): Abnormal cathepsin B activity in Batten disease. Am J Med Genet, Suppl 5:209-220.

Dyken PR (1988): Reconsideration of the classification of the neuronal ceroid-lipofuscinoses. Am J Med Genet, Suppl 5:69-84.

Hall NA, Patrick AD (1988): Accumulation of dolichol-linked oligosaccharides in ceroid-lipofuscinosis (Batten disease). Am J Med Genet, Suppl 5:221-232.

Halliwell B, Gutteridge JMC (1984): Oxygen toxicity, oxygen radicals, transition metals and disease. Biochem J 219:1-14.

Halliwell B, Gutteridge JMC (1984): Lipid peroxidation, oxygen radicals cell damage, and antioxidant therapy. Lancet 1:1396-1397.

Heiskala H, Gutteridge JMC, Westermarck T, Alanen T, Santavuori P (1988): Bleomycin-detectable iron and phenanthroline-detectable copper in the cerebrospinal fluid of patients with neuronal ceroid-lipofuscinoses. Am J Med Genet, Suppl 5:193-202.

Ivy GO, Schottler F, Wenzel J, Baudry M, Lynch G (1984): Inhibitors of lysosomal enzymes: accumulation of lipofuscin-like dense bodies in the brain. Science 226:985-987.

Jolly RD, Shimada A, Craig AS, Kirkland KB, Palmer ND (1988): Ovine ceroid-lipofuscinosis II: pathologic changes interperted in light of biochemical observations. Am J Med Genet, Suppl 5:159-170.

Kohlschütter A, Laabs R, Albani M (1988): Juvenile neuronal
    ceroid lipofuscinosis (JNCL): quantitative
    description of its clinical variability. Acta
    Paediatr Scand 77:867-872.

Krusius T, Viitala J, Palo J, Maury CP (1986): Enrichment of
    high mannose-type glycans in nervous tissue
    glycoproteins in neuronal ceroid-lipofuscinosis. J
    Neurol Sci 72:1-10.

Palmer DN, Martinus RD, Barns G, Reeves RD, Jolly RD (1988):
    Ovine ceroid-lipofuscinosis I: lipopigment
    composition is indicative of a lysosomal proteinosis.
    Am J Med Genet, Suppl 5:141-158.

Philippart M (1988): Diagnosis and treatment of typical and
    atypical forms of lipopigment storage disorders. Am J
    Med Genet, Suppl 5:291-298.

Pullarkat RK, Kim KS, Sklower SL, Patel VK (1988):
    Oligosaccharyl diphosphodolichols in the ceroid-
    lipofuscinoses. Am J Med Genet, Suppl 5:243-252.

Rapola J, Santavuori P, Heiskala H (1988): placental
    pathology and prenatal diagnosis of infantile type of
    neuronal ceroid-lipofuscinosis. Am J Med Genet Suppl
    5:99-103.

Santavuori P, Westermarck T, Rapola J, Lappi M, Moren R,
    Vuonnala U (1985): Antioxidant treatment in
    Spielmeyer-Sjögren's disease. Acta Neurol Scand
    71:136-145.

Santavuori P (1988): Neuronal ceroid-lipofuscinoses in
    childhood. Brain Dev 10:80-83.

Santavuori P, Heiskala H, Westermarck T, Sainio K, Moren R
    (1988): Experience over 17 years with antioxidant
    treatment in Spielmeyer-Sjögren disease. Am J Med
    Gen, Suppl 5:265-274.

Westermarck T, Santavuori P, Heiskala H (1988): Antioxidant
    therapy in neuronal ceroid lipofuscinoses (NCL). In:
    Handbook of Free Radicals and Antioxidants in
    Biomedicine, Vol II, CRC Pres Inc. Boca Baton, Eds
    Miquel J, Weber H & Quintanilha A, 281-287.

Westermarck T, Heiskala H, Johansson E, Lindh U, Santavuori P
    (1989): Iron, copper and zinc concentrations in
    mononuclear cells of patients with ceroid-
    lipofuscinoses. Elsevier (in press)

Wisniewski KE, Rapin I, Haney-Kieras J (1988): Clinico-
    pathological variability in the childhood neuronal
    ceroid-lipofuscinoses and new observations on
    glycoprotein abnormalities. Am J Med Genet, Suppl
    5:27-46.

Wolfe LS, Gauthier S, Haltia M, Palo J (1988): Dolichol and dolichyl phosphate in the neuronal ceroid-lipofuscinoses and other diseases. Am J Med Genet, Suppl 5:233-242.

Zeman W, Donohue S, Dyken P, Green J (1970): The neuronal ceroid-lipofuscinoses (Batten-Vogt syndrome): in Vinken PJ, Bruyn GW (eds): "Handbook of Clinical Neurology, Vol 10 (Leukodystrophies and Poliodystrophies)." Amsterdam: North Holland Publ. Co., pp 588-679.

DISCUSSION

PORTA: I suppose that you have been treating your patients with antioxidants because you may have at least some evidence that the morphological and/or functional alterations are due to an increase in prooxidants or a decrease of antioxidants. Do you have any data showing that this has been the case, such as levels of antioxidants or levels of common products of lipid peroxidation measured by TBA reaction, diene conjugation analysis, ethane evolution, chemiluminescence, etc.? If so, did the treatment modify these values?

WESTERMARCK: We have tried to use some of these procedures, like ethane evolution, but so far we have not been successful. The antioxidant capacity of the CSF was actually decreased, but we have not determined the results of treatment in relation to parameters that you mentioned.

HALL: In some of the Swedish patients with juvenile Batten's disease, the clinical course is somewhat different from that in Britain. Do you have any evidence to show that the clinical course of the untreated patients in the Finnish series is similar to that in the German patients? The second question is: were the same physicians involved in scoring the patients disabilities?

WESTERMARCK: We have been collaborating with different professionals in the evaluation. It is admittedly difficult to say what the benefits of the antioxidant treatment are, but at least, in the Finnish patients without treatment, the disabilities were obviously worse than in the treated ones.

HALL: How big was that group?

WESTERMARCK: We have a total of 53 patients.

PALMER: The first thing to worry about iron in the CSF is blood contamination. The second and more serious problem relates to the use of deferoxamine which may have side effects. Would you like to comment on that?

WESTERMARCK: During one year use of deferoxamine the results were not satisfactory, and we have abandoned it. I agree about the side effects.

GOEBEL: You seem to have a heterogenous group of patients. The patient you showed that had an IQ of over 100 at 24 years and being blind, seems to be a protracted form of juvenile ceroid-lipofuscinosis rather than the classical form. We described two siblings who were blind for about 20 years before dementia occurred, and they died in their mid-thirties; so, that somewhat obscures your protocol. Did you actually compare scores in the same patients with control patients?

WESTERMARCK: We compared retrospectively the scores in the same patients several times.

THE ROLE OF CEROID IN LUNG AND

GASTROINTESTINAL DISEASE IN HERMANSKY-PUDLAK SYNDROME

Carl J. Witkop[1], DeWayne Townsend[2], Peter B. Bitterman[2], Keith Harmon[2]

Departments of Oral Science[1] and Medicine[2]
Health Sciences Center, University of Minnesota
515 Delaware Street, S.E.
Minneapolis, Minnesota 55455
U.S.A.

SUMMARY

Studies of ceroid associated lesions in Hermansky-Pudlak syndrome (HPS) indicate that restrictive lung disease and granulomatous gastrointestinal lesions are among the most frequent and account for 60% of the deaths of the patients. No defects in the immune system in HPS were found. Histological, ultrastructural and chemical studies show accumulation of non-biodegradable ceroid in tissue cells and associated macrophages of HPS patients. There is no known degradative pathway for ceroid. Ceroid is eliminated from cells by exocytosis. Wild type and pale eared mice treated with leupeptin, which inhibits exocytosis, accumulate ceroid in organ cells in the same sequence seen in HPS. Young HPS patients without significant pulmonary function deficits were lavaged, the macrophages examined by TEM and tested for platelet derived growth factor. Macrophages contained ceroid and 7/12 patients had 27±42 units of PDGF bioactivity compared to zero activity in controls. Purified ceroid was fed to macrophages lavaged from the lungs of non-smoking control subjects. Prior to feeding, less than 5% of cells contained one or two small yellow-orange autofluorescent granules resembling ceroid. After feeding, approximately 20% of control cells had ingested ceroid, but PDGF was not increased. The immunologic and histologic studies and the production of PDGF by macrophages which precedes lung fibrosis all point to a central role of the macrophage in these lesions. These studies did not distinguish whether the macrophages ingested ceroid from other cells, or whether ceroid is produced intrinsically by the HPS macrophage.

Key Words: Hermansky-Pudlak syndrome, ceroid, pulmonary fibrosis, colitis, macrophage.

Supported by NIH Grants: GM PO1-22167 and Clinical Program Grant PR400.

INTRODUCTION

Hermansky-Pudlak syndrome (HPS)[1,2] is an autosomal recessively inherited disorder consisting of tyrosinase-positive oculocutaneous albinism, a bleeding diathesis due to storage-pool deficient platelets lacking dense bodies[3] and ceroid storage disease[4]. The most frequent ceroid-associated disorders in HPS are fibrotic restrictive lung disease,[2,4-6] granulomatous enteropathic lesions,[2,4,5,7] granulomatous gingivitis,[4] and kidney failure.[8-10] The enteropathic and gingival granulomatous lesions are nearly identical to those seen in Crohn's disease.[4,11,12]

A study of 484 HPS patients in Puerto Rico[10] found that among 44 recently deceased patients that 23 (52%) died from fibrotic restrictive lung disease between ages 35 and 54 years and that 4 (9%) died from sequelae of granulomatous entero-pathic disease between ages 19 and 44 years. Some of these patients had combinations of these disorders as well as kidney failure at the time of death.

Granulomatous Enteropathic Disease

The familial aggregation, the clinical appearance and course of the enteropathic disease and the histologic appearance of the lesions in HPS are so similar to those in Crohn's disease that they elicit a diagnosis of Crohn's disease by both gastroenterologists and pathologists who encounter these patients. The etiology and pathogenesis of Crohn's disease is poorly understood.[13] A large number of factors have been purported to act in the pathogenesis of Crohn's disease. Among those proposed are immunologic factors, particularly antibody-depend cell-mediated cytotoxicity, cell-mediated delayed hypersensitivity, disturbances in the complement system and elevation of IgG and IgM to the detriment of IgA. It is thought that the formation of granulomas is evidence of the absorption of insoluble antigens.[13]

Pulmonary Fibrosis

In pulmonary fibrosis, mesenchymal cells and connective tissue accumulate within the alveolar wall.[14] One hypothesis posits that these structural changes result from the release of growth factors from unregulated alveolar macrophages.[15] The concept that the release of mesenchymal growth factors from alveolar macrophages precedes the fibrosis was derived from a study of at-risk relatives of patients with autosomal dominant familial idiopathic pulmonary fibrosis.[15] However, in this study,[15] there was no means of identifying the relatives who were carrying the gene for pulmonary fibrosis.

Project Studies on the Enteropathic and Pulmonary Lesions in HPS

Studies under the Hermansky-Pudlak Project have been directed toward elucidating the factors in the pathogenesis of the enteropathic and pulmonary disease in HPS.

Immunology

A study of 15 HPS patients,[16] 4 with and 11 without granulomatous lesions, found no abnormalities in immunoglobulin
levels, complement, lymphocyte subsets, natural killer and
lymphokine-activated cytotoxicity, mixed lymphocyte responses,
lectin-induced transformation, neutrophile function including
luminol-dependent chemiluminescence, chemotaxis and aggregation. There was no lymphocyte proliferative response to an
isolated ceroid preparation. Serum protein electrophoresis
and quantitative IgA, IgG and IgM levels were normal and each
patient had total hemolytic complement (CH_{50}) assays and
levels of complement (C_3 and C_4) within normal limits.[16]
Rheumatoid factor and immunofluorescence assays for anti-
thyroid, anti-smooth muscle, anti-mitochondrial, and anti-
nuclear antibodies were negative in all patients.

Ceroid

The exact chemical composition of ceroid which accumulates
in HPS is unknown. It shares the features of yellow
autofluorescence under ultraviolet illumination, stains acid
fast and positive with the Armed Forces Institute of Pathology
lipofuscin stain, is insoluble in strong acids and alkalis and
has ultrastructural similarities with ceroid in the ceroid-
lipofuscin diseases such as Batten disease.[4,5,17] It does not
appear to be derived from polyunsaturated fats.[18] Dietary
intake of saturated or unsaturated fat does not influence urinary ceroid output.[18] Ceroid in HPS is stored intracellulary
in lysosomes.[4,5,11,17-19] Ceroid accumulates first and in
largest quantities in the proximal tubules of the kidney, in
the reticuloendothelial system, and in the liver.[5,8,19]
Moderate amounts are found in lungs, gastrointestinal tract,
and cardiac muscle[4,5,8,19] and lesser amounts in other
organs.[5,19] There is no known degradative biochemical pathway
for ceroid.[17] It is believed to be eliminated from cells by
exocytosis.[4,17] Blocking the exocytic mechanism by leupeptin
causes accumulation of ceroid in the tissues of mice in the
same sequence as that seen in HPS patients.[4]

Alveolar Macrophages and Growth Factors[20]

Clinical, radiographic and pulmonary function tests of 22
young non-smokers with HPS (13m, 9f) found that 13 of 22 were
normal and 9 had mild restrictive lung disease. Twelve of the
13 patients with normal tests were lavaged. Differential cell
counts were normal: (91±5.0% alveolar macrophages; 7±5.0%
lymphocytes; 2±1.6% neutrophiles) indicating no unusual
inflammatory component. The release of one prototype growth
factor, platelet derived growth factor (PDGF) was assayed in a
standard serum-free bioassay. PDGF was released in 7 of the
12 (HPS = 27±42 units; normals undetectable). This study
indicated that release of growth factors by the alveolar
macrophages precedes the fibrosis.[20]

As the immunologic, pulmonary alveolar macrophage and
ultrastructural studies suggested a central role for macrophages in HPS, this study was designed to investigate the role of
ceroid in macrophages in the enteropathic and pulmonary

disease of HPS patients who can be identified in infancy,[3,4] who have no detected immunologic defects,[16] who nearly invariably develop fibrotic restrictive lung disease by the fourth or fifth decade of life,[6,10] and who have a high risk of developing granulomatous enteropathic disease.[5,7,10]

METHODS

Patients who gave informed consent for the procedures used in this study were selected from albino patients diagnosed as having HPS on the basis of absent platelet dense bodies[3] from among subjects studied in the Puerto Rican population[10] and from HPS patients from continental north America. Subjects ranged in age from 16 to 48 years of age.

Tissue specimens were obtained from patients having diagnostic biopsies, from surgical specimens and from autopsies. Biopsies were obtained from patients with no clinical or colonoscopic evidence of disease other than from areas of the colon that on colonoscopy showed a color change. These areas were distinguished by a yellow linear streak in the mucosa. Biopsies were also obtained from patients with granulomatous changes without ulceration and from patients with ulcerated mucosa. These were fixed in either 10% formalin for histology or in 3% cold gluteraldehyde for electron microscopy. Alveolar macrophages from pulmonary lavage were prepared and fixed in a similar manner. In addition to the preparation of routine hemotoxylin and eosin stained tissues, sections were stained with the Armed Forces Institute of Pathology stain for ceroid-lipofuscin.[21] Unstained deparaffinized sections were observed for yellow-orange autofluorescent ceroid using a Zeiss photoscope with ultra violet illumination with a broad spectrum UG1 excitor filter and 50 or 53 barrier filters. Tissues for electron microscopy were fixed in 3% cold gluteraldehyde and processed in the usual manner.

Pulmonary function tests, stress tests and chest radiographs were performed on all patients and control subjects who were subsequently lavaged for alveolar macrophages. Tests included spirometry, lung volumes by body plethysomography, airway resistance, and single breath capacity using Medical Graphics Corporation Systems 1070 and 1085. Only those subjects (12) who had normal chest radiographs and no deficits on function or stress tests were subsequently included in the analysis of platelet derived growth factor (PGDF).

Ceroid was prepared from 24h urine specimens from HPS patients. Urine was centrifuged for 15 min at 5,000 rpm in a Beckman J2-21 centrifuge. The supernatant discarded and the pellet was washed with distilled water and resuspended and centrifuged sequentially in: 0.1 N HCl, distilled water, 4mM EDTA, distilled water, 0.1 M borate, pH 10.5, distilled water. The remaining pellet was extracted with chloroform:methanol 1:2, centrifuged, dried and kept dessicated at 30°C until used. Prior to adding the ceroid to cultures of alveolar macrophages from non-HPS patients, the ceroid was heated to 80°C for $\frac{1}{2}$h to kill organisms, suspended in sterile saline and sonicated for 45 sec to a uniform suspension of approximately a 1-3µ size and checked for autofluorescence. Two hundred µl of suspended ceroid was added to 5x10[6] alveolar macrophages suspended in 2.5 ml of serum-free media and incubated at 37˚C for 14h in a gently shaking water bath.

Alveolar macrophages were lavaged from HPS and control subjects in sterile saline, the cells counted, centrifuged and resuspended in serum-free media and cell concentration adjusted to 2x10[6]/ml. Slides for differential counts and ceroid autofluorescence were prepared by placing 0.5 ml of the cell suspension into the chamber of a Sheldon Elliott Cytospin and centrifuged on to a microscopic slide to provide a con- sistent cell density. The slides were fixed in 95% ethanol, dehydrated in absolute ethanol and were used unstained in xylol for ceroid fluorescence or stained with Wright's stain for differential cell counts.

PDGF was assayed by the dot blot technique on isolated ceroid prepared as above, and on cultures of lavaged macropha- ges from HPS patients, and on cultures of cells of unaffected control subjects before and after feeding macrophages with ceroid prepared as above.

RESULTS

Enteropathic Findings

Surgical specimens of enteropathically unaffected patients showing only yellow streaks and were grossly uninvolved with granulomatous changes showed that the first detectable altera- tion was the deposition of ceroid in plaques in the lamina propria (Fig. 1). At this point, there was little change in the tissue adjacent to these ceroid plaques, and the overlying mucosa was intact (Fig. 1). Some other asymptomatic patients had small granulomas seen on colonoscopy but intact epithelial mucosa. These areas contained small non-caseating granulomas, consisting of central multinucleated giant cells surrounded by macrophages containing autofluorescent granules which stained positive for ceroid with the AFIP ceroid stain (Fig. 2). Multinucleated giant cells in the granulomas frequently con- tained AFIP stain positive autofluorescent material (Fig. 2). Electron microscopy showed ceroid material in macrophages stored in membrane-bound structures resembling lysosomes (Fig. 3). The gut wall adjacent to the granulomas was morbidly thickened with proliferating cellular fibroblasts (Fig. 4). In symptomatic patients, the number and extent of the granulo- mas increased in areas subjacent to regions of necrosis and ulceration of the mucosa (Fig. 5).

Lung Findings

Biopsies and surgical specimens of lung tissue from normal areas and areas of fibrosis in HPS patients showed large accu- mulations of AFIP ceroid-lipofuscin stain positive, yellow- orange autofluorescent material in alveolar and bronchial cells, in alveolar spaces and in alveolar macrophages (Fig. 6).

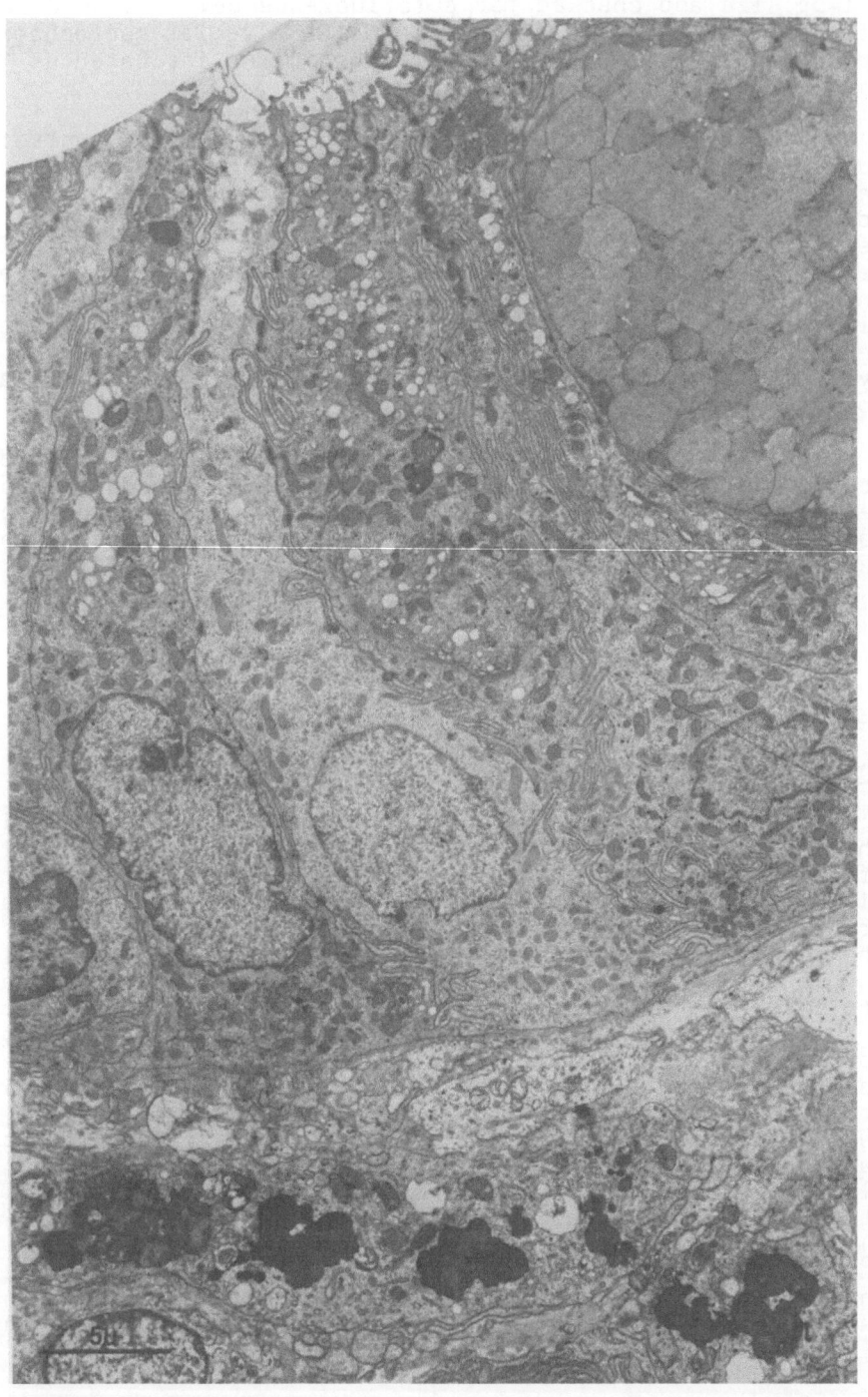

Figure 1. Biopsy from a patient without symptoms of enteropathic disease showing the earliest detectable change is the
accumulation of ceroid plaques in the lamina propria with
relatively little change in the surrounding cells. The
overlying mucosal cells are intact with normal cells junctions.

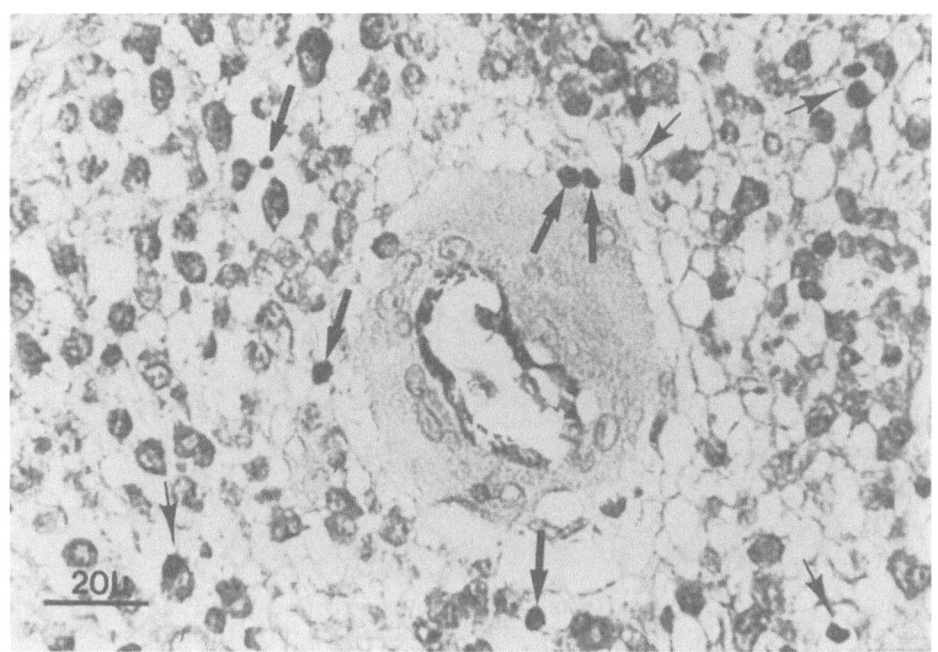

Figure 2. Granuloma from the colonic wall showing AFIP stain positive ceroid in the multinucleated giant cell and surrounding macrophages (arrows). Small granulomas may occur in patients with or without clinical symptoms of enteropathic disease (AFIP ceroid-lipofuscin stain).

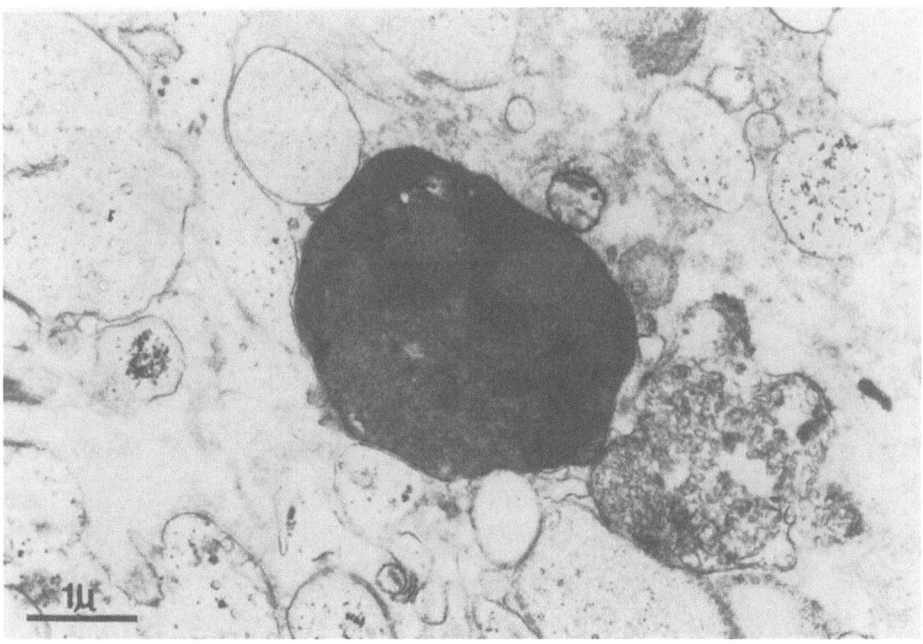

Figure 3. Electron dense material in membrane-bound lysosomal-like structures from a macrophage from the colon resembles the ceroid seen in bone marrow and pulmonary macrophages (see Figure 7).

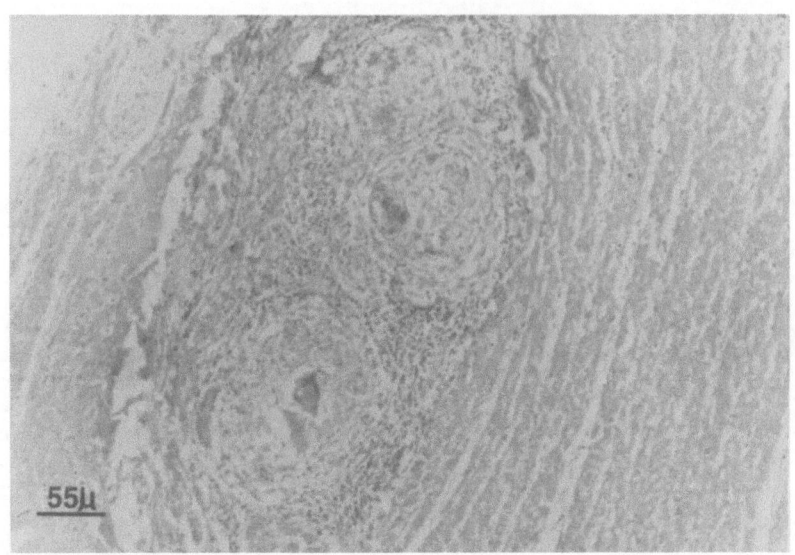

Figure 4. Small granulomas from the colonic wall of a patient with mild enteropathic symptoms of 1 year duration showing thickening of the colonic wall by proliferation of fibroblasts surrounding the granulomas. The epithelial mucosa overlying this lesion was intact. (H E Stain.)

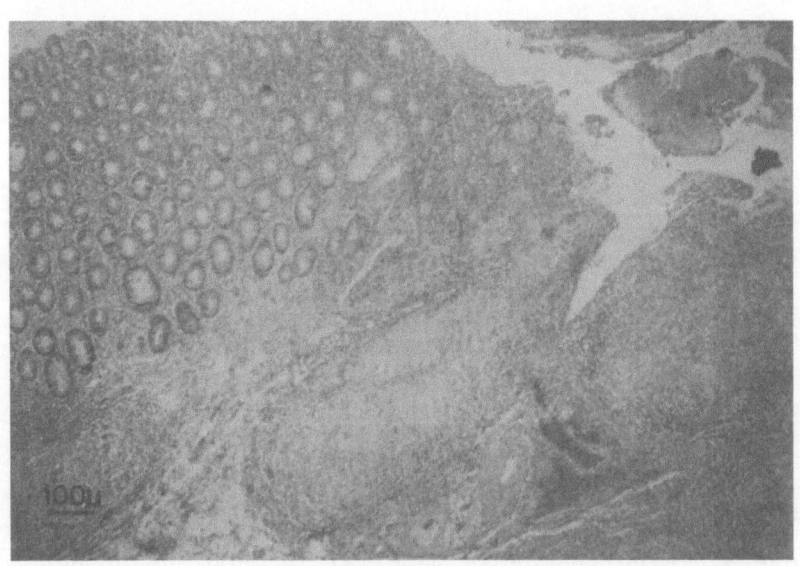

Figure 5. Area of ulceration from a patient with symptoms of 16 years duration. Note numerous large granulomatous areas beneath ulcerated mucosa. (H E Stain.)

Alveolar macrophages lavaged from HPS patients showed numerous electron dense inclusions similar to the ceroid inclusions seen in the colonic, gingival and bone marrow lesions of HPS patients, as well as laminated surfactant material (Fig. 7).

Nearly all alveolar macrophages lavaged from HPS patients contained numerous autofluorescent ceroid granules (Fig. 8).

Lavaged alveolar macrophages from unaffected non-smoker control subjects had less than 5% of cells with only one or two autofluorescent granules (Fig. 9). Freshly lavaged alveolar macrophages from control subjects had undetectable amounts of PDGF, and isolated ceroid alone gave no detectable amounts of PDGF activity. Cell preparations of alveolar macrophages from control subjects after ceroid feeding for 14h showed that approximately 20% of cells had ingested autofluorescent material (Figs. 10 and 11). None of the cell preparations from ceroid-fed control macrophages achieved the ceroid-positive cell density as seen in the macrophages of HPS patients.

The assay of PDGF in the post-fed alveolar macrophages did not show a definite increase in PDGF levels over those of the unfed cultures.

DISCUSSION

Histologic and ultrastructural evidence from the entero-pathic disease indicate that the first detectable alteration in tissue is the accumulation of ceroid in plaques in the lamina propria below an intact mucosa with little alteration in the surrounding tissue. In patients who on colonscopy have a few small granulomatous areas, macrophages and giant cells with ingested ceroid accumulate in the gut wall under an intact mucosa. These granulomas become progressively surrounded by thick layers of proliferating fibroblasts. In symptomatic patients with colonoscopic evidence of ulceration, the gut wall had numerous large granulomas and the overlying mucosa was necrotic and ulcerated. The role of the ceroid-containing macrophage as a possible source of fibroblast stimulating growth factors in the gut lesion is less amenable to experimental manipulation than is the alveolar macrophage. However, the evidence is clear that plaques of ceroid in the lamina propria precede the formation of granulomas and fibroblast proliferation is subsequent to the formation of granulomas with ceroid-containing macrophages.

In the pulmonary tissue, these experiments demonstrate that the release of growth factors for mesenchymal cells pre-cedes the fibrosis. These cells contain large amounts of ceroid. However, we have failed to demonstrate whether the source of the ceroid is either from exocytosis or dying cells with ingestion by the macrophages, or whether ceroid is generated de novo within the alveolar macrophage. We have failed to demonstrate that ceroid as isolated and prepared in this experiment is ingested in large quantities by normal macrophages and whether it is ceroid or other factors that act to release mesenchymal growth factors. One problem may be the method of isolating preparing ceroid in this experiment.

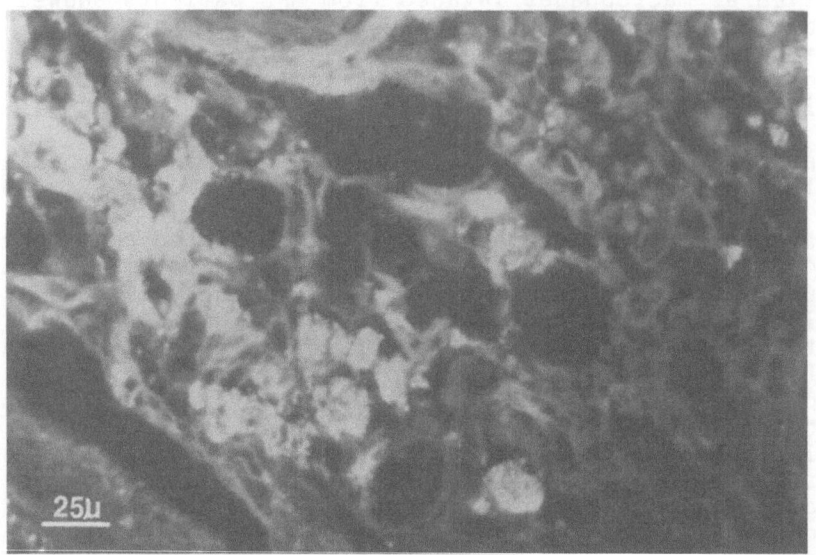

Figure 6. Biopsy of radiodense area of lung showing fibrosis and autofluorescent material in alveolar cells, macrophages and lying free in the alveoli. (Unstained section, UV illumination.)

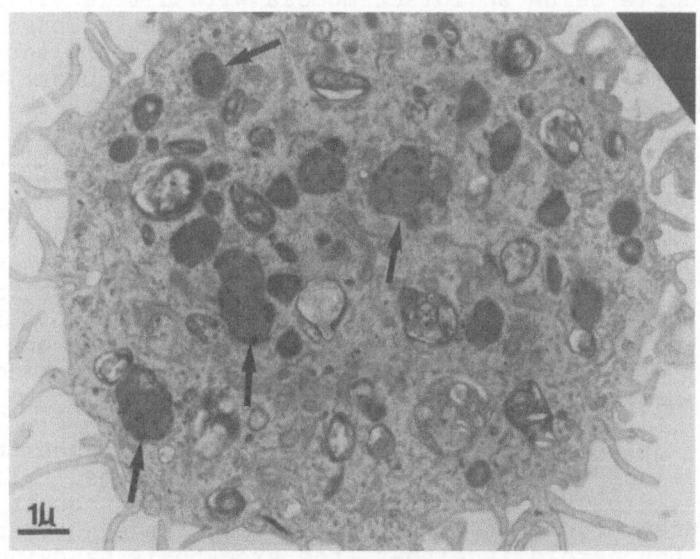

Figure 7. Alveolar macrophage lavaged from an 18-year-old HPS patient with no detectable deficits in pulmonary function tests or chest radiographs. Note electron dense material in lysosomal-like structures (arrows) similar to those seen in colonic macrophages (Figure 3). Laminated material is surfactant.

292

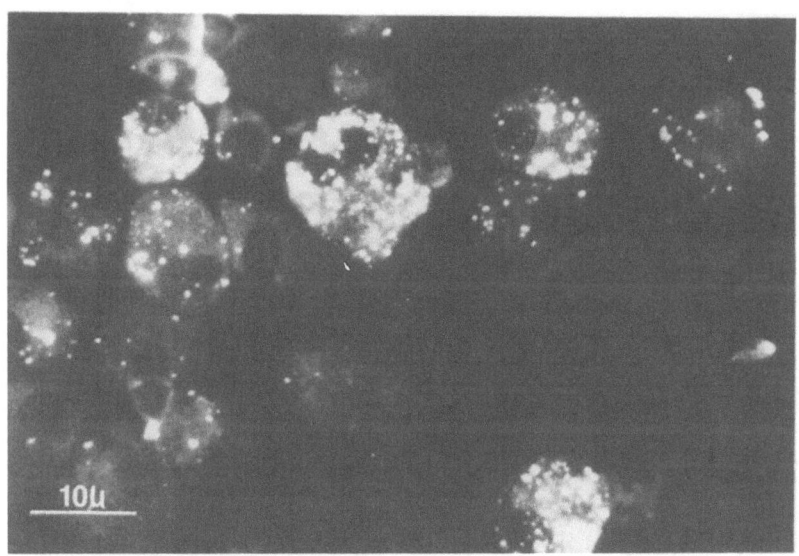

Figure 8. Nearly all alveolar macrophages lavaged from the patient in Figure 7 showed numerous yellow-orange autofluorescent granules characteristic of ceroid. (Unstained cytospin, UV illumination.)

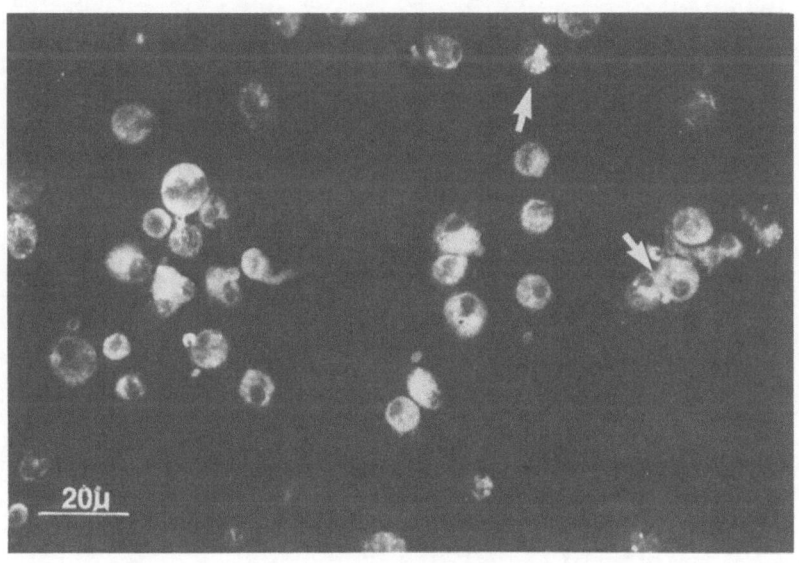

Figure 9. Lavaged alveolar macrophages from a non-smoker control subject aged 46 years. Only a few cells have a few small autofluorescent granules (arrows). (Unstained cytospin, UV illumination).

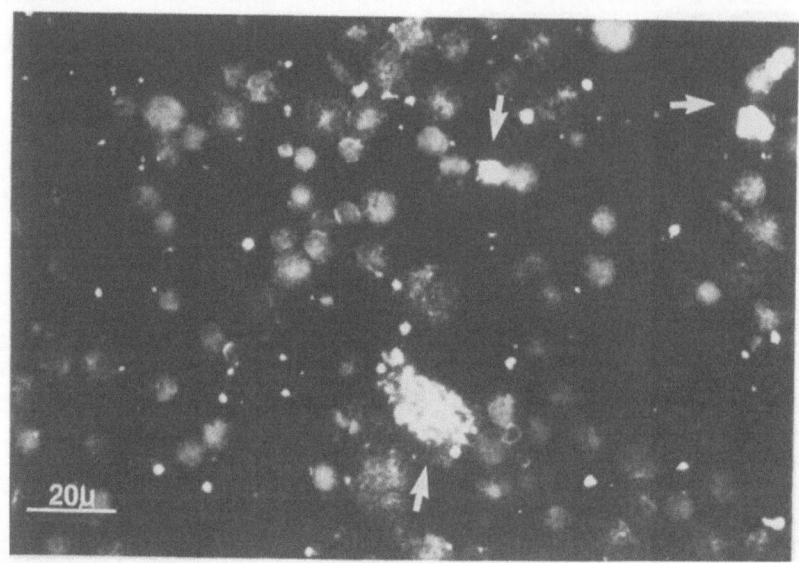

Figure 10. Alveolar macrophages from culture of cells in patient in Figure 9 incubated with sonicated ceroid prepared from urine sediment show that only about 20% of cells had ingested the autofluorescent material. (Unstained cytospin, UV illumination.)

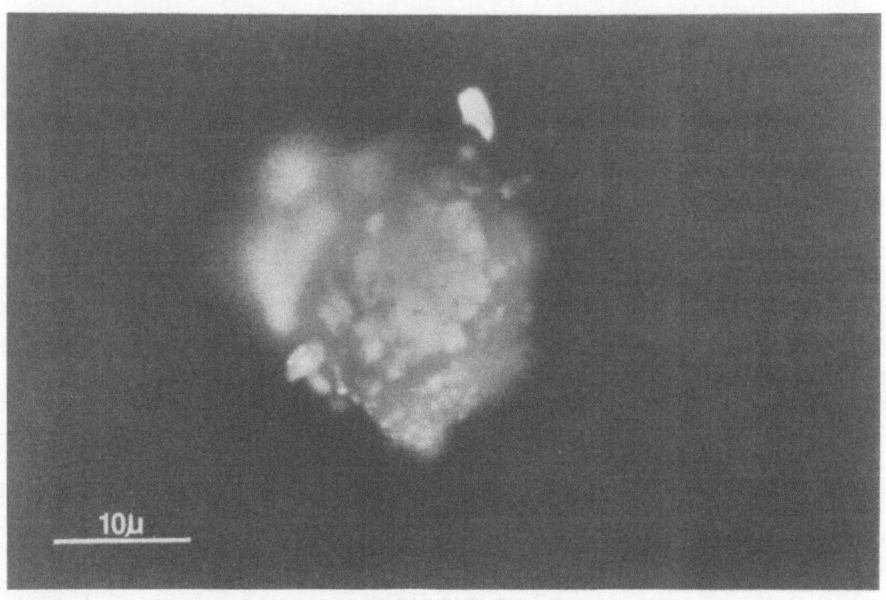

Figure 11. Large macrophage with ingested ceroid from culture of macrophages fed prepared ceroid. (Unstained cytospin, UV illumination.)

Macrophages have a propensity for ingesting lipid materials. The chloroform:methanol step in the purification of ceroid as used herein probably has removed lipid components. We now are preparing ceroid in a more natural state directly from lavaged alveolar macrophages to conduct similar feeding experiments.

REFERENCES

1. F. Hermansky and P. Pudlak, Albinism associated with hemorrhagic diathesis and unusual pigmented reticular cells in the bone marrow: report of two cases with histochemical studies, Blood 14:162 (1959).

2. C.J. Witkop, Jr., W.C. Quevedo, Jr., T.B. Fitzpatrick, and R.A. King, Albinism, in: "The Metabolic Basis of Inherited Disease," 6th ed., C.R. Scriver, A.L. Baudet, W.S. Sly, and D. Valle, eds., McGraw-Hill, New York (1989) pp. 2905-2947.

3. C.J. Witkop, Jr., M. Krumweide, H. Sedano, and J.G. White, Reliability of absent platelet dense bodies as a diagnostic criterion for Hermansky-Pudlak syndrome, Am J Hematol 26:305 (1987).

4. C.J. Witkop, Jr., J.G. White, D. Townsend, H.O. Sedano, S.X. Cal, M. Babcock, M. Krumwiede, K. Keenan, J.E. Love, and L.S. Wolfe, Ceroid storage disease in Hermansky-Pudlak syndrome: induction in animal models, in: "Lipofuscin-1987: State of the art, I," Zs-Nagy, ed., Elsevier, Amsterdam (1988), pp. 413-436.

5. R.A. Schinella, M.A. Greco, S.N. Garay, J. Lackner, S.R. Wolman, and E.P. Frazzini, Hermansky-Pudlak syndrome: a clinicopathologic study, Hum Pathol, 16:336 (1985).

6. S.M. Garay, J.E. Gardella, E.P. Frazzini, and R.M. Goldring, Hermansky-Pudlak syndrome: pulmonary manifestations of a ceroid storage disorder, Am J Med 66:737 (1979).

7. R.A. Schinella, M.A. Greco, B.L. Colbert, L.W. Denmark, and R.P. Cox, Hermansky-Pudlak syndrome with granulomatous colitis. Ann Intern Med 92:20 (1980).

8. C.J. Witkop, Jr., Inherited disorders of pigmentation, Clin Dermatol 3:70 (1985).

9. J.S. Bomalski, D. Green, and F. Carone, Oculocutaneous albinism, platelet storage pool disease, and progressive lupus nephritis, Arch Int Med 143:809 (1983).

10. C.J. Witkop, Jr., M.N. Babcock, G.H.R. Rao, F. Gaudier, C.G. Sommers, F. Shanahan, K.R. Harmon, D. Townsend, R.A. King, H.O. Sedano, S.X. Cal, M. Krumwiede, C. Almadovar, H. Cruz, B. Pinero, and J.G. White, Albinism and Hermansky-Pudlak syndrome in Puerto Rico, Bol Assoc Med PR (in press).

11. C.J. Witkop, Jr., R.A. King, and D. Townsend, Human albinism and animal models of albinism, Pig Cell Res Suppl 1:88 (1988).

12. H. Thompson, The role of rectal biopsy in inflammatory bowel disease, in: "Inflammatory bowel disease: some international data and reflections," F.T. de Dombal, J. Myren, I.A.D. Bouchier, and G. Watkins, eds., Oxford University Press, Oxford (1986) pp. 134-160.

13. Z. Maratka, Pathogenesis and aetiology of inflammatory bowel disease, in: "Inflammatory bowel disease: some international data and reflections," F.T. de Dombal, J. Myren, I.A.D. Bouchier, and G. Watkins, eds., Oxford University Press, Oxford (1986) pp. 29-65.

14. R.C. Crystal, P.B. Bitterman, S.I. Rennard, A.J. Hance, and B.A. Keogh, Interstitial lung disease of unknown cause: Disorders characterized by chronic inflammation of lower respiratory tract (first of two parts), New Eng J Med 310:154 (1984).

15. P.B. Bitterman, S.I. Rennard, B.A. Keogh, M.D. Wewers, S. Adelberg, and R.G. Crystal, Familial idiopathic pulmonary fibrosis: Evidence of lung inflammation in unaffected family members, New Eng J Med 314: 1343 (1986).

16. F. Shanahan, L. Randolf, R. King, R. Oseas, M. Brogan, C. Witkop, J. Rotter, and S. Targan, Hermansky-Pudlak syndrome: an immunologic assessment of 15 cases. Am J Med 85:823 (1988).

17. L.S. Wolfe, G.O. Ivy, and C.J. Witkop, Jr., Dolichols, lysosomal membrane turnover and relationships to the accumulation of ceroid and lipofuscin in inherited disease, Alzheimer's disease and aging. Twelfth Nobel Conference: Structure, biosynthesis and function of isoprenoid compounds in eukaryotic cells, Sodegarn, Sweden, May 25-28, 1986. Chemica Scripta 27:79 (1986).

18. C.J. Witkop, Jr., L.S. Wolfe, S.X. Cal, J.G. White, D. Townsend, K.M. Keenan, Elevated urinary dolichol excretion in the Hermansky-Pudlak syndrome: indicator of lysosomal dysfunction, Am J Med 82:463.

19. A. Takahashi and T. Yokoyama, Hermansky-Pudlak syndrome with special reference to lysosomal dysfunction: a case report and review of the literature, Virch Arch (Pathol Anat) 402:247 (1984).

20. K.R. Harmen, L.S. Snyder, R.A. King, C.J. Witkop, Jr., J.G. White, J. Tashjian, and P.B. Bitterman, Alveolar macrophage release of growth factors precedes lung fibrosis in the Hermansky-Pudlak syndrome. Clin Res 36:505A (1988).

21. L.G. Luna, "Manual of histologic staining methods of the Armed Forces Institute of Pathology," ed. 3, McGraw-Hill, New York (1960).

DISCUSSION

TAKAHASHI: You showed a granuloma in the wall of the gastrointestinal tract. Have you examined it bacteriologically in the necrotic center?

WITKOP: We used acid-fast stain for the detection of ceroid, but we have not seen any acid-fast positive organisms like tuberculosis, etc. The cultures we have made from these lesions all remain negative.

TAKAHASHI: Chronic granulomatosis in children has a very similar picture to the one you showed in Hermansky-Pudlak syndrome.

WITKOP: There are several granulomatous processes with disseminated granulomas and fibrosis of the lung. For instance, in asbestosis and sarcoidosis, you see the same thing. When you have a nonbiodegradable material in macrophages, this material probably stimulates the production of mesenchymal growth factors, and it doesn't make any difference whether is asbestos, ceroid, or something else.

TAKAHASHI: All these patients have albinism. What is the cause for this?

WITKOP: It starts out from tyrosine all the way down to melanin. In this syndrome, the pathway is intact. There is no defect in tyrosinase or in other enzymes down in the pathway. What happens is the membrane bound tyrosine reductase, a free radical trap, working outside the melanocytes and keratinocytes, traps free radicals, such as those generated by U.V. light. When it works inside the cell, it acts by reducing tyrodoxine, and thus you get a marked reduction in tyrosinase activity. If you take these patients out and put them in the sun, they freckle but not really tan.

PORTA: You showed some laminated bodies in the surfactant. Have you seen any transition between these laminated structures and the more heterogenous granules of ceroid?

WITKOP: No. I have seen the sort of light gray bodies with small little white dots in it. These get darker and darker, not totally, uniformly dark,but they become finally very dark, and when we isolate them, these are the the ones that fluoresce orange, while the new ones fluoresce yellow.

ABNORMAL LIPOPIGMENTS AND LYSOSOMAL RESIDUAL BODIES

IN METACHROMATIC LEUKODYSTROPHY

Hans H. Goebel and H. Busch

Division of Neuropathology, University of Mainz
Langenbeckstraße 1, 6500 Mainz/FRG

SUMMARY

Ultrastructurally, metachromatic leukodystrophy (MLD) is marked by
characteristic features such as herringbone, prismatic and tufaceous
patterns which are typically encountered within oligodendrocytes of the
central nervous system (CNS) and in Schwann cells (PNS). These patterns
can be documented in late infantile, juvenile, and adult forms. In the
latter, aging of the ailing individual adds another component, the accumulation of lipopigments which are marked by an opaque supposedly lipid
droplet and a granular component. While MLD-specific lysosomal residual
bodies occur in myelinforming cells, lipopigments accrue in neurons and
to a lesser degree in astrocytes.
MLD represents a unique example in which these two separate lysosomal
storage processes combine to form a wide spectrum of ultrastructurally
divergent MLD-lipopigments affecting several cell type in the CNS and PNS.
Lipopigments and MLD-specific lysosomal inclusions also assemble in sweat
gland epithelial cells again combined within the same residual body and
in Schwann cells of non-myelinated axons which may also regularly display
lipopigments, contrary to Schwann cells of myelinated axons. Comparative
studies on sweat glands in childhood and adult forms confirm the earlier
observations that composite MLD-lipopigments are frequent in adult MLD.
Thus, although lipopigment formation and MLD are different processes,
their common occurrence in MLD provides evidence of the mutual morphogenic
influence which results in a diversified population of MLD-lipopigments.
Whether there is absolute increase of lipopigment formation beyond the
age-related level remains to be clarified by quantitative and morphometric
data. The presence of lipopigments in childhood MLD in sweat gland epithelial cells favors the concept of accelerated lipopigment formation while
the lysosomal compartiment is already stimulated by an inborn error of
lysosomal metabolism. Whether this principle also prevails in adult forms
of other non-MLD lysosomal disorders remains to be elucidated by respective ultrastructural investigations.

Key words: Electron microscopy - cytosomes - lipopigments - lysosomal
          disease - metachromatic leukodystrophy - storage

INTRODUCTION

Lipopigments accumulate as lipofuscin due to aging or wear and tear or as ceroid due to rapid, often experimentally induced formation. Among human diseases that are marked by abnormal amounts of lipopigments, the neuronal ceroid-lipofuscinoses (NCL) (Goebel et al., 1979) and vitamin E deficiency, the latter of various acquired and seemingly idiopathic (Burck et al., 1980) types, are well known examples of pathological increment of intracellular lipopigments represented by abnormal lysosomal residual bodies, thus rendering NCL - and less obviously vitamin E deficiency - lysosomal disorders defined by morphological, i.e. ultrastructural criteria. On the other hand, there are, in a purer sense due to lysosomal enzyme defects, lysosomal diseases, sphingolipidoses, mucopolysacchari-doses (MPS), mucolipidoses, type II glycogenosis, and others, some of which run a brief course of illness, others a more protracted one. Some such conditions have been found to also show increased amounts of lipo-pigments, as MPS III (Martin et al., 1979; Oldfors and Sourander, 1981; Wisniewski et al., 1985), aspartylglucosaminuria and Salla disease (Sourander, 1985), the GM_1 (Lowden et al., 1981, 1982), and GM_2 (Jellinger et al., 1982) gangliosidoses, and sialidosis, the former cherry-red spot myoclonus syndrome (Rapin et al., 1978). Ultrastructurally, lipopigments of the regular lipofuscin type consist of a rather electron lucent opaque component, considered by many to be lipid, and an electron dense granular matrix. In addition, at high magnification, lamellae might occasionally be seen incorporated into the electron dense matrix, especially inside oligo-dendrocytic (Schlote und Boellaard, 1983) and intraneuronal human lipo-pigments (Boellaard and Schlote 1986). Conversely, the ultrastructure of the lipopigments in the NCL differ as to their fine structure, e.g. the curvilinear profiles and fingerprint patterns in the late infantile, early juvenile, and juvenile forms. In infantile NCL, the lipopigments are gra-nular without an opaque component (Siegismund et al., 1982) while in the adult or Kufs type lipopigments in the central nervous system resemble regular lipofuscin with some lamellae encased (Goebel et al., 1982). Lipopigments observed in conjunction with lysosomal storage in above men-tioned lysosomal disorders also seem to be of a regular lipofuscin type, consisting of the granular electron dense matrix and the opaque component, thus differing from those lipopigments seen in childhood forms of NCL. Metachromatic leukodystrophy (MLD) is a lysosomal disease marked by a lysosomally based myelin disorder of the central and peripheral nervous system, thus foremost affecting oligodendrocytes and Schwann cells which normally produce myelin. These cell types are normally not known for con-spicuous lipopigment formation.
The study presented here reports aspects of ultrastructural complexes which combine the divergent components of lipopigments and MLD-specific lysosomal residual bodies in MLD.

MATERIAL AND METHODS

Tissues available for this ultrastructural study were derived from the central (CNS) and peripheral (PNS) nervous system of a patient with adult MLD (Seidel et al., 1980; Goebel et al., 1980), the CNS and PNS of a pa-tient afflicted with late infantile MLD, skin, obtained by biopsy, from two patients with adult MLD and from another one with late infantile MLD. Furthermore, sural nerve biopsy specimens from children with late infan-tile and juvenile MLD were also studied. CNS and PNS tissues from the patient with adult MLD as well as the biopsied skin and sural nerve spe-cimens were fixed in buffered glutaraldehyde, the CNS and PNS tissue from the patient afflicted with late infantile MLD was primarily fixed in formalin. After washing in buffer, osmication, dehydration in increasing concentrations of ethanol, and embedding in epoxy resins, ultrathin

sections were studied with the electron microscope. The correct nosological identification of the patients' MLD was obtained by showing biochemical arylsulfatase A deficiency and/or ultrastructural features
considered typical of MLD such as herringbone or prismatic and tufaceous
patterns.

RESULTS

While MLD-specific herringbone or prismatic profiles could be observed
within membrane-bound lysosomal residual bodies of the CNS in adult MLD
(Fig. 1a), pure lipopigments in which occasional wavy lamellae appeared
quite distinct from membranous profiles (Fig.1b) could be observed as well
as composite membrane-bound inclusions combining lipopigment and membranous features (Fig.1c). In other membrane-bound cytosomes, numerous opaque
components were scattered among densely packed often parallel running membranes or a less well defined matrix (Fig. 2a). In the spinal ganglia, an
electron dense component, appeared added to the opaque and membranous constituents of membrane-bound lysosomal residual bodies (Fig. 2b). In spinal
anterior horn neurons of infantile MLD, both pure membranous inclusions,
resembling membranous cytoplasmic bodies (Fig. 3a) as well as others with
an additional opaque component (Fig. 3b) were seen. Already in this early
childhood form, composite lipopigment-membranous bodies (Fig. 3c) of sometimes bizarre arrangement of the membranes (Fig. 3d) were present.
Schwann cells of unmyelinated axons, studied in sural nerve specimens of
juvenile MLD (Fig. 4a) and late infantile MLD (Fig. 4b) also contained
membrane-bound lamellar inclusions that featured additional opaque components or a granular matrix (Fig. 4b). Sweat gland epithelial cells in
adult MLD showed a variegated spectrum of membrane-bound inclusions, some
of which were of the zebra body type (Fig. 5a) separate from lipopigments
or of a rather complex membranous-opaque ultrastructure (Fig.5b). An occasional sweat gland epithelial cell in late infantile MLD also displayed
lipopigments, but with a rather large opaque component and only a few parallel lamellae within the appendix-like granular matrix (Fig. 5b, inset).
Thus, in addition to MLD-specific cytosomes and pure lipopigments, combined lipopigment-MLD forms were encountered in cells of the CNS and the
PNS as well as in sweat gland epithelial cells. The ratio of lipopigment:
MLD-components varied between little lipopigments and abundant MLD-materi-
al on one side of the spectrum and a reversed ratio at the opposite end.
Both lipopigments and MLD-inclusions were more conspicuous in sweat gland
cells of adult MLD than in similar cell types of late infantile MLD.

DISCUSSION

First of all, our ultrastructural study on the CNS in late infantile
MLD confirmed the presence of intraneuronal membranous inclusions, earlier
seen (Peng and Suzuki, 1987). These lamellar inclusions resembled membranous cytoplasmic bodies, typical of the gangliosidoses rather than MLD-
specific features, prismatic or tufaceous patterns which are usually seen
in oligodendrocytes and Schwann cells indicating that the type of cell,
neuron versus glial cell, molds the ultrastructural pattern of the respective lysosomal residual bodies.
While regular lipopigment bodies, consisting of an opaque component and a
granular matrix within a trilaminar membrane were noted in the adult MLD
CNS, investigative emphasis provided a motley gamut of composite cytosomes
that harbored both lipopigment and membranous features. Such composite
lipopigment-MLD-cytosomes were encountered in adult and late infantile
MLD, though more frequently in the former clinical form. Not infrequently,
the opaque component which is considered by many to be neutral lipid - but
this is contested by others because its lipid nature has not been

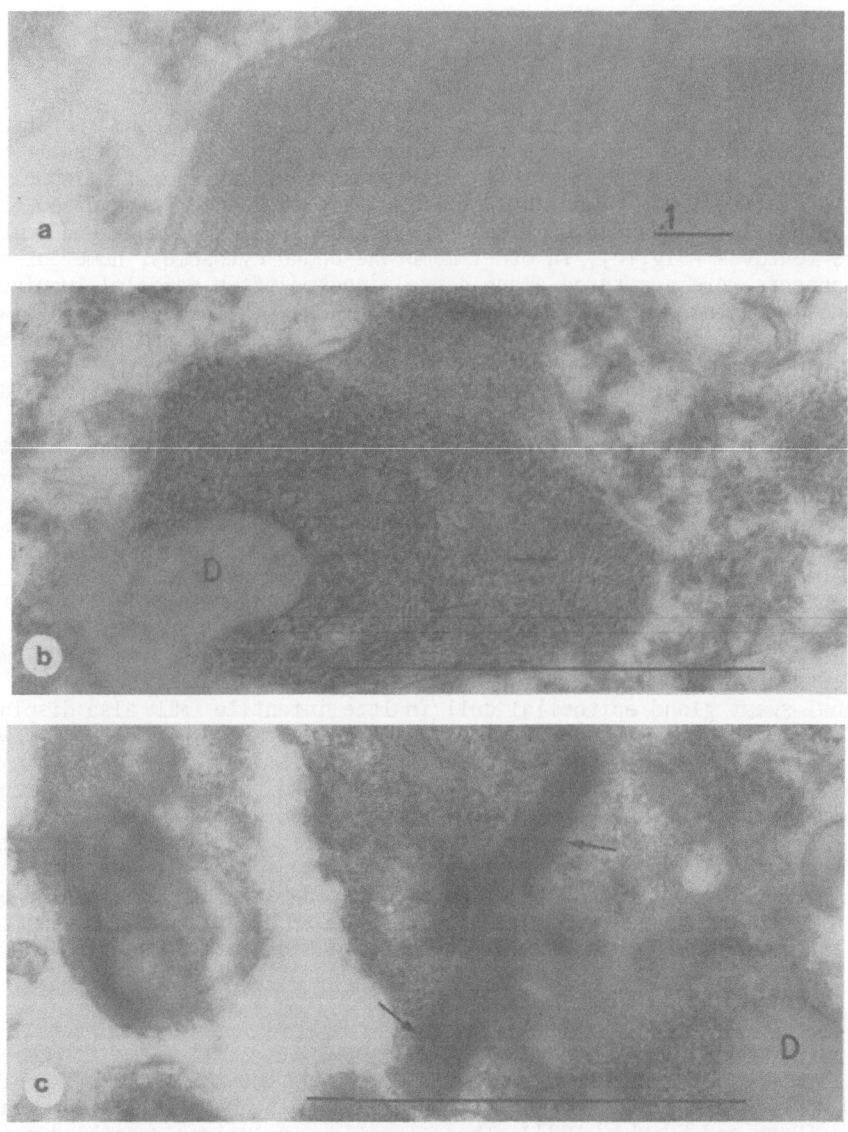

Fig. 1. Adult metachromatic leukodystrophy (MLD), central nervous
system (CNS):
  a) Disease-specific herringbone or prismatic profiles within a
     membrane-bound inclusion body.
  b) A non-specific lipopigment body contains an opaque droplet
     (D), and a matrix replete with wavy profiles, (arrows), but
     no MLD-specific components.
  c) A combined lipopigment MLD inclusion harbors granular
     matrix, opaque droplets (D), and parallel lamellae (arrows).

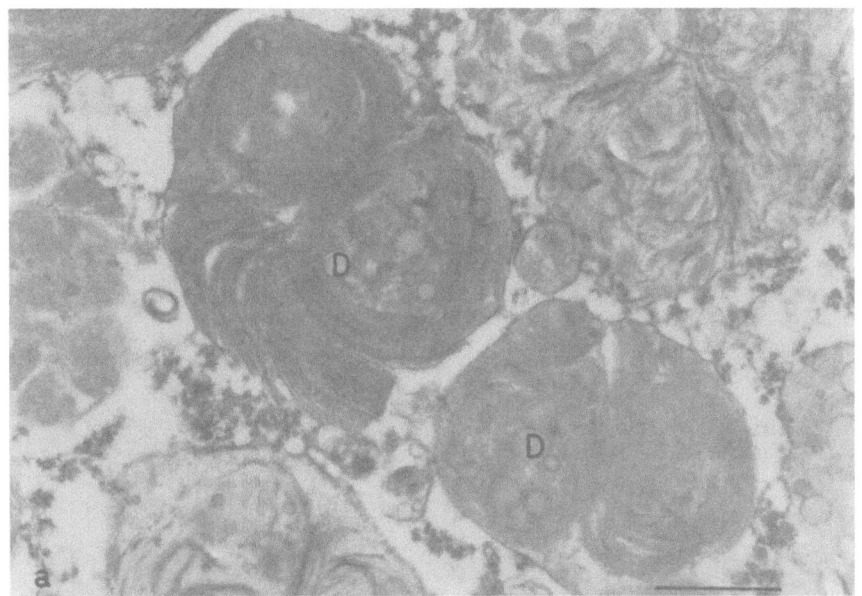

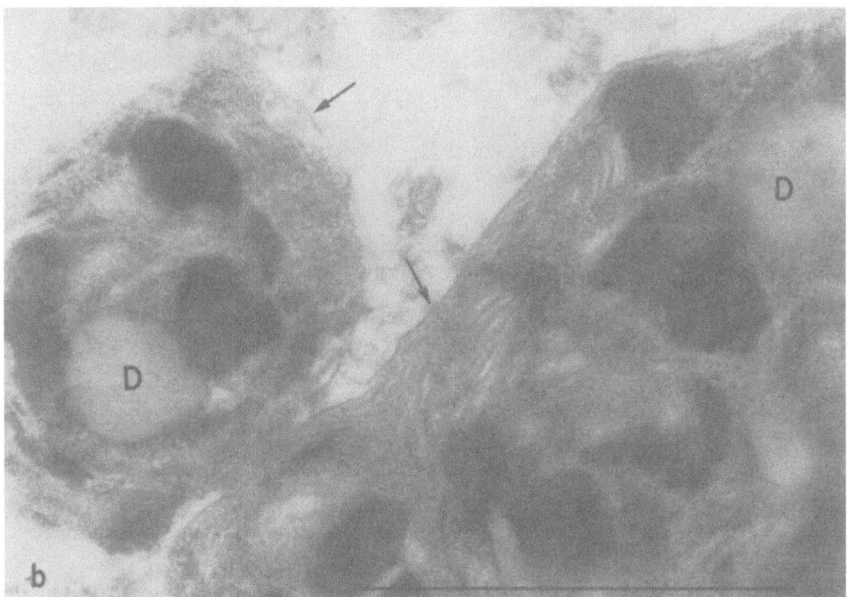

Fig. 2. Adult metachromatic leukodystrophy.
a) CNS: large individual inclusions with both opaque droplets
(D) and parallel lamellae.
b) Spinal Ganglion: two membrane-bound (arrows) inclusions
contain both opaque droplets (D) and membranous profiles
and an additional electron dense component.

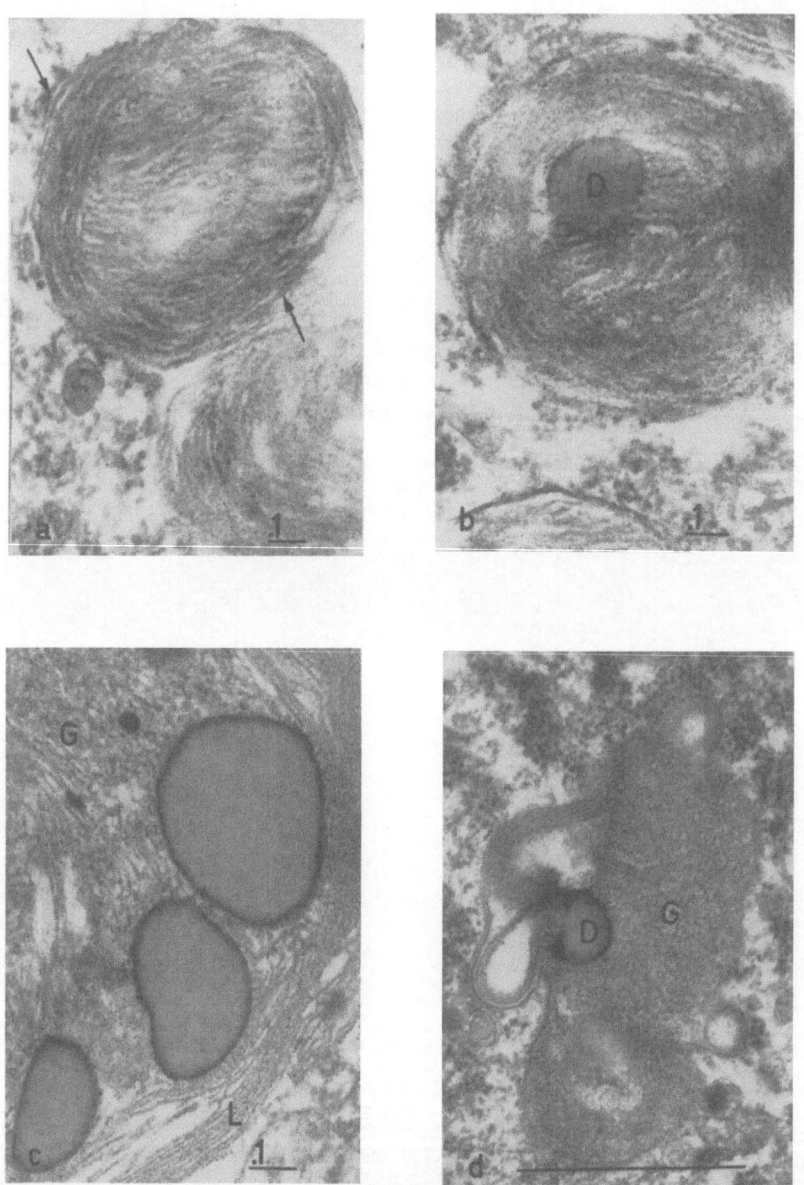

Fig. 3. Late infantile metachromatic leukodystrophy.
   a) Spinal anterior horn: a membrane-bound (arrows) inclusion
      resembles a membranous cytoplasmic body.
   b) Spinal anterior horn: Another similar inclusion also harbors
      an opaque droplet (D).
   c) Lateral geniculate body: several opaque droplets, parallel
      lamellae (L) and a granular (G) component form this
      lysosomal residual body.
   d) Stellate Ganglion: another composite inclusion consists of a
      granular (G) matrix, an opaque droplet (D), and parallel
      membranes.

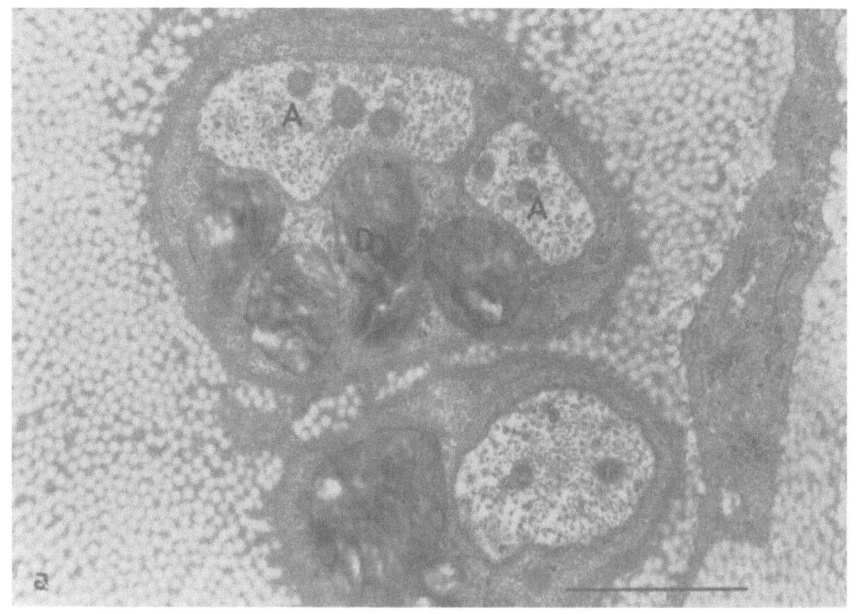

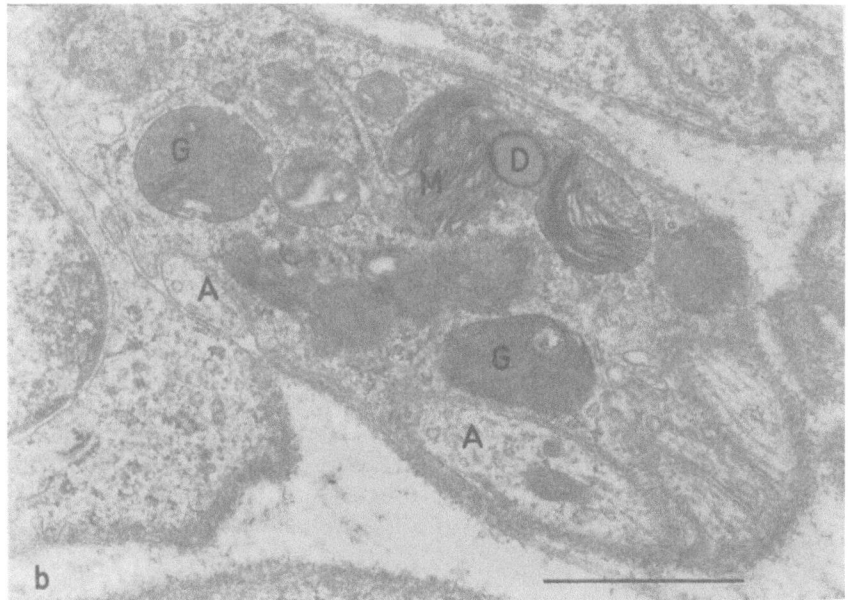

Fig. 4. Sural nerve.
  a) Juvenile MLD: a Schwann cell not only contains two
     unmyelinated axons (A) but also numerous membrane-bound
     lamellar bodies with an occasional opaque droplet (D).
  b) Late infantile MLD: a Schwann cell covering two unmyelinated
     axons (A) contains membrane-bound inclusions some of
     which harbor a granular (G) matrix, membranes (M), and an
     occasional opaque droplet (D).

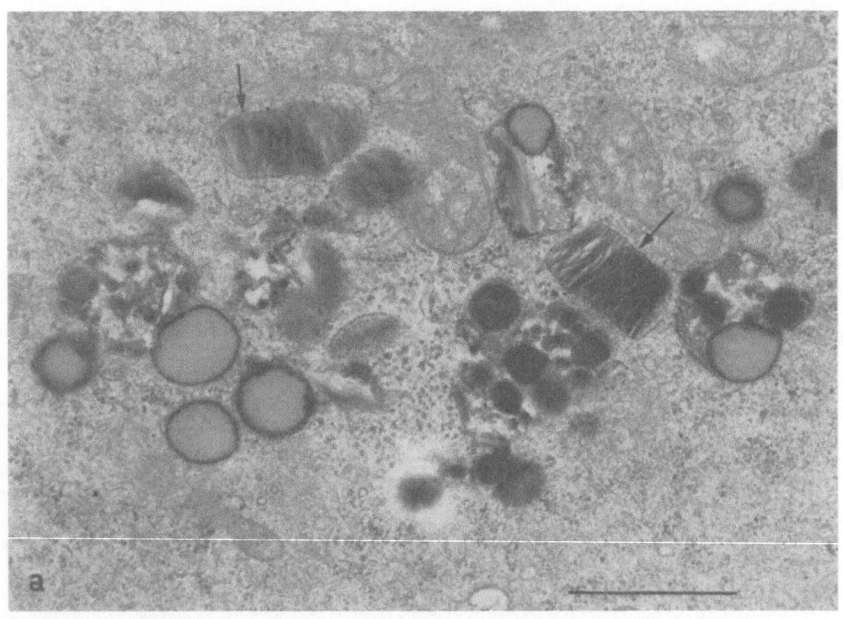

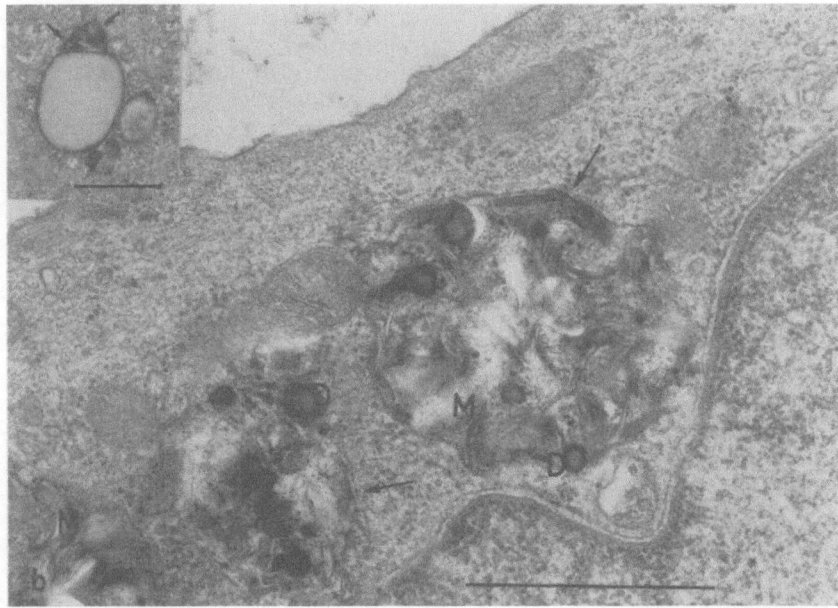

Fig. 5. Sweat gland epithelial cells.
  a) Adult MLD: membrane-bound (arrows) membranous (M) bodies
     and lipopigments.
  b) Adult MLD: Three large membrane-bound (arrows) inclusions
     with combined opaque droplets (D) and membranous (M)
     profiles.
Inset: Late infantile MLD: the electron dense component of lipo-
       pigment bodies contains a few parallel membranes (arrows).

sufficiently proven (Jolly et al., 1989) - often appeared as the sole non-MLD component in the respective cytosomes in adult and late infantile MLD, and both within the CNS and the PNS.

Lipopigments accrue with age and thus are more frequently seen in adult MLD. The composite lipopigment-MLD-inclusions are also more frequent although similar examples may already be observed in late infantile MLD. This combination of lipopigments and membranes has not been specifically described or illustrated in an earlier study confined to the CNS of late infantile MLD (Peng and Suzuki, 1987), while our electron microscopic examination surfaced such combined lipopigment-membranous features in the lateral geniculate body and the stellate, i.e. sympathetic ganglion. As lipopigments increase with age of the individual it is not surprising to realize that lipopigments were also more abundant in adult MLD. This observation is in agreement with those made in other lysosomal diseases with lipopigment accumulation which also occurred in late onset or protracted forms of such entities as MPS III (Oldfors and Sourander, 1981), GM_1-gangliosidosis (Lowden et al., 1981, 1982), GM_2-gangliosidosis (Jellinger et al., 1982) or the cherry-red spot-myoclonus syndrome/sialidosis (Rapin et al. 1978). Furthermore, composite inclusions consisting of lipopigments and the respective disease-specific components were likewise encountered in sialidosis (Rapin et al., 1978), MPS III (Martin et al., 1979), or juvenile (Suzuki et al., 1970) and adult (Jellinger et al., 1982) GM_2 gangliosidoses. Membranes arranged in a fingerprint-like pattern within composite lipopigment disease-specific inclusions seem to be evidence of the enzyme deficiency-defined lysosomal disorder rather than evidence of NCL or Batten's disease (Carpenter, 1982).

While accumulation of both disease-specific lysosomal storage components and lipopigments in those lysosomal disorders which are marked by neuronal storage is not surprising, our study adds another aspect to this problem in that the combination of lipopigments and disease-specific lysosomal storage material has occurred in a group of lysosomal disorders, the MLD of adult, juvenile and late infantile forms, which is primarily recognized as a myelin disorder affecting the neuronal population only to a minor degree. However, our observation enlarges this general principle beyond the group of pure neuronal lysosomal storage disorders. On the other hand, whether all of the membrane containing lysosomal residual bodies in MLD neurons represent sulfatides, the disease-specific stored material in MLD, requires biochemical analysis of such isolated inclusions as, for instance, in NCL zebra bodies and membranous bodies may be a distinct, albeit rare feature (Goebel et al., 1979). Furthermore, fingerprint profiles identical to those seen in NCL may also be found in circulating lymphocytes in certain MPS (Goebel et al., 1981) complicating the interpretation of which fine structural pattern belongs to which biochemical material in the respective lysosomal disorders combined with accumulation of lipopigments.

Whether there is an actual increase in accumulation of lipopigments beyond the respective age level in late onset or protracted lysosomal disorders marked by conspicuous lipopigment formation remains a matter for future quantitative studies.

ACKNOWLEDGEMENTS

Financial support by the Deutsche Forschungsgemeinschaft 477/1500/89 is gratefully acknowledged.
Thanks are also due to Mrs. I. Warlo for electron microscopy, to Mr. W. Meffert for photography, and to Mrs. M. Messerschmidt for editorial assistance.

REFERENCES

Boellaard, J.W., and Schlote, W., 1986, Ultrastructural heterogeneity of neuronal lipofuscin in the normal human cerebral cortex. Acta Neuropathol. (Berl.) 71:285-294

Burck, U., Goebel, H.H., Kuhlendahl, H.D., Meier, C., and Goebel, K.M., 1981, Neuromyopathy and vitamin E deficiency in man. Neuropediatrics 12:267-27819.

Carpenter, S., 1982, Type 2 GM_1 gangliosidosis and neuronal ceroid lipofuscinosis. Neurology 32:575-576

Goebel, H.H., Zeman, W., Patel, V.K., Pullarkat, R.K., and Lenard, H.G., 1979, On the ultrastructural diversity and essence of residual bodies in neuronal ceroid-lipofuscinosis. Mech. Ageing Dev. 10:53-70

Goebel, H.H., Argyrakis, A., Shimokawa, K., Seidel, D., and Heipertz, R., 1980, Adult metachromatic leukodystrophy. IV. Ultrastructural studies on the central and peripheral nervous system. Eur. Neurol. 19:294-307

Goebel, H.H., Ikeda, K., Schulz, F., Burck, U., and Kohlschütter, A., 1981, Fingerprint profiles in lymphocytic vacuoles of mucopolysaccharidoses I-H, II, III-A, and IIIB. Acta Neuropathol. (Berl.) 55:247-249

Goebel, H.H., Braak, H., Seidel, D., Doshi, R., Marsden, C.D., and Gullotta, F., 1982, Morphologic studies on adult neuronal-ceroid lipofuscinosis (NCL). Clin. Neuropathol. 1:151-162

Jellinger, K., Anzil, A.P., Seemann, D., and Bernheimer, H., 1982, Adult GM_2 gangliosidosis masquerading as slowly progressive muscular atrophy: motor neuron disease phenotype. Clin. Neuropathol. 1:31-44

Jolly, R.D., Slack, P.M., Palmer, D.N., Dalefield, R.R., and Hartley, W.J., 1989, Ceroid and lipofuscin pigments in veterinary pathology. Abstract, Third International Symposium on Lipofuscin and Ceroid Pigments, Maui, Hawaii/USA, 1989.

Lowden, J.A., Callahan, J.W., Gravel, R.A., Skomorowski, M.A., Becker, L., and Groves, J., 1981, Type 2 GM_1 gangliosidosis with long survival and neuronal ceroid lipofuscinosis. Neurology 31:719-724

Lowden, J.A., Callahan, J., Gravel, R., Skomorowski, M.A., Becker, L., and Groves, J., 1982, Reply from the authors: Type 2 GM_1 gangliosidosis and neuronal ceroid lipofuscinosis. Neurology 32:576

Martin, J.J., Ceuterick, C., Van Dessel, G., Lagrou, A., and Dierick, W., 1979, Two cases of mucopolysaccharidosis type III (Sanfilippo). An anatomopathological study. Acta Neuropathol. (Berl.) 46:185-190

Oldfors, A., and Sourander, P., 1981, Storage of lipofuscin in neurons in mucopolysaccharidosis. Report on a case of Sanfilippo's syndrome with histochemical and electronmicroscopic findings. Acta Neuropathol. (Berl.) 54:287-292

Peng, L., and Suzuki, K., 1987, Ultrastructural study of neurons in metachromatic leukodystrophy. Clin. Neuropathol. 6:224-230

Rapin, I., Goldfischer, S., Katzman, R., Engel, J. Jr., and O'Brien, J.S., 1978, The cherry-red spot-myoclonus syndrome. Ann. Neurol. 3:234-242

Schlote, W., and Boellaard, J.W., 1983, Role of lipopigment during aging of nerve and glial cells in the human central nervous system. In: Cervos-Navarro J, Sarkander HJ (eds) Brain aging: Neuropathology and Neuropharmacology (Aging, Vol 21) Raven Press, New York, pp 27-74

Seidel, D., Heipertz, R., Goebel, H.H., Duensing, I., and Pilz, H., 1980, Adult metachromatic leukodystrophy. III. Clinical course, final stages and first biochemical results. Eur. Neurol. 19:288-293

Siegismund, G., Goebel, H.H., and Löblich, H.J., 1982, Ultrastructure and visceral distribution of lipopigments in infantile neuronal ceroid-lipofuscinosis. Path. Res. Pract. 175:335-347

Sourander, P., 1985, Increased neuronal lipofuscin as an unspecific marker of genetically undetermined lysosomal diseases. Comm. to International Workshop on Age pigments: Biological markers in aging and environmental stress? Vico Equense, Italy. Arch Biol (Bruxelles) 96:373

Suzuki, K., Suzuki, K., Rapin, I., Suzuki, Y., and Ishii, N., 1970, Juvenile GM$_2$-gangliosidosis. Clinical variant of Tay-Sachs disease or a new disease. Neurology 20:190-204

Wisniewski, K., Rudelli, R., Laure-Kamionowska, M., Sklower, S., Houck, G.E. Jr, Kieras, F., Ramos, P., Wisniewski, H.M., and Braak, H., 1985, Sanfilippo disease. Type A with some features of ceroid lipofuscinosis. Neuropediatrics 16:98-205

DISCUSSION

ELLEDER: I have arrived at the same conclusion as you have regarding the combined lysosomal storage processes, but I would also think that the lysosomal storage is due to some sort of "low" enzymopathic defect. The storage material appears probably in a lysosome predestined to produce lipopigment, and both processes may appear simultaneously. However, if the storage is very intense, it may easily suppress the production of lipofuscin. At least, in the liver of adult people with lysosomal storage disease, the storaged material can almost totally suppress the normal production of lipofuscin.

GOEBEL: Yes, quantitative data may actually show that there is less lipofuscin formation.

IVY: Do you know which neurons in particular in the brain accumulate the lipopigment?

GOEBEL: Neurons in the thalamus, in peripheral tissue and in particular in the motor neurons of the anterior spinal cord.

IVY: Have you looked at the hippocampus or cerebral cortex?

GOEBEL: I have looked at the cerebral cortex, and I have not seen very convincing examples, but I have not looked at the entire brain.

PORTA: Have these "herring-bone" bodies been isolated and biochemically studied? What is known about their chemical composition?

GOEBEL: I think they are mainly sulphatides. This material has been actually isolated and also fed to tissue cultures, and the storage material was very similar ultrastructurally to the original material. Whether this is pure sulphatide or with some other material is difficult to determine at the ultrastructural level.

IN VITRO SYSTEMS OF LIPOPIGMENT FORMATION

ON THE ORIGIN OF LIPOFUSCIN; THE IRON CONTENT OF RESIDUAL BODIES, AND THE RELATION OF THESE ORGANELLES TO THE LYSOSOMAL VACUOME. A STUDY ON CULTURED HUMAN GLIAL CELLS

U.T. Brunk

Department of Pathology, Linköping University, S-581 85 Linköping, Sweden

SUMMARY

Cultured human glial cells constitute a suitable model system for the study of lipofuscinogenesis *in vitro*. These cells, although not post-mitotic, can be kept for several months in stable monolayers due to their display of very pronounced density-dependent inhibition of cell growth. Residual bodies, or lipofuscin pigment granules, accumulate over time in this "pseudo" post-mitotic cell system.

I.   In early dense cultures, exposed to purified rat liver mitochondriae, it was possible to follow the uptake of mitochondriae and their degradation, which was found to be incomplete and result in the formation of numerous residual bodies containing lipofuscin-type material. It was concluded that incomplete degradation of mitochondriae may be an important origin of lipofuscin.

II.   Dense, older cultures exposed to electron dense marker particles (colloidal thorium dioxide) accumulated these markers within endosomes, and later in secondary lysosomes of various types, including residual bodies. It was concluded that residual bodies constitute an integral part of the lysosomal vacuome system.

III.   Phase III glial cells were cultured on formvar-coated gold EM-grids and studied by whole cell transmission electron microscopy using TEM and STEM techniques in combination with energy dispersive X-ray microanalysis. It was found that residual bodies contained iron. This fact was taken as a further indication that lipofuscin has its origin in autophagocytosed mitochondriae and ER-material rich in metallo-enzymes. Due to their high concentration of iron, residual bodies may constitute unstable structures within the cells. Since iron is a well known catalyst of various peroxidative processes, the surrounding lysosomal membrane might be damaged, e.g. by oxidative stress, with risk for leakage of degradative lysosomal enzymes into the cell sap.

INTRODUCTION

The degradation of material contained in phagosomes takes place when these organelles have fused with primary or secondary lysosomes. Phagosomes can be classified under two main headings; heterophagosomes containing exogenous material absorbed by the cell during endocytosis, and autophagosomes containing the cell's own cytoplasmic components. In the course of the degradation some material may be found to be indigestible. Such substances, if not subjected to exocytosis, will remain within the lysosomal vacuome, forming the content of residual bodies, or lipofuscin pigment granules. These organelles are found in abundance within aged post-mitotic cells such as

KEY WORDS: Aging; Lipofuscin; Residual bodies; Lysosomes; Glial cells (human); Cell culture.

Lipofuscin and Ceroid Pigments
Edited by E. A. Porta
Plenum Press, New York, 1990

neurons and cardiac muscle cells (1,2,3,4) and have also been demonstrated in cultured cells in phase III, and in phase II after prolonged periods in non-dividing state (5,6,7,8,9,10). Lipofuscin is thought to be the accumulation of non-degradable residues of partially digested lipids and proteins (11), which initially may have been introduced into the lysosomal vacuome by autophagocytosis. This theory of origin for lipofuscin, however, is based mainly on circumstantial evidence (1,12,13).

Residual bodies, or lipofuscin pigment granules, constitute a special type of secondary lysosomes; they are large, compared to other lysosomes, irregularly shaped, and contain an electron-dense, osmiophilic, heterogeneous matrix, membranous structures, ferritin-like grains, and often lipid globules (14). Fluorescence and cytochemical studies have revealed that lipofuscin is rich in polymerized and peroxidized protein and lipid residues (15,16).

Residual bodies seem to form continuously during the life-span of man and animal (1). Lipofuscin pigment granules found in cultured human glial cells and neonatal rat myocytes have been proved to be analogous to those found in post-mitotic cells *in vivo* (6, 17).

The functional significance of residual bodies remains obscure. The granules have mainly been considered as inert bags for waste products. The validity of this theory is difficult to assess in a complex *in vivo* system, but might be more easily studied in well-controlled cell culture systems, especially on cultured human glial cells. These cells show well-developed contact inhibition of movement and division, and thus can be maintained in stable monolayers for several months without significant mitotic activity or cell loss (5).

Since the bulk of the material within autophagic vacuoles are mitochondriae, it is an obvious possibility that incompletely degraded mitochondriae will consistute a major part of the "raw material" for lipofuscin. This possibility will be commented upon in this article. Moreover, the problem of whether residual bodies really are inert, or whether they participate in functional cellular metabolism, as do other secondary lysosomes, will be addressed as well as whether cultured human glial cells have the capacity to extrude such organelles. Moreover, electron probe X-ray microanalysis has been performed on residual bodies in cultured human glial cells in phase III, grown on formvar-coated gold EM-grids and studied as whole cell preparations using 3-D TEM and STEM techniques.

EXPERIMENTS

Cell lines and culture conditions

All experiments were performed on cultured, diploid, normal human glial cells in phase II and III. The cells were derived and kept in culture by described methods (5). These cells have a large nucleus, are relatively cytoplasm-rich, display an astrocyte-like morphology *in vitro*, and produce the neuroectoderm-specific protein S100. These findings, along with the absence of morphologic features of other likely cell types, make an astrocyte stem cell the most probable cell of origin.

I.   Confluent monolayers were exposed to a suspension of rat liver mitochondriae (18). Cultures were fixed for SEM and TEM at intervals between 5 min and 14 days following exposure to mitochondria.

II.   Other confluent cultures were exposed to Thorotrast® particles (colloidal thorium dioxide) for 12 hours, then washed free of the electron-dense marker particles, and further cultivated before fixation for TEM after periods varying between 15 min and 5 months (19).

III.   Phase III cells, rich in residual bodies, were subcultivated onto formvar coated gold EM-grids and allowed to settle and stretch for 24 h before preparation for elemental analysis (20).

Preparation for TEM, SEM, STEM and electron probe X-ray microanalysis

For morphologic studies, the cells were directly immersed in 2 % glutaraldehyde in 0.1 M cacodylate buffer with 0.1 M sucrose at 4°C for 1 hour (pH 7.2, total osmolality = 510 mOsm., vehicle osmolality = 300 mOsm.), then carefully rinsed in 0.1 M cacodylate buffer with 0.1 M sucrose, and postfixed in 1% OsO_4 in 0.15 M cacodylate buffer (pH

7.2) for 90 minutes at 4°C. Cultures for TEM examination were dehydrated, embedded in Epon 812 cut and stained as reported (21).

The preparations to be studied by SEM were, after GA and OsO_4 fixation, dehydrated in acetone and critical point-dried from carbon dioxide. They were then sputter-coated with a 5-10 nm thick layer of a gold-palladium alloy and studied in a JEOL 100-C microscope equipped with a side entrance goniometer and a scanning attachment.

Phase III cells, growing on gold grids, were fixed in the GA-fixative described above for 60 min at 4°C , rapidly rinsed in distilled water, freeze-dried, and carbon-coated. Analytic electron microscopy was performed with a Princeton Gammatech (PGT) energy-dispersive X-ray spectrometer, in combination with a JEOL 100-C electron microscope (20).Areas of cytoplasm with residual bodies and adjacent cytoplasmic areas with or without mitochondriae were analyzed. The electron beam had a diameter of 0.3-0.5 μm. To prevent contamination and minimize evaporation of material from the specimens the "cold trap" and the "cold finger" were used. Photographs of the regions under analysis were taken in both TEM and STEM mode. PGT programs Alfa-16 was used for data processing.

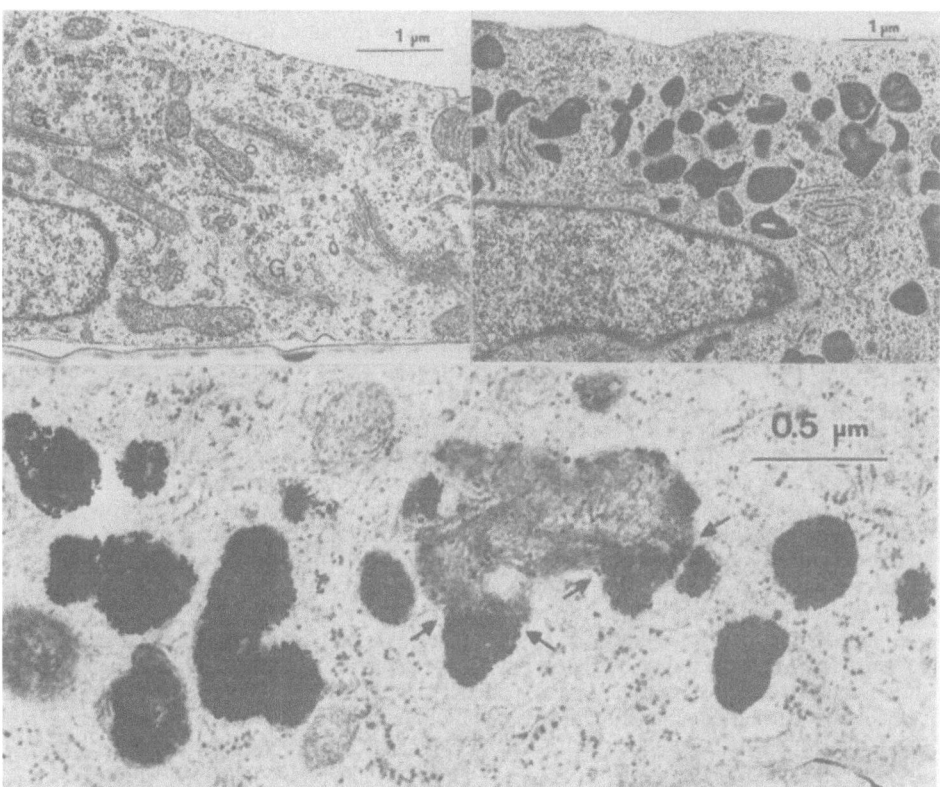

Fig. 1 Phase II glial cell (upper left) in logarithmic growth with almost no pigment granules.
Growth inhibited phase II glial cell after 3 weeks in dense state (upper right) showing numerous lipofuscin pigment granules (residual bodies).
Phase III glial cell (lower micrograph) three days after exposure to thorium dioxide particles. Note labeled residual bodies.

RESULTS AND DISCUSSION

I.   Glial cells exposed to mitochondriae, and studied by inverted phase-contrast microscopy, were at first found to be completely covered with these organelles. When studied in the SEM, after 4 hours of exposure, the cells showed numerous elongated microvilli on their upper surfaces. Membrane activity could also be present at such sites in the form of ruffle-like structures of variable size. Mitochondriae were present almost exclusively in these areas. These mitochondriae were apparently so strongly attached to the cell surface that they remained throughout the multiple rinsings of the preparative procedure. The mitochondriae were engulfed by ruffle-like structures, or trapped in local indentations. After endocytosis, the liver mitochondriae were found in membrane-bound vacuoles, *i.e.*, in phagosomes. Immediately after phagocytosis these mitochondriae had a similar morphology to those found outside the cell. Mitochondrial breakdown occurred relatively rapidly but was not complete. Osmiophilic and membraneous materials accumulated within the secondary lysosomes, resulting in an increase in the occurrence of lysosomes of the residual body type. Thus, incomplete mitochondrial digestion seems to contribute to the formation of these structures. Our findings support the theory that phagocytosis of cellular organelles, as occurs normally during autophagocytosis, may result in material which accumulates in secondary lysosomes some of which are gradually transformed into residual bodies.

It has previously been suggested that residual bodies accumulate as a result of continuously progressing autophagocytosis in combination with inability of the cells to rid themselves of the residues of digestion to a sufficient degreee (1,5, 6). This would explain the finding of residual bodies *in vivo* mainly in non-dividing cells which are unable to dilute their supply of these bodies by way of division. It would thus appear that the accumulation is dependent on (a) the rate of autophagocytosis or cellular uptake of material which cannot be completely degraded; (b) the efficiency of the lysosomal enzymes in the cell under consideration; (c) the rate of elimination of undigestible residues (either through exocytosis of the enclosed material, extrusion of the residual bodies, or dilution of the residual bodies during cell division).

II.   The results from the Thorotrast® labelling experiments showed that cells from late passages engulfed the marker particles in the same manner, as did cells from early passages. The particles were thus rapidly found in coated and uncoated pits and brought into the cells within endocytotic vacuoles. One hour following the administration of the marker substance, the particles were found in conventional secondary lysosomes of small and medium size while the residual bodies contained only scattered granules. Following longer periods of time there was also accumulation of marker particles within the residual bodies which appeared to be accomplished through of fusion with conventional secondary lysosomes loaded with thorium dioxide particles. In the cells which were harvested 72 h after the initial exposure to the marker, all residual bodies were rich in particles. As long as 5 months after the initial exposure of the cells to the marker particles, labeled secondary lysosomes were found, strongly suggesting that the rate of extrusion of secondary lysosomes, or exocytosis of their content must be very low, if occurring at all. Whether this applies generally to normal *in vitro* cultured cells is unknown and may be difficult to find out because few cell lines can be kept in a stable, contact-inhibited layer without degeneration for such long periods of time as the human glial cells. *In vivo,* this problem is hard to approach because of labelling difficulties as well as possible reutilization of marker substances. It is, however, probable that the capacity for extrusion and/or exocytosis differs between individual cell types. There is thus chemical as well as morphological evidence that lysosomes of liver parenchymal cells discharge their contents into the bile canaliculi and space of Disse. On the other hand, the increasing accumulation of residual bodies during aging in neurons and cardiac muscle cells suggests that these cells do not have this capacity to any sufficient degree. The implications of the findings of the labelling experiments are that secondary lysosomes form part of a vacuome system which by fusion and fission allows material within membrane-limited vacuoles to spread throughout the system, although it is at any

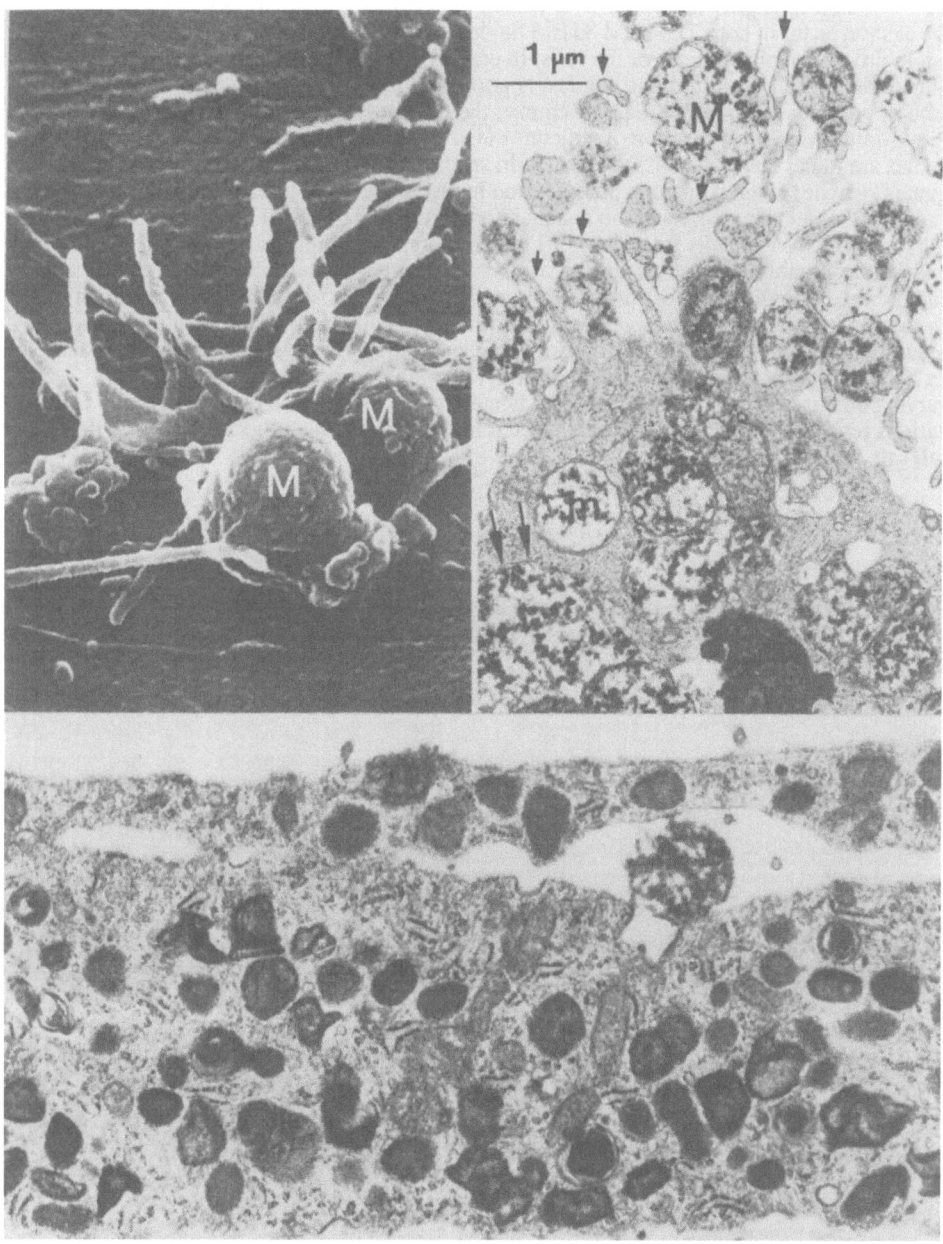

Fig. 2 Glial cells recently exposed to mitochondriae as seen in SEM (upper left) and
TEM (upper right) and after another 14 days in TEM (lower). Note numerous
residual bodies in the lower fig. with resemblance to phase II cells after prolonged
periods of confluence.

one moment discontinuous. This view is in accordance with de Duve's exoplasmic vacuome concept. The results further indicate that residual bodies form an integral part of the lysosomal vacuome system and regularly receive lytic enzymes by fusion with other types of lysosomes.

III.   Whole cells growing on formvar coated gold EM-grids can be well penetrated by the electon beam in both TEM and STEM modes at 100 kV if the dehydration is carried out in order to avoid collapse of the cells. In critical-point dried cells, fixed with both glutaraldehyde and osmium, the contrast is good, and most details may be studied with ease as previously described (22). Of course, the thickness of the cells prevents high resolution, and superimposition of structures also complicates the interpretation. Residual bodies are rather large and easily detected. In specimens fixed only briefly in glutaraldehyde, without ensuing post-fixation in OsO_4 and then freeze-dried the endogenous electron density of the residual bodies suffices to reveal these organelles for X-ray analysis. Most cells from phase III glial cultures contain large numbers of residual bodies, presumably because a majority of the cells from such cultures are non-dividers and do not dilute their number of these organelles during mitosis.This notion is supported by the finding that density-dependent growth-inhibited cells from phase II cultures also accumulate residual bodies (5). All residual bodies examined with microanalysis revealed an X-ray emission peak characteristic of iron, although this metal obviously exists in varying amounts in different bodies, as indicated by the differences in the spectra of various residual bodies. No iron was detected in the nucleus, the mitochondriae, or the cell sap.

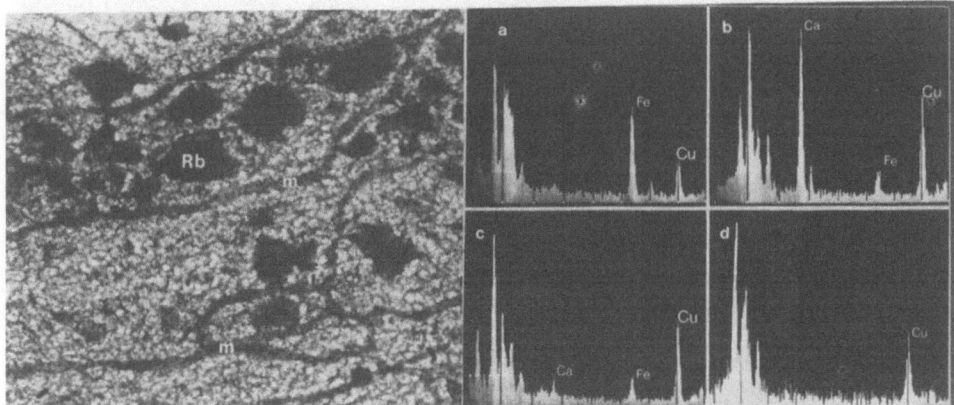

Fig. 3 STEM micrograph of a whole freeze-dried, unstained cell with elongated
mitochondriae and clumpsy residual bodies. Energy dispersive X-ray analysis of
three residual bodies, **a - c**, and surrounding cytoplasm, **d**, resulted in shown
spectra.

Since mitochondriae and the ER are rich in iron-containing metallo-enzymes, it is reasonable to assume that iron should be a component of the non-degradable material within the residual bodies if mitochondriae and ER are believed to be the material of origin for lipofuscin within residual bodies. This assumption is supported by morphologic findings (ferritin-like granules) and cytochemistry (23,24,25). Such methods are, however, rather non-specific and in fact neither prove the presence of iron nor indicate its amount. Energy dispersive X-ray microanalysis, however, is an objective and reasonably sensitive technique. It is well known that iron is a potent catalyst of peroxidation processes, which to some extent seem to be a normal event in living cells. Damage by peroxidation is normally minimized by protecting enzyme systems, as well as by the occurrence of substances capturing reactive oxygen metabolites. It is tempting to assume, however, that under certain circumstances the peroxidative level might increase and cause damage, especially to membranes in direct contact with iron. Since residual bodies also contain active hydrolytic enzymes (1,26) one may speculate about their possible role as weak structures of the cell. Damage to their membranes with subsequent leakage of hydrolytic enzymes would lead to degenerative alterations within cells rich in residual bodies, such as aged neurons and heart muscle cells. This hypothesis should be further tested by studies of the stability of residual bodies and the latency of their hydrolytic enzymes as compared with other types of lysosomes.

REFERENCES

1. U.Brunk and J.L.E. Ericsson, Electron microscopical studies on rat brain neurons: Localization of acid phosphatase and mode of formation of lipofuscin bodies. *J. Ulstrastruct . Res.* 38:1 (1972).
2. A.B.Novikoff, Lysosomes in nerve cells. In *The Neuron*, H.Hydén, ed., Elsevier, Amsterdam, p.319 (1967).
3. W.Reichel, Lipofuscin pigment accumulation and distribution in five rat organs as a function of age. *J. Gerontol.*, 23:145 (1968).
4. D.F.Travis and A.Travis, Ultrastructural changes in the left ventricular rat myocardial cells with age. *J. Ultrastruct. Res.*, 39:124 (1972).
5. U.Brunk, J.L.E.Ericsson, J.Pontén and B.Westermark, Residual bodies and "aging" in cultured human glial cells: effect of entrance into phase III and prolonged periods of confluence. *Exp. Cell. Res.*, 79:1 (1973).
6. V.P.Collins and U.Brunk, Characterization of residual bodies formed in phase II cultivated human glial cells. *Mech. Ageing Dev.*, 5:193 (1976).
7. V.P.Collins and U.Brunk, Quantitation of residual bodies in cultured human glial cells during stationary and logarithmic growth phases. *Mech. Ageing Dev.*, 8:139 (1978).
8. J.Lipetz and V.J.Cristofalo, Ultrastructural changes accompanying the aging of human diploid cells in culture. *J. Ultrastruct. Res.*, 39:43 (1972).
9. K.Nandy and H.Schneider, Lipofuscin pigment formation in neuroblastoma cells in culture. In *Neurobiology of Aging*, R.D.Terry and S.Gershon, eds., Raven Press, New York, p.245 (1976).
10. E.Robbins, E.M.Levine and H.Eagle, Morphologic changes accompanying senescence of cultured human diploid cells. *J. Exp. Med.*, 131:1211 (1970).
11. A.L.Tappel, Lipid peroxidation and flurescent molecular damage to membranes. In *Pathology of Cell Membrane*, B.F.Trump and A.V.Arstila, eds., Academis Press, New York, p.145 (1975).
12. J.L.E.Ericsson, Mechanism of cellular autophagy. In *Lysosomes in Biology and Pathology*, J.T.Dingle and H.B.Fell, eds, North Holland, Amsterdam, Vol. II, p. 345 (1969).
13. S.E.Toth, The origin of lipofuscin age pigments. *Exp. Gerontol.*, 3:19 (1968).
14. E.Holzman, Lysosomes: A Survey. *Cell Biology Monographs*, Springer Verlag, New York, Vol. 3 (1976).
15. B.L.Strehler, On the histochemistry and ultrastructure of age pigment. *Adv. Gerontol. Res.*, 1:343-354 (1964).
16. A.L.Tappel, Lysosomal enzymes and other components. In *Lysosomes in Biology and Pathology*, J.T.Dingle and H.B.Fell, eds., North Holland, Amsterdam, Vol. 2, pp. 207-244 (1969).
17. R.S.Sohal, M.R.Marzabadi, D.Galaris and U.T.Brunk, Effect of ambient oxygen concentration on lipofuscin accumulation in cultured rat heart myocytes - a novel in vitro model of lipofuscinogenesis. *Free Rad. Biol. Med.*, 6:23 (1989).

18. V.P.Collins, B.Arborgh, U.Brunk and J.P.M.Schellens, Phagocytosis and degradation of rat liver mitochondria by cultivated human glial cells. *Lab. Invest.*, 42:209-216 (1980).
19. U.Brunk, Distribution and shifts of ingested marker particles in residual bodies and other lysosomes. *Exp. Cell Res.*, 79:1 (1973).
20. E.Blomquist, B.-A.Fredriksson and U.Brunk, Electron probe X-ray microanalysis of residual bodies in aged cultured human glial cells. *Ultrastruct. Pathol.*, 1:11 (1980).
21. U.Brunk, J.L.E.Ericsson, J.Pontén and B.Westermark, Specialization of cell surfaces in contact-inhibited human glia-like cells *in vitro*. *Exp. Cell Res.*, 67:407 (1971).
22. V.P.Collins, U.Brunk, B.-A.Fredriksson and B.Westermark, The fine structure of growing human glia and glioma cells. Whole cell preparations. *Acta Pathol. Microbiol. Scand. (A)*, 87:29 (1979).
23. H.Miyawaki, Histochemistry and electron microscopy of iron-containing granules, lysosomes and lipofuscin in mouse mammary glands. *J. Natl. Cancer Inst.*, 34:601 (1965).
24. S.Goldfischer, H.Villaverde and R.Forschirm, The demonstration of acid hydrolase, thermostable reduced diphosphopyridine nucleotide-tetrazolium reductase and peroxidase activities in human lipofuscin pigment granules. *J. Histochem. Cytochem.*, 14:641 (1966).
25. A.Brun and U.Brunk, Heavy metal localization and age related accumulation in the rat nervous system: A histochemical and atomic absorption spectrophotometric study. *Histochemie*, 34:333-342 (1973).
26. H.E.Hirsch, Enzyme levels of individual neurons in relation to lipofuscin content. *J. Histochem. Cytochem.*, 18:268 (1970).

DISCUSSION

ELLEDER: I was really impressed by the induction of lipofuscin-like granules by mitochondria. Did you check whether these mitochondria were autofluorescent before reaching the interior? There is a possibility that they might have been autofluorescent before.

BRUNK: Yes, they were to some extent and showed slight bluish autofluorescence, but not the yellowish one that we get when they have been there for some time. It might have been better, of course, to purify the mitochondria anaerobically, because they certainly become peroxidized to some degree during contact with the atmosphere, but it is extremely difficult to purify in a sterile way mitochondria anaerobically, because we have to use special equipment, centrifuges, etc.

ELLEDER: You see? It is really a problem with the autofluorescence, because there is an increasing body of evidence that it may not be given by chemical structure, but by physical states of the molecules. Probably, if you treated the mitochondrial pellet by various ways, you might induce other spectral characteristics. Was it possible to see some persistent mitochondrial enzyme activities in the residual bodies?

BRUNK: We did demonstrate succinic dehydrogenase activity cytochemically afterwards, but I must admit that we did not do those experiments very carefully, and so I am not able to tell whether this activity finished out completely, since we followed it for no more than 3 or 4 days. What we have seen was that when we gave mitochondria to the cells, the mitochondria changed their form. I don't claim that mitochondria were normal when they were taken up by the cells.

Zs.-NAGY: I wonder why you used iron 3 because it is well known that iron 2 in these systems is much more efficient in generating free radicals. Floyd and coworkers found that if you complex iron 2 with ATP or ADP, you can keep the iron in divalent state at pH 7.4 which is in the biological range.

BRUNK: We actually also used both divalent and trivalent iron complexed with ADP, but we got exactly the same results. We do think, however, that the use of trivalent iron is preferred because then we also have the possibility to study the reduction process within the lysosomes from trivalent to divalent iron, and which we think is a physiological event.

HALL: I think that you developed two beautiful systems to study the formation of lipofuscin. Whether you are observing the same phenomena that occurs normally in vivo is unlikely. My question is: how will you obtain real biochemical evidence to support your pathological evidence?

BRUNK: Well, what we are doing is to see whether cells with a high load of lipofuscin have diminished capacity for autophagocytosis by feeding radioactive L-lysine to the medium. Then, by changing experiments, it is possible to study this load which indicates the degree of phagocytosis.

KATZ: You did not show any fluorescent micrographs or any spectra of the fluorophores that you were quantitating. What evidence do you have that the fluorophores that you are generating, particularly with iron, are the same as the age-pigment fluorophores?

BRUNK: We have done these spectral analysis, but I did not show the pictures in my presentation. They do show an emission of about 430 nm.

PORTA: I want to congratulate you for your elegant and imaginative in vitro studies which may provide important clues on the physiologic and pathologic aspects of iron metabolism as well as lipopigment formation. In vivo conditions, however, practically all our cells in front of iron overloading conditions, such as hemosiderosis and hemochromatosis, have the mechanisms to maintain the free iron in the form of ferritin and hemosiderin that prevent cell damage. In fact, in experimental iron overload in rats, for instance, it takes a relatively long time and about 200-fold increase over the normal hepatic levels of iron before microsomal lipid peroxidation can be detected. My question then is whether you have determined or at least studied histochemically the content of ferritin and hemosiderin in cardiac myocytes, as well as the contents of antioxidant components of these cells.

BRUNK: Your question is very relevant, and we are well aware of these studies on experimental hemochromatosis. I completely agree that iron is kept in stable conditions and is not catalyzing peroxidative events until very late stages. Obviously, also, there are some kinds of iron chelators in the lysosomal vacuome, the chemical nature of which is still imprecise. We have not studied the oxidative defenses of these cells. I think that the amount of iron that we added was very large and is probably overwhelming the regulating iron mechanisms within the lysosomes, because we can see enormous amounts of iron. So, probably this is a rather artificial model.

PORTA: If I understood correctly, you also add vitamin E in your system, and it seems to me that the added amounts were not quite effective in preventing lipopigment formation. Could you clarify this?

BRUNK: What we did was to add tocopherol acetate in water soluble form at the concentration of 10 nmoles, and then we observed slight inhibition of lipopigment formation. There was, however, no statistical difference at the 12 day level, but there was some difference at the 7 day level.

SHIMASAKI: I want to ask just a technical question. How long the cells lived in your cultures?

BRUNK: Human astrocytes, for instance, went into phase 3 in about 45 passages, but you can keep them in a dish without touching them for a very long period of time. We have actually kept cells in a density dependent stage of inhibition for more than 1 1/2 years, at which moment they were very heavily loaded with lipopigments. The myocytes, however, don't survive for more than a maximum of 35 - 40 days, except when you use some collagen or other type of matrix, in which case they survive for more prolonged periods.

ELLEDER: Since you loaded the cultures with large amounts of iron salts, have you considered the possibility that iron itself deposited in lysosomes might inhibit the pigment autofluorescence? This is usually the case in human tissues with excessive hemosiderin, and, in order to detect the pigment's autofluorescence, it is necessary first to remove the iron.

BRUNK: I think you are right; iron really quenches considerably the autofluorescence.

LIPID PEROXIDATION AND STORAGE OF FLUORESCENT PRODUCTS BY MACROPHAGES

IN VITRO AS A MODEL OF CEROID-LIKE PIGMENT FORMATION

H. Shimasaki, R. Maeba and N. Ueta

Department of Biochemistry
Teikyo University School of Medicine
Itabashi-ku, Tokyo 173, Japan

SUMMARY

Storage of fluorescent products, ceroid-like pigments was observed in P388D$_1$ cells, an established macrophage-like cell line, when the cells were cultured in the presence of rat liver phosphatidylcholine liposomes containing polyunsaturated fatty acids. Some fluorescent products accumulated in the cells were extractable in organic solvents, ethanol/ether (3:1, v/v), while others were insoluble in organic solvents, but soluble in detergent. The fluorescent products dissolved in organic solvents or in detergent had a fluorescence maximum at 430 nm when excited at 360 nm. The formation of the ceroid-like pigments was inhibited, at least in part, by antioxidants, such as alpha-tocopherol and butylated hydroxy-toluene (BHT). The fluorescence intensity of the pigments was quenched in alkaline media and restored by adjustment of pH to neutrality. These findings indicate that liposome uptake by macrophages causes the formation of ceroid-like pigments, and that the fluorescent chromophores of the pigments are Schiff base structures derived from the reaction of lipid peroxides from the exogenous phospholipids and subcellular amino compounds.

INTRODUCTION

Ceroid pigments, as well as lipofuscin, are generally believed to be end products of free radical-induced lipid peroxidation of subcellular components. The term ceroid describes the pathologically induced autofluorescent granules found in vitamin E-deficient animals or in humans with neuronal ceroid-lipofuscinosis, while lipofuscin is used to denote the pigmented autofluorescent granules that accumulate as a function of age in humans or animals (1,2). The mechanisms for the formation of ceroid or lipofuscin are not well understood (2-5), but products with similar fluorescence characteristics are produced in model systems involving the reaction of malonaldehyde with amino compounds. Tappel and co-workers (6,7) have proposed that the fluorophore structures of lipofuscin or ceroid are modified-Schiff bases. The products have a fluorescence maximum in the region of 430-440 nm when excited at 360 nm. The fluorescence intensity of the products is quenched at alkaline pH and restored by adjustment

KEY WORDS: Ceroid, lipid peroxidation, Schiff base, fluorescent chromophore,
    alpha-tocopherol, macrophage, P388D$_1$ cell

of the pH to neutrality (8). We have shown that the age-related fluorescent substances isolated from the testes and kidneys of old rats have the typical fluorescence characteristics of Schiff bases (9,10).

On the other hand, Kikugawa and Ido (11) have recently shown that the fluorescent products derived from the reaction of malondialdehyde with primary amines are 1,4-dihydropyridine-3,5-dicarbaldehydes, which have a fluorescence maximum of 462 nm and an excitation maximum of 403 nm. The fluorescence intensity of these products is quenched in acidic media below pH 4, indicating the fluorophore structure differs from the Schiff bases. We have also shown that the fluorescence intensity of compounds derived from the reaction of malondialdehyde with glycine is quenched at acid pH, while fluorescent products formed during lipid peroxidation of rat liver phosphatidylcholine liposomes containing glycine as an amino compound, are reduced in intensity at alkaline pH (12). Moreover, the excitation and emission maxima of the products formed in the rat liver phosphatidylcholine liposomes are different from those of the compounds derived from the reaction of malondialdehyde with glycine. The earlier story suggested that malondialdehyde was formed from lipid hydroperoxides and then reacted with amino compounds to form Schiff bases. Our data, however, have shown that malondialdehyde is not the precursor of the fluorescent products that arise during lipid peroxidation of liposomes (12,13). We also suggested that glycerophospholipids containing mono-aldehydes, which are formed via β-scission cleavage of alkoxyl radicals in acyl branches of oxidized phospholipids, may react with amino compounds to form fluorescent Schiff bases (12). Recently, Iio and Yoden reported the existence of fuorescent products located in the acyl branch of glycerophopholipids (14,15). In this paper, we report that ceroid-like pigments accumulate in P388D$_1$ cells cultured with rat liver phosphatidylcholine liposomes and that the fluorescence characteristics of the pigments are similar to those of Schiff base fluorescent products formed during the lipid peroxidation.

MATERIALS AND METHODS

Multilamellar liposomes were prepared as described previously (13) with a slight modification. A lipid mixture dissolved in chloroform was pipetted into a sterilized flask, and the solvent was removed by a rotary evaporator at 20 °C. The dry lipid film was suspended in 0.3 M glucose with a vortex mixer under nitrogen atomosphere passed through a filter (millipore, 0.22 µm). The lipid composition of the liposomes was rat liver phosphatidylcholine, bovine brain phosphatidylserine, cholesterol, and dicetylphosphate (molar ratio, 1 : 1 : 2 : 0.2). The liposomes (3 µmol of total phospholipid) were added to P388D$_1$ cells suspended in freshly-prepared culture medium (5×10^6 cells/10ml/dish).

P388D$_1$ cells were incubated with rat liver phosphatidylcholine liposomes in RPMI 1640 medium (Flow Laboratories, Irvine, U.K.) containing 10% fetal calf serum (FCS) in 60 mm diameter plastic culture dishes under 5% CO_2 at 37 °C. After incubation for up to 8 days, the cells were washed with phosphate-buffered saline (PBS; 10 mM, pH 7.4), and examined for auto-fluorescence under a fluorescent microscope (Olympus BH-2) equipped with a mercury vapor lamp and a Y495 barrier filter. The unstained cells were also examined under a light microscope equipped with a W-lamp. Microphotographs from the fluorescent and light microscopes were taken at magnification x 200.

Fluorescent products accumulated in P388D$_1$ cells were extracted with 2 ml of ethanol/ether (3:1, v/v) by vigorous mixing on a vortex mixer, followed by centrifugation. Aliquots (1.5 ml) of the supernatant were collected to measure the fluorescence spectra of organic solvent-soluble products, and material sedimented at the bottom of the tube was dissolved in 2 ml of 15% SDS-PBS solution. The fluorescent spectra of the organic solvent-soluble and -insoluble (detergent soluble) products were measured with a fluorescence spectrophotometer (Hitachi 204). Fluorescence intensity of 0.1 μM quinine sulfate in 0.1 N H_2SO_4 solution was measured with a fluorescence maximum of 448 nm and an excitation maximum of 352 nm, and taken to be 100 as a standard.

P388D$_1$ cells (5×10^5 cells/dish) were also cultured in the presence of rat liver phosphatidylcholine liposomes containing 0.2 μCi ^{3}H-cholesterol for up to 7 days. The total lipids were extracted with chloroform/methanol/water (1:2:0.8, v/v/v) by the method of Bligh and Dyer. The lipid extracts were subjected to thin layer chromatography developed in hexane/ether/acetic acid (70:30:1, v/v/v). The spots of neutral lipids were identified from their Rf values by comparison with authentic standards, and scraped directly into vials. Radioactivities of ^{3}H-cholesterol and ^{3}H-cholesterol ester were measured by counting with a Scintillation Spectometer (Aroka LSC-700) in a toluene solution of 0.4% 2,5-diphenyloxazole (PPO) and 0.01% 1,4-bis-2-(5-phenyloxazolyl)-benzene (POPOP) obtained from Dojin Chemical Institute (Osaka, Japan).

RESULTS AND DISCUSSION

I. Liposome uptake by P388D$_1$ cells

When P388D$_1$ cells were cultured in the presence of multilamellar liposomes consisting of rat liver phosphatidylcholine, bovine brain phosphatidylserine, ^{3}H-cholesterol and dicetylphosphate, in a molar ratio of 1 : 1 : 2 : 0.2, uptake of liposome by the cells was observed for up to 4 days. Uptake was determined by the cellular content of ^{3}H-labeled cholesterol esters. The esterification of ^{3}H-cholesterol incorporated with liposomes into the cells may be catalyzed by lysosomal enzymes in the cells. The time course for ^{3}H-cholesterol ester formation (liposome uptake) is shown in Figure 1. The accumulation of ceroid-like pigments appears after liposome uptake by the macrophages. In contrast, the rat liver phosphatidylcholine liposomes without phosphatidylserine are poorly incorporated into cells and show only slight pigment formation under the same conditions.

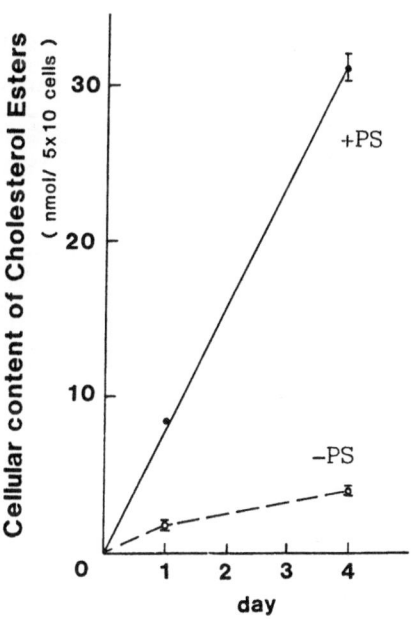

Figure 1. Time course of ^{3}H-cholesterol ester formation (liposome uptake) by P388D$_1$ cells.

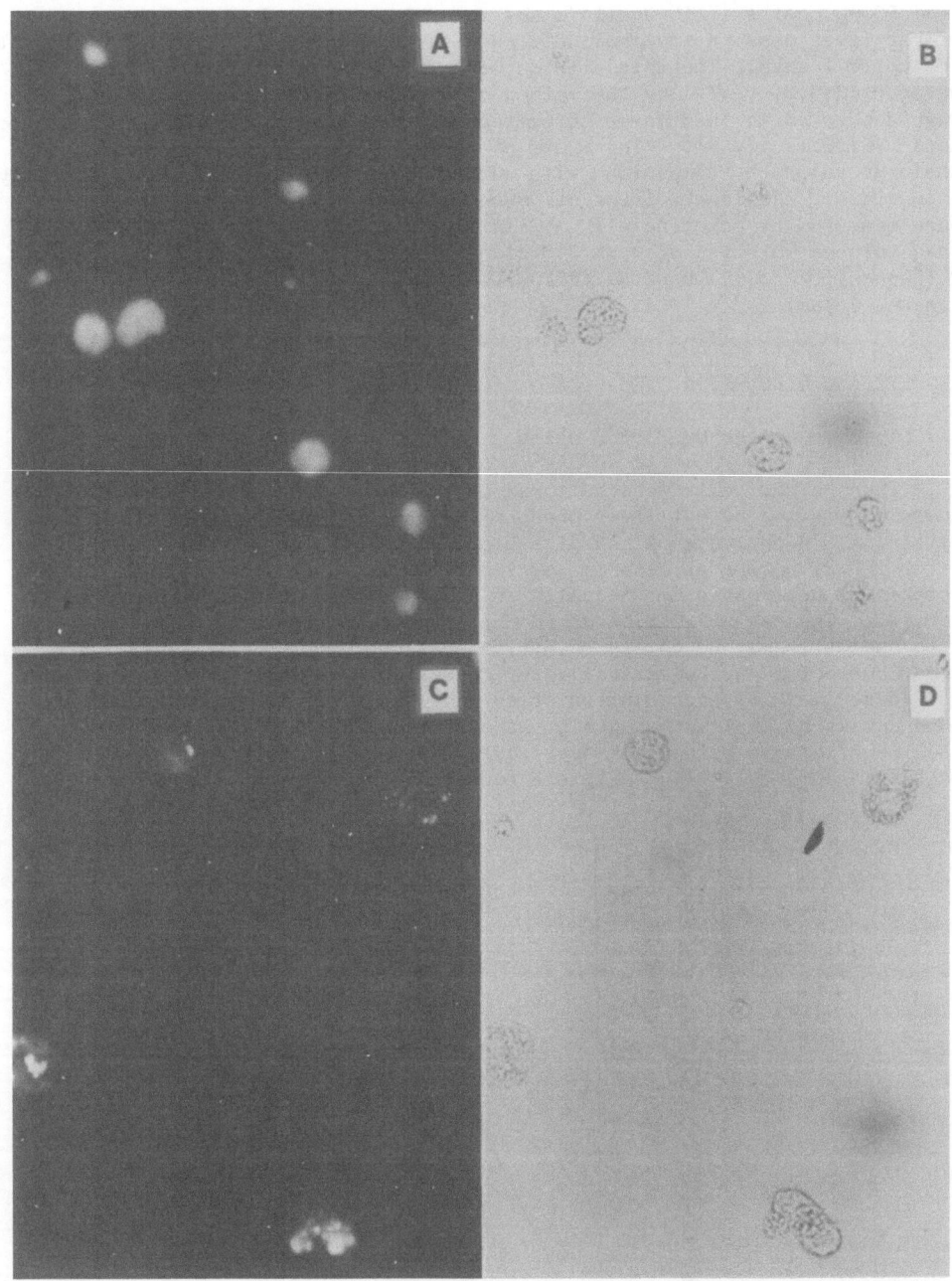

Fig. 2 Fluorescent and light microscopy. Photomicrographs from
P388D$_1$ cells cultured with rat liver phosphatidylcholine
liposomes (A and B), or liposomes containing alpha-
tocopherol (C and D).

II. Fluorescent and light microscopy

P388D$_1$ cells cultured in the presence of rat liver phosphatidylcholine liposomes, or liposomes containing alpha-tocopherol were observed by fluorescent and light microscopy (Fig. 2). The ceroid-like pigments and lipid droplets accumulated in the cells and increased continuously up to 8 days (Fig.2 A and B). The cells incubated without liposomes showed no appreciable fluorescence or lipid droplet formation under the same conditions. The ceroid-like pigments in P388D$_1$ cells cultured with liposomes appeared yellow in color without stain under the fluorescent microscope. The fluorescence of the pigments was similar to that of ceroid pigments accumulating in several tissues and organs of rats fed a vitamin E-deficient diet for one year (16).

III. Extraction and fluorescence measurement

Some fluorescent products in P388D$_1$ cells incubated with rat liver phosphatidylcholine liposomes are extractable in organic solvents, such as ethanol/ether (3:1, v/v), while others are insoluble in organic solvents, but soluble in detergent. The organic solvent-soluble fluorescent products were extracted from the cells with 2 ml of ethanol/ether (3:1, v/v) and, after centrifugation, the supernatants were collected for fluorescence measurement. Materials sedimented at the bottom of the tube were washed twice with 10 ml of the solvent mixture and then dissolved in 2 ml of 15 % SDS-PBS solution. Both the organic solvent-soluble and the detergent-soluble products exhibited a fluorescence maximum at 430 nm when excited at 360 nm (Fig. 3). The fluorescence spectra are similar to those of Schiff base structures, which have emission maxima in the region of 430-440 nm and excitation maxima in the region of 350-365 nm (7,8).

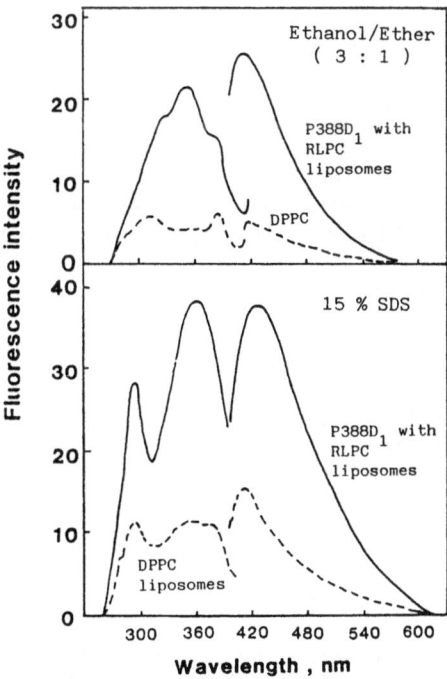

Figure 3. Fluorescence spectra of organic solvent-soluble and and detergent-soluble fluorescent products accumulated in P388D$_1$ cells.

Table I   Effects of antioxidants on liposome uptake and
formation of fluorescent products by P388D$_1$ cells

Antioxidants	µM	Liposome uptake (^{3}H-choresterol ester formation)		Fluorescence Intensity
		n mol	%	%
none	–	35.6	100	100
α-tocopherol	0.4	35.6	100	53
	2	37.8	106	52
BHT*	10	36.4	102	53
	50	31.9	90	28

*Butylated hydroxytoluene

IV. Effect of antioxidants on fluorescent product formation

The accumulation of ceroid-like pigments in P388D$_1$ cells was partially
inhibited by the presence of alpha-tocopherol, an antioxidant, in the rat
liver phosphatidylcholine liposomes.   The accumulation of lipid droplets,
however, appeared to be the same as when liposomes without alpha-tocopherol
were used (Fig.2 C and D).   Alpha-tocopherol, as well as butylated hydroxytoluene (BHT), had no effect on liposome uptake by the cells, but suppressed
the formation of fluorescent products.   The amount of liposome uptake
and the fluorescence intensity of detergent-soluble products accumulated
in P388D$_1$ cells are shown in Table I.   It should be note that the accumulation of ceroid-like pigments by P388D$_1$ cells was also inhibited when the
cells were cultured with liposomes containing dipalmitoyl phosphatidylcholine instead of rat liver phosphatidylcholine (Fig. 3).   The fatty
acids in the acyl branches of dipalmitoyl phosphatidylcholine are completely
saturated, while rat liver phosphatidylcholine contains polyunsaturated
fatty acids (13); thus the results indicate that the ceroid-like pigments
are, at least in part, lipid peroxidation products whose formation can
be inhibited by antioxidants, and that lipid peroxidation of liposomal
phospholipids occurs after uptake by the cells.

V. Fluorescence characteristics of products

There are currently two schemes to explain the formation of fluorescent
compounds from lipid peroxides and amino compounds.   In early studies,
Tappel and colleagues (6-8) proposed a model for the formation of the fluorescent components of lipofuscin, a Schiff base.   However, Kikugawa and
his co-workers (11,17) have recently reported an alternate theory, which
is that dihydropyridines are formed in the reaction of malondialdehyde
with primary amines.   The fluorescence characteristics of the dihydropyridines differ from those of the Schiff base products.   The fluorescence
intensity of the dihydropyridines is reduced at acid pH, while that of
the Schiff base products is quenched by alkaline pH.   The fluorescence
of both species can be restored by readjusting the pH to neutrality.
The fluorescence intensity of the ceroid-like pigments accumulated in P388D$_1$
cells cultured in the presence of rat liver phosphatidylcholine liposomes
was reduced at alkaline pH and restored by adjusting the pH to neutrality
(18).   This result indicates that the ceroid-like pigments in P388D$_1$ cells

have properties indicative of Schiff base fluorescent chromophores. More-over, the fluorescence properties are similar to those of the fluorescent substances that accumulate in rat testes and kidney with increasing age (10).

The earlier theory for the formation of fluorescent compounds was that malondialdehyde reacts with amino compounds to form Schiff base fluore-scent chromophores. Recent studies, however, have shown that the precursor of the fluorescent products formed during lipid peroxidation is not malondi-aldehyde (12-15,17). It is suggested that glycerophospholipids containing mono-aldehydes in their acyl branches formed during lipid peroxidation of glycerophospholipids may react with amino compounds to form Schiff base fluorescent chromophores. The possible mechanisms for lipid peroxide cleavage and fluorescent product formatiom are shown in Figure 4. Although the structure and mechanism of formation of ceroid pigments have not been clarified, the findings presented here suggest that the ceroid-like pigments accumulated in P388D$_1$ cells cultured with liposomes contain, at least in part, the Schiff base fluorescent chromophores derived from the reaction between peroxidized glycerophospholipids and subcellular amino compounds.

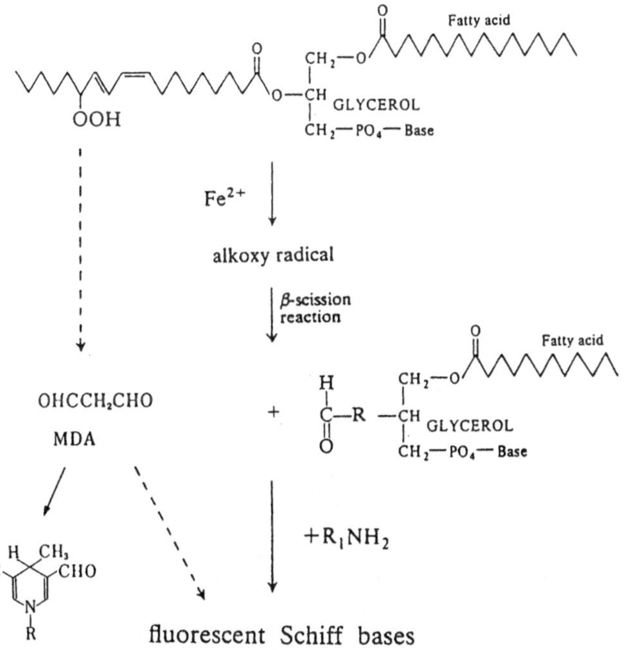

Figure 4. Possible mechanisms for hydroperoxide cleavage and fluore-scent product formation during lipid peroxidation of liposomal phospholipids.

REFERENCES

1. E.A. Porta, W.S. Hartroft, Lipid pigments in relation to aging and dietary factors (Lipofuscins), in "Pigments in Pathology," M. Wolman, ed., Academic Press, New York (1969).
2. C. Oliver, Lipofuscin and ceroid accumulation in experimental animals, in: "Age Pigments," R.S. Sohal, ed., Elsevier/North-Holland Biomedical Press, Amsterdam (1981).

3. J. Miquel, J. Oro, K.G. Bensch, J.E. Johnson, Jr., Lipofuscin: Fine-structural and biochemical studies, in "Free Radicals in Biology," vol. III, W.A. Pryor, ed., Academic Press, New York (1977).

4. D.N. Palmer, D.R. Husbands, P.J. Winter, J.W. Blunt, R.D. Jolly, Ceroid lipofuscinosis in sheep. I. Bis(monoacylglycero)phosphate, dolichol, ubiquinone, phospholipids, fatty acids, and fluorescence in liver lipopigment lipids, J. Biol. Chem. 261:1766 (1986).

5. D.N. Palmer, G. Barns, D.R. Husbands, R.D. Jolly, Ceroid Lipofuscinosis in sheep. II. The major component of the lipopigment in liver, kidney, pancreas, and brain is low molecular weight protein, J. Biol. Chem. 261:1773 (1986).

6. K.S. Chio, A.L. Tappel, Synthesis and characterization of the fluorescent products derived from malonaldehyde and amino acids, Biochemistry 8:2821 (1969).

7. A.L. Tappel, Measurement of and protection from in vivo lipid peroxidation, in "Free Radicals in Biology," vol. IV, W.A. Pryor, ed., Academic Press, New York (1980).

8. V.G. Malshet, A.L. Tappel, V.M. Burns, Fluorescent products of lipid peroxidation: II. Methods for analysis and characterization, Lipids 9:328 (1974).

9. H. Shimasaki, T. Nozawa, O.S. Privett, W.R. Anderson, Detection of age-related fluorescent substances in rat tissues, Arch. Biochem. Biophys. 183:443 (1977).

10. H. Shimasaki, N. Ueta, O.S. Privett, Isolation and analysis of age-related fluorescent substances in rat testes, Lipids 15:236 (1980).

11. K. Kikugawa, Y. Ido, Studies on peroxidized lipids. V. Formation and characterization of 1,4-dihydropyridine-3,5-dicarbaldehydes as model of fluorescent components in lipofuscin, Lipids 19:600 (1984).

12. H. Shimasaki, N. Hirai, N. Ueta, Comparison of fluorescence characteristics of products of peroxidation of membrane phospholipids with those of products derived from reaction of malonaldehyde with glycine as a model of lipofuscin fluorescent substances, J. Biochem. 104:761 (1988).

13. H. Shimasaki, N. Ueta, H. Mowri, K. Inoue, Formation of age pigment-like fluorescent substances during peroxidation of lipids in model membranes, Biochim. Biophys. Acta 792:123 (1984).

14. T. Iio, K. Yoden, Hydrolysis of a fluorescent substance formed from an oxidized phospholipid and an amino compound by phospholipase A$_2$, Lipids 23:937 (1988).

15. T. Iio, K. Yoden, Fluorescence formation from hydroperoxide of phosphatidylcholine with amino compound, Lipids 23:65 (1988).

16. T. Sato, Y. Tokoro, H. Tauchi, K. Kohtani, T. Mizuno, H. Shimasaki, N. Ueta, Morphometrical and biochemical analysis on autofluorescent granules in various tissues and cells of the rats under several nutritional conditions, Mech. Aging Dev. 43:229 (1988).

17. K. Kikugawa, Fluorescent products derived from the reaction of primary amines and components in peroxidized lipids, Adv. Free Rad. Biol. Med. 2:389 (1986).

18. R. Maeba, H. Shimasaki, N. Ueta, K. Inoue, Accumulation of ceroid-like pigments in macrophages cultured with phosphatidylcholine liposomes in vitro, Biochim. Biophys. Acta in press.

DISCUSSION

KATZ: The main data to support your conclusion that these autofluorescent pigments in the cells were Schiff bases were the effects of pH on the autofluorescence. I think, however, that a lot of different fluorophores can probably have fluorescence quenched either at acid or base pH, and there is not sufficient evidence that these were Schiff base compounds. It is rather a nonspecific criteria.

SHIMASAKI: Yes, our data on the effect of pH may provide an orientation, but may not be sufficient to prove these are Schiff base products.

WITKOP: I think you mentioned that you take macrophages and culture them for a period of time with liposomes, and then you check the accumulation of autofluorescent material. Is that correct?

SHIMASAKI: Yes, that's correct.

WITKOP: Do you have any ultrastructural evidence that the liposomes were actually phagocytized by the macrophages?

SHIMASAKI: We add cholesterol to liposomes, and then when liposomes enter the cell, they are esterified and by determining this, we obtain proof of autophagocytosis.

KATZ: Was there any reason why you did not add phosphatidyl ethanolamine in your liposomes, because this was capable of forming Schiff bases?

SHIMASAKI: I used both phosphatidyl ethanolamine and phosphatidyl serine, but with phosphatidyl serine, the phagocytic activity was much higher than with phosphatidyl ethanolamine.

KATZ: But you were using mixed phospholipids in one of your liposomes that contained phosphatidyl choline, phosphatidyl serine, as well as cholesterol, but you could have just included phosphatidyl ethanolamine. What I am interested to know is whether some of the fluorophores degenerated simply by oxidation of phosphatidyl ethanolamine, so it could be that some of your organic solvent extractable fluorophores were in fact oxidation products of phosphatidyl ethanolamine.

SHIMASAI: Yes, I see the point, but the liposomes with added phosphatidyl ethanolamine were not phagocytized by the macrophages.

MODULATION OF CEROID ACCUMULATION IN MACROPHAGES IN VITRO

Keri L.H. Carpenter, Richard Y. Ball, Nigel P. Carter, Susan E. Woods, Sarah L. Hartley, Sheena Davies, Jacqui H. Enright and Malcolm J. Mitchinson

Department of Pathology
University of Cambridge
Tennis Court Road
Cambridge CB2 1QP U.K.

SUMMARY

Mouse resident peritoneal macrophages (MPM) cultured with artificial lipoprotein consisting of cholesteryl linoleate complexed with bovine serum albumin (CL/BSA) rapidly accumulate ceroid in the form of rings. Experiments with various phenolic radical scavenger antioxidants and derivatives showed that the radical scavengers which are strongly lipophilic, and possess a free (i.e. non-esterified) phenolic hydroxyl group are inhibitors of ceroid ring formation. Time-course experiments with MPM and CL/BSA in which either or both of the components of the artificial lipoprotein have been oxidised before feeding showed that such oxidation accelerated ceroid accumulation, and suggested that oxidation of the lipoprotein is rate-determining in ceroid accumulation. Copper appeared to be a good catalyst for this. Agents able to activate the respiratory burst production of reactive oxygen species appeared to have no accelerating effect on ceroid accumulation from CL/BSA by MPM in a time-course. A novel method has been attempted for quantitating ceroid in MPM by means of its autofluorescence, using a Fluorescence-activated Cell Sorter (FACS). The results from FACS agree qualitatively with those from alcohol-xylene treatment followed by oil-red-o staining (AX/ORO). MPM cultured with CL/BSA for up to 4 days showed a 2.7 - 4.6-fold increase in mean fluorescence (at wavelengths greater than 490nm) over MPM cultured with cholesteryl oleate/BSA (CO/BSA), with CL/BSA/butylated hydroxytoluene (CL/BSA/BHT), with CL/BSA/probucol, and with no artificial lipoprotein. The implications of the findings with respect to human atherosclerosis are discussed.

INTRODUCTION

Ceroid is a constituent of many of the foam cells (i.e. fat-filled cells) in all human atherosclerotic lesions (Mitchinson et al., 1985). The foam cells have been shown to be macrophages (Aqel et al., 1984, 1985; Klurfeld, 1985; Jonasson et al., 1986; Gown et al., 1986) and we have used an in vitro model for foam cells by culturing MPM with artificial lipoproteins consisting of emulsions of lipid with BSA

KEY WORDS: Macrophages, ceroid, artificial lipoproteins, radical scavengers, lipid oxidation, respiratory burst activators, autofluorescence, Fluorescence-activated Cell Sorter.

(Ball et al., 1987a; Carpenter et al., 1988). When the lipid is polyunsaturated (such as CL, or cholesteryl arachidonate, or trilinolein) the MPM accumulate ceroid and soluble lipid, but when the lipid is monounsaturated (such as CO or triolein) the cells accumulate soluble lipid but not ceroid (Ball et al., 1987a; Carpenter et al., 1988). The lipid species we have chosen for our studies on ceroid modulation is CL (a major constituent of serum low density lipoprotein (LDL)). CL/BSA leads to the appearance of ceroid rings within the MPM foam cells resembling both histologically and ultrastructurally the ceroid found in human atherosclerosis (Carpenter et al., 1988; Ball et al., 1988). In order to attempt further to elucidate the mechanisms of ceroid formation we have tested the effects of various substances for their ability to inhibit or accelerate ceroid accumulation in MPM incubated with CL/BSA, using AX/ORO to visualise ceroid and counting the percentage of cells containing ceroid rings. In the course of these studies it became apparent that a more quantitative method for ceroid would be highly desirable, and so we have begun to develop a method, which measures the autofluorescence intensity of ceroid, using a FACS.

Knowledge of substances with the ability to inhibit ceroid formation may be of clinical interest in the field of atherosclerosis for a number of reasons. First, inhibition of ceroid formation may lead to a decrease in production of certain potentially toxic oxidised lipid species that appear to be produced simultaneously with ceroid (Ball et al., 1987b). Secondly, inhibition of ceroidogenesis may prevent the auto-allergic reaction to ceroid which sometimes occurs as a complication of advanced atherosclerosis (Parums et al., 1986). Thirdly, as ceroid is essentially insoluble and can occur as an outer skin trapping a droplet of soluble lipid, inhibition of ceroidogenesis may help prevent the progression of the lesion to irreversibility (Ball et al., 1987b).

Culture conditions for MPM and artificial lipoprotein preparation technique were as described previously (Ball et al., 1987a), except that RPMI 1640 without phenol red was substituted for DMEM with phenol red in the later experiments (see Table 1). The reason for this was that in our experiments on production of soluble oxidised lipids by MPM, RPMI gave a lower, more reproducible background level of lipid oxidation in the absence of cells, and the omission of phenol red gave cleaner lipid extracts.

ATTEMPTS TO INHIBIT CEROID ACCUMULATION

Ceroid accumulation in MPM exposed to CL/BSA for up to 3 days can partially be inhibited by incorporating the phenolic antioxidant radical scavenger BHT into the artificial lipoprotein prior to incubation with the cells (Ball et al., 1987; Carpenter et al., 1988). This abolished the formation of ceroid rings, but by 3 days almost all of the cells contained some ceroid in the form of numerous small granules or a diffuse cytoplasmic blush (staining with AX/ORO). A structurally-related radical scavenger, butylated hydroxyanisole (BHA) had a similar effect. This supported our hypothesis that ceroid formation involved free radicals, which would explain why polyunsaturated lipids lead to ceroid in our system, whereas monounsaturates do not, since the bisallylic methylene groups in polyunsaturates can easily lose a hydrogen atom to produce a delocalised radical, which readily picks up molecular oxygen to form peroxides, then hydroperoxides; these intermediates can then rearrange, fragment, polymerise, or participate in chain reactions with other lipid or protein species. Such reactions could lead to the formation of a complex polymer such as we believe ceroid to be.

We therefore set out to investigate the effects in our in vitro system of several other radical scavengers and derivatives. These compounds were incorporated into the artificial lipoprotein at a molar ratio of 60:1:2.7 CL:BSA:radical scavenger as used for CL/BSA/BHT and CL/BSA/BHA (Ball et al., 1987a), except where stated otherwise. Results are summarised in Table 1, and are described below. It is important to note that some experiments were done only once but with clear results. BHT and BHA are

Table 1 Effects of various substances on ceroid accumulation in mouse peritoneal macrophages, as judged by cell counts after AX/ORO staining.

Addition to medium	Medium	Ceroid rings within cells 1 and 3 days	Effect on time course 0-24h	No. of expts
CL/BSA/BHT	DMEM	-	ND	4
CL/BSA/BHA	DMEM	-	ND	2
CL/BSA/probucol	RPMI	-	ND	2
CL/BSA/propyl gallate	RPMI, DMEM	+	ND	4
CL/BSA/octyl gallate	RPMI	±	ND	2
CL/BSA/lauryl gallate	RPMI	±	ND	2
CL/BSA/α-tocopherol	RPMI	-	ND	4
CL/BSA/α-tocopherol acetate	RPMI	+	ND	1
CL/BSA/α-tocopherol acid succinate	RPMI	+	ND	1
CL/BSA + α-tocopherol phosphoric acid ester disodium salt	DMEM	+	ND	2
O_2-oxCL/BSA	DMEM	+	>	3
CL/Cu^{2+}-oxBSA	DMEM	+	>	2
Cu^{2+}-ox[CL/BSA]	DMEM	+	>	1
CL/BSA + Cu^{2+}	DMEM	+	>	2
γ-IFN 24h, then CL/BSA + γ-IFN	RPMI	+	0	2
PMA 6h, then CL/BSA	DMEM	+	0	1
PMA + A23187 6h, then CL/BSA	DMEM	+	0	1
A23187 6h, then CL/BSA	DMEM	+	0	1
LPS 6h, then CL/BSA	DMEM	+	0	1

Notes: CL = cholesteryl linoleate, BSA = bovine serum albumin, BHT = butylated hydroxytoluene, BHA = butylated hydroxyanisole, γ-IFN = human recombinant γ-interferon 10 000 U/ml, PMA = phorbol myristate acetate 10ng/ml, A23187 = calcium ionophore A23187 0.1 μM, LPS = lipopolysaccharide (from *Salmonella abortus*) 10 μg/ml. In column 3, + denotes ceroid rings in 85% or more of the cells, - denotes ceroid rings in 10% or fewer of the cells, ± denotes partial inhibition of ceroid rings, with diminished staining intensity. In column 4, ND = not determined, > = acceleration of ceroid accumulation, and 0 = no acceleration of ceroid accumulation. Time of addition of CL/BSA = time 0 of time-course. All experiments were concurrent with normal MPM + CL/BSA controls.

antioxidants widely used in the food industry (Pokorny, 1987). Another class of phenolic antioxidant radical scavengers used similarly are the gallates. These include propyl gallate (PG) (3-carbon side chain), octyl gallate (OG) (8-carbon side chain) and lauryl gallate (LG) (12-carbon side chain). Although all are soluble in fats and organic solvents such as acetone, the longer the side-chain, the more lipophilic the gallate (lauryl>octyl>propyl). Propyl gallate is slightly water-soluble, whereas BHT and BHA are insoluble in water. On the basis of their molecular structures, we would expect octyl gallate and lauryl gallate to be more hydrophilic than BHT or BHA. When tested in our MPM-CL/BSA system for up to 3 days, we found that PG was ineffective at inhibiting ceroid ring formation, whereas OG and LG were partially effective (Table 1), with the ceroid rings somewhat less strongly staining (AX/ORO) than for CL/BSA or CL/BSA/PG. However such partial inhibition is extremely difficult to assess visually, especially as the intensity of staining can show some degree of variation even for normal CL/BSA incubations. A more quantitative, less subjective technique for estimating ceroid accumulation is required. These results tentatively suggest that the more lipophilic the radical scavenger, the more effective it is at inhibiting ceroid ring formation. These findings imply that lipid oxidation is the crucial process in ceroidogenesis. This is compatible with our previous work on MPM cultured with lipid-polyaminoacid complexes (Carpenter et al., 1988), where the nature of the lipid, but not of the polyaminoacid, appeared to be the determinant of the formation of ceroid and its morphology.

An antioxidant phenolic radical-scavenging drug which has recently come to the forefront of interest in the field of atherosclerosis is probucol, which structurally resembles two linked molecules of BHT. This drug appears to have anti-atherogenic properties independent of its mild cholesterol-lowering effect in serum, when tested on Watanabe heritable hyperlipidemic (WHHL) rabbits, which are a strain lacking the LDL receptor and which succumb early in life to widespread atherosclerotic lesions (Kita et al., 1987; Carew et al., 1987). These authors attributed probucol's anti-atherogenicity to its radical-scavenging property, as in both humans and rabbits it renders serum LDL less susceptible to oxidation (Kita et al., 1987; Parthasarathy et al., 1986).

We therefore tested probucol in our MPM-CL/BSA system. Probucol was used at half the molar concentration of the other radical scavengers because it contains two phenolic groups per molecule instead of one. Probucol inhibited ceroid ring formation (AX/ORO staining) without visibly diminishing lipid uptake (ORO staining). This finding is the first instance in which a pharmacological agent capable of slowing the progression of experimental atherosclerosis has been shown to impair ceroid formation within macrophages in vitro. In addition we have found that both probucol and BHT almost completely abolished the appearance of cholesterol oxidation products cholest-5-en-3β,7α-diol and cholest-5-en-3β,7β-diol both within the cells and in the extracellular medium in the MPM-CL/BSA system (Carpenter, unpublished results). The radical scavengers were incorporated into the artificial lipoprotein as previously described, prior to feeding. This correlated with the inhibition of ceroid, supporting the idea that ceroid formation and lipid oxidation are related processes.

A well-known naturally-occurring dietary antioxidant is Vitamin E (α-tocopherol). It is practically insoluble in water, and freely soluble in lipid solvents. In the MPM-CL/BSA system, dl-α-tocopherol, when incorporated into the artificial lipoprotein at a molar ratio of 60:1:2.7 CL:BSA:α-tocopherol prior to feeding, led to an inhibition of ceroid ring formation (AX/ORO staining) without visibly affecting lipid uptake (ORO staining). Higher and lower concentrations of α-tocopherol were then tested. A ratio of CL:BSA:α-tocopherol of 60:1:0.27 was ineffective against ceroid ring formation, although the intensity of the staining was somewhat decreased. Ratios of 60:1:27 and 60:1:270 inhibited ceroid ring formation without appearing toxic to the MPM; in addition the numbers of ORO-positive lipid droplets within the cells were considerably diminished. Several derivatives of α-tocopherol proved ineffective against ceroid ring formation (Table 1). α-Tocopherol acetate is practically insoluble in water but freely soluble in lipid solvents. The acetate and acid succinate were incorporated into the artificial lipoprotein prior to feeding as for α-tocopherol, but the phosphoric acid ester disodium salt is readily water-soluble, so was added to the cultures as an aqueous solution. The inactivity of these 3 derivatives implies that a free phenolic group is required for effective radical scavenging, and that hydrolysis does not proceed rapidly enough within the system to de-esterify the tocopherol quickly enough to inhibit ceroid ring formation. In addition, the acid succinate and phosphoric acid ester disodium salt are more hydrophilic derivatives of α-tocopherol, and presumably this hydrophilicity disfavours close association with the very hydrophobic cholesteryl linoleate. This is consistent with the evidence above that the more lipophilic the antioxidant, the more effective it is as an inhibitor of ceroid ring formation, in this system.

ATTEMPTS TO ACCELERATE CEROID ACCUMULATION

Since MPM contain abundant ceroid by just one day of incubation with CL/BSA, we needed to utilise a time-course over the first 24h of exposure, running each "activated" experiment simultaneously with a normal MPM-CL/BSA control. Typically cells were fixed at 2,4,8,12 and 24h and stained with AX/ORO as before.

It has been suggested that ceroid could result from oxidation of lipoprotein particles by the macrophages' microbicidal mechanism, which is triggered by phagocytosis (Mitchinson, 1983). This "respiratory burst" involves the membrane-bound NADPH-oxidase, which catalyses the conversion of oxygen to superoxide radical anion, which then undergoes further reaction to give hydrogen peroxide and hydroxyl radicals (Fantone and Ward, 1982; Babior 1987). γ-interferon (γ-IFN) is now accepted as being "macrophage activation factor" and has numerous well-documented effects on both macrophages and macrophage cell lines (Nathan et al., 1983). γ-IFN is reported to activate NADPH-oxidase in a monocytic cell line (Andrew et al., 1987), and stimulates human monocyte-macrophages in their ability to kill schistosomes (Cottrell et al., 1989). However, in our experiments human recombinant γ-IFN did not appear to accelerate ceroid accumulation in our MPM-CL/BSA 24h time-course. MPM were cultured in the presence of human recombinant γ-IFN, 10 000 units per ml, from the time of plating through to the time of fixing, thus receiving 24h priming before the addition of CL/BSA. Neither did the same γ-IFN appear to have any enhancing effect on the ability of human monocyte-macrophages to accumulate ceroid when cultured with CL/BSA (Carpenter, unpublished observations).

Activation is a term generally applied to macrophages when they show enhanced production of reactive oxygen species, i.e. superoxide radical anion, hydrogen peroxide, hydroxyl radical, and hypochlorous acid (Fantone and Ward, 1982), or enhanced cytotoxicity, and the latter is often expressed as the macrophages' ability to kill, for example, tumour cells. A number of substances besides γ-IFN have been reported to bring about macrophage activation. These include phorbol myristate acetate (PMA) (Johnston et al., 1978; Nathan and Root, 1977), PMA plus calcium ionophore A23187 (Somers et al., 1986), calcium ionophore A23187 (Somers et al., 1986), and bacterial lipopolysaccharide (LPS) (Doe and Hensen, 1978). These substances are all activators of the respiratory burst (Fantone and Ward, 1982), but it is not always known whether increased production of reactive oxygen species is actually responsible for the increased cytotoxicity. PMA or A23187 used separately had no accelerating effect on ceroid accumulation, just as, with the priming conditions used (Table 1), they have been reported to have minimal effect individually on cytotoxicity (Somers et al., 1986). The combination of PMA with A23187 is reported to enhance cytotoxicity (Somers et al., 1986) but, using the same priming conditions, we observed no acceleration of ceroid accumulation (Table 1). Likewise, LPS was reported to increase macrophages' cytotoxicity (Doe and Hensen 1978), but we observed no acceleration of ceroid accumulation (Table 1). The lack of effect of any of the above activators, or of γ-IFN, on ceroid formation in our experiments (Table 1) might imply that the respiratory burst reactive oxygen species are not involved in the rate-determining step in ceroid formation in our system. Such negative results always need interpreting with caution. The concentrations of the activators and the priming regimes used were comparable with those from the literature (Cottrell et al., 1989; Somers et al., 1986; Doe and Hensen, 1978 and refs. therein). Higher levels or longer priming may be needed in our system in order to affect ceroid production. The apparent lack of effect of human recombinant γ-interferon on ceroidogenesis in our MPM-CL/BSA system could be due to a species incompatibility. Thus our suggestion that reactive oxygen species may not be involved in the rate determining step of ceroid formation is as yet only very tentative. However, it is interesting to note that even when LDL is damaged considerably by superoxide and hydroxyl radicals (generated using a cobalt source) this LDL is nevertheless not taken up readily by macrophages (Bedwell et al., 1989), whereas copper-oxidised LDL is avidly taken up leading to production of "foam cells" rich in both soluble lipids and ceroid (Ball et al., 1986).

The suggestion that reactive oxygen species may not be involved in the rate determining step of ceroidogenesis is not incompatible with the finding that ceroidogenesis is impaired by radical scavengers, since the polyunsaturated lipid radicals described earlier do not require reactive oxygen species for their formation. The abstraction of a hydrogen atom from a bisallylic methylene group in a polyunsaturated fatty acid chain could be the rate-determining step, and so that any substance which catalyses this, such as a transition metal, would thus accelerate ceroid formation. In the plant lipoxygenase-catalysed oxidation of linoleic acid to hydroperoxides, the rate-determining step is indeed the abstraction of a hydrogen atom from the bisallylic methylene group of the linoleic acid, and iron plays an essential role in the enzyme's catalytic activity (Vliegenthart, 1979). An alternative explanation of the lack of effect of respiratory burst activators on the rate of ceroid formation could simply be that reactive oxygen species are already produced in excess in the "non-activated" control, triggered by the phagocytosis of CL/BSA particles. Activation of the cell by an agent such as γ-IFN would give an increase in reactive oxygen species, but if this was simply swelling the excess there would be no effect on ceroid formation kinetics.

Thus our attempts to activate MPM failed to accelerate ceroid accumulation over 24h. However, oxidation of CL with oxygen at 37°C prior to emulsifying with BSA (O_2-oxCL/BSA) and feeding to MPM did increase their rate of ceroid accumulation over the same time-course (Ball et al., 1987a). Similarly oxidation of the CL/BSA artificial lipoprotein (using 800μM copper (II) chloride for 6h at room temperature, followed by dialysis to remove copper chloride) (Cu^{2+}-ox[CL/BSA]) prior to incubation with MPM, and oxidation of BSA (using 400 and 800μM copper (II) chloride, conditions as above) prior to emulsifying with non-oxidised CL (CL/Cu^{2+}-oxCL/BSA)and then feeding to MPM both accelerated ceroid accumulation over 24h (Table 1), though CL/Cu^{2+}-oxBSA showed some toxicity to MPM. Addition of copper (II) chloride to the incubation medium at final concentrations of 40μM and 0.01μM accelerated ceroid accumulation in MPM exposed to CL/BSA, and increased the intensity of AX/ORO staining relative to control MPM-CL/BSA incubations with no added copper. In each of the above examples, the time taken for 85% or more of the cells to contain ceroid was reduced to 2/3 or less relative to that of the MPM-CL/BSA controls. This suggests that oxidation of lipoprotein may be rate-limiting for ceroidogenesis.

QUANTITATION OF CEROID ACCUMULATION

The above studies amply illustrate that a major limitation to our work on ceroid modulation has been the inability to measure ceroid accumulation quantitatively. Simply counting the percentage of cells that contain AX/ORO-positive rings is laborious and is a very crude measure of ceroid since it tells us nothing about the amount of ceroid within each cell, and sometimes we have had to qualify the numerical data with qualitative judgements of the relative intensity of staining. Early work attempted to grade ceroid within cells using a scoring system (Ball et al., 1984) but this proved

Table 2 Mean autofluorescence intensity and side scattered light intensity measured by FACS, for mouse peritoneal macrophages cultured for 4 days with and without artificial lipoproteins.

Addition to medium	Harvesting method	FL	$\dfrac{FL}{FL\ no\ adds.}$	SSC	$\dfrac{SSC}{SSC\ no\ adds.}$
CL/BSA	Trypsin/EDTA	37.57	4.63	18.21	2.94
No additions	Trypsin/EDTA	8.12	1.00	6.19	1.00
CO/BSA	Trypsin/EDTA	8.18	1.01	7.08	1.14
CL/BSA	Scrape	30.37	4.12	17.64	2.59
No additions	Scrape	7.37	1.00	6.81	1.00
CO/BSA	Scrape	8.34	1.13	7.14	1.05

Notes: FL = mean autofluorescence intensity per cell in arbitrary units. SSC = mean side-scattered light intensity per cell in arbitrary units. These values were calculated by converting the means in Fig. 1, and the corresponding means for the scraped cells, from a log scale to a linear scale. CL = cholesteryl linoleate. CO = cholesteryl oleate. BSA = bovine serum albumin.

unwieldy and too subjective, and so was abandoned. Thus up to now we have been limited to studying gross effects, and truly quantitative results, which are highly desirable when studying inhibitors and activators, have hitherto not been possible.

We have therefore recently carried out a pilot study of a novel method for ceroid quantitation, using a FACS. The main features of this instrumentation are described by Herzenberg et al., 1976. We have attempted to utilise the yellowish autofluorescence of ceroid (Ball et al., 1987a) as a means of quantitating it in MPM. MPM cultured with lipid/BSA artificial lipoproteins were analyzed using a FACStar Plus (Becton Dickinson Diagnostic Systems) flow cytometer equipped with a Spectra Physics 2025 argon ion laser. The laser was tuned to emit 300W of light in the ultraviolet wavelengths 351.1-363.8nm. Intact cells were identified using forward light scatter, and fluorescence was measured above 490nm using a long-pass filter in front of the detector. Light intensities were measured in arbitrary units.

MPM cultured with CL/BSA for up to 4 days gave a 2.7-4.6 - fold increase in mean fluorescence per cell over MPM cultured with no artificial lipoprotein in the medium (Tables 2 and 3). These FACS data were all obtained using fixed cells. MPM cultured with CO/BSA, CL/BSA/probucol, and CL/BSA/BHT gave the same mean fluorescence per cell as MPM cultured with no artificial lipoprotein (Tables 2 and 3). Examples of fluorescence intensity population distribution curves and additional side-scattered (SSC) light intensity population distribution curves (a measure of cellular granularity) are shown in Fig. 1. In this system, SSC light intensity (which is a measure of "granularity" within the cell) appeared to correlate with fluorescence intensity. MPM incubated with CO/BSA contain abundant lipid droplets visible as phase-bright droplets on phase-contrast microscopy, but give little SSC, in keeping with the non-granular appearance, whereas MPM incubated with CL/BSA appear more dark and granular on phase-contrast microscopy, and give pronounced SSC. The autofluorescence measured by FACS correlates well with AX/ORO staining.

Autofluorescence intensity measurement on FACS promises, from these very preliminary data, to be a suitable method for ceroid quantitation. It has the potential for giving much finer discrimination than simple AX/ORO-positive counts, for quantitating any type or combination of ceroid morphology. It also has the following advantages

Table 3 Effect of radical scavengers on mean autofluorescence intensity per cell, measured by FACS, of mouse peritoneal macrophages cultured with artificial lipoproteins.

Addition to medium	Duration	FL	$\dfrac{\text{FL}}{\text{FL no adds.}}$
CL/BSA	2 days	125.72	2.72
CL/BSA/probucol	2 days	52.48	1.13
CL/BSA/BHT	2 days	49.58	1.07
No additions	2 days	46.27	1.00
CL/BSA	4 days	159.59	2.74
CL/BSA/probucol	4 days	54.29	0.93
CL/BSA/BHT	4 days	51.16	0.88
No additions	4 days	58.33	1.00

Notes: CL = cholesteryl linoleate. BSA = bovine serum albumin. BHT = butylated hydroxytoluene. FL = mean fluorescence intensity per cell in arbitrary units. The 256 detector channels were arranged linearly 0-256. Cells were harvested by scraping. Each FL value is the average FL of 2 simultaneous incubations.

over ordinary spectrofluorimetric measurements. (1) Broken and clumped cells can be easily excluded from FACS results by using appropriate population selection as mentioned above, whereas they would have to be physically removed in order to be excluded from measurement on a spectrofluorimeter. (2) There is no need accurately to know the number of cells in each sample beforehand, as the FACS makes correlated measurements on individual cells. (3) Simple spectrofluorimetry would be difficult to do quantitatively on a cell suspension because of lack of homogeneity of sample. Ceroid itself would be difficult to extract as it is essentially insoluble. (4) The FACS gives the fluorescence intensity distribution within the cell population generated from individual measurements (which is in itself useful information), from which the FACS data system can automatically calculate the mean fluorescence intensity of the population. Ordinary spectrofluorimetry would simply give total intensity of fluorescence, which would have to be divided by number of cells (determined separately using a haemocytometer or Coulter Counter).

FACS is a powerful and sensitive technique for quantitating fluorescence in cells. Its only real limitations are that it cannot be used for scanning (i.e. plotting a spectrum), and that there are restrictions in the choice of wavelengths, governed by the laser (for excitation).

A potential drawback of the technique for ceroid quantitation by FACS is that it may be measuring soluble autofluorescent lipids as well as insoluble autofluorescent ceroid. It would in principle be possible to treat the cells with alcohol-xylene (or some other suitable lipid solvent) before FACS measurements, so obtaining data on insoluble material only. However the co-detection of soluble autofluorescent substances may not necessarily be a disadvantage, since the production of such substances may be intimately linked with ceroidogenesis. Indeed, in our experience of fluorescence microscopy of fixed MPM (prior to AX/ORO treatment) we have never observed significant autofluorescence without subsequent histochemical detection of ceroid.

Autofluorescence   Side-scatter

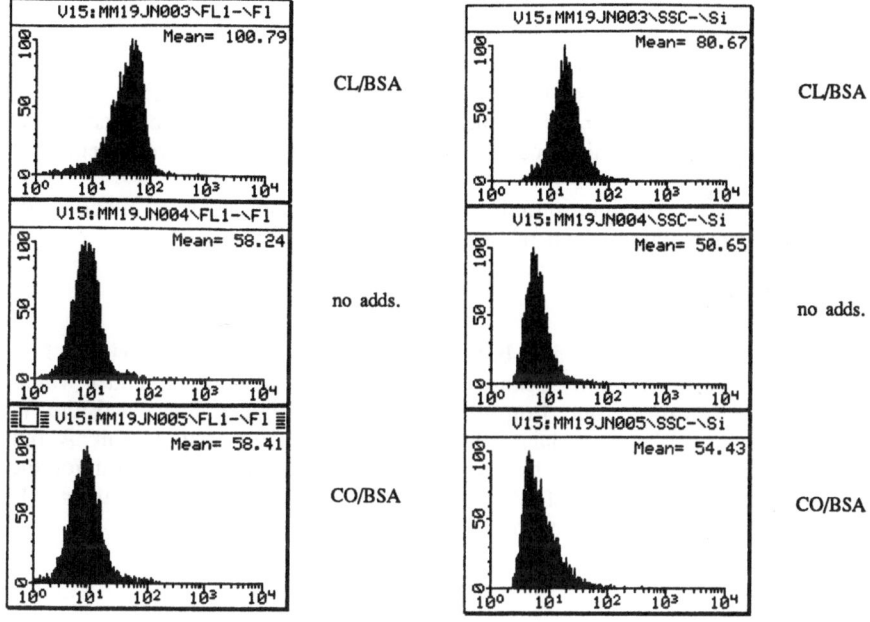

Figure 1   Autofluorescence intensity and side-scattered light intensity population
distributions measured by FACS, for mouse peritoneal macrophages
cultured for 4 days with and without artificial lipoproteins.

Notes: Autofluorescence intensity and side scattered light intensity measured in arbitrary units (x-axis).
y axis = number of cells. The 256 detector channels were arranged logarithmically over 4 decades of
intensity (0 - 10 000). CL = cholesteryl linoleate. CO = cholesteryl oleate. BSA = bovine serum
albumin. No adds. = no artificial lipoprotein. Cells were harvested by trypsinisation. Similar results,
but with rather wider population distributions, were obtained for cells harvested by scraping. Results
are tabulated in Table 2.

CONCLUSIONS

Our previous work on ceroid (Carpenter et al., 1988 and refs. therein) established
a convenient in vitro model, using MPM incubated with lipid/BSA artificial lipoproteins,
and gave evidence that ceroid formation involved free-radical oxidation of
polyunsaturated lipids. The present studies have given further insight into ceroidogenesis.

More evidence has been found that lipid oxidation is the key to ceroid
formation, as the more strongly lipophilic radical scavengers appear to be more
effective than the less lipophilic ones at inhibiting ceroid. One of these strongly
lipophilic radical scavengers which inhibited ceroid was the drug probucol, which is
antiatherogenic in an animal model (Kita et al., 1987; Carew et al., 1987).

It has previously been suggested that ceroid in atherosclerotic foam cells arises
as a result of the macrophages' oxidative microbicidal mechanism being triggered by
phagocytosis of lipoprotein particles (Mitchinson, 1983). We found no evidence,
however, that agents capable of activating the respiratory burst accelerate ceroid
accumulation in our MPM-CL/BSA incubations.   This does not mean that reactive
oxygen species play no role in ceroidogenesis, but it might tentatively suggest that they
are not involved in the rate-determining step. Alternatively, reactive oxygen species
might be involved in the rate-determining step but might already be produced in excess

in the non-activated MPM-CL/BSA control, so that any further increase in their production as a result of an activator such as γ-IFN would have no effect on the rate of ceroid formation.

Acceleration of ceroid accumulation in the MPM-CL/BSA system can be achieved by oxidising either or both of the components of the artificial lipoprotein prior to incubation with MPM, or by supplementing the medium with copper. These results imply that lipoprotein oxidation may be the rate limiting step in ceroidogenesis.

The FACS technique promises to improve subsequent estimations of ceroid accumulation and the assessment of its inhibitors.

ACKNOWLEDGEMENTS.

K.L.H.C. and J.H.E. are both supported by the British Heart Foundation. K.L.H.C. and R.Y.B. gratefully acknowledge past financial support, under which part of this research was carried out, from the John Lucas Walker Fund and the Wellcome Trust respectively. We also thank the East Anglian Regional Health Authority for financial support, Mr. B. Potter for technical assistance, Dr. B.J. Cottrell for the gift of γ-IFN, and Dr. R. Rosenberg of Merrell Dow Pharmaceuticals Ltd., Staines, Middlesex, U.K. for providing us with probucol.

REFERENCES

Andrew, P. W., Robertson, A. K., Lowrie, D. B., Cross, A. R., and Jones, O. T. G., 1987, Induction of the synthesis of the components of the hydrogen peroxide-generating oxidase activation of the human monocytic cell line by interferon-G, Biochem. J., 248:281.

Aqel, N. M., Ball, R. Y., Waldmann, H., and Mitchinson, M. J., 1984, Monocytic origin of foam cells in human atherosclerotic plaques, Atherosclerosis, 53:265.

Aqel, N. M., Ball, R. Y., Waldmann, H., and Mitchinson, M. J., 1985, Identification of macrophages and smooth muscle cells in human atherosclerosis using monoclonal antibodies, J. Pathol., 146:197.

Babior, B. M., The respiratory burst oxidase, 1987, Trends Biochem. Sci., 12:241.

Ball, R. Y., Brodley, H., Brooks, P.N., and Mitchinson, M. J., 1984, The production of ceroid by mouse peritoneal macrophages in vitro, Br. J. exp. Pathol., 65:719.

Ball, R. Y., Bindman, J. P., Carpenter, K. L. H., and Mitchinson, M. J., 1986, Oxidised low density lipoprotein induces ceroid accumulation by murine peritoneal macrophages in vitro, Atherosclerosis, 60:173.

Ball, R. Y., Carpenter, K. L. H., Enright, J. H., Hartley, S. L., and Mitchinson, M. J., 1987a, Ceroid accumulation by murine peritoneal macrophages exposed to artificial lipoproteins, Br. J. exp. Pathol., 68:427.

Ball, R. Y., Carpenter, K. L. H., and Mitchinson, M. J., 1987b, What is the significance of ceroid in human atherosclerosis?, Arch. Path. Lab. Med., 111:1134.

Ball, R. Y., Carpenter, K. L. H., and Mitchinson, M. J., 1988, Ceroid accumulation by murine peritoneal macrophages exposed to artificial lipoproteins: ultrastructural observations, Br. J. exp. Pathol., 69:43.

Bedwell, S., Dean, R. T., and Jessup, W., 1989, The action of oxygen-centred free radicals on human LDL, Biochem. J., 262:707.

Carew, T. E., Schwenke, D.C., and Steinberg, D., 1987, Antiatherogenic effect of probucol is unrelated to its hypocholesterolemic effect: evidence that antioxidants in vivo can selectively inhibit LDL degradation in macrophage-rich fatty streaks and slow the progression of atherosclerosis in the WHHL rabbit, Proc. Natl. Acad. Sci. U.S.A., 84:7725.

Carpenter, K. L. H., Ball, R. Y., Ardeshna, K. M., Bindman, J. P., Enright, J. H., Hartley, S. L., Nicholson, S., and Mitchinson, M.J., 1988, Production of ceroid and oxidised lipids by macrophages in vitro, in: "Lipofuscin - 1987: State of the Art," I. Zs.-Nagy, ed., Akademiai Kiado, Budapest and Elsevier Science Publishers, Amsterdam.

Cottrell, B., Pye, C., and Butterworth, A., 1989, Cytotoxic effects in vitro of human monocytes and macrophages on schistosomula of *Schistosoma mansoni*, Parasite Immunol., 11:91.

Doe, W. F., and Hensen, P. M., 1978, Macrophage stimulation by bacterial lipopolysaccharides I. Cytolytic effect on tumour target cells, J. Exp. Med., 148:544.

Gown, A. M., Tsukada T., and Ross R., 1986, Human atherosclerosis II. Immunocytochemical analysis of the cellular composition of human atherosclerotic lesions, Am. J. Pathol., 125:191.

Fantone, J. C., and Ward, P. A., 1982, Role of oxygen-derived free radicals and metabolites in leucocyte-dependent inflammatory reactions, Amer. J. Pathol., 107:397.

Herzenberg, L. A., Sweet, R. G., and Herzenberg, L. A., (1976), Fluorescence-activated cell sorting, Scientific American 234:108.

Johnston, R. B., Godzik, C. A., and Cohn, Z. A., 1978, Increased superoxide anion production by immunologically activated and chemically elicited macrophages. J. Exp. Med., 148:115.

Jonasson, L., Holm, J., Skalli, I., Bondjers, G., and Hansson G. K., 1986, Regional accumulation of T cells, macrophages and smooth muscle cells in the human atherosclerotic plaque, Arteriosclerosis, 6:131.

Kita, T., Nagano, Y., Yokode, M., Ishii, K., Kumi, N., Ooshima, A., Yoshida, H., and Kawai, C., 1987, Probucol prevents the progression of atherosclerosis in Watanabe heritable hyperlipidemic rabbit, an animal model for familial hypercholesterolemia, Proc. Natl. Acad. Sci. U.S.A., 84:5928.

Klurfeld, D. M., 1985, Identification of foam cells in human atherosclerotic lesions as macrophages using monoclonal antibodies, Arch. Path. Lab. Med., 109:445.

Mitchinson, M. J., 1982, Insoluble lipids in human atherosclerotic plaques, Atherosclerosis, 45:11.

Mitchinson M. J., 1983, Macrophages, oxidised lipids and atherosclerosis, Med. Hypotheses, 12:171.

Mitchinson, M. J., Hothersall, D. C., Brooks, P. N., and de Burbure, C. Y., 1985, The distribution of ceroid in human atherosclerosis, J. Pathol., 145:177.

Nathan, C. F., and Root, R. K., 1977, Hydrogen peroxide release from mouse peritoneal macrophages, J. Exp. Med., 146:1648.

Nathan, C.F., Murray, H. W., Wiebe, M. E., and Rubin, B. Y., 1983, Identification of interferon-gamma as the lymphokine that activates human macrophage oxidative metabolism and antimicrobial activity, J. Exp. Med., 158:670.

Parthasarathy, S., Young, S. G., Witztum, J. L., Pittmann, R. C., and Steinberg, D., 1986, Probucol inhibits oxidative modification of LDL, J. Clin. Invest., 77:641.

Parums, D. V., Chadwick, D. R., and Mitchinson, M.J., 1986, The localisation of immunoglobulin G in chronic periaortitis, Atherosclerosis, 61:117.

Pokorny, J., 1987, Major factors affecting the autoxidation of lipids, in: "Autoxidation of Unsaturated Lipids," H. W-S. Chan, ed., Academic Press Ltd., London.

Somers, S. D., Weiel, J. E., Hamilton, T. A., and Adams, D. O., 1986, Phorbol esters and calcium ionophore can prime murine peritoneal macrophages for tumour destruction, J. Immunol., 136:4199.

Vliegenthart, J. F. G., 1979, Enzymic and non-enzymic oxidation of polyunsaturated fatty acids, Chem. Ind., 241.

FLUORESCENT AND CROSS-LINKED PROTEINS FORMED BY FREE RADICAL AND ALDEHYDE SPECIES GENERATED DURING LIPID OXIDATION

Kiyomi Kikugawa, Tetsuta Kato, Masatoshi Beppu
and Akira Hayasaka

Tokyo College of Pharmacy
1432-1 Horinouchi, Hachioji
Tokyo 192-03, JAPAN

SUMMARY

One of the mechanisms for the formation of lipofuscin-like fluorescent substances is considered to be related to lipid oxidation of tissues. Induction of lipid oxidation in tissues or cells produces cross-links and borohydride-reducible functions together with fluorescence in proteins. In order to elucidate the structures of fluorophores, cross-links and borohydride-reducible functions produced in proteins by lipid oxidation, the reactions of a lipid peroxy free radical with amino acids and proteins, and those of an aldehyde with primary amines were investigated. We demonstrated here two possible types of the reactions that produce the modified proteins. A peroxy free radical generated during lipid oxidation may attack the tyrosine residue in proteins to form the tyrosine radical which may be in turn dimerized into fluorescent and cross-linked tyrosine dimer. Aldehyde species formed by degradation of the peroxy free radical may be polymerized into dimer, trimer, tetramer and so on, which may react with the amino groups of protein to produce fluorescence, cross-links and borohydride-reducible functions. Cross-links can be produced by the formation of Schiff base between the tetrameric dialdehyde and the amino groups of proteins.

INTRODUCTION

Lipofuscin-like fluorescent substances increase with age of tissues or cells. One of the mechanisms for the formation of these substances is considered to be related to lipid oxidation of tissues, since it has been shown that incubation of microsomes, mitochondria and lysosomes of rat liver under aerobic conditions produce characteristic fluorescent substances[1]. Several mechanisms for the formation of these fluorescent substances with respect to lipid oxidation have been proposed[2-4]. Induction of lipid oxidation in tissues or cells produces cross-links and borohydride-reducible functions together with fluorescence in proteins[2-4].

Unsaturated fatty acids undergo oxidation in the presence of oxygen

KEY WORDS: Lipofuscin, lipid oxidation, peroxy free radical, aldehyde, fluorescence, cross-link, borohydride-reducible function

via free radical initiation, propagation and termination[5]. Initial products are lipid hydroperoxides. Decomposition of the hydroperoxides proceeds by the formation of alkoxy or peroxy free radicals. These radicals undergo carbon-carbon cleavage to form breakdown products including many aldehyde species. Lipid hydroperoxides produce fluorescence and cross-links by reaction with proteins, and the fluorescence characteristics of the products are similar to those of lipofuscin-like substances. Malonaldehyde that may be generated during lipid oxidation produces fluorescence, cross-links and borohydride-reducible functions in proteins[2], but the fluorescence characteristics of the products are not always similar to those of lipofuscin-like substances[3,4]. Hence, the reaction of malonaldehyde is of little significance in the formation of lipofuscin-like fluorescent substances. It remains unclear what kinds of reaction are involved in the formation of fluorescent and cross-linked proteins with borohydride-reducible functions during lipid oxidation.

We will report here two possible reactions that produce fluorescence, cross-links and borohydride-reducible functions in proteins. One is the reaction induced by lipid peroxy free racicals and, the other is the reaction of aldehyde species other than malonaldehyde.

RESULTS

Formation of fluorescence, cross-links and borohydride-reducible functions in proteins by reaction with a lipid peroxy free radical

It is known that lipid hydroperoxides breakdown into the corresponding peroxy free radicals by interaction with ferric complexes[6]. 13-Hydroperoxylinoleic acid (13-LOOH) may produce its peroxy free radical by interaction with methemoglobin as shown below.

$$13\text{-LOOH} + \text{Methemoglobin (Fe}^{3+})$$
$$\downarrow$$
$$13\text{-LOO}\cdot + H^+ + \text{Hemoglobin (Fe}^{2+})$$

Formation of fluorescence in the reaction of several amino acids with 13-LOOH in the presence of methemoglobin was investigated. A mixture of 10 mM each amino acid, 5 mM 13-LOOH and 50 µM methemoglobin in 0.2 M borate buffer (pH 9.5) was incubated at 37°C for 24 hr (Table 1). Control reaction mixture without amino acids produced a signigicant amount of fluorescence probably due to the reaction of the peroxy radical with the globin moiety of methemoglobin. The reaction mixtures with glycine, lysine, arginine, histidine and phenylalanine produced much higher fluorescence with excitation maxima at 345-370 nm and emission maxima at 420-435 nm, which was decreased to the control level on treatment with borohydride. The reaction mixture with methionine produced no increased fluorescence. The reaction mixture with cystein produced much lower fluorescence than the control, indicating that sulfhydryl group scavenged the free radical and thus suppressed the formation of fluorescence of the control. The reaction mixture with tryptophan showed its intrinsic high fluorescence and did not reveal newly formed fluorescence. It is noted that the reaction mixture with tyrosine showed much higher fluorescence than that formed in the reaction mixtures with other amino acids. Fluorescence spectra of the reaction mixture with tyrosine showed an excitation maximum at 298 nm and an emission maximum at 388 nm; the wavelengths being rather shorter than those observed with the reaction mixtures with other amino acids. The fluorescence could not be destroyed on treatment with borohydride. Similar results were obtained with the reactions performed at pH 7.5.

When the reaction with the amino acids with α-amino groups blocked by carbobenzoxy (Z) was performed similarly, the reaction mixtures with Nα-Z-lysine and Z-tyrosine produced significant fluorescence and the other derivatives did not (Table 1). The fluorescence from Z-tyrosine was not destroyed but that from Nα-Z-lysine was destroyed on treatment with borohydride. Hence, it is evident that two types of fluorescence were produced in the reaction of amino acids with the peroxy free racical: one is the fluorescence from α-amino groups of various amino acids and ε-amino group of lysine which was destroyed on treatment with borohydride, and the other is the fluorescence that from phenolic group of tyrosine which was not destroyed on treatment with borohydride.

A major fluorescent reaction product of tyrosine with the peroxy free radical was isolated by use of high pressure liquid chromatography (HPLC). The retention time of the product was identical with that of authentic tyrosine dimer (Fig. 1) prepared by the reaction of tyrosine with hydrogen peroxide and horseradish peroxidase[7-9]. Ultraviolet absorption and fluorescence spectra of the product were coincided with those of tyrosine dimer. Fluorescence spectra of tyrosine dimer were dependent on the solvent employed (Fig. 2). At neutral pH (phosphate buffer) it showed excitation maximum at 320 nm and an emission maximum at 412 nm with relative molar intensity of 20% of quinine sulfate. At alkaline pH (carbonate buffer and sodium hydroxide solution) it showed similar spectra

Table 1. Relative Fluorescence Intensity of 13-LOOH/Methemoglobin-Amino Acid Reaction Products

Mixture	Fluorescence of Reaction mixture			Fluorescence of reaction mixture after borohydride-reduction		
	Ex max (nm)	Em max (nm)	Relative fluorescence intensity	Ex max (nm)	Em max (nm)	Relative fluorescence intensity
Control	355	432	24	349	427	29
Glycine	352	430	53	341	420	29
Lysine	353	430	59	348	425	33
Arginine	355	432	37	345	425	26
Histidine	367	423	57	356	428	29
Cystein	348	438	4	334	415	6
Methionine	354	435	24	345	425	25
Phenylalanine	356	430	46	350	428	28
Tryptophan	283	362	8244	283	363	6617
Tyrosine	298	388	399	298	386	385
Control	354	430	27	347	425	26
Nα-Z-Lysine	351	428	69	346	422	36
Z-Arginine	355	430	20	347	426	21
Z-Histidine	354	430	20	349	427	23
Z-Phenylalanine	353	430	26	347	424	25
Z-Tyrosine	297	384	591	298	384	595

A mixture of 10 mM each amino acid, 5 mM 13-LOOH and 50 μM methemoglobin in 0.2 M sodium borate buffer (pH 9.5) was incubated at 37°C for 24 hr. The reaction mixture was treated with 88 mM sodium borohydride. Fluorescence spectra and fluorescence intensities were measured after dilution into the same buffer. Relative fluorescence intensities were obtained against 0.1 μM quinine sulfate in 0.1 N sulfuric acid. Z: carbobenzoxy.

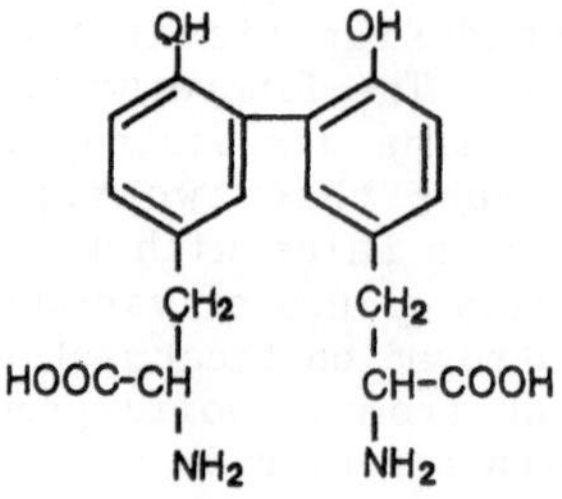

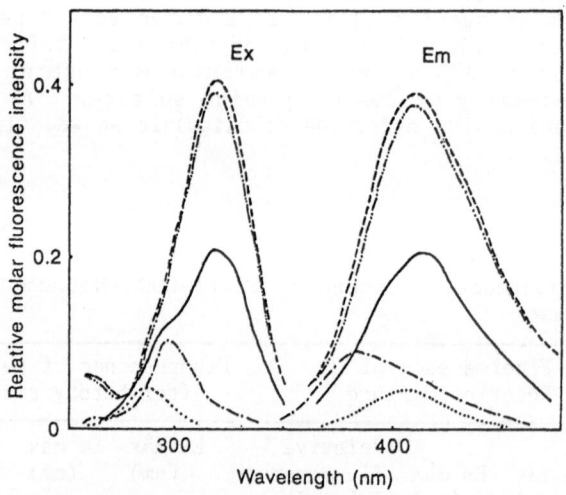

Fig. 1. Tyrosine dimer: fluorescent compound derived from the reaction of tyrosine and 13-LOOH/methemoglobin.

Fig. 2. Fluorescence spectra of tyrosine dimer.
Spectra were taken in 0.1 M phosphate buffer (pH 7.0)(————), 0.2 M carbonate buffer (pH 9.5)(— — — —), 0.2 M borate buffer (pH 9.5)(—·—·—), NaOH solution (pH 10.0)(—··——··—), and HCl solution (pH 2.0)(·········). Molar intensity relative to quinine sulfate is expressed.

with much higher intensity. In the borate buffer, it showed maximum excitation and emission with shorter wavelengths and much lower intensity probably owing to the chelation of borate with amino groups. At acidic pH (hydrochloric acid solution), its fluorescence intensity was much lowered. The compound was not destroyed on treatment with borohydride.

The reaction of α-crystallin from bovine eye lens with the peroxy free radical was performed. A mixture of 0.25 mM crystallin, 11 mM 13-LOOH and 1.3 uM methemoglobin was incubated at 37°C for 24 hr in 0.2 M borate buffer. SDS-polyacrylamide gel electrophoresis of the reaction mixture revealed that high-molecular weight polymers were produced, and thus cross-links were produced in the protein. Fluorescence spectrum of the reaction mixture showed an excitation maximum at 348 nm and an emission maximum at 419 nm (Fig. 3A). On reduction with borohydride, fluorescence intensity

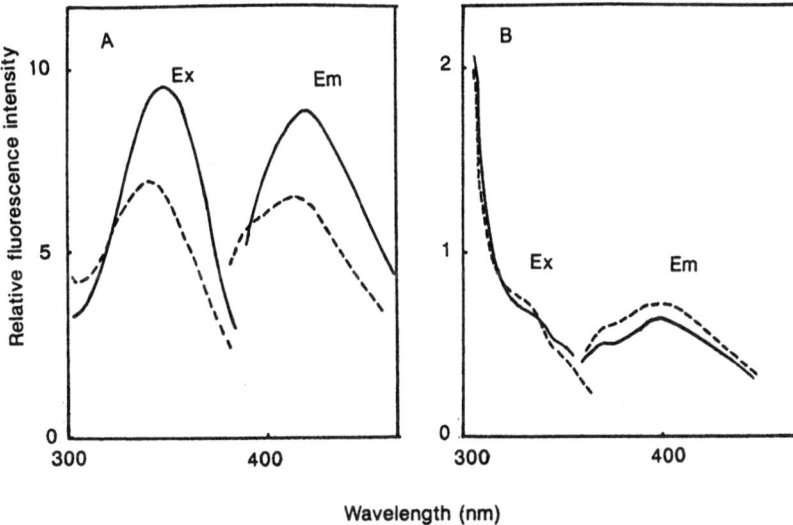

Fig. 3. Fluorescence spectra of the reaction mixtures of A: 0.25 mM α-crystallin, 11 mM 13-LOOH and 1.3 μM methamoglobin and B: 0.25 mM α-crystalline, 59 mM H_2O_2 and 2.5 μM horseradish peroxidase, at 37°C for 48 hr in 0.2 M borate buffer. Spectra were taken before (———) and after (— — —) treatment with 5.0 mM sodium borohydride. Fluorescence intensity relative to 0.1 μM quinine sulfate is shown.

of the reaction mixture was reduced to 76% of the initial one but the excitation and emission maxima remained little changed. The reaction mixture was dialyzed against water, and subjected to acid hydrolysis with 6 N HCl for amino acid analysis. The fluorescence spectrum of the hydrolysate in carbonate buffer showed an excitation maximum at 326 nm and an emission maximum at 415 nm. The fluorescence lost on treatment with borohydride may be due to the fluorophores generated at the 𝛿-amino groups of lysyl residues and α-amino groups of the N-terminal amino acid. The fluorescence resistant to borohydride may be due to the formation of tyrosine dimer, whereas the spectrum was slightly different from that of tyrosine dimer.

Bondaness and coworkers[10,11] treated crystallin with hydrogen peroxide in the presence of heme peptide from cytochrome C and demonstrated that fluorescence and cross-links due to tyrosine residue developed. It has been shown that hydrogen peroxide generates superoxide anion O_2^-, hydrogen peroxy radical $HO_2\cdot$ and hydroxyl radical $OH\cdot$.[12]. A mixture of 0.25 mM crystallin, 59 mM hydrogen peroxide and 2.5 μM horseradish peroxidase were treated at 37°C for 24 hr in 0.2 M borate buffer. Cross-links were formed as analyzed by SDS-polyacrylamide gel electrophoresis. Fluorescence spectrum of the reaction mixture revealed an excitation maximum at 325 nm and an emission maximum at 400 nm (Fig. 3B). On treatment with borohydride, the fluorescence of the reaction mixture did not alter significantly. The fluorescence and cross-links may be due to tyrosine dimer.

Amino acid analysis on amino acid analyzer and HPLC of the acid hydrolysates of borohydride-treated reaction mixtures of 13-LOOH- and hydrogen peroxide-modified crystallin revealed that both the reactions produced tyrosine dimer in the protein molecules (Fig. 4). Its contents were estimated to be 232 pmol/mg protein (amino acid analyzer) or 255 pmol (HPLC) for 13-LOOH-modified crystallin and 59 pmol (amino acid analyzer) or 61 pmol (HPLC) for hydrogen peroxide-modified crystallin.

When human serum γ-globlin and bovine serum albumin were treated with 13-LOOH/methemoglobin and hydrogen peroxide/hoseradish peroxidase, both the proteins produced fluorescence and cross-links by these agents. Amino acid analysis indicated that significant amount of tyrosine dimer was produced in these modified proteins.

The peroxy free radical from 13-LOOH may react with proteins to form fluorescence and cross-links due to the formation of tyrosine dimer. However, another kinds of fluorescence, cross-links and borohydride-reducible functions may be produced due to the reaction of ε-amino groups of lysine residues and α-amino groups of N-terminal amino acid residues of proteins. The relevance of amino groups of lysyl residues to the formation of fluorescence and cross-links has been well documented[13]. Thus, the reaction of polylysine with 13-LOOH gave two fluorescent peptides, one is a monomer and the other is a di- or trimer of the peptide. This reaction may not be due to the radical but to its degradation products such as aldehyde species.

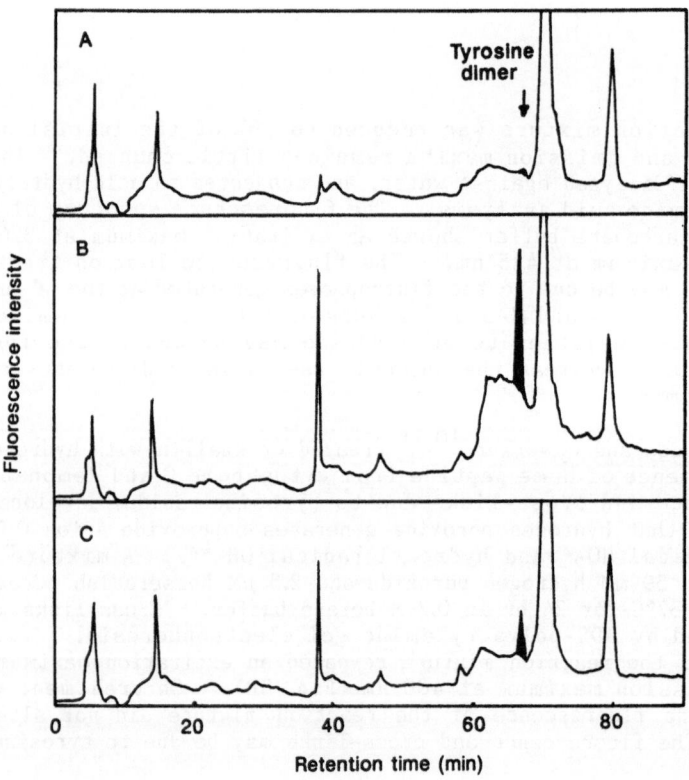

Fig. 4. Analysis of the acid hydrolysate of the borohydride-treated crystallin by use of an amino acid analyzer. A: control, B: 13-LOOH/ methemoglobin-modified and C: H_2O_2/horseradish peroxidase-modified crystallin.

Possible formation of fluorescence, cross-links and borohydride-reducible functions in proteins by reaction with aldehyde species

Various aldehyde species other than malonaldehyde are considered to be more important products to induce the protein damage[3,4]. We have shown that various monofunctional aldehydes produce fluorescence, cross-links and borohydride-reducible functions in proteins[13-16]. Since polylysine undergoes the damage with formation of fluorescence and cross-links by reaction with monofunctional aldehydes[13], the reaction of amino groups of proteins with these aldehydes may play an important role in the protein damage. Hence, we investigated what components are produced by the reaction of primary amines and monofunctional aldehydes in order to elucidate the reaction mechanisms for the formation of fluorescence and cross-links in proteins by these monofunctional aldehydes.

We have previously shown that the reaction of 1-butanal with methylamine gave 2-ethyl-2-hexenal as a result of aldol condensation of the aldehyde, and that several fluorophores were produced by the reaction of 1-butanal and 2-ethyl-2-hexenal with methylamine[17]. One of the fluorescent substances showed an excitation maximum at ~360 nm and an emission maximum at ~430 nm. The fluorophore was not destroyed on treatment with borohydride. This time, we attempted to elucidate the structures of the fluorescent products and characterize the reaction products.

A mixture of 200 mM 1-butanal and 100 mM methylamine in 70% methanol/ 0.1 M phosphate buffer (pH 7) was incubated at 37°C for 48 hr. On treatment with borohydride, the mixture showed fluorescence with shorter excitation at 345 and emission maxima at 408 nm. It was found that four fluorescent products (1-4) were formed in the reaction as analyzed by HPLC (Fig. 5), two (1 and 2) of which remained unchanged, and two (3 and 4) of which were converted into another fluorescent products (R1 and R2) on treatment with borohydride. Major fluorescent compound R1 produced on treatment with borohydride was isolated and was identified as 2,4-diethyl-2,5-dihydrofuran (Fig. 6). It exhibited fluorescence with an excitation maximum at 345 nm and an emission maximum at 406 nm with 1.0% molar fluorescence intensity relative to quinine sulfate. The product may be formed

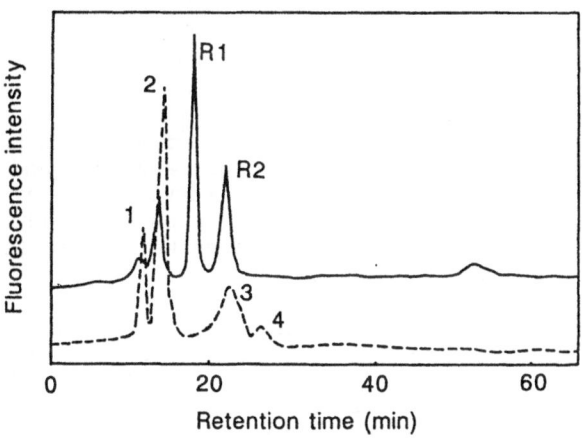

Fig. 5. HPLC of the reaction mixture of 200 mM 1-butanal with 100 mM methylamine in 0.1 M phosphate buffer (pH 7.0) containing 70% methanol at 37 C for 48 hr. The mixture before (-----) and after (——) treatment with borohydride was subjected to an Inertsil ODS column and eluted with methanol/0.01 M triethylamine bicarbonate (pH 7.6) (1:1).

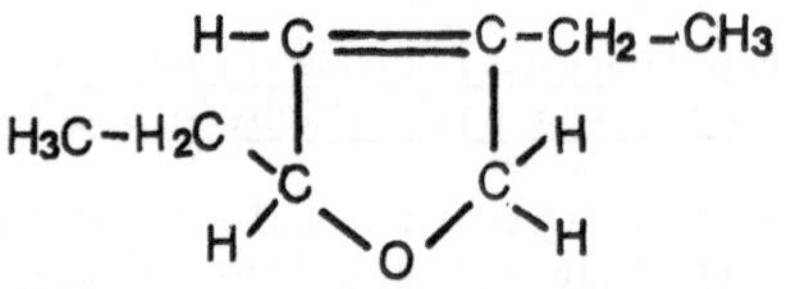

Fig. 6. 2,4-Diethyl-2,5-dihydrofuran (R_1): Fluorescent compound derived from the reaction of 1-butanal and methylamine and subsequent treatment with borohydride.

by condensation of 2 molecules of 1-butanal in the presence of methylamine and subsequent borohydride treatment. It is interesting to note that a self-condensation product of the aldehyde exhibited fluorescence whose spectrum was similar to those of the lipid hydroperoxide-modified proteins. However, the product may be specific to the reaction of methylamine, since the reactions of 1-hexylamine and benzylamine with 1-butanal did not produce the 2,5-dihydrofuran but produced other fluorescent products whose fluorescence was lost on treatment with borohydride. Non-fluorescent products composed of 2:1, 3:1 and 4:1 of 1-butanal and methylamine were isolated from the reaction mixtures of methylamine and 1-butanal, indicating that 1-butanal readily condensed into high molecular weight substances under the reaction conditions. Formation of various fluorescent products may involve self-condensation of 1-butanal in the presence of methylamine.

We have then investigated the reaction of 1-butanal and benzylamine. A mixture of 200 mM 1-butanal and 100 mM benzylamine (or 100 mM 2-ethyl-2-hexenal) in 90% methanol/0.1 M phosphate buffer (pH 7) was incubated at 37° C for 48 hr. Fluorescence spectra of the reaction mixtures showed excitation maxima at $\sim$ 365 nm and emission maxima at $\sim$ 400 nm. HPLC of the reaction mixture revealed a fluorescent peak and a large number of ultraviolet-absorbing peaks. Any attempts to isolate the fluorescent substances were unsucessful. HPLC of the reaction mixture after treatment with borohydride indicated that the fluorescent peak completely disappeared but most of the ultraviolet-absorbing peaks remained unchanged.

Total ion monitoring in GC-MS analysis of the chloroform extracts of the borohydride-treated reaction mixtures indicated the presence of various products (Fig. 7). The ion peak at a retention time of 1.1 min showed the spectrum corresponding to 2-ethyl-2-hexenol (BR_1). The ion peak at a retention time of 3.3 min showed the spectrum of N-butylbenzylamine (BR_2), the reduction product of the Schiff base of 1-butanal and benzylamine. The ion peak at 9.1 min showed the spectrum corresponding to N-(2-ethyl-2-hexyl)benzylamine (BR_3), the reduction product of the Schiff base of 2-ethyl-2-hexenal and benzylamine. These products must be the major borohydride-reducible condensation products that suffer borohydride reduction.

Among the many other products appeared on GC-MS analysis, compounds BR_4, BR_5 and BR_6 were purified by preparative HPLC. Mass spectra of these compounds showed the same molecular ion peak at 430 m/z. High resolution mass spectra indicated the formulae of these compounds were identical and found to be $C_{30}H_{42}N_2$. These compounds were also observed in the reaction mixtures before reduction with borohydride, and thus they were not the compounds produced on borohydride-treatment. The contents of carbon and

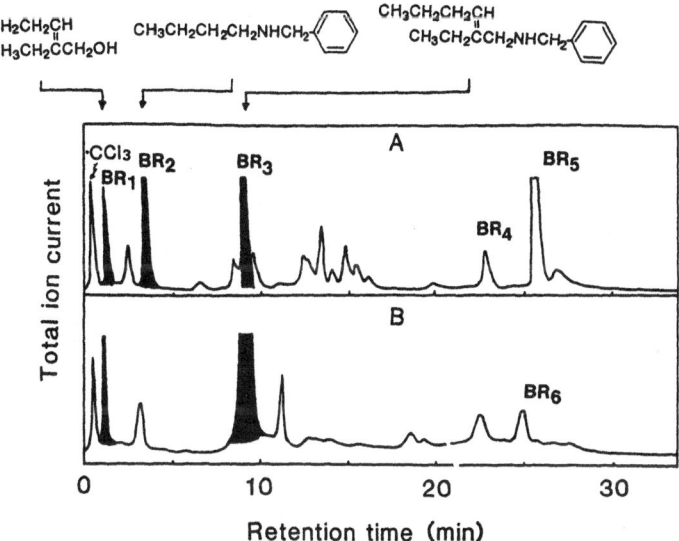

Fig. 7. GC-MS of the chloroform extract of the borohydride-treated reaction mixture of 200 mM 1-butanal (A) and 100 mM 2-ethyl-2-hexenal (B) with 100 mM benzylamine at 37°C for 48 hr in 0.1 M phosphate buffer (pH 7.0) containing 90% methanol.

The chloroform extract was subjected to GC-MS with a Silicone OV-101 column and helium as a carrier gas.

nitrogen atoms suggested that these compounds were derived from 4 molecules of 1-butanal (or 2 molecules of 2-ethyl-2-hexenal) and 2 molecules of benzylamine.

The reaction may proceed as illustrated in Fig. 8; two molecules of 1-butanal were condensed by aldol condensation and dehydrated to produce 2-ethyl-2-hexenal, and the dimer aldehyde may undergo Michael addition with one molecule of 1-butanal and subsequent aldol condensation with one molecule of 1-butanal to form the tetrameric dialdehyde. Alternatively, 2-ethyl-2-hexenal may be converted into the tetrameric dialdehyde by Michael addition to itself. The formulae of compounds BR_4, BR_5 and BR_6 were consistent with the Schiff base structure of the tetrameric dialdehyde and benzylamine as shown in Fig. 8, but they could not be reduced on treatment with borohydride, and their NMR spectra did not supported the structure. NMR spectral analysis of the compounds suggested the structures are the cyclized forms as shown in Fig. 9. Formation of these compounds strongly supported the idea that the reaction of 1-butanal and benzylamine proceeded via the formation of the tetrameric dialdehyde shown in Fig. 8.

Formation of the tetrameric dialdehyde from 1-butanal under mild conditions has an important significance in the protein cross-linking by 1-butanal and related aldehydes. The formation of the tetrameric dialdehyde can be induced in the presence of primary amines, amino acids and proteins.

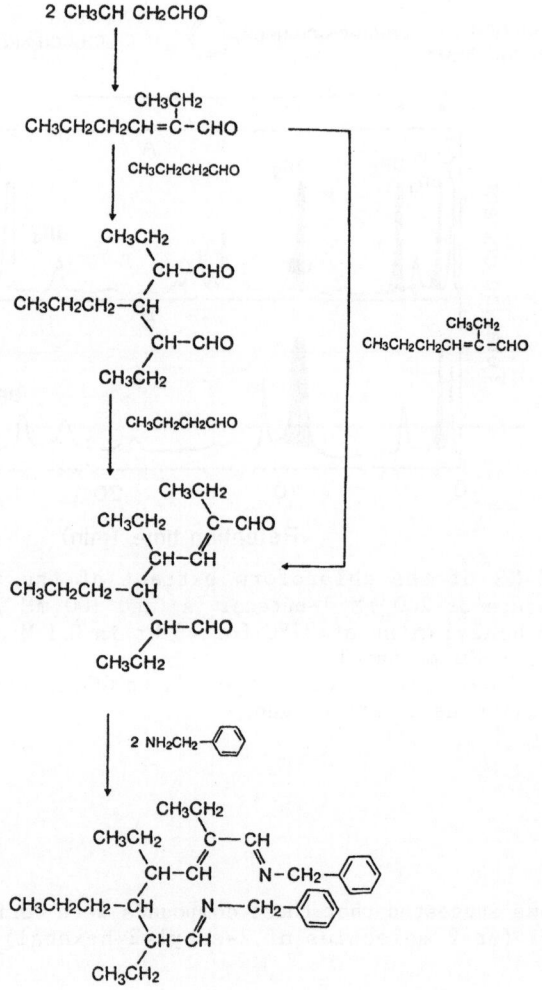

Fig. 8.   Possible reaction scheme of 1-butanal and benzylamine.

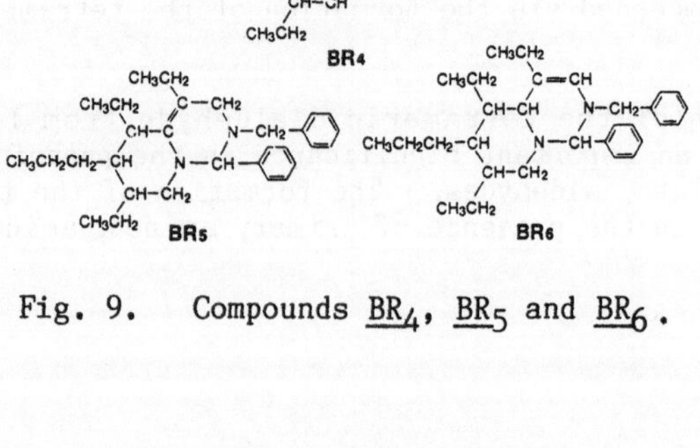

Fig. 9.   Compounds $\underline{BR}_4$, $\underline{BR}_5$ and $\underline{BR}_6$.

In the reaction of proteins with 1-butanal, the tetrameric dialdehyde may be initially produced by the catalytic action of proteins, and the tetrameric dialdehyde may produce unstable Schiff bases between the amino groups of the protein as illustrated in Fig. 10. The Schiff bases may be transformed into stable adducts similar to compounds $\underline{BR_4}$, $\underline{BR_5}$ and $\underline{BR_6}$. Thus, the stable cross-links can be formed in the proteins.

From the model experiments on the reactions of 1-butanal with primary amines, mechanisms for the reaction between proteins with 1-butanal were suggested. Major borohydride-reducible functions produced in the proteins may be the Schihff bases between amino groups of proteins and 1-butanal or its polymer aldehydes. The fluorophores formed in the protein molecules have not yet been identified, but they are susceptible to borohydride treatment. Cross-links in the protein molecules may be due to the reaction of amino groups of the proteins with the tetrameric dialdehyde formed by catalysis of the proteins.

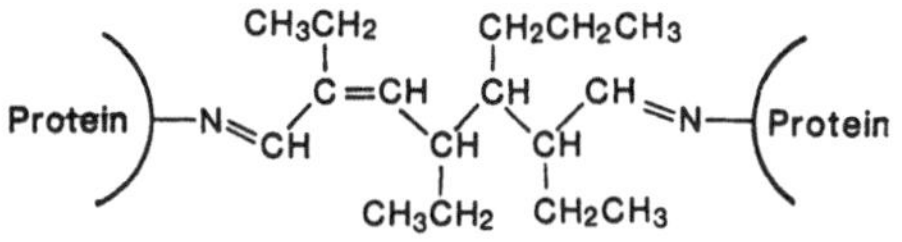

Fig. 10. Possible cross-links in the protein by 1-butanal.

DISCUSSION

Mechanisms for the formation of fluorescence, cross-kinks and borohydride-reducible functions in proteins by interaction with lipid peroxy free radicals are suggested (Fig. 11). One of the sites for the formation of fluorescence and cross-links was tyrosine residues of proteins. Tyrosine radical may be formed and dimerized into fluoresent and cross-linking tyrosine dimer. Another sites for the formation of fluorescence

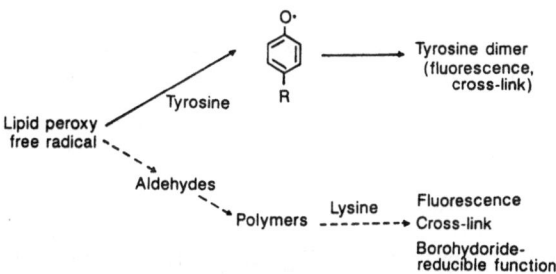

Fig. 11. Possible mechanisms for the formation of fluorescence, cross-links and borohydride reducible functions in proteins treated with lipid peroxy free radical.

and cross-links may be the amino groups of lysyl and N-terminal amino acid residues. Lipid peroxy free radicals may be decomposed into various kinds of aldehydes which may be polymerized by the catalytic action of the proteins. Polymeric aldehyde species may react with the amino groups of proteins to form fluorescence, cross-links and borohydride-reducible functions.

REFERENCES

1. K.S. Chio. R. Reiss, B. Fletcher, and A.L. Tappel, Peroxidation of subcellular organelles: formation of lipofuscinlike fluorescent pigments, Science 166: 1535 (1969).
2. A.L. Tappel, Measurement of and protection from in vivo lipid peroxidation, in: "Free Radicals on Biology", Vol. 4, W.A. Pryor, ed., pp 1, Academic Press (1980).
3. K. Kikugawa, Fluorescent products derived from the reaction of primary amines and components in peroxidized lipids, Adv. Free Radical Biol. Med. 2: 389 (1986).
4. K. Kikugawa, and M. Beppu, Involvement of lipid oxidation products in the formation of fluorescent and cross-linked proteins, Chem. Phys. Lipids 44: 277 (1987).
5. H. Esterbauer, Aldehydic products of lipid peroxidation, in: "Free Radicals, Lipid Peroxidation and Cancer", D.C.H. McBrien, and T.F. Slater, eds, pp 101, Academic Press (1982).
6. B. Halliwell, and J.M.C. Gutteridge, Oxygen toxicity, oxygen radicals, transition metals and disease, Biochem. J. 219: 1 (1984).
7. R. Amado, R. Aeschbach, and H. Neukom, Dityrosine: in vitro production and characterization. Methods Enzymol. 107: 377 (1984).
8. A.J. Gross, and I.W. Sizer, The oxidation of tyramine, tyrosine, and related compounds by peroxidase, J. Biol. Chem. 234: 1611 (1959).
9. S.O. Andersen, Covalent cross-links in a structural protein, resilin, Acta Physiol. Scand. 66, Supple. 263: 9 (1966).
10. R.S. Bondaness, and J.S. Zigler, Jr., The rapid H_2O_2-mediated nonphotodynamic crosslinking of lens crystallins generated by the heme-undecapeptide from cytochrome c: potential implications for cataracto-geneses in man, Biochem. Biophys. Res. Commun. 113: 592 (1983).
11. R.S. Bondaness, M. Leclair, and J.S. Zigler, Jr., An analysis of the H_2O_2-mediated crosslinking of lens crystallins catalyzed the heme-undecapeptide from cytochrome c, Arch. Biochem. Biophys. 231: 461 (1984).
12. H. Kohler, and H. Jenzer, Interaction of lactoperoxidase with hydrogen peroxide, formation of enzyme intermediates and generation of free radicals, Free Radical Biol. Med. 6: 323 (1989).
13. K. Kikugawa, K. Takayanagi, and S. Watanabe, Polylysines modified with malonaldehyde, hydroperoxylinoleic acid and monofunctional aldehydes, Chem. Pharm. Bull. 33: 5437 (1985).
14. M. Beppu, K. Murakami, and K. Kikugawa, Fluorescent and cross-linked proteins of human erythrocyte ghosts formed by reaction with hydroperoxylinoleic acid, malonaldehyde and monofunctional aldehydes, Chem. Pharm. Bull. 34: 781 (1986).
15. M. Beppu, K. Murakami, and K. Kikugawa, Detection of oxidized lipid-modified erythrocyte membrane proteins by radiolabeling with tritiated borohydride, Biochim. Biophys. Acta 897: 169 (1987).
16. K. Kikugawa, A. Iwata, and M. Beppu, Formation of cross-links and fluorescence in polylysine, soluble proteins and membrane proteins by reaction with 1-butanal, Chem. Pharm. Bull. 36: 685 (1988).
17. K. Kikugawa, and A. Sawamura, Formation of fluorescent substances in reaction of aliphatic aldehydes and methylamine, J. Am. Oil Chem. Soc. 64: 1156 (1987).

DISCUSSION

Zs.-NAGY: I want to congratulate you for your interesting presentation and just add that it has been known for a long time that the old collagen gives a fluorescence which is not present in the young one. It has been suspected that this fluorescence was due to the presence of tyrosine or tryptophan. Do you have any data showing that tyrosine dimers can be also formed in the protein chain?

KIKUGAWA: Yes, we have identified tyrosine dimers in proteins.

Zs.-NAGY: Yes, but in collagen, for instance, do you have any data?

KIKUGAWA: I don't have data on collagen.

ELLEDER: When I discussed with my colleagues what to do in the future, I proposed to try to see whether sodium borohydride would influence the intensity of autofluorescence in NCL deposits. Do you think it would make sense to follow this line of histochemical studies on the basis of your findings?

KIKUGAWA: Yes, I think it would be interesting to explore that.

ELLEDER: My next question is related to the fact that elastin fluoresces intensely in normal human tissues. Could you tell us anything about the fluorogenes in elastin?

KATZ: The experiments you described are based on the assumption that lipofuscin contains blue-emitting fluorophores, as for example these tyrosine dimers you showed and which are blue emitters. However, as far as I am aware, there is no evidence at all that lipofuscin mainly contains blue-emitting fluorophores. Could you comment on this?

KIKUGAWA: Lipofuscin-like pigments formed in cells and tissues exhibit yellow, green, orange and blue fluorescence. However, isolated lipofuscin fluoresces blue. I don't know the reason for this. What I can only say is that our model is one that gives this blue fluorescence.

KATZ: The blue fluorophores that you are mentioning are those extractable by chloroform-methanol as described originally by Tappel. What you need to do is to use corrected fluorescent spectrum. The spectra that Tappel recorded when he found the blue fluorophores were uncorrected. Spectrofluorometers have typically a high bias for the blue, and the fotomultipliers generally used for this type of measurements are practically blind for the red end of the visible spectrum. You can fool yourself if you don't correct the spectra for this instrumental bias.

POSTER PRESENTATIONS

A COMPUTER-ASSISTED MORPHOMETRIC STUDY OF SYNAPTIC MITOCHONDRIA IN YOUNG RATS FED A VITAMIN E DEFICIENT DIET. C. Bertoni-Freddari, P. Fattoretti, T. Casoli, W. Meier-Ruge and J. Ulrich. Center for Surgical Research, INRCA Research Department, Via Birarelli 8, 60121 Ancona, Italy and Neuropathology Division, University of Basel, Schoenbeinstrasse 40, CH-4003 Basel, Switzerland.

A morphometric study has been carried out on synaptic mitochondria in the cerebellar glomeruli of young, adult, old normal rats and young vitamin E-deficient rats. In tissue samples, prepared according to conventional electron microscopic techniques, the following parameters were measured by means of an ASBA (Wild + Leitz. AG) computer-assisted image analyzer properly programmed by ourselves: the numerical (Nv) and volumetric (Vv) densities of the mitochondria, that is the number and volume occupied by the organelles in a unit volume of tissues; the average volume (V) and skeleton (Sk) of the mitochondria i.e. a measurement of their average size and shape. We found no difference in Vv among the groups investigated. Nv significantly increased between young and adult animals, whereas decreased in the old group in a comparison both with young and adult rats. In vitamin E deficient animals Nv was significantly decreased as compared to normally fed rats of different ages. V and Sk significantly decreased in adult group and increased in old animals. Vitamin E deficient rats vs. control littermates showed a significant increase of both V and SK, however, comparing the vitamin E deficient group with the old control one, Sk was significantly higher and V significantly lower, respectively. In our previous work (Bertoni-Freddari et al. Mech. Age. Dev. 24:225-232, 1984) we demonstrated that synaptic junctions from the cerebellar glomeruli of young rats fed the same vitamin E deficient diet undergo serious alterations of their ultrastructural features which are very similar to those found in normally fed old rats. The present findings appear to support that: a) age-related morphological adaptations of mitochondrial ultrastructure may underly the reported alterations of synaptic junctions; b) vitamin E deficiency resulted in aging-like changes of mitochondrial morphology.

DISCUSSION

GOEBEL: Did your rats develop the spinocerebellar degeneration that may be seen in prolonged vitamin E deficiency?

BERTONI-FREDDARI: No, the animals were fed the vitamin E deficient diet for only 10 months and did not develop spinocerebellar degeneration.

GOEBEL: The changes that you observed in the synaptic mitochondria were only localized in cerebellar glomerular layer or was it a more generalized phenomenon?

BERTONI-FREDDARI: We have reasons to believe that this may be a generalized phenomenon, and we are planning to make similar determinations in other areas of the brain.

GOEBEL: Were the changes you described in mitochondria the only ones or did you find other changes such as in the shape of cristae or in the presence of inclusion bodies, etc.?

BERTONI-FREDDARI: Even if we have seen some other mitochondrial changes, our computer-assisted program for quantitative analysis was not prepared for these other determinations.

GOEBEL: Did the cells that supplied the synaptic areas where you measured the mitochondrial changes accumulate lipofuscin?

BERTONI-FREDDARI: We did not explore this aspect.

GOEBEL: Do you have any evidence or information whether the administration of inhibiting enzymes or other substances would harm the mitochondria in this area?

BERTONI-FREDDARI: We don't have any personal data on that.

GOEBEL: Are the mitochondrial changes that you described specific for vitamin E deficiency, or are they simply nonspecific reactions?

BERTONI-FREDDARI: This is a good question, but difficult to answer. Somewhat similar mitochondrial changes to those found in our vitamin E deficient animals have been found at least in the cerebellar Purkinje cells of rats after prolonged alcohol consumption by Tavares and Paula-Barbosa (J. Submicrosc. Cytol., 15:713, 1983). It is my impression that the changes we observed in the mitochondria of our animals are adaptive rather than pathological ones.

EFFECT OF IRON ON THE SENSITIVITY OF LIPOFUSCIN-CONTAINING RAT CARDIAC MYOCYTES TO H₂O₂-INDUCED OXIDATIVE STRESS

Anders Brunmark
Department of Pathology, University of Linköping, S-581 85 Sweden.

An important question in the field of lipofuscin research is whether its accumulation in post-mitotic cells affects their cellular functions, including resistance against oxidative stress. Among several properties of lipofuscin, with potential importance for the effect on cellular viability, is the sequestration of transition metals, especially iron, in this pigment. Transition metals in the lysosomal vacuome may influence its stability and thus capability of the cells to resist oxidative conditions, such as those following post-ischemic reoxygenation.

In this study we used neonatal rat myocytes plated on collagen-coated plastic dishes and cultured for 12 days in Eagles Minimal Essential Medium supplemented with 10% newborn calf serum in the presence or absence of Fe^{3+} (30μM). The latter treatment enhances the lipofuscin content and iron concentration in the lysosomal vacuome of rat cardiac myocytes. Sensitivity to oxidative stress was studied by exposure of the cells to hydrogen peroxide. Cell viability was determined as release of lactate dehydrogenase.

Cells grown in the presence of Fe^{3+} were found to exhibit significantly higher sensitivity to oxidative stress imposed by hydrogen peroxide. It is possible that the iron present in lipofuscin-containing cells may affect their viability under oxidative conditions. The mechanism may involve peroxidation of the lysosomal membrane with resulting leakage of lytic enzymes. It can be speculated that cardiac myocytes in the aged organism, possessing high amounts of lipofuscin, are more sensitive to e.g. ischemia due to their lysosomal transition metal content.

DISCUSSION

PORTA: You mentioned that the viability of the cardiac myocytes exposed to hydrogen peroxide stress was determined by the release of lactic dehydrogenase. Is that correct? Did you measure the release of any lysosomal enzyme?

BRUNMARK: No. Actually, the cell viability was monitored by dye exclusion test. I want to add that this may be a good model not only for metal toxicity, but also for ischemic reperfusion damage.

GOEBEL: I am not very familiar with tissue culture of cardiac myocytes. Did you perform the experiments in explanted surviving cells, or did you get growth or proliferation from any dormant satellite cells as they are in the skeleton muscle?

BRUNMARK: No. They were actually primary explants.

SPECIFIC KIDNEY INJURY OF RATS BY LIPOFUSCIN FORMATION UNDER VITAMIN E DEFICIENCY AND GSH DEPLETION. Tomio Ichikawa and Kiyokazu Hagiwara. National Institute of Nutrition, Shinju-ku-ku, Tokyo 162, JAPAN.

Four-week old Wistar male rats were fed either a VE deficient or a sufficient diet for 6 weeks and then intraperitoneally injected with BSO at the dosage of 1 mmol/kg body weight once a day for 3 days. GSH depletion by BSO treatment in VE defi-cient rats caused tissue injury, especially in the kidneys. It was observed that the epithelial cells of the renal tubu-les were severely injured and showed necrosis and desquama-tion. No injury was detected in kidneys of 10E and 10E-BSO groups. TBA value in kidneys of the 0E-BSO group was lower than that of the 0E group, but lipofuscin in kidneys of the 0E-BSO group was 10 times higher than that of the 0E group.

In order to elucidate the mechanisms involved in the marked lipofuscin formation in the 0E-BSO group, the TBA value and lipofuscin levels of kidney homogenates were determined in vitro by $AsA-Fe^{2+}$. The TBA value increased chronologically with incubation time, but lipofuscin levels did not change. The polyunsaturated fatty acid content decreased in kidney homogenates of the 0E-BSO group. GSH-Px activity was low and GSSG-Rd activity tended to be low in the 0E-BSO group. Increased GST activity was observed. It is known that GSH depletion in various cells increases sensitivity to oxygen stress. Concerning the nutritive state of rats, decreased intake of sulfur amino acids in the VE deficient situation may increase the risks for renal injury.

DISCUSSION

PORTA: Your studies on the effects of glutathione depletion in vitamin E deficient rats are very interesting, but I am very much intrigued by the exclusive localization of necrotic lesions in the kidneys. Wouldn't you expect to find degenerative and necrotic lesions in other organs like, for instance, the liver?

ICHIKAWA: We have examined all the organs, and we only found kidney lesions.

PORTA: In your photographs, the lesions appear particularly prominent in the proximal convoluted tubules of the kidney. Do you have any explanation for this particular selectivity?

ICHIKAWA: I don't have at this moment any good explanation for this organ selectivity. We also found increased lipofuscin in the kidneys.

PORTA: Well, frankly, looking at your photographs, I could not see any lipofuscin or ceroid pigment accumulation. How did you estimate the amount of lipofuscin or ceroid?

ICHIKAWA: We have solubilized the pigment and measured lipofuscin autofluorescence at 435 nm emission.

KITANI: So, you have measured, in fact, the blue fluorescence of the lipid soluble extracts. Is that correct?

ICHIKAWA: Yes, that's correct.

PORTA: Well, if this was the case, I would like to suggest that you should use more acceptable methods of evaluation. The so-called organic solvent soluble lipofuscin pigments are not presently considered to represent the pigments that accumulate in tissues, and which typically display yellow-orange autofluorescence in situ.

WHAT ARE THE TBA-REACTIVE SUBSTANCES IN TISSUE HOMO-GENATE? Hiroko Kosugi,* Takashi Kojima,** and Kiyomi Kikugawa.** *Ferris University, Yokohama, Japan. **Tokyo College of Pharmacy, Hachioji, Tokyo, Japan.

The thiobarbituric acid (TBA) assay has been used for determination of lipid peroxidation of biological samples. Several methods of the TBA assay have been developed for determination of lipid peroxidation of liver homogenate or microsomes and of blood plasma. A red pigment with an absorption maximum at 532 nm produced in the TBA assay has been considered to be specific to malonaldehyde. However, it has been shown recently that determination of malonaldehyde by alternative methods showed that malonaldehyde contents in various samples are much lower than those estimated by the TBA assay. We have shown previously that red pigment is formed from alkenals (Anal. Biochem. 165, 456-464 1987) and alkadienals (Lipids 23, 1024-1031 1988), and its formation is extensively enhanced by organic hydroper-oxides (J. Japan Oil Chem. Soc. 38, 224-230 1989). We investigated which components of rat liver homoge-nate are TBA-reactive by use of four different methods of the TBA assay. Major differences of the reaction conditions of these methods were those of the solvents employed: trichloroacetic acid-hydrochloric acid was used in method A as a solvent, phosphoric acid in method B, acetate adjusted at pH 3.5 and sodium dodecyl sulfate in method C, and diluted acetic acid alone in method D. Development of the pigment greatly depended on the methods and on the pH values of the reaction mixtures. Method C, whose reaction mixture was adjusted at pH 3.5, was most sensitive to the TBA-reactive substances. Amounts of the pigments produced were dramatically increased by introduction of 2 mM t-butyl hydroperoxide in the reaction mixtures of the methods. These characteristics of the pigment formation from the homogenate were similar to those from alkadienals and essentially different from those from malonaldehyde. Alkadienals were likely candidates for the TBA-reactive substances in the homogenate.

DISCUSSION

KITANI: For many years, we have been using the TBA reaction, but some investigators have reservations about its significance. Yet, it is still widely used, and therefore, it would be of interest to hear comments from the audience.

Zs.-NAGY: I was always convinced that this is a wrong and useless method. The chemical environment where the TBA reaction takes place may greatly influence the results. We just recently observed that even minor amounts of serine may produce as much as 10 times absorption in the same wavelength. This is falsifying everything; obviously, we cannot have always the same chemical environment for this reaction. Another point is that many people apply the TBA reaction expecting to find what has happened 2 - 4 weeks before. The products that react with TBA cannot wait in the tissues for one month. I think that people should be discouraged to use this method.

KITANI: Can you then propose the use of something better?

Zs.-NAGY: Yes, one more precise method could be spin-trapping, and another could be HPLC. These methods are, admittedly, more elaborate than the simple TBA reaction, but are much better. Perhaps, Dr. Porta may like to comment on this.

PORTA: I suppose that we all have some justified concerns about the adequacy and usefulness of the TBA reaction in the evaluation of in vivo lipid peroxidation in animal specimens. Obviously, there is still a great deal of confusion about the real significance of this reaction, and the method is frequently misused, and its results misinterpreted. This doesn't mean that the method is totally useless and should be abandoned. First, this method is not intended to detect free radicals, as has been apparently implied in the comments of Dr. Zs.-Nagy, but to detect some of the late decomposition products of lipid peroxidation. It should not be, therefore, compared with spin-trappping or other methods used for that purpose. Second, as Dr. Kosugi and other investigators have shown, the TBA reaction is not specific for malonaldehyde since, at the absorption maxima of 532 nm; it not only detects malonaldehyde, but particularly other important products of lipid peroxidation, such as alkenals and alkadienals. Third, due to its simplicity and sensitivity, the TBA reaction is still widely used as a reliable measure of lipid peroxidation in well defined in vitro chemical systems under standardized and controlled conditions. However, when the method is applied to homogenates or cell fractions of animal tissues, the situation is more complex due to the numerous factors influencing the reaction. In this case, I agree that the results should not be interpreted and equated with the eventual occurrence of in vivo lipid peroxidation. There are other methods, such as diene conjugation analysis, loss of PUFA from membrane phospholipids, measurements of ethane, etc., which provide a better measure of in vivo lipid peroxidation. The in vitro production of TBA reactive substances occurs not only in animal specimens with lipids that underwent peroxidation in vivo, but practically in all normal and abnormal tissues and cell fractions aerobically incubated. The reason for the common use of the TBA reaction in animal specimens by many experts in the field is that it does provide useful information about the "lipoperoxidative potentials," resulting from the overall balance between the various prooxidant and antioxidant factors present in tissues.

BRUNMARK: Dr. Zs.-Nagy, you mentioned the usefulness of spin-trapping for the detection of free radicals in vivo, but I think that this would be extremely difficult to perform.

Zs.-NAGY: Well, I was referring to the in vitro systems. Although there are some possibilities for the in vivo studies, these have not been yet completely utilized.

KOSUGI: The TBA assay may not be the best method to study the problem of in vivo lipid peroxidation or to measure malonaldehyde, but still we consider it a very useful and practical method in the whole context of lipid peroxidation.

THE VALUE OF ULTRASTRUCTURAL LYMPHOCYTE LYSOSOMAL ANALYSIS FOR DIAGNOSING CEROID LIPOFUSCINOSIS, Stephen A. Smith, University of Minnesota, Minneapolis, Minnesota, USA.

The clinical diagnosis of neuronal ceroid lipofuscinosis (NCL) is made by finding abnormal lysosomes containing ceroid lipofuscin storage material in body tissues. Lymphocytes, skin biopsies, and conjunctival biopsies are commonly studied. The accuracy of the morphologic investigations for diagnostic purposes, including exclusion of NCL, has been of concern. Increased excretion of urinary dolichols helps confirm the diagnosis of NCL.

This study analyzes 108 blood specimens submitted to look for lymphocyte lysosomal inclusions to screen for NCL. Lymphocytes normally contain no abnormal lysosomal inclusions. In NCL 50% or more of lymphocytes may contain ceroid lipofuscin material. In the infantile form of NCL granular inclusions are most common. Late infantile disease shows primarily curvilinear profiles. Fingerprint inclusions are typically present in juvenile NCL. The technique requires a fresh, EDTA preserved blood sample, that is promptly processed. After centrifugation the buffy coat is infiltrated with 4:1 E/M fixative followed by post-fixation with osmium. Lymphocyte rich areas are selected for thin sectioning and electron microscopic examination. 100 lymphocytes are examined ultrastructurally at a magnification of 58,000x. All suspicious and clearly positive lymphocytes are reproduced on photographs at magnifications of 57,000 to 87,000x. The number of abnormal lymphocytes is counted, and the types of inclusions are described.

Positive identification ultrastructurally of abnormal lysosomes was found in 12 patients (11%), and negative results were present in 96 (89%). Fingerprint inclusions were most commonly present (10 of 12 patients), followed by curvilinear bodies (2 of 12 patients). The type of inclusion correlated with the clinical subtype of NCL. There were no examples of infantile or late juvenile/early adulthood onset disease in this study. Positive diagnoses correlated with urinary dolichol excretion when examined for and autopsy findings when available. No false negative results were identified when case records were analyzed.

Utilizing this technique of processing a freshly obtained blood specimen, it is concluded that ultrastructural examination of lymphocytes for abnormal lysosomes containing ceroid lipofuscin material is an accurate way to diagnose NCL.

DISCUSSION

BRUNK: There are a few things that I didn't quite understand in your presentation. Do you mean that only 11% of the patients were diagnosed using this method?

SMITH: The 11% is simply the percentage of 108 patients that were examined.

BRUNK: And so, do they all have this disease? Does it mean that the method gives only 11% positive results?

SMITH: No, what I said is that in 108 patients with a wide variety of neurological disorders, including cerebral palsy, seizures, etc., we had only 11% of the cases with CL.

KITANI: Then, which is the real percentage of positiveness with this method?

SMITH: As far as I can tell, it is virtually 100%. We have not yet found any false positive.

ELLEDER: Would it make sense to correlate the electron microscopic findings with histochemical or autofluorescent studies in smears?

SMITH: Although we have not done that, it seems theoretically possible.

ELLEDER: I wonder if these tiny deposits are detectable by autofluorescence. When I received a biopsy specimen for diagnosis of CL, the first thing I do is to look for autofluorescence. If it is negative, then it will never be positive for electron microscopy. So, in this way, you can save time by avoiding the relatively laborious procedures for electron microscopy.

SMITH: I don't have experience in examining samples for autofluorescence.

GOEBEL: I may, perhaps, answer this question. There is one study where the authors claimed that the diagnosis of CL could be made in smears by fluorescent microscopy. I have a few questions. Did you encounter any adult cases?

SMITH: This study was only done in children.

GOEBEL: Did you see in any single patient combined forms, such as lymphocytes with curvilinear bodies as well as with vacuoles with fingerprints?

SMITH: No.

GOEBEL: Did the presence of inclusions correlate always with urinary dolichols?

SMITH: No, there were 2 or 3 cases where the urinary sediment was negative for dolichols, but positive by lymphocytic analysis.

GOEBEL: Did you analyze each of your patients once, or did you repeat the analysis?

SMITH: We have repeated the analysis in some cases, but there were no differences with the results of the original analysis.

HALL: The clinical presentation of the juvenile form is really very different from that of the late infantile or the early juvenile form. In the juvenile form of the disease, you can invariably see vacuolated lymphocytes simply by light microscopy. It seems to me that if you suspect a juvenile form, you don't really need electron microscopy. However, in the suspicion of a late infantile or early juvenile form, it may be worth to do electron microscopy.

SMITH: Yes, I agree with that.

GOEBEL: I wouldn't rely only in light microscopy for the detection of vacuolated lymphocytes as seen in the juvenile form, because many other lesions unrelated to CL may produce vacuolations in lymphocytes, such as swollen mitochondria and other lysosomal disorders.

NEURONAL CEROID-LIPOFUSCINOSIS IN INFANTILE OSTEOPETROSIS.

Kiyoshi Takahashi, Makoto Naito, and Fumie Yamamura. Kumamoto University Medical School, Second Department of Pathology, Kumamoto, Japan.

Neuronal ceroid-lipofuscinosis and infantile osteopetrosis are rare autosomal recessive diseases. We have recently experienced an autopsy case of neuronal ceroid-lipofuscinosis occurring in infantile osteopetrosis and examined it pathologically, histochemically, electron-microscopically and biochemically. This case was a 1-year and 11-month-old girl who exhibited clinical pictures of manifestations. In the terminal stage, she developed frequent convulsions, cerebral and optic atrophy, retinal degeneration, loss of voluntary movement, and pathological reflexes. Her elder brother died of the same disease, and her parents were a consanguinous marriage. Autopsy revealed generalized sclerosis and thickening of cortical and spongy bones, formation of mineralized cartilagenous tissues, narrowing of the marrow cavities and decreased hematopoietic cell components. Spindle-shaped osteoclasts were found around the thickened bone trabecles. Electron-microscopically, the osteoclasts had numerous cytoplasmic vacuoles and showed the disappearance of ruffled borders and clear zone along their cell membrane facing the bone matrix surface. All osteopetrosis. In the current case, the most striking findings were in the central nervous system. The brain was markedly atrophic with loss of neuronal cells and focal gliosis, and the remaining neuronal cells showed cytoplasmic ballooning with accumulation of brown granular pigments. The pigments emitted autofluorescence and were confirmed histochemically to be ceroid or lipofuscin. Electron-microscopically, they showed a granular or amorphous, osmiophilic appearance, but no curvilinear tubular structures or lamellar concentric materials. Similar deposits were observed in foamy brain macrophages. These findings were consistent with those of infantile neuronal ceroid-lipofuscinosis. To our knowledge, this case is the first autopsy case of neuronal ceroid-lipofuscinosis complicating in infantile osteopetrosis, except for a suggestive autopsy case reported by Ambler et al.

DISCUSSION

PORTA: I want to especially congratulate you, because your poster display was very well organized, very clear and well illustrated with excellent tables and photography. I only have a question. Was the hepatosplenomegaly due to extramedullary hematopoiesis?

TAKAHASHI: Yes, that's correct.

GOEBEL: Since infantile ceroid lipofuscinosis is a ubiquitous disease, as far as the morphological distribution of lipopigments is concerned, I wonder if you have seen abnormal lipopigments outside the brain, as for instance in the liver, to say that this is a generalized disorder.

TAKAHASI: In this case, the pigment was confined to the brain.

GOEBEL: You indicated that a sister died. Was there any evidence that she had this combination, or did she just have osteopetrosis?

TAKAHASHI: Unfortunately, we have not done the brain autopsy, and therefore, we don't know.

IVY: Could you, please, say something about neuronal death in the brain? Was the presence of the pigment related to cell death anywhere in the brain?

TAKAHASHI: Deposition of ceroid-lipofuscin pigment was found in brain cortex, basal ganglia, cerebellum, pond and spinal cord, but cell death was only prominent in the cerebellar Purkinje cells.

PORTA: Now, we would like to request the opinions of the participants about future lines of investigations on lipofuscin and ceroid pigments.

JOLLY: Something that has not been touched here, except by Dr. Brunk this morning, is the idea of age-pigment being nothing more than sequestration of heavy metals in tertiary lysosomes. We cannot ignore that some metals are sequestered there and may react with proteins and other substances.

Zs.-NAGY: If we think that lipofuscin is just sequestration of heavy metals, we have to consider that heavy metals have faster turnover in young organisms. Why are they sequestered in the old, but not in the young? I personally don't believe that the presence of some heavy metals in lysosomes indicates just sequestration.

KATZ: I want to point out an error that some people continue to make despite the data that Dr. Eldred has presented at the two previous symposia. This is about the fluorescent spectra data that have been still presented in the uncorrected form, and which, therefore, have little meaning in order to properly characterize lipofuscin. Numerous papers are still giving the emission maxima in the range of 400 - 470 nm, and which is still in the blue region of the visible spectrum, while lipofuscin is not mainly containing blue fluorophores but rather yellow and orange ones.

HARMAN: Looking at the voluminous literature on lipopigments, one is intrigued at the relatively few studies done on the isolation procedures and further biochemical analysis of isolated pigment material. It would be important in future meetings to dedicate one entire session to discuss and present new procedures and data that can be applied to the understanding of the lipopigment problem.

ELLEDER: If somebody would ask me what I would like to know about these pigments, I would say that I want to know how to really define them. The autofluorescence is a reliable criteria to define lipopigments of any type. Some people rely on the chemical nature of the fluorogens; some others denied this, and think that this is only a reflection of physical nature that has nothing to do with the chemical composition. Another thing I would like to know is whether all the autofluorescent pigments accumulate in lysosomes simply because all of them are indigestable material, or whether, as I suspect, in some instances the accumulation is due to enzymatic defects or enzymopathies. Studies along this line would be rewarding.

JOLLY: Since very little is known about the functional implications due to lipopigment accumulation, I think that we need more functional studies.

IVY: Just a brief comment on what Dr. Jolly just said. It would be quite convenient, I agree, to have more functional studies in aged animals in which, for instance, the brain accumulates great amounts of lipofuscin. Unfortunately, the type of functional studies that can be done are almost always confounded by secondary or tertiary effects of either aging or drug treatments.

JOLLY: I was thinking not necessarily in terms of in vivo tests, but in some histochemical studies in neurons, for instance.

HALL: I think that another aspect of lipofuscin that perhaps should be investigated in the future, is the fact that lipofuscin is associated with lysosomes and, therefore, I think that if we want to look at the functional effects of lipofuscin, the first thing to explore would be whether lysosomes are functionally normal.

PORTA: I will totally agree with the suggestions of Drs. Hall and Jolly, and particularly with that of Dr. Elleder, that one of the most pressing and probably most rewarding studies should be those directed to determine possible functional alterations, especially through the detailed histochemical or biochemical analysis of lysosomal enzymes. I feel that there is not enough data on this aspect, and the available ones are usually conflictive. Although it has been mentioned here that during aging there is a decline in the activity of some cathepsins, there are still many discrepant data, and these have been attributed to the large variations in the contribution of the lysosomes from diverse cell populations present in whole organs or tissue homogenates.

BRUNK: I think that we should as much as possible consider the possibility of utilizing cytochemical technics. I have the feeling that when we grind a tissue and measure its enzymatic activities, we may be looking at something that is totally unphysiological.

ELLEDER: I just want to make a brief comment. If we are dealing with deposits that are segregated in the lysosomal system, I think that it would be very convenient to know whether they are primarily indigestible for any type of normally equipped cell, because this would mean that the primary lesion is probably in the extra lysosomal system. I feel, however, that in certain cases, the primary defect is in the lysomal system.

CONTRIBUTORS

ALHO, H., University of Tampere, Finland

ALOJ-TOTARO, E., University of Calabria, and Zoological Station, Naples, Italy

BALL, R.Y., University of Cambridge, England

BAYLISS, S.L., Massey University, Palmerston North, New Zealand

BEPPU, M., Tokyo College of Pharmacy, Japan

BERRA, A., University of Buenos Aires, Argentina

BERTONI-FREDDARI, C. INCRCA Research Department, Ancona, Italy

BITTERMAN, P.B., University of Minnesota, Minneapolis, MN, USA

BORGMAN, R., Alcon Laboratories, Ft. Worth, TX, USA

BRUNMARK, A., University of Linkoping, Sweden

BRUNK, U.T., University of Linkoping, Sweden

BUSCH, H., University of Mainz, Fed. Rep. Germany

CARPENTER, K.L.H., University of Cambridge, England

CARTER, N.P., University of Cambridge, England

CASOLI, T., INRCA Research Department, Ancona, Italy

DALEFIELD, R.R., Massey University, Palmerston North, New Zealand

DAVIS, S., University of Cambridge, England

DAWSON, G., University of Chicago, IL, USA

DAWSON, S.A., University of Chicago, IL, USA

ELLEDER, M., Charles University, Prague, Czechoslovakia

ENRIGHT, J.H., University of Cambridge, England

FATORETTI, P., INRCA Research Department, Ancona, Italy

FEARNLEY, I.M., Medical Research Council Lab. Molecular Biol., Cambridge, England

GOEBEL, H.H., University of Mainz, Fed. Rep. Germany

HAGIWARA, K., National Institute of Nutrition, Tokyo, Japan

HALL, N.A., Institute of Child Health, London, England

HARMAN, D., University of Nebraska, Omaha, NB, USA

HARMAN, K., University of Minnesota, Minneapolis, MN, USA

HARTIKAINEN, K., University of Tampere, Finland

HARTLEY, S.L., University of Cambridge, England

HATANPÄÄ, University of Tampere, Finland

HAYASAKA, A., Tokyo College of Pharmacy, Japan

HEISKALA, A., University of Helsinki, Finland

HERVONEN, A., University of Tampere, Finland

ICHIKAWA, T., National Institute of Nutrition, Tokyo, Japan

IVY, G.O., University of Toronto, Ontario, Canada

JAATINEN, P., University of Tampere, Finland

JOHANSSON, E., University of Tampere, Finland

JOLLY, R.D., Massey University, Palmerston North, New Zealand

KANAI, S., Tokyo Metropolitan Inst. Gerontology, Japan

KATO, T., Tokyo College of Pharmacy, Japan

KATZ, M.L., University of Missouri, Columbia, MO, USA

KIKUGAWA, K., Tokyo College of Pharmacy, Japan

KITANI, K., Tokyo Metropolitan Inst. Gerontology, Japan

KOBAYASHI, M. Tokyo Metropolitan Inst. Gerontology, Japan

KOISTINAHO, J., University of Tampere, Finland

KOJIMA, T., Tokyo College of Pharmacy, Japan

KOSUGI, H., Ferris University, Yokohama, Japan

LAAKSONEN, H.M., University of Tampere, Finland

LAKE, B.D., Institute of Child Health, London, England

LUCADAMO, L., Zoological Station, Naples, Italy

LUSTIK, Gy., Verzar Intern. Lab. for Experimental Gerontology, Debrecen, Hungary

MAEBA, R., Teikyo University School of Medicine, Tokyo, Japan

MARTINUS, R.D., Massey University, Palmerston North, New Zealand

MEDD, S.M., Medical Research Council Lab. Molecular Biol., Cambridge, England

MEIER-RUGE, W., University of Basel, Switzerland

MITCHINSON, M.J., University of Cambridge, England

MONSERRAT, A.J., University of Buenos Aires, Argentina

NAITO, M., Kumamoto University, Japan

NOKUBO, M., Tokyo Metropolitan Inst. Gerontology, Japan

OHTA, M., Tokyo Metropolitan Inst. Gerontology, Japan

OTSUBO, K., Tokyo Metropolitan Inst. Gerontology, Japan

PALMER, D.N., Massey University, Palmerston North, New Zealand

PATRICK, A.D., Institute of Child Health, London, England

PISANTI, F.A., Zoological Station, Naples, Italy

PORTA, E.A., University of Hawaii, Honolulu, HI, USA

RUBIO, M.C., Institute of Pharmacol. Investigations, CONICET, Buenos Aires, Argentina

SANTAVUORI, P., University of Helsinki, Finland

SATO, Y., Tokyo Metropolitan Inst. Gerontology, Japan

SHIMASAKI, H., Teikyo University School of Medicine, Tokyo, Japan

SIAKOTOS, A.N., Indiana University School of Medicine, IN, USA

SMITH, G., University of Toronto, Ontario, Canada

SMITH, S.A., University of Minnesota, Minneapolis, MN, USA

SOHAL, R.S., Southern Methodist University, Dallas, TX, USA

TAKAHASHI, K.M., Kumamoto University Medical School, Japan

TOWNSEND, D., University of Minnesota, Minneapolis, MN, USA

UETA, N., Teikyo University School of Medicine, Tokyo, Japan

ULRICH, J., University of Basel, Switzerland

WALKER, J.E., Medical Research Council Lab. Molecular Biol., Cambridge, England

WESTERMARCK, T.N., University of Helsinki, Finland

WITKOP, C.J., University of Minnesota, Minneapolis, MN, USA

WOLFE, L.S., Montreal Neurological Institute, McGill Univ., Montreal, Quebec, Canada

WOODS, S.E., University of Cambridge, England

YAMAMURA, F., Kumamoto University Medical School, Japan

Zs.-NAGY, I., Verzar Intern. Lab. for Experimental Gerontology, Ancona, Italy

Zs.-NAGY, V., Verzar Intern. Lab. for Experimental Gerontology, Ancona, Italy

INDEX

Ceroid (continued)
    in renal proximal tubules, 175-176, 180-181
Ceroid-lipofuscin (see also Ceroid, Ceroid-lipofuscinosis, Lipofuscin, Lipofuscin-like pigments and Lipopigments)
    autofluorescence, 78, 163
    cathepsins, 85
    dolichols, 225
    fluorescent emission 79
    formation, 84
    in kupffer cells, 86
    intracellular distribution, 85
    solubility, 164
Ceroid-lipofuscinosis, 4, 9, 163-164, 211-219, 225-240, 259-269, 273-279, 366 (see also Batten's disease, Ceroid and Ceroid-lipofuscin)
    canine, 263-264
    carbohydrate content, 234
    clinical course, 277
    concanavalin A-binding components, 230, 236, 246, 248
    DCCD binding protein subunit, 163
        amino acid sequences, 215
        genes for DCCD, 216-217
    dolichol-linked oligosaccharides, 212, 225-240
    extraction properties, 237
    fingerprint inclusions, 300, 366
    glycoconjugates, 225-240
    glycoprotein turnover, 274
    incidence, 212
    in infantile osteopetrosis, 368
    lymphocyte lysosomal analysis, 366
    lysosomal phospholipases, 262-263, 275
    ovine, 211-212, 250
    storage bodies, 213
    urinary dolichols, 366
Chloroquine, 38, 161
Cholesterol, 6, 177, 180, 182
Copper, 335
Cysteine proteinase inhibitors, 170
Cythiolone (n-acetyl-homocysteine-thiolactone), 136

DCCD (dicyclohexylcarbodiimide)
    binding subunit, 163, 211, 213-217, 227,260
Deferoxamine, 21, 282
DNA repair processes, 3
Dolichols, 6,8, 78, 83-84, 168, 175, 178, 225, 242, 366
Dolichol-linked oligosaccharides, 212, 225-240

E-64C protease inhibitor, 33, 38, 41, 77
Electron microscopy of lipopigments, 5, 21, 23, 35-39, 81, 150-151, 231, 289, 300-307, 314-317, 366
Ethanol, 23
Extractability of lipopigments, 200

Fenton reaction, 21
Fingerprint profiles, 56, 259, 300

Fluorescent-activated cell sorter, 334
Fluorophores, 6, 15, 24, 46, 112-114, 117, 158, 170, 331, 351, 357, 369
Free radicals, 4, 6, 28, 42, 121, 130, 135, 142, 143

Glutathione, 121, 127-131
    depletion, 363
Glutathione peroxidase, 9, 121, 144, 366
G_M gangliosidosis, 227, 237, 246, 248
Glycoproteins, 243-256

Heart, 4, 35, 40, 146, 148, 177
Hemoceroid, 158
Hepatocytes, 40, 75-88, 99, 190
Hermansky-Pudlak syndrome, 46, 168, 283-295
    albinism, 284, 297
    alveolar macrophages, 285, 292
    granulomatous enteropathic disease, 284, 287, 290
    immunologic changes, 285
    pulmonary fibrosis, 284, 287-290
Herringbone profiles, 301, 309
4-hydroxynonenal, 9, 261, 266

Interceroid, 158
Intracellular monovalent electrolytes, 93-104
Ionizing radiation, 135-140
Iron, 9, 20, 24, 318-319, 321, 362

Keshan's disease, 155
Kidney, 35, 40, 177-178, 180, 182, 363
Kufs disease, 4, 212, 248
Kupffer cells, 40, 79, 81, 86, 169-184

Lectin histochemistry, 244
Leupeptin, 7, 9, 33-37, 75-89, 113, 116, 161, 169-185
    bromosulphalein test, 83
    cathepsin, 88
    clearance, 85
    dolichols, 87, 180, 182
    functional alterations, 86
    ouabain excretion, 82
    serum cholesterol, 180, 182
    serum enzymes, 180, 182
Lipid binding subunit of mitochondrial ATP synthase, 163, 227
Lipid peroxidation, 6-7, 17, 22, 130, 135, 143, 158, 170, 203, 262, 323-324, 345-356, 364
Lipofuscin, 3-6, 225, 246, 254 (see also Age pigment, Ceroid, Ceroid-lipofuscin, Lipopigments and Lipofuscin-like pigments)
    and aging, 17, 25, 32
    autofluorescence, 5, 18, 20, 53, 113, 237, 322
    carbohydrates, 41, 234
    chemical composition, 6
    distribution, 4, 69
    electron microscopy, 5-6, 21, 50, 55-56
    exocytosis, 316
    ferritin-like granules, 318
    fluorophores, 6, 113-114